Springer
New York
Berlin
Heidelberg
Barcelona
Hong Kong
London
Milan
Paris
Singapore
Tokyo

Universitext

Editors (North America): S. Axler, F.W. Gehring, and K.A. Ribet

(continued after index)

Kenji Matsuki

Introduction to
the Mori Program

With 61 figures

Springer

Kenji Matsuki
Department of Mathematics
1395 Mathematical Science Building
Purdue University
West Lafayette, IN 47907-1395
USA

Mathematics Subject Classification (2000): 15-01, 15A04, 15A69, 16-01

Library of Congress Cataloging-in-Publication Data
Matsuki, Kenji, 1958–
    Introduction to the Mori Program / Kenji Matsuki.
        p. cm. — (Universitext)
    Includes bibliographical references and index.

    1. Algebraic varieties—Classification theory. I. Title.
    QA564 .M38 2001
    516.3'53—dc21   00-067917

Printed on acid-free paper.

Printed in the United States of America.

9 8 7 6 5 4 3 2 1

ISBN 978-1-4419-3125-2

Springer-Verlag   New York Berlin Heidelberg
A member of BertelsmannSpringer Science+Business Media GmbH

Preface

This book started as a collection of personal notes that I made to help me to understand what we call the Mori program, a program that emerged in the last two decades as an effective approach toward the biregular and/or birational classification theory of higher-dimensional algebraic varieties. In some literatures the Mori program restrictively refers to an algorithm, called the minimal model program, to produce minimal models of higher-dimensional algebraic varieties. (Classically, the construction of minimal models was known only for algebraic varieties of dimension less than or equal to 2.) Here in this book, however, we use the word in a broader sense to represent a unifying scheme that is a fusion of the minimal model program and the so-called Iitaka program.

As such, I had no hesitation to, or rather even made elaborate efforts to, extract nice arguments from the existing literature whenever it seemed appropriate to fit them into a comprehensible development of the theory, even to the extent of copying them literally word by word. I am particularly aware of the original sources of the subject matters in the following list:

Chapter 1.

Barth–Peters–Van de Ven [1], Beauville [1], Clemens–Kollár–Mori [1], Griffiths–Harris [1], Hartshorne [3], Iitaka [5], Kawamata–Matsuda–Matsuki [1], Kodaira [2][3][4][5], Kollár [5], Mori [2][3], Reid [9], Shafarevich [1], Wilson [1]

Chapter 2.

Iitaka [2] [3] [5], Kawamata [1][2], Vojta [1]

Chapter 3.

Clemens–Kollár–Mori [1], Corti [1], Corti–Pukhlikov–Reid [1], Iitaka [1] [4][5], Kawamata [4][5][6][12], Kawamata–Matsuda–Matsuki [1], Kollár [5], Kollár et al [1], Miyaoka–Mori [1], Mori [2][3][4][5], Reid [2][3][6][7][8], Sarkisov [3], Shokurov [1][2]

Chapter 4.

Artin [1][2][5], Brieskorn [1], Kawamata [2][9], Mori [4], Mumford [3], Reid [2][3][7]

Chapter 5.

Clemens–Kollár–Mori [1], Esnault–Viehweg [1], Kawamata [3]. Kodaira [1], Kollár [3][4][7], Viehweg [2]

Chapter 6.

Angehrn–Siu [1], Ein–Lazarsfeld [1], Kawamata [4][6], Kawamata–Matsuda–Matsuki [1], Shokurov [1]

Chapter 7.

Clemens–Kollár–Mori [1], Kawamata [5], Kollár [1], Mori [2]

Chapter 8.

Kawamata [5], Kawamata–Matsuda–Matsuki [1], Mori [2]

Chapter 9.

Kawamata [9], Kawamata–Matsuda–Matsuki [1], Mori [5], Shokurov [1]

Chapter 10.

Kawamata [5][13] Kollár [1] [12] Miyaoka–Mori [1] Mori [1] [2]

Chapter 11.

Iitaka [2][3][5], Kawamata–Matsuda–Matsuki [1] Shokurov [2]

Chapter 12.

Kawamata [9], Kollár [8], Reid [3]

Chapter 13.

Corti [1], Reid [6], Sarkisov [3], Takahashi [1][2][3]

Chapter 14.

Danilov [1][2], Fulton [2], Oda [1][2], Reid [5]

I learned about the Mori program from my teachers, S. Iitaka, Y. Kawamata, J. Kollár, S. Mori, M. Reid, and V. V. Shokurov. My indebtedness to them goes far beyond what I can express in words both mathematically and personally.

My personal notes grew into the present form in the process of communicating what I learned about the Mori program through seminars and classes held at Brandeis University and Purdue University. My thanks are due to D. Arapura, J. Lipman, T. Matsusaka, and the students who clarified many ambiguities and mistakes in my presentation of this beautiful theory. Special thanks go to D. Eisenbud, without whose encouragement this book would have never converged into the present form.

I claim no originality except for simplification or reworking of some classical results (e.g., the analysis of rational double points as canonical singularities and the description of log terminal singularities as hyperquotients in Chapter 4) and for some explicit presentation of folklore-type results that are well known to experts but have never appeared in the literature (e.g., the toric Sarkisov program in Chapter 14). My priority is to present the Mori program in as easy and digested a form as possible, with the degree of motivational background that many have wished for but has not hitherto been revealed, and I apologize to those researchers who feel that the beauty of their original ideas has been deformed or lost in the process. Prerequisites are such that a graduate student who has read Hartshorne [3] or Iitaka [5] should have no difficulty understanding the material. (That is how much background knowledge I had when I started learning the subject.) My preference leans toward the geometric side rather than the purely algebraic, and this has made me feel free to switch back and forth between the analytic category and algebraic category.

This book consists of what I feel should roughly be the basics for an understanding of the global picture of the Mori program, leaving the hard-core analysis of 3-dimensional terminal singularities and the proof of the existence of flips to the original research literature. But this does not mean that the reader has to read the entire 400 plus pages to get a rough idea of what the Mori program is all about. The introduction in Chapter 0 and the overview in Chapter 3 should provide an easy guide for the remaining chapters, each of which could be read separately depending on the interest of the reader.

After circulating the first draft of this book in 1997, I received detailed suggestions and corrections from several people after some very careful readings, despite the numerous mistakes contained in the draft. Thanks are due to the attentions of D. Abramovich, V. Alexeev, B. Hassett, K. Hunt, S. Kovács and V. Masek. Especially the notes of K. Oguiso and J. Sakurai from the seminars held at Tokyo University were extremely helpful in revising the book.

A stay at Warwick University was crucial for the book at the final stage of its preparation. M. Reid provided linguistic, mathematical, and philosophical help to the author with warm hospitality. S. Mukai gave private tutoring sessions to the author on several topics through stimulating conversations in the dark winter nights of England.

I would like to thank Ina Lindemann of Springer-Verlag, who kindly and patiently waited for the completion of the manuscript, which was prepared and typeset using $\mathcal{A}_{\mathcal{M}}\mathcal{S}$-TEX.

I would also like to thank David Kramer, the copyeditor, who made an enormous effort to make this book readable.

The calligraphy on the front cover is done by Yumiko Takahashi.

This book was conceived and written during a most turbulent and happiest time, when our first baby, Mark Takamichi Matsuki, was born. I would like to dedicate this book to my wife, Dina Matsuki, known as Dinochka, who brought me a cup of tea every night after attending to crying Mark.

Contents

List of Notation

(Note that the notation for the groups in the table of Corollary 4-6-16 on Page 212 is different from that in the table of Theorem 4-6-20 on Pages 219 and 220.)

Introduction: The Tale of the Mori Program

The purpose of this book is to give an introductory and comprehensible account of what we call the **Mori program**, a program that emerged in the last two decades as an effective approach toward the biregular and/or birational classification theory of higher-dimensional algebraic varieties.

It may be said that one of the ultimate goals of algebraic geometry is to classify all the projective varieties $X \subset \mathbb{P}^n$ over $\mathbb{C}$ up to isomorphism.

The Mori program consists of the following four strategic schemes, MP 1 through MP 4, toward this goal, based upon a point of view in **birational geometry**:

| MP 1 | Find a good representative in a fixed birational equivalence class. |

| MP 2 | Study the properties of the good representative. |

| MP 3 | Study the (birational) relation among possibly many choices of the good representatives. |

| MP 4 | Construct the moduli space (fixing some discrete invariants like genus or Chern classes but varying the birational equivalence classes). |

We would like to demonstrate these strategic schemes in the classical case of algebraic curves, i.e., algebraic varieties of dimension 1, as a prototype of the higher-dimensional case.

Mori Program for Algebraic Curves

| MP 1 | Even if we fix a birational equivalence class, i.e., an isomorphism class of the function fields, we still have many projective varieties in the class with various singularities. (For example, consider a smooth rational curve, a cuspidal rational curve, a nodal rational curve, etc., all with the same function field.) But if we take the normalization, then it is nonsingular and unique in the fixed birational equivalence

class. We take this as a good representative for the varieties of dimension one, i.e., algebraic curves.

MP 2   Once we choose the good representative, a nonsingular projective curve C, our next object is to study its properties. The **canonical divisor** K_C, which defines the sheaf of regular holomorphic 2-forms $\mathcal{O}_C(K_C) \cong \Omega_C^2$, provides the decisive information, as in the following table:

C	$\mathbb{P}^1$	elliptic	hyperbolic
$g(C) = h^0(\mathcal{O}_C(K_C))$	0	1	≥ 2
$\deg K_C$	< 0	$= 0$	> 0
Universal Cover	$\mathbb{P}^1$	$\mathbb{C}$	the unit ball
$\mathrm{Aut}(C)$	$\mathrm{PGL}_1(\mathbb{C})$	$C \times$ finite grp	finite grp
Moduli	pt.	$\mathbb{A}^1$	a variety of dimension $3g(C)-3$

MP 3   The distinctive feature of dimension one is that there is a *unique* nonsingular projective representative in a fixed birational equivalence class. So there is no need to study the relation.

MP 4   Fixing the genus $g(C) \geq 2$, we construct the moduli space embedding good representatives into a projective space by the (pluri-)canonical maps and then apply geometric invariant theory to the corresponding (subspace of the) Hilbert scheme. When $g(C) = 1$, the moduli of elliptic curves can be studied based on their specific features as Abelian varieties, while we have the moduli consisting of a single point representing $\mathbb{P}^1$ if $g(C) = 0$.

Mori Program for Algebraic Surfaces

Chapter 1 is devoted to presenting the **Enriques classification** of algebraic surfaces in the framework of the Mori program. Though in the same basic spirit and principle as in dimension 1, our strategic schemes MP 1 though MP 4 become more involed in dimension 2 and start revealing their true features as we move toward higher dimensions. Eventually, we will observe that the Mori program provides a simpler and more unified way toward the classification than its classical counterparts.

MP 1   First, as in the case of dimension 1 for algebraic curves, we take a nonsingular projective representative by **resolution of singularities**. The major difference between dimensions 1 and 2 is that there are many nonsingular projective representatives even in a fixed birational equivalence class, due to the operation called **blowing up**. In order to avoid this ambiguity in choosing a nonsingular projective representative, one is naturally led to the inverse operation, called **blowing down**, via **Castelnuovo's contractibility criterion** of (-1)-curves. This will lead to the **minimal model program** (called **MMP** for short) in dimension 2. It can be considered a black box that given an input of an arbitrary nonsingular projective surface in a birational equivalence class produces either a **minimal surface** or a **ruled surface** as output by successive contractions of (-1)-curves. (A minimal surface is a minimal model in dimension 2, whereas a ruled surface is a Mori fiber space in dimension 2.) We take this output of MMP to be our good representative.

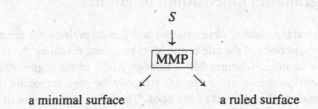

$$S$$
$$\downarrow$$
$$\boxed{\text{MMP}}$$

a minimal surface   a ruled surface

Note that the decisive factor in MMP is played by the canonical divisor K_S of the surface S, e.g., giving the characterization of a (-1)-curve C by its self-intersection $C^2 = -1$ and the intersection with the canonical divisor $K_S \cdot C = -1$.

$\boxed{\text{MP 2}}$ Just as the canonical divisor dictates the process of MMP, it also dictates the behavior of the good representative, a minimal surface or a ruled surface.

The importance of the characterization of a minimal surface, the canonical divisor being nef, is only reinforced by the **abundance theorem**, which asserts that the pluricanonical system is base point free and hence defines a morphism (onto the **canonical model**). The theorem plays an essential role in our classification schemes, in which we observe a fusion of the **Iitaka program** with the MMP.

A ruled surface has a rather rigid structure, since it is covered by rational curves. In dimension 2, via Tsen's lemma, they are birationally equivalent to a product of a curve and a rational curve $\mathbb{P}^1$.

$\boxed{\text{MP 3}}$ If the **Kodaira dimension** of a surface is greater than or equal to 0, then the outcome of MMP is necessarily a minimal surface, and it is uniquely determined by its birational equivalence class. Thus in this case there is no need to study the relation. On the other hand, if the Kodaira dimension of a surface is $-\infty$, then MMP produces many different outcomes. Here arises the necessity to study the relation among them. The classical **Castelnuovo–Noether theorem**, decomposing a given birational map between two different ruled surfaces into elementary transformations, is one of the most important descriptions of the relation. We present the theorem in the modern framework of the **Sarkisov program**.

$\boxed{\text{MP 4}}$ Though we will not discuss the moduli problem of surfaces (of general type) in Chapter 1 or in the book, the pluricanonical maps (which provide the major tool for the construction of the moduli) are discussed in some detail, leading to the basic properties of the canonical models and the canonical bundle formula for elliptic fibrations.

Chapter 2 is devoted to advertising **Iitaka's philosophy** on the **logarithmic category** (another important idea of Iitaka aside from the Iitaka program). Several key ideas of the Mori program, e.g., the celebrated Kawamata–Viehweg vanishing theorem, would look rather technical and unnatural if not for their logarithmic interpretations.

Mori Program in Dimension 3 or Higher

Having discussed the cases of dimensions 1 and 2 as its prelude, we are now ready to go into the main body of the tale of the Mori program, revealing the remarkable features of the strategic schemes MP 1 through MP 3 in the higher-dimensional setting. The fourth scheme, MP 4, though arguably the most important of all, will not be discussed at all in the rest of the book. There is a vast amount of literature on the classical subject of the moduli space of curves (See, e.g., Harris–Morrison [1] for references). We refer the reader only to the original research papers such as Alexeev [2][3] Kollár [2][8], Kollár–Shepherd–Barron [1], and the book by Viehweg [5] on the subject of the moduli of higher-dimensional varieties.

Chapter 3 takes the reader on a quick roller coaster ride of the Mori program to show him what kind of entertainment is awaiting in this Disney world, without going into the details of the proofs. We hope that the reader will have a glimpse of the global picture at this early stage.

Chapters 4 through 14 are dedicated to presenting the details of MP 1 through MP 3.

MP 1   First, as in the case of dimension 1 or 2, we take a nonsingular projective representative X by resolution of singularities. Second, we try to make X "smaller" by contracting some appropriate subvarieties analogous to the (-1)-curves on surfaces in dimension 2. The whole procedure is packaged in the so-called minimal model program.

The two main machineries in the package, the **cone theorem** and the **contraction theorem**, will be discussed with proofs in Chapters 7 and 8. Our presentation is based upon Kawamata–Reid–Shokurov's cohomological approach via the **Kawamata–Viehweg vanishing theorem**, the **nonvanishing theorem of Shokurov** and the **base point freeness theorem**.

We emphasize the point that the Kawamata–Viehweg vanishing theorem is nothing but the logarithmic generalization of the **Kodaira vanishing theorem** based upon Iitaka's philosophy of Chapter 2, and we present its proof from this point of view in Chapter 5.

The base point freeness theorem is discussed in Chapter 6 as one of the features of the adjoint linear systems in the spirit of Ein–Lazarsfeld [1]. Combined with the study of the locus of "just" log canonical singularities, the Kawamata–Viehweg vanishing theorem finds its powerful applications in the base point freeness theorem and nonvanishing theorem of Shokurov.

It should be noted that the above cohomological approach, centered on the analysis of the divisors, is dual to the original approach of Mori, centered on the analysis of the curves and their deformations. We discuss his original approach in Chapter 10 via the method of **bend and break**, which led him to the notion of **extremal rays**.

One of the new features of MMP in dimension 3 or higher is that we are inevitably and naturally forced to extend the category from that of nonsingular varieties to the one consisting of varieties with some specific **singularities**. A systematic study of the basic properties of these specific singularities, originated by Reid, is the subject of Chapter 4. Most important are the notions of **terminal and canonical singularities** together with their logarithmic generalizations, log terminal and log canonical singularities. We give the complete classification in dimension 2.

We discover (after Mori) another new feature, "**flip**," whose existence and termination will be the key to operating MMP and will be the subject of Chapter 9. The existence of flip is a deep theorem of Mori in dimension 3, whose full proof would require a book of its own. We leave this intricate and beautiful story, though central to our study, to the original research literature. We will be content merely to find the basic relation of the existence to the **finite generation conjecture of the (relative) canonical ring**.

With all these key ingredients in hand, we finally obtain the black box MMP, which produces either a **minimal model** or a **Mori fiber space**:

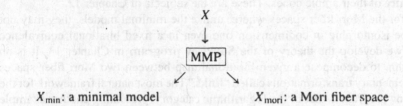

X_{min}: a minimal model X_{mori}: a Mori fiber space

MP 2   The canonical divisor dictates the behavior of the good representative. The Kodaira dimension κ, the fundamental birational invariant associated to the canonical divisor, determines the outcome of MMP (the **hard dichotomy conjecture**): It is a minimal model (respectively a Mori fiber space) iff $\kappa \geq 0$ (respectively $\kappa = -\infty$).

For a minimal model, whose characterization is given by the canonical divisor being nef, we have the **abundance conjecture**, which claims that the pluricanonical system is base point free. Via the conjecture, the classical **Iitaka fibration** given by the pluricanonical system is not only a rational map but a morphism onto a variety of (log) general type, called the canonical model, whose general fiber is a variety whose canonical divisor is zero (up to multiple). This leads to the **classification principle** that via the Iitaka fibration the study of algebraic varieties with $\kappa \geq 0$ is reduced to that of varieties of (log) general type and that of varieties whose canonical divisors are zero (up to multiple). For the varieties of (log) general type, we try to solve the moduli problem by embedding them into the projective space via the pluri(log)canonical maps. If the varieties whose canonical divisors are zero (up to multiple) have irregularity nonzero, then we can utilize the Albanese maps to analyze their structures. Thus, at the end, the study of varieties whose canonical divisors are zero (up to multiple) and irregularity zero, called the

Calabi–Yau varieties, remains a problem of central importance. The problem is, unfortunately, one of the subjects that we do *not* discuss at all in the book.

We look at the other end. One of the main features of a variety with $\kappa = -\infty$ (and hence whose representaive in its birational class is a Mori fiber space) is that it is covered by rational curves and hence is **uniruled**. This is proved in Chapter 10 as an application of Mori's bend and break method. The subtle difference between the notions of "ruled" and "uniruled" in higher dimensions is discussed briefly in Chapter 3 concerning the rationality question.

$\boxed{\text{MP 3}}$ Going to dimension 3 or higher, even for a birational equivalence class that produces a minimal model through MMP, we may not have a unique representative, i.e., we may have many different minimal models, though they are isomorphic in codimension one. This is in clear contrast to the case of lower dimensions 1 and 2, where we have the unique minimal model in a fixed birational equivalence class, and it is a distingushed feature in higher dimensions. The "local" property of the relation among any *two* of these various minimal models is that they are connected by a sequence of "**flops**," the codimension two operations similar to and in close connection with "flips." The more "global" property concerning the relation among *all* the minimal models is described in terms of the **chamber structure** of their ample cones. These are the subjects of Chapter 12.

As for the Mori fiber spaces where, unlike the minimal models, they may not even be isomorphic in codimension one even in a fixed birational equivalence class, we develop the theory of the **Sarkisov program** in Chapter 13. It is an algorithm to decompose a given birational map between two Mori fiber spaces into elementary transformations called "**links**." The most natural framework for the Sarkisov program is that of the logarithmic category, presenting another example of Iitaka's philosophy. We include Takahashi's work providing a new proof for the classical result on the structure of the automorphism group of the affine 2-space $\mathbb{A}^2$ as an application of the (log) Sarkisov program.

Chapter 14 is intended as a coffee break, where all the conjectures in the general setting of the Mori program have very clear interpretations in terms of the geometry of convex cones and hence are established theorems concerning toric varieties. We hope that the reader can play with the toric varieties and see all the ingredients in the previous chapters (if he has survived through them) in a transparent way.

It may be worthwhile to take a second look at the original goal (question) of classifying all the projective varieties $X \subset \mathbb{P}^n$ over $\mathbb{C}$ up to isomorphism, now that we have explained briefly the contents of each chapter of the book. There are no a priori reasons for the restrictions we put on the objects to classify: the varieties being projective, irreducible, and reduced over the base field, which is the field $\mathbb{C}$ of complex numbers.

The reason to put the projectivity restriction is rather methodological to use the techniques of the Hilbert schemes and various vanishing theorems of cohomology concerning the ample divisors.

If a given scheme is reducible, then we decompose it into the irreducible components. If a given scheme is not reduced, then we take the reduced structure with the same underlying topological space. Thus we consider the irreducible and reduced schemes (separated and of finite type over $\mathbb{C}$), i.e., those that are defined to be **varieties** in this book, as the building blocks to analyze the original reducible and/or nonreduced schemes.

Our main reason to stick to the category over $\mathbb{C}$ is again a methodological one to use resolution of singularities, the Kodaira vanishing theorem, and some analytic and/or transcendental methods such as Hodge theory. Most of our results are valid over an algebraically closed field of characteristic zero via the Lefschetz principle or some direct arguments, while establishing the Mori program in positive characteristic remains an obvious but difficult problem.

What Is Missing from the Book?

We try to be self-contained, proving most of the basic theorems from scratch. We list the major exceptions to this principle below:

1. We use resolution of singularities without proof. We refer the reader to Hironaka [2], Bierstone–Milman [1], Encinas–Villamayor [1], Villamayor [1], and Abramovich–De Jong [1], Bogomolov–Pantev [1], De Jong [1], Paranjape [1], for the recent developments.
2. The proof of (existence of flip) in dimension 3 by Mori [5], based upon a detailed structure theorem of terminal singularities in dimension 3 of Reid [2][3][7] and Mori [4].
3. The proofs of (existence of log flip) and (termination of log flips) in dimension 3 by Shokurov [3], Kawamata [15], Kollár et al. [1] (cf. Takagi [1]).
4. The proof of the abundance theorem in dimension 3 by Miyaoka [1][2][3], Kawamata [12], (cf. Kollár et al. [1]), and that of the log abundance theorem by Keel–Matsuki–McKernan [1] (cf. Kawamata [14], Kollár et al. [1]).
5. The proof of the boundedness of the $\mathbb{Q}$-Fano 3-folds with Picard number 1 by Kawamata [11] and that of S_3 (local) by Alexeev [1][2] (cf. Kollár et al. [1]). These facts are used to show termination of the Sarkisov program in dimension 3 in Chapter 13 following Corti [1].
6. The entire theory concerning the construction of the moduli spaces, especially that of varieties of (log) general type, in the starategic scheme of MP 4. We only refer the reader to the recent exciting developments of Alexeev [1][2][3], Karu [1], Kollár [2][9], Kollár–Shepherd–Barron [1], Viehweg [5].
7. Further discussion on the classification principle via the Iitaka fibration, especially the Iitaka conjecture on the (sub-)additivity of the Kodaira dimensions and the current developments on subject of Calabi–Yau varieties.

We hope that the reader, after looking at this book, will be well prepared to read, with good motivation, the original research papers for the missing material.

Prerequisites

We end this introduction by saying what is expected as the prerequisites for reading this book.

We expect that the reader should have mastery of the basic knowledge of algebraic geometry at the level of Hartshorne [3] or Iitaka [5] and complex analytic geometry at the level of Griffiths–Harris [1]. But this does *not* mean that the reader should finish reading all of these books before starting to read this book. We rather hope that the occasional reference to any of these books when needed should more than suffice to provide the confidence in rigor and understanding on the part of the reader. We often write (**Exercise! Why?**) whenever we feel that it would be better if the reader himself fills in the details of the argument. The level of difficulty of these exercises varies from one to another, but usually the solution is nothing more than an application or review of the materials discussed in that chapter or section.

Though the book is about algebraic varieties and hence on the subject of algebraic geometry, my preference leans toward the geometric side rather than the purely algebraic one, and therefore I felt free to switch back and forth between the algebraic category and the complex analytic category. Especially, the reader should be aware that all the cohomology groups are taken and computed in the analytic category with the usual analytic topology. For example, the exponential sequence

$$0 \to \mathbb{Z} \to \mathcal{O}_X \to \mathcal{O}_X^* \to 0$$

is valid only in the analytic category, and the cohomology groups with $\mathbb{Z}$-coefficients $H^i(X, \mathbb{Z})$ computed in the analytic category coincide with the usual singular cohomology groups, while the ones computed in the Zariski topology may not. Of course, for coherent sheaves on projective varieties, the cohomology groups computed in the algebraic category and the analytic category coincide via Serre's GAGA [1] and should cause no confusion.

CHAPTER 1

Birational Geometry of Surfaces

The purpose of this chapter is to present the main features of birational geometry of surfaces in the framework of the **Mori program**, following the main strategic schemes MP 1 through MP 4 as presented in the introduction.

The first three sections discuss the strategic scheme MP 1 for choosing a good representative in a birational equivalence class in dimension 2, which is drastically different from the one in dimension 1.

In dimension 1 there is a unique nonsingular projective curve, once we fix a birational equivalence class, as we discussed in Chapter 0. In dimension 2, however, after finding a nonsingular projective surface in a given birational equivalence class via resolution of singularities, we can create another in the same birational equivalence class through the operation of **blowing up**. It creates a distinctive curve called a (-1)-**curve**. Thus, in search of a good representative suggested in MP 1, we try to eliminate this ambiguity in the choice of nonsingular projective representatives by detecting a (-1)-curve and "blowing it down" via **Castelnuovo's contractibility criterion**. This leads to our crude form of the **minimal model program** (MMP for short) in dimension 2 in Section 1-1.

The key criterion for our crude form of MMP is this: Is there a (-1)-curve? Though natural and classical, the criterion does not provide much global information for an end result of the program. (We know only that it does not have any (-1)-curve.) One of Mori's brilliant ideas is to replace this criterion with the one dictated by the **canonical divisor** (canonical bundle) K_S of the surface S: Does the canonical divisor K_S have nonnegative intersection with any curve C on S, i.e., $K_S \cdot C \geq 0$? In other words, is K_S **nef**?

We show in Section 1-2 that when K_S is not nef, there is an **extremal contraction**. The notion of an extremal contraction turns out not only to generalize Castelnuovo's contractibility criterion but also to provide decisive information on the global structures of the end results of the program.

So have we established the MMP in dimension 2 with this new criterion? Not yet!

Section 1-3 gives the characterization of an extremal contraction in terms of the geometry of the convex cone generated by effective 1-cycles in the space of the numerical classes of curves: An extremal contraction corresponds to an **extremal ray** of the cone.

The **cone theorem** and **contraction theorem** finally establish the complete form of the MMP in dimension 2 at the end of Section 1-3, which provides a good representaive required in MP 1 in the form of a **minimal model** or a **Mori fiber space** as an end result of the program.

According to the strategic scheme MP 2, we now try to study the properties of these end results. Section 1-4 discusses the basic properties of the Mori fiber spaces in dimension 2, whereas Section 1-5 discusses those of the minimal models in dimension 2.

The **abundance theorem**, one of the key properties of the minimal models in dimension 2 discussed in Section 1-5, asserts that the pluricanonical system is base point free and hence that the associated rational map is actually a morphism. Section 1-6 studies this morphism, called the **Iitaka fibaration**, onto the **canonical model** of (the minimal model of) a surface. Though the strategic scheme MP 4 is hardly touched upon in this book, the embedding of (the canonical models of) surfaces of general type by the pluricanonical systems is crucial for the construction of the moduli.

Our study of birational geometry of surfaces culminates in Section 1-7, on the **Enriques classification** of surfaces, a proof of which is thus given now in the unified framework of the Mori program.

Section 1-8 discusses the birational relation among surfaces. We give a proof of the classical factorization theorem of birational maps in dimension 2 into blowups and blowdowns. The rest of the discussion is carried out according to the strategic scheme MP 3. Uniqueness of the minimal model in dimension 2 in a fixed birational equivalence class being the main feature of the birational relation on the one hand, we describe the relation among Mori fiber spaces in dimension 2 on the other via the modern interpretation of the classical **Castelnuovo–Noether theorem** as the **Sarkisov program** in dimension 2.

1.1 Castelnuovo's Contractibility Criterion

We start with the discussion of the operation of **blowing up** a point on a nonsingular projective surface with its basic properties.

Definition-Proposition 1-1-1. *Let $p \in T$ be a point on a nonsingular projective surface T. We take an analytic open neighborhood U of p with local coordinates (x, y). We define $\tilde{U}$ to be*

$$\tilde{U} = \{(s : t) \times (x, y) \in \mathbb{P}^1 \times U : xt - ys = 0\} \to U.$$

Observe the isomorphism

$$\tilde{U} - E \xrightarrow{\sim} U - \{p\},$$

where $E = \mathbb{P}^1 \times (0,0)$ is the inverse image of the origin $p = (0,0)$ under the natural projection. Thus we obtain the morphism

$$
\begin{array}{ccc}
S & \xrightarrow{\mu} & T \\
\| & & \| \\
\tilde{U} \cup (T - \{p\}) & \longrightarrow & U \cup (T - \{p\})
\end{array}
$$

where $\tilde{U}$ and $(T - \{p\})$ are amalgamated via the above-mentioned isomorphism to obtain S.

The morphism $\mu : S \to T$, called the **blowup** of T at p, has the following basic properties:

(i) *The blowup can be described as*

$$\mu : S := \mathrm{Blp}_p T = Proj \oplus_{d \geq 0} m_p{}^d \to T,$$

   and hence S is a nonsingular projective surface.

(ii) $\mu : S - \mu^{-1}(p) \xrightarrow{\sim} T - p.$

(iii) $\mu^{-1}(p) = e \cong \mathbb{P}^1$ *and* $E^2 = -1.$

(iv) $\mathrm{Pic}(S) = \mu^* \mathrm{Pic}(T) \oplus \mathbb{Z} \cdot E$ *and* $\dim_{\mathbb{R}} H^2(S, \mathbb{R}) = \dim_{\mathbb{R}} H^2(T, \mathbb{R}) + 1.$

(v) *We have the ramification formula $K_S = \mu^* K_T + E$ between the canonical divisors K_S of S and K_T of T.*

PROOF. For a proof, we refer the reader to Hartshorne [3], Section V.3, page 386–387. □

Our aim is to "detect" the blowup in terms of the distinctive properties of the (exceptional) curve E it creates.

Definition 1-1-2. *An irreducible and reduced curve E on a nonsingular projective surface S is called a (-1)-curve if*

$$E \cong \mathbb{P}^1 \quad and \quad E^2 = -1.$$

One of the main tools of our analysis throughout this chapter is the following.

Theorem 1-1-3 (Riemann–Roch Theorem on a Surface). *If D is any divisor on a nonsingular projective surface S, then*

$$\chi(\mathcal{O}_S(D)) = \frac{1}{2}(D - K_S) \cdot D + \chi(\mathcal{O}_S).$$

PROOF. For a proof, see Hartshorne [3] Section V.1 page 362–363. □

As an example of how we utilize the Riemann–Roch theorem, we prove the following characterization of (-1)-curves.

Lemma 1-1-4. *An irreducible and reduced curve E on a nonsingular projective surface S is a (-1)-curve iff*

$$K_S \cdot E < 0 \quad and \quad E^2 < 0.$$

PROOF.   From the Riemann–Roch theorem

$$\chi(\mathcal{O}_S(-E)) = \frac{1}{2}(K_S + E) \cdot E + \chi(\mathcal{O}_S).$$

By considering the exact sequence

$$0 \to \mathcal{O}_S(-E) \to \mathcal{O}_S \to \mathcal{O}_E \to 0,$$

we compute

$$\chi(\mathcal{O}_E) = h^0(E, \mathcal{O}_E) - h^1(E, \mathcal{O}_E)$$
$$= \chi(\mathcal{O}_S) - \chi(\mathcal{O}_S(-E)) = -\frac{1}{2}(K_S + E) \cdot E.$$

Since $h^0(E, \mathcal{O}_E) = 1$ for an irreducible reduced curve E, we derive the **arithmetic genus formula**.

$$h^1(E, \mathcal{O}_E) = \frac{1}{2}(K_S + E) \cdot E + 1.$$

We use the notation $(\Rightarrow)$ and $(\Leftarrow)$ to indicate the "only if" and "if" parts of the proof, respectively.

$(\Rightarrow)$ Since $E \cong \mathbb{P}^1$ and $E^2 = -1$,

$$0 = h^1(E, \mathcal{O}_E) = \frac{1}{2}(K_S + E) \cdot E + 1,$$

which implies $K_S \cdot E = -1$.

$(\Leftarrow)$ We have

$$0 \le h^1(E, \mathcal{O}_E) = \frac{1}{2}(K_S + E) \cdot E + 1.$$

Thus the inequalities imply

$$K_S \cdot E = E^2 = -1 \quad and \quad H^1(E, \mathcal{O}_E) = 0.$$

Now the claim is an easy consequence of the following exercise

Exercise 1-1-5. *Suppose E is a projective, irreducible, and reduced curve (but maybe singular a priori). Show that $E \cong \mathbb{P}^1$ if $H^1(E, \mathcal{O}_E) = 0$.*

[*Hint*. We have only to show that E is nonsingular or equivalently normal since $\dim E = 1$, since it is well known that a nonsingular projective curve E with $g(E) = \dim H^1(E, \mathcal{O}_E) = 0$ is isomorphic to $\mathbb{P}^1$. (Exercise! Check this "well-known" fact.)]

Let $\pi : \tilde{E} \to E$ be the normalization. Then we have the exact sequence

$$0 \to \mathcal{O}_E \to \pi_* \mathcal{O}_{\tilde{E}} \to \text{Coker} \to 0$$

and the cohomology sequence

$$H^0(E, \mathcal{O}_E) \to H^0(E, \pi_* \mathcal{O}_{\tilde{E}}) \to H^0(\text{Supp}(\text{Coker}), \text{Coker})$$
$$\to H^1(E, \mathcal{O}_E).$$

Since $H^1(E, \mathcal{O}_E) = 0$ by assumption and

$$H^0(E, \mathcal{O}_E) \tilde{\to} H^0(E, \pi_* \mathcal{O}_{\tilde{E}}) = H^0(\tilde{E}, \mathcal{O}_{\tilde{E}}) \cong \mathbb{C},$$

we conclude that since Supp (Coker) is a finite number of points,

$$H^0(\text{Supp}(\text{Coker}), \text{Coker}) \cong \text{Coker} = 0.$$

Therefore, the map $\mathcal{O}_E \hookrightarrow \pi_* \mathcal{O}_{\tilde{E}}$ is actually an isomorphism and hence so is π. Thus $E \cong \tilde{E}$ is normal.   □

The next theorem provides the key machinery for a crude form of MMP in dimension 2 to work.

Theorem 1-1-6 (Castelnuovo's Contractibility Criterion). *Let S be a nonsingular projective surface, $E \subset S$ a (-1)-curve. Then there exists a morphism called the contraction of E,*

$$\mu : S \to T,$$

onto another projective surface T such that

(i) *$\mu(E) = \text{pt.}(= p \in T)$ and $\mu : S - E \tilde{\to} T - p$,*
(ii) *T is nonsingular.*

In fact, $\mu : S \to T$ is the blowup of T at p.

PROOF. Note first that any morphism from S to a projective space

$$\mu : S \to \mathbb{P}^N \supset H$$

must be of the form

$$\mu : S \quad \to \qquad \mathbb{P}^N$$
$$\psi \qquad\qquad\qquad \psi$$
$$q \quad \to \quad \mu(q) = (\xi_0(q) : \xi_1(q) : \cdots : \xi_N(q)),$$

where H is a hyperplane of the projective space $\mathbb{P}^N$ and $\{\xi_0, \xi_1, \ldots, \xi_N\}$ are taken from $H^0(S, \mathcal{O}_S(\mu^* H))$ so that the sections $\xi_0, \ldots, \xi_N$ do not have any common zeros. (We say then that $|\mu^* H|$ has no base points.) When $\{\xi_0, \xi_1, \ldots, \xi_N\}$ are taken to be a basis of $H^0(S, \mathcal{O}_S(\mu^* H))$, we denote μ by $\Phi_{|\mu^* H|}$.

So finding the desired μ is equivalent to finding an appropriate divisor $\mu^* H = L$ (without a priori knowing μ). It is easy to see that in order for μ to satisfy the condition of the theorem, L has to satisfy the following numerical conditions:

(a) L is **nef**, i.e.,

$$L \cdot C \geq 0 \text{ for any curve } C \subset S.$$

(b) For any curve $C \subset S$

$$L \cdot C = 0 \Leftrightarrow C = e.$$

Our strategy is first to find a divisor L with the properties (a) and (b) and then to verify that such an L actually leads to the desired contraction.

Step 1. Construction of L.

Choose a very ample divisor A on S and set

$$L := A + kE,$$

where $k = A \cdot E$.

It is straightforward to check the conditions (a) and (b) for this choice of L.

Step 2. $H^0(S, \mathcal{O}_S(L))$ has no common zeros, i.e., $|L|$ has no base points, and thus

$$\Phi_{|L|} : S \to \mathbb{P}^N$$

is a morphism defined everywhere.

Since

$$|L| \supset |A| + kE,$$

where A is very ample, the base points of L, if any, can be only on E. Thus we have only to show that $|L|$ has no base point on E.

Consider the exact sequence

$$0 \to \mathcal{O}_S(L - E) \to \mathcal{O}_S(L) \to \mathcal{O}_E(L) \to 0,$$

which gives rise to the long cohomology sequence

$$H^0(S, \mathcal{O}_S(L)) \to H^0(E, \mathcal{O}_E(L)) \cong H^0(\mathbb{P}^1, \mathcal{O}_{\mathbb{P}^1})$$
$$\to H^1(S, \mathcal{O}_S(L - E)).$$

Since $H^0(\mathbb{P}^1, \mathcal{O}_{\mathbb{P}^1}) \cong \mathbb{C}$ has a section that does not vanish anywhere on $\mathbb{P}^1 \cong E$, it suffices to show that this section can be lifted to some section in $H^0(S, \mathcal{O}_S(L))$. This follows if we can show that

$$H^1(S, \mathcal{O}_S(L - E)) = H^1(S, \mathcal{O}_S(A + (k - 1)E)) = 0,$$

which will be justified in Step 5.

Step 3. $\mu(E) = pt(= p)$ and $\mu : S - E \xrightarrow{\sim} T - p$.

Again, since

$$|L| \supset |A| + kE,$$

where A is very ample, $\mu = \Phi_{|L|}$ is an isomorphism outside of E, since the sections corresponding to $|A|$ already separate points and tangents. Now

$$0 = L \cdot E = H \cdot \mu_* E$$

implies that $\mu(E)$ must be a point on T.

Step 4. T is nonsingular (at p).

This is the hardest and most subtle part of the proof.

We expect that

$$\mu : E \subset S \to p \in T$$

should be, in a neighborhood of p, analytically isomorphic to the blowup of the origin of A^2,

$$\mathrm{Bl}_{(0,0)} A^2 = \left\{ (s : t) \times (x, y) \in \mathbb{P}^1 \times A^2 ; xt - ys = 0 \right\} \to A^2.$$

The local coordinates (x, y) around the origin p should be given by the ratios of 3 sections $\xi_0, \xi_1, \xi_2 \in H^0(S, \mathcal{O}_S(L))$,

$$x = \frac{\xi_1}{\xi_0} \text{ and } y = \frac{\xi_2}{\xi_0},$$

where ξ_0 does not vanish anywhere along E, ξ_1 and ξ_2 do vanish on E as x and y vanish at p. Moreover, they should satisfy the equation

$$\frac{\xi_1}{\xi_0} \cdot t - \frac{\xi_2}{\xi_0} \cdot s = 0,$$

though the equation is meaningless at this point without specifying the homogeneous coordinates s and t in terms of the ξ_i.

Once this is understood, finding ξ_0, ξ_1, ξ_2 and proving T nonsingular at p is straightforward (though subtle) as follows.

Consider the exact seqeuence

$$0 \to \mathcal{O}_S(L - 2E) \to \mathcal{O}_S(L - E) \to \mathcal{O}_E(L - E) \cong \mathcal{O}_{\mathbb{P}^1}(1) \to 0,$$

which induces the cohomology sequence

$$H^0(S, \mathcal{O}_S(L - E)) \to H^0(E, \mathcal{O}_E(L - E))$$
$$\to H^1(S, \mathcal{O}_S(L - 2E)) = H^1(S, \mathcal{O}_S(A + (k - 2)E)) = 0,$$

where the vanishing of the last term will be justified in Step 5.

Then lift a basis

$$s, t \in H^0(E, \mathcal{O}_E(L - E)) \cong H^0(\mathbb{P}^1, \mathcal{O}_{\mathbb{P}^1}(1))$$

to

$$\xi_1, \xi_2 \in H^0(S, \mathcal{O}_S(L - E)).$$

Also consider the exact sequence

$$0 \to \mathcal{O}_S(L - E) \to \mathcal{O}_S(L) \to \mathcal{O}_E \to 0,$$

which induces the cohomology sequence

$$H^0(S, \mathcal{O}_S(L)) \to H^0(E, \mathcal{O}_E) \cong \mathbb{C}$$
$$\to H^1(S, \mathcal{O}_S(L - E)) = H^1(S, \mathcal{O}_S(A + (k - 1)E)) = 0.$$

Then take $\xi_0 \in H^0(S, \mathcal{O}_S(L))$, which maps to a nonzero constant in $\mathbb{C} \cong H^0(E, \mathcal{O}_E)$.

Observe that ξ_1 and ξ_2 considered as elements in $H^0(S, \mathcal{O}_S(L))$ through the injection $H^0(S, \mathcal{O}_S(L - E)) \hookrightarrow H^0(\mathcal{O}_S(L))$ map to $0 \in H^0(E, \mathcal{O}_E)$ and hence vanish along E.

Since the morphism μ is given by

$$\mu = \Phi_{|L|} : S \to T \subset \mathbb{P}^N$$
$$\cup \qquad \cup$$
$$q \mapsto (X_0 : X_1 : X_2 : \cdots) = (\xi_0(q) : \xi_1(q) : \xi_2(q) : \cdots),$$

the open neighborhood $S_0 = \{\xi_0 \neq 0\}$ of E in S maps entirely into $\mathbb{A}^N \cong \{X_0 \neq 0\} \subset \mathbb{P}^N$ and hence to $\mathbb{A}^2$ by the projection

$$\mathbb{A}^N \qquad \to \qquad \mathbb{A}^2$$
$$\cup \qquad\qquad\qquad \cup$$
$$\left(\frac{X_1}{X_0}, \frac{X_2}{X_0}, \cdots\right) \quad \mapsto \quad \left(\frac{X_1}{X_0}, \frac{X_2}{X_0}\right).$$

By construction the induced map $S_0 \to \mathbb{A}^2$ factors through the blowup of $\mathbb{A}^2$ at the origin

$$E \subset S_0 \qquad \xrightarrow{\mu} \qquad p \in T_0 \subset \mathbb{A}^N$$
$$\tau \downarrow \qquad\qquad\qquad \eta \downarrow$$
$$\mathbb{P}^1 \times (0,0) \subset \mathrm{Blp}_{(0,0)}\mathbb{A}^2 \quad \xrightarrow{\nu} \quad (0,0) \in \mathbb{A}^2$$

with τ defined as

$$(s : t) \times (x, y) = (\xi_1 : \xi_2) \times \left(\frac{\xi_1}{\xi_0}, \frac{\xi_2}{\xi_0}\right),$$

since then

$$xt - ys = \frac{\xi_1}{\xi_0} \cdot \xi_2 - \frac{\xi_2}{\xi_0} \cdot \xi_1 = 0.$$

Claim 1 of Step 4. *There exist analytic open neighborhoods $V \subset S_0$ of E and $W \subset \mathrm{Blp}_{(0,0)}\mathbb{A}^2$ of $\mathbb{P}^1 \times (0,0)$ such that the map τ induces analytic isomorphisms*

$$E \subset V \xrightarrow{\sim} \mathbb{P}^1 \times (0,0) \subset V.$$

Note first that, since ξ_1 and ξ_2 vanish along E, τ maps E to $\mathbb{P}^1 \times (0,0)$. Since

$$\left(\frac{\xi_2}{\xi_0}, \frac{\xi_1}{\xi_2}\right) \text{ or } \left(\frac{\xi_1}{\xi_0}, \frac{\xi_2}{\xi_1}\right) \left(\text{respectively } \left(y, \frac{x}{y}\right) \text{ or } \left(x, \frac{y}{x}\right)\right)$$

gives local coordinates on E (respectively on $\mathbb{P}^1 \times (0,0)$), τ is a local analytic isomorphism at each point $e \in E$, i.e., there exist open neighborhoods V_e of e in S_0 and $W_{\tau(e)}$ of $\tau(e)$ in $\mathrm{Blp}_{(0,0)}\mathbb{A}^2$ such that τ induces an analytic isomorphism between them:

$$e \in V_e \xrightarrow[\tau]{\sim} \tau(e) \in W_{\tau(e)}.$$

Since $\mathbb{P}^1 \cong \mathbb{P}^1 \times (0,0)$ is simply connected, we conclude that τ induces an isomorphism E onto $\mathbb{P}^1 \times (0,0)$:

$$\tau : E \xrightarrow{\sim} \mathbb{P}^1 \times (0,0).$$

Set

$$V' = \bigcup_{e \in E} V_e \quad \text{and} \quad W' = \bigcup_{e \in E} W_{\tau(e)}.$$

Take an open neighborhood V_0 of E so that the closure $\overline{V_0}$ is compact and $\overline{V_0} \subset V'$.

Consider the locus in $\overline{V_0}$ where τ fails to be injective, i.e.,

$$S = \{z \in \overline{V_0}; \exists z' \in \overline{V_0} \text{ s.t. } z' \neq z \text{ and } \tau(z) = \tau(z')\}.$$

It is easy to see that S is closed. In fact, if not, then there would exist a sequence $\{z_i\}$ in S,

$$z_i \to \tilde{z}, \text{ where } \tilde{z} \notin S \text{ is the limit of the sequence,}$$

while by taking a subsequence we may assume that the sequence $\{z_i'\}$, z_i' being as described in the definition of S, also converges:

$$z_i' \to \hat{z} \text{ for some } \hat{z} \in \overline{V_0},$$

since $\overline{V_0}$ is compact. Observe that $\tau(\tilde{z}) = \tau(\hat{z})$, since τ is continuous.

If $\tilde{z} \neq \hat{z}$, then by setting $\tilde{z}' = \hat{z}$, we conclude $\tilde{z} \in S$, a contradiction!

If $\tilde{z} = \hat{z} \in V_e$ for some $e \in E$, then for $i \gg 0$ we have

$$z_i, z_i' \in V_e, z_i \neq z_i' \text{ and } \tau(z_i) = \tau(z_i'),$$

contradicting the injectivity of τ restricted to V_e.

Thus S is closed.

Observe also that

$$\tau^{-1}(\mathbb{P}^1 \times (0,0)) \cap V' = e.$$

Since τ induces an isomorphism from E to $\mathbb{P}^1 \times (0,0)$, S is disjoint from E.

Now the claim is immediate, once we set

$$V = V_0 - S \quad \text{and} \quad W = \tau(V).$$

Claim 2 of Step 4. *There exists an open neighborhood U of $(0,0)$ such that $\eta : T_0 \to \mathbb{A}^2$ induces an analytic isomorphism*

$$p \in \eta^{-1}(U) \xrightarrow[\eta]{\sim} (0,0) \in U.$$

Take an open neighborhood U of $(0,0)$ such that

$$(0,0) \in U \subset \mathbb{A}^2 - \nu(\text{Blp}_{(0,0)}\mathbb{A}^2 - W).$$

Then by the previous claim it is easy to see that $\eta : \eta^{-1}(U) \to U$ is proper and bijective (and hence a homeomorphism) and that

$$\eta : \eta^{-1}(U) - \{p\} \to U - \{(0,0)\}$$

is an analytic isomorphism. Thus in order to show that

$$\eta : \eta^{-1}(U) \to U$$

is an isomorphism between the analytic spaces (as ringed spaces), we have only to show that

$$\Gamma(U', \mathcal{O}_{\mathbb{A}^2}) \overset{\eta^*}{\to} \Gamma(\eta^{-1}(U'), \mathcal{O}_{T_0}) \text{ for any open } U' \subset U$$

is an isomorphism, since their topologies coincide via the homeomorphism η. The map η^* is injective as a map of continuous functions. It is also surjective, since any holomorphic function $h \in \Gamma(\eta^{-1}(U'), \mathcal{O}_{T_0})$, considerd as a holomorphic function over $\eta^{-1}(U') - \{p\} \overset{\sim}{\to} U' - \{(0,0)\}$, extends to a holomorphic function over the entire U' by Hartogs's theorem.

This completes the proof of Step 4.

Step 5. Justification of

$$H^1(S, \mathcal{O}_S(L - E)) = H^0(S, \mathcal{O}_S(A + (k-1)E)) = 0,$$
$$H^1(S, \mathcal{O}_S(L - 2E)) = H^0(S, \mathcal{O}_S(A + (k-2)E)) = 0$$

by replacing A by its high multiple (and hence also replacing L by its high multiple).

Replacing A with its high multiple, we may assume

$$H^1(S, \mathcal{O}_S(A)) = 0$$

by the following.

Theorem 1-1-7 (Serre's Criterion for Ampleness: Serre's Vanishing Theorem).
A divisor A is ample on S iff any of the conditions below holds:

(i) *For any coherent sheaf $\mathcal{F}$ on S*

$$H^i(S, \mathcal{F} \otimes \mathcal{O}_S(nA)) = 0 \quad i > 0$$

for $n \gg 0$.

(ii) *For any coherent sheaf $\mathcal{F}$ on S, $\mathcal{F} \otimes \mathcal{O}_S(nA)$ is generated by its global sections.*

(iii) *For any line bundle $\mathcal{L}$ on S, $\mathcal{L} \otimes \mathcal{O}_S(nA)$ is (very) ample for $n \gg 0$.*

We claim that

$$H^1(S, \mathcal{O}_S(A + iE)) = 0 \text{ for } i = 0, 1, \ldots, k+1,$$

which we show by induction in i.

For $i = 0$, we have

$$H^1(S, \mathcal{O}_S(A + 0 \cdot E)) = 0$$

by our choice of A.

For $i > 0$, consider the exact sequence

$$0 \to \mathcal{O}_S(A + (i-1)E) \to \mathcal{O}_S(A + iE) \to \mathcal{O}_E(A + iE) \cong \mathcal{O}_{\mathbb{P}^1}(k-i) \to 0,$$

which induces the cohomology sequence

$$H^1(S, \mathcal{O}_S(A + (i-1)E)) \to H^1(S, \mathcal{O}_S(A + iE)) \to H^1(\mathbb{P}^1, \mathcal{O}_{\mathbb{P}^1}(k - i)),$$

where the third term vanishes for $k - i > -2$. (Exercise! Why?) Now the assertion is clear by induction.

This completes the proof of Step 5 and hence that of Theorem 1-1-6. $\square$

Remark 1-1-8.

(i) It is worth noting that Step 5 is essential in the above proof, though it may look technical at first sight. As a consequence of the theorem, we expect that $L = \mu^* H$ for some ample divisor H on T and that $-E$ and $-K_S$ are both μ-ample, since $-E \cdot E = -K_S \cdot E = 1 > 0$. Therefore, for sufficiently large $l \in \mathbb{N}$, we have both

$$lL - E - K_S$$

and

$$lL - 2E - K_S$$

ample. (We leave it as an exercise for the reader to show only from the assumptions of the theorem that both divisors are ample using, e.g., the Nakai–Moishezon criterion for ampleness (cf. Theorem 1-2-5), without referring to the consequence of the theorem.) By replacing lL with L, we conclude as in Step 5 that

$$H^1(S, \mathcal{O}_S(L - E)) = H^1(S, \mathcal{O}_S(L - E - K_S + K_S)) = 0,$$
$$H^1(S, \mathcal{O}_S(L - 2E)) = H^1(S, \mathcal{O}_S(L - 2E - K_S + K_S)) = 0,$$

by the Kodaira vanishing theorem (cf. Theorem 1-2-13). Here we observe a prototype of the idea of showing the existence of some contraction morphism (a suitable base point free linear system) via the Kodaira vanishing theorem (or its generalizations), which will be further developed in the later chapters.

(ii) The key point of Step 4 is to find (regular) local coordinates (x, y) of T at p. We can expect that (x, y) maps a neighborhood of p in T to a neighborhood of the origin in $\mathbb{A}^2$ only via an *analytic* isomorphism but *not* via an *algebraic* isomorphism (unless T is rational!!). Thus it is inevitable to go to the analytic category or argue in the étale topology. Accordingly, Hartshorne [3] carries out the argument using formal functions. His argument has the advantage of being valid even in positive characteristic, while ours, at the cost of being restricted to the field of complex numbers, may be more elementary.

Exercise 1-1-10.

(i) Define the notion of "blowing up" a higher-dimensional nonsingular variety along some nonsingular subvariety called the center of the blowing up.

(ii) By the same argument as in Theorem 1-1-6, prove the following contractibility criterion of $E \cong \mathbb{P}^{n-1}$ in a nonsingular projective variety X of dimension n with $\mathcal{O}_E(E) \cong \mathcal{O}_{\mathbb{P}^{n-1}}(-1)$: There exists a contraction morphism

$$\mu : X \to Y$$

such that

(a) $\mu(E) = \mathrm{pt}. = p \in Y$ and $\mu : X - E \widetilde{\to} Y - p$,
(b) Y is nonsingular projective.

In fact, $\mu : X \to Y$ is the blowup of Y at p.

(iii) Contractibility in the *projective* category of a $\mathbb{P}^l$-bundle that appears to arise from blowing up a nonsingular center is very subtle, as opposed to Castelnuovo's criterion or the above contractibility criterion: Construct an example of a $\mathbb{P}^1$-bundle $r : E \to C$ over a nonsingular projective curve C embedded in a nonsingular projective 3-fold X such that the normal bundle has the property

$$\mathcal{N}_{E/X}|_{r^{-1}(p)} \cong \mathcal{O}_X(E)|_{r^{-1}(p)} \cong \mathcal{O}_{\mathbb{P}^1}(-1)$$

for each fiber $r^{-1}(p) \cong \mathbb{P}^1$ of r but such that E can never be contracted to a projective variety, i.e., there is *No* morphism $\phi : X \to Y$ with the following properties:

(a) $\phi : E \to \phi(E)$ coincides with $r : E \to C$,
(b) $\phi : X - E \widetilde{\to} Y - \phi(E)$, and
(c) Y is normal and *projective*.

Using Castelnuovo's contractibility criterion, we describe our crude form of the minimal model program in dimension 2 in Flowchart 1-1-11.

Flowchart 1-1-11.

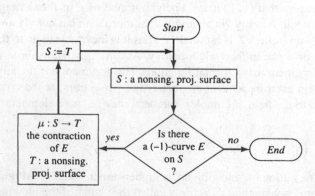

Remark 1-1-12.

(i) The program has to come to an *End* after finitely many executions of it, since every time we contract a (-1)-curve the second Betti number drops by 1:

$$0 \le \dim_{\mathbb{R}} H^2(T, \mathbb{R}) = \dim_{\mathbb{R}} H^2(S, \mathbb{R}) - 1.$$

(ii) The crude form of MMP in dimension 2 as above provides little global information as to the end results, except for telling us that they do not have any (-1)-curves. (The end results of this crude MMP are classically called "relatively minimal.") This is one of the reasons why the global analysis of the relatively minimal models toward the classification of surfaces tends to be a collection of ad hoc methods if we stick to this crude form of MMP in dimension 2.

1.2 Surfaces Whose Canonical Bundles Are Not Nef I

Throughout Section 1-2 the letter S denotes a nonsingular projective surface.

Recall that the key criterion for our crude form of MMP in dimension 2 is,

Is there a (-1)-curve on S?

One of the brightest ideas of Mori is to replace this with the following new criterion on the canonical bundle K_S:

Is K_S nef?

At first sight, this new and simple criterion does not seem to bring much improvement to MMP over our crude form. In fact, it seems we would be left clueless if the answer to the above question were *no*. (The analysis of the surfaces when the answer to the above question is *yes* is the main subject of Section 1-5.) Here Mori saves us by providing decisive information:

If K_S is not nef, then there exists an "extremal contraction."

The purpose of this section is to explain what an extremal contraction is and to verify the above assertion via the

rationality theorem, and the
base point freeness theorem.

We follow the approach developed by Kawamata–Reid–Shokurov–Kollár, who studied the behavior of divisors, dual to the original appoach of Mori, who studied the behavior of curves.

Then we present our provisional form of MMP in dimension 2, which will be completed in the next section via the

cone theorem and the
contraction theorem.

Theorem 1-2-1 (Extremal Contraction). *If K_S is not nef, then there exists a morphism*

$$\phi : S \to W$$

called an **extremal contraction** *such that*

(i) *ϕ is not an isomorphism,*

(ii) *for a curve $C \subset S$*

$$\phi(C) = pt \implies K_S \cdot C < 0,$$

(iii) *all the curves contracted by ϕ are numerically proportional, i.e.,*

$$\phi(C) = pt \ and \ \phi(D) = pt \implies [D] = c[C] \ in \ H_2(S, \mathbb{R}) \ for \ some \ c \in \mathbb{Q}_{>0},$$

(iv) *ϕ has connected fibers with W normal and projective.*

One remark should be made to explain informally the definition of an extremal contraction, which may look somewhat technical at first, so that it will sink into the mind of the reader as a natural notion.

Condition (i) is to avoid the identity morphism from the extremal contractions, since it does not provide any information on S in the analysis of its structure.

Condition (ii) is also natural, since an extremal contraction should occur exclusively when K_S is *not* nef. When we want to emphasize the reference to the canonical bundle K_S, we call it an extremal contraction "**with respect to K_S.**"

Now we would like an extremal contraction to be a building block to construct the entire picture of the structure of the surface S. As such, we would like each building block ϕ to contract *as little as possible*: Observe that for any morphism $\psi : S \to V$ onto a projective variety, if two curves C and D are numerically proportional, then for an ample divisor H on V

$$\psi(C) = pt \iff \psi^*H \cdot C = 0 \iff \psi^*H \cdot D = 0 \iff \psi(D) = pt$$

That is to say, once a curve is contracted, all the curves in its numerical class (up to multiple) must also be contracted. Condition (iii) requires in return that those be the *only* curves contracted by an extremal contraction.

For ϕ to contract "as little as possible," W has to be "as big as possible." Even after requiring condition (iii) and specifying the numerical class of the curves to be contracted, the biggest W has to be characterized by condition (iv) as the Stein factorization (cf. Proposition 1-2-16).

Brief Review on the Intersection Theory on Surfaces

First we review briefly the intersection theory on a nonsingular projective surface S, though it has already been used in Section 1-1.

There is an intersection pairing between the line bundles and the 1-cycles on S,

$$\text{Pic}(S) \times Z_1(S) \to \mathbb{Z},$$

induced from the natural pairing

$$H^2(S, \mathbb{Z}) \times H_2(S, \mathbb{Z}) \to \mathbb{Z},$$

through the homomorphisms

$$\mathrm{Pic}(S) \to H^2(S, \mathbb{Z}),$$
$$Z_1(S) \to H_2(S, \mathbb{Z}).$$

Exercise 1-2-2. *Prove that the above pairing coincides with the following pairings defined in two different ways:*

(i) $$\mathrm{Pic}(S) \times Z_1(S) \to \mathbb{Z}$$

given by

$$L \cdot C = \deg_{\tilde{C}} \mu^* L,$$

where L is a line bundle and C is a curve with the normalization $\mu : \tilde{C} \to C$,

(ii) $$\mathrm{Pic}(S) \times Z_1(S) \to \mathbb{Z}$$

given by

$$L \cdot D = \chi(\mathcal{O}_S) - \chi(L^{-1}) - \chi(\mathcal{O}_S(-D)) + \chi(L^{-1} \otimes \mathcal{O}_S(-D))$$

for a line bundle L and $D \in Z_1(S)$ *considered a divisor on S, by showing that all these pairings satisfy the characterization*

1. $\mathcal{O}_S \cdot D = 0 \quad \forall D \in Z_1(S)$,
2. $\mathcal{O}_S(D_1) \cdot D_2 = \mathcal{O}_S(D_2) \cdot D_1 \quad \forall D_1, D_2 \in Z_1(S)$,
3. $L \cdot (a_1 D_1 + a_2 D_2) = a_1 L \cdot D_1 + a_2 L \cdot D_2 \quad \forall L \in \mathrm{Pic}(S), D_1, D_2 \in Z_1(S)$ *and* $a_1, a_2 \in \mathbb{Z}$, *and*
4. $L \cdot C = \deg_C L$ *for a nonsingular curve* $C \subset S$ *and* $L \in \mathrm{Pic}(S)$.

We extend this pairing to the one over $\mathbb{R}$,

$$N^1(S) := \{\mathrm{Pic}(S)/ \equiv\} \otimes_{\mathbb{Z}} \mathbb{R} \times N_1(S) := \{Z_1(S)/ \equiv\} \otimes_{\mathbb{Z}} \mathbb{R} \to \mathbb{R},$$

where $\equiv$ denotes **numerical equivalence**.

We note that the natural map

$$\tau : \mathrm{Pic}(S) \to H^2(S, \mathbb{R})$$

induces a surjective map

$$\mathrm{Pic}(S)/\ker(\tau) \to \mathrm{Pic}(S)/ \equiv,$$

since (co) homological equivalence implies numerical equivalence by definition. Therefore, the **Picard number** $\rho(S)$ is finite:

$$\rho(S) = \dim_{\mathbb{R}} N_1(S) = \dim_{\mathbb{R}} N^1(S) \leq \dim_{\mathbb{R}} H^2(S, \mathbb{R}) < \infty.$$

We use the notation

$$\mathrm{NE}(S) := \text{the convex cone generated by effective 1-cycles,}$$
$$\overline{\mathrm{NE}}(S) := \text{the closure of NE}(S) \text{ in } N_1(S).$$

Remark 1-2-3. (i) Actually, (co)homological equivalence and numerical equivalence coincide: In order to check that a 1-cycle D is (co)homologous to 0, one has to check that its intersection with any class in $H^2(S, \mathbb{R})$ is 0, thanks to Poincaré duality. But since D is algebraic and hence its Poincaré dual sits in $H^{1,1}(S, \mathbb{R})$, we have only to test the intersection with all the classes in $H^{1,1}(S, \mathbb{R}) \subset H^2(S, \mathbb{R})$. Now thanks to the Lefschetz theorem, which asserts that every (1,1)-class

$$\gamma \in \dot{H}^{1,1}(S, \mathbb{R}) \cap H^2(S, \mathbb{Z})$$

is analytic, i.e., there is a line bundle $L \in \mathrm{Pic}(S)$ such that

$$c_1(L) = \tau(L) = \gamma$$

(cf. Griffiths–Harris [1] page 163), we conclude that we have only to test the intersection with all line bundles. Thus (co)homological equivalence coincides with numerical equivalence.

Therefore, we have the inclusions

$$N^1(S) \hookrightarrow H^2(S, \mathbb{R}) \quad \& \quad N_1(S) \hookrightarrow H_2(S, \mathbb{R}).$$

(ii) By abuse of notation, we often do not distinguish a divisor D from its associated line bundle $\mathcal{O}_S(D)$. Accordingly, we speak not only of the intersection of line bundles with curves but also that of divisors with curves.

Definition 1-2-4 (The notion of "nef"). *A line bundle L on S is* **nef** *if*

$$L \cdot C \geq 0 \text{ for any curve } C \subset S,$$

which is equivalent to saying

$$L \geq 0 \text{ on } \mathrm{NE}(S),$$

or

$$L \geq 0 \text{ on } \overline{\mathrm{NE}}(S).$$

There are two criteria for ampleness that play an important role in what follows.

Theorem 1-2-5 (Kleiman's Criterion for Ampleness). *A line bundle L on S is ample iff*

$$L > 0 \text{ on } \overline{\mathrm{NE}}(S).$$

We remark that the closure "–" in the criterion is crucial, since

$$L \text{ ample} \Rightarrow L > 0 \text{ on } \mathrm{NE}(S)$$

holds true, but

$$L \text{ ample} \Leftarrow L > 0 \text{ on } \mathrm{NE}(S)$$

does not hold in general.

PROOF. See Kleiman [1] or Hartshorne [2] for a proof. In Hartshorne [2], one also finds an example, by Mumford, showing the failure of the last implication in the above remark. □

Theorem 1-2-6 (Nakai–Moishezon Criterion for Ampleness). *A line bundle L on S is ample if*

$$L^2 > 0 \text{ and } L \cdot C > 0 \text{ for any curve } C \subset S.$$

PROOF. See Hartshorne [3] Section V.1, Theorem 1.10, page 365. □

Strategy to prove Theorem 1-2-1

How should we find an extremal contraction $\phi : S \to W$?

We analyze what we should expect assuming that we have already found such a ϕ. If we take an ample divisor H on W and its pullback $L = \phi^* H$ on S, Kleiman's criterion tells us that the hyperplane

$$L^\perp = \{z \in N_1(S); L \cdot z = 0\}$$

"touches" $\overline{NE}(S)$, since it is nef but not ample. If we take

$$A = L - rK_S \text{ for some small } r \in \mathbb{Q}_{>0},$$

then $A^\perp$ will then be away from $\overline{NE}(S)$, implying that A is ample.

Figure 1-2-7.

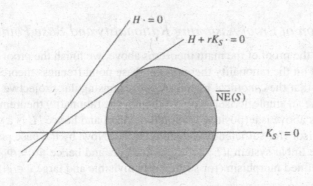

As in the proof of Castelnuovo's contractibility criterion, in order to find ϕ, it suffices to find L. We backtrack the process above.

Take an arbitrary ample divisor A on S. Then we would like to perturb A by some multiple of K_S until the hyperplane associated to it touches $\overline{NE}(S)$.

There are two fundamental questions we have to ask in order to make this approach legitimate:

Question 1 (Rationality). *Set $r = \sup\{t; A + tK_S \text{ is nef}\}$.*

Is r a rational number?

By setting $L = A + rK_S$ formally, $L^\perp$ touches $\overline{\text{NE}}(S)$ as required, no matter whether r is rational or not. But if r is not rational, then L (or its multiple) is not a well-defined integral divisor, and hence we cannot hope to use L to construct ϕ. Thus the affirmative answer to Question 1 is crucial for our approach.

Suppose r is rational as we desire. We still have to face one more question:

Question 2 (Base-Point Freeness). *Suppose r is rational. Then does L (or its sufficiently divisible and large multiple lL) give a base point free linear system $|lL|$?*

The affirmative answer to Question 2 is again crucial to our approach, since we would like to construct $\phi = \Phi_{|lL|}$, as a morphism associated to the linear system $|lL|$, well-defined everywhere on S, since $|lL|$ is base point free. (Note that actually in the proof that follows we construct ϕ from $\Phi_{|lL|}$ and we do not quite assert that $\phi = \Phi_{|lL|}$ is an extremal contraction, due to the technical possibility that $L^\perp$ might touch $\overline{\text{NE}}(S)$ along a face of dimension 2 or more, and hence $\Phi_{|lL|}$ might contract curves not numerically proportional to each other.)

The affirmative answers to Questions 1 and 2 will be the two main theorems of this section:

rationality theorem,
base point freeness theorem.

Completion of Proof Assuming Rationality and Base Point Freeness

Postponing the proof of the main theorems above, we finish the proof of Theorem 1-2-1 based on the rationality theorem and base point freeness theorem.

Suppose that the canonical bundle K_S of a nonsingular projective surface S is not nef. Take an ample divisor A on S. Then by the rationality theorem the number r defined as above is a (positive) rational number, and hence lL is a well-defined integral divisor (for sufficiently divisible $l \in \mathbb{N}$). Now, by the base point freeness theorem the linear system $|lL|$ is base point free, and hence $\Phi = \Phi_{|lL|} : S \to V$ is a well-defined morphism (for sufficiently divisible and large $l \in \mathbb{N}$).

> Case: $\dim V = \dim S = 2$

Since $\Phi : S \to V$ is not an isomorphism, we can take an irreducible curve E on S such that $\Phi(E) = $ pt. (Exercise! Why?) Then $K_S \cdot E < 0$, since $0 = L \cdot E = (A + rK_S) \cdot E = 0$ with $r > 0$, A being ample and $E^2 < 0$. Thus E is a (-1)-curve by Lemma 1-1-4. We have only to set $\phi = \mu : S \to T = W$ to be the contraction of E. We note that $E^2 < 0$ is a consequence of the Hodge index theorem or negativity of self-intersection of contractible curves.

Theorem 1-2-8 (Hodge Index Theorem). *The intersection form on S has the signature $(1, \rho(S) - 1)$, where $\rho(S) = \dim_{\mathbb{R}} N^1(S)$. In particular, if a nonzero divisor E has intersection $D \cdot E = 0$ for a divisor with $D^2 > 0$, then $E^2 < 0$.*

PROOF.   See Hartshorne [3] Section V.1, Theorem 9, Page 364.   □

We note that in our case for an ample divisor H on V and $D = \Phi^* H$ we have $D \cdot E = 0$ and $D^2 > 0$.

Theorem 1-2-9 (Negativity of Self-Intersection of Contractible Curves). *Let $\Phi : S \to V$ be a proper birational morphism from a nonsingular surface S. Then for any nonzero divisor E on S with $\Phi(E) = pt$ we have $E^2 < 0$.*

PROOF.   We refer the reader to Theorem 4-6-1 in Section 4-6.   □

We go back to the proof of Theorem 1-2-1.

Case: dim $V = 1$

We take the Stein factorization of $\Phi : S \to V$ and use the same notation Φ for it by abuse of notation.

Subcase: There is a reducible fiber $\Phi^{-1}(p)$ for some $p \in V$.

Take an irreducible component $E \overset{\neq}{\subset} \Phi^{-1}(p)$. Since E is contained in a fiber, $L \cdot E = (A + r K_S) \cdot E = 0$, which implies $K_S \cdot E < 0$. The next lemma shows that $E^2 < 0$ and thus E is a (-1)-curve. Again we have only to set $\phi = \mu : S \to T = W$ to be the contraction of E.

Lemma 1-2-10. *Let $\Phi : S \to V$ be a morphism with connected fibers from a nonsingular projective surface S onto a nonsingular projective curve V. Then for any divisor D whose support lies in fibers, we have*

$$D^2 \leq 0.$$

Moreover,

$$D^2 = 0 \Leftrightarrow D \text{ is a (rational) multiple of full fibers.}$$

PROOF.   Take a very ample divisor H on V. Then $\Phi^* H \cdot D = 0$ and $(\Phi^* H)^2 = 0$. Thus the Hodge index theorem implies $D^2 \leq 0$. Suppose $D^2 = 0$. This implies $D_p^2 = 0$, where D_p is the part of D contained in a fiber $\Phi^{-1}(p)$. Thus we may assume that D itself is contained in a fiber $\Phi^{-1}(p)$. If D is not a multiple of $F = \Phi^{-1}(p)$, then we can take $a \in \mathbb{Q}$ (Exercise! Why?) such that

$$D + aF \text{ is nonzero effective and } \mathrm{Supp}(D + aF) \overset{\neq}{\subset} \mathrm{Supp} F.$$

Since F is connected, we can take an irreducible component $G \subset \mathrm{Supp} F$ such that $(D + aF) \cdot G > 0$. Then since $(D + aF)^2 = D^2 = 0$, we have for $0 < x \ll 1$

$$0 \geq \{(D + aF) + xG\}^2 = 2x(D + aF) \cdot G + x^2 G^2 > 0,$$

a contradiction!   □

Subcase: There is no reducible fiber.

Then all the fibers are algebraically equivalent being parametrized by the curve V and thus numerically equivalent. Therefore, we have only to set $\Phi = \phi : S \to V = W$.

$\boxed{\text{Case: } \dim V = 0}$

In this case, we necessarily have

$$A + rK_S \equiv 0 \text{ i.e., } A \equiv -rK_S.$$

If $\rho(S) = \dim_{\mathbb{R}} N_1(S) = 1$, then all the curves on S are numerically proportional, and we have only to set $\Phi = \phi : S \to V = W$.

If $\rho(S) > 1$, then take another ample divisor A' not numerically proportional to A. (Exercise! Why can we take such an A'?) Then the contraction morphism $\Phi' : S \to V'$ attached to $L' = A' + r'K_S$, constructed similarly as before, must necessarily be in one of the previous cases (otherwise, we would end up with $A' \equiv -r'K_S$, a contradiction to the choice of A').

This completes the proof of Theorem 1-2-1 assuming the rationality theorem and base point freeness theorem.

It remains to prove the rationality theorem and the base point freeness theorem as the affirmative answers to Questions 1 and 2. In dimension 2, their proofs can be obtained easily just using the Kodaira vanishing theorem and the intersection theory on surfaces, which we have already been utilizing.

Theorem 1-2-11 (Rationality Theorem in Dimension 2). *Let S be a nonsingular projective surface whose canonical bundle is not nef, A an ample divisor on S.*

Then the positive number $r := \sup\{t; A + tK_S \text{ is nef}\}$ is rational, i.e., $r \in \mathbb{Q}$.

PROOF.   We first consider an easy case where $|kK_S| \neq \emptyset$ for some $k \in \mathbb{N}$ and hence we can take a member $\Sigma d_i D_i \in |kK_S|$. Then

$A + tK_S$ nef

$\Leftrightarrow (A + tK_S) \cdot C \geq 0$ for any curve $C \subset S$

$\Leftrightarrow \left(A + t \left(\frac{1}{k} \Sigma d_i D_i \right) \right) \cdot C \geq 0$ for any curve $C \subset S$

$\Leftrightarrow \left(A + t \left(\frac{1}{k} \Sigma d_i D_i \right) \right) \cdot D_j \geq 0$ for $\forall j$.

Therefore, in this case

$$r = \min \left\{ t_j; \left(A + t_j \left(\frac{1}{k} \Sigma d_i D_i \right) \right) \cdot D_j = 0 \right\}$$

is a rational number.

The same argument proves the following.

Lemma 1-2-12. *Suppose there exists a rational number $r_0 \geq r$ such that $|k(A + r_0 K_S)| \neq \emptyset$ for some $k \in \mathbb{N}$. Then r is a rational number.*

Now we prove the general case. Assuming r is irrational, we would show that there should exist r_o as above (which would indicate that r is rational, leading to a contradiction!).

Consider the following quadratic polynomial in x and y:

$$P(x, y) := \chi(\mathcal{O}_S(xA + yK_S))$$
$$= \frac{1}{2}(xA + yK_S - K_S) \cdot (xA + yK_S) + \chi(\mathcal{O}_S),$$

If r is irrational, it follows from elementary number theory that there are infinitely many pairs (u, v) of positive integers such that

$$0 < \frac{v}{u} - r < \frac{1}{3u}.$$

For any of these pairs (u, v), we observe that the polynomial $P(ku, kv)$ is quadratic in k and that

$$P(ku, kv) = 0 \text{ as a polynomial in } k \Leftrightarrow P(x, y) \text{ divisible by } (xv - yu).$$

Choose a pair (u_0, v_0) such that

$$0 < \frac{v_0}{u_0} - r < \frac{1}{3u_0}$$

and such that $(xv_0 - yu_0)$ does not divide $P(x, y)$ and thus

$$P(ku_0, kv_0) \neq 0 \text{ as a polynomial in } k.$$

Then, for $k = 1, 2,$ or 3, we should have

$$0 \neq P(ku_0, kv_0) = \chi(\mathcal{O}_S(ku_0 A + kv_0 K_S))$$
$$= h^0(S, \mathcal{O}_S(ku_0 A + kv_0 K_S)) - h^1(S, \mathcal{O}_S(ku_0 A + kv_0 K_S))$$
$$+ h^2(S, \mathcal{O}_S(ku_0 A + kv_0 K_S)).$$

Since

$$ku_0 A + kv_0 K_S = K_S + ku_0 \left(A + \frac{kv_0 - 1}{ku_0} K_S \right),$$

where

$$ku_0 \left(A + \frac{kv_0 - 1}{ku_0} K_S \right) \text{ is ample as } 0 \leq \frac{kv_0 - 1}{ku_0} < r,$$

the Kodaira vanishing theorem gives

$$h^i(S, \mathcal{O}_S(ku_0 A + kv_0 K_S)) = h^i(S, \mathcal{O}_S(K_S + ku_0 \left(A + \frac{kv_0 - 1}{ku_0} K_S \right))) = 0$$

for $i = 1, 2$. But then

$$h^0(S, \mathcal{O}_S(ku_0 A + kv_0 K_S)) = P(ku_0, kv_0) \neq 0,$$

and thus we have obtained

$$r_0 = \frac{ku_0}{ku_0} > r \text{ with } |ku_0(A + r_0 K_S)| \neq \emptyset,$$

which would imply that r is rational by Lemma 1-2-12, a contradiction!

Theorem 1-2-13 (Kodaira Vanishing Theorem). *Let* S *be a nonsingular projective variety and* A *an ample divisor on* S. *Then*

$$H^i(S, \mathcal{O}_S(K_S + A)) = 0 \text{ for } i > 0.$$

We refer the reader to Chapter 5 for a discussion of the Kodaira vanishing theorem.

Theorem 1-2-14 (Base Point Freeness Theorem in Dimension 2). *Let* L *be a (rational) divisor on* S *of the form*

$$L = A + r K_S,$$

where A *is an ample divisor,* r *a nonnegative rational number.*
    Assume that L *is nef.*
    Then $|lL|$ *is base point free for sufficiently (divisible and) large* $l \in \mathbb{N}$ *(so that* lL *is an integral divisor).*

PROOF.   First we give a "fake" proof, which demonstrates the core of the idea utilizing the Kodaira vanishing theorem, though it is not rigorous as it is. It can be modified and made rigorous via the Kawamata–Viehweg vanishing theorem. See Section 6-2 for details.

"Fake" Proof: Suppose $p \in S$ is a base point of the linear system $|lL|$ for sufficiently (divisible and) large $l \in \mathbb{N}$. Let $\nu : \tilde{S} \to S$ be the blowup of S at p with the exceptional curve $E = \nu^{-1}(p)$.

Consider the exact sequence

$$0 \to \mathcal{O}_{\tilde{S}}(\nu^* lL - E) \to \mathcal{O}_{\tilde{S}}(\nu^* lL) \to \mathcal{O}_E(\nu^* lL) \cong \mathcal{O}_E \to 0,$$

which induces the cohomology sequence

$$H^0(\tilde{S}, \mathcal{O}_{\tilde{S}}(\nu^* lL)) \to H^0(E, \mathcal{O}_E) \to H^1(\tilde{S}, \mathcal{O}_{\tilde{S}}(\nu^* lL - E)).$$

If

$$H^1(\tilde{S}, \mathcal{O}_{\tilde{S}}(\nu^* lL - E)) = 0,$$

then a nonzero constant in $\mathbb{C} \cong H^0(E, \mathcal{O}_E)$ lifts to a section in $H^0(\tilde{S}, \mathcal{O}_{\tilde{S}}(lL)) \cong H^0(S, \mathcal{O}_S(lL))$. This section does not vanish along E and hence not at p. This is a contradiction, since we chose p to be a base point.

Thus it suffices to show that

(*) $H^1(\tilde{S}, \mathcal{O}_{\tilde{S}}(\nu^* lL - E)) = H^1(\tilde{S}, \mathcal{O}_{\tilde{S}}(\nu^*(lL - K_S) - 2E + K_{\tilde{S}})) = 0.$

Observe that $lL - K_S$ is ample on S and $-2E$ is ν-ample.

Therefore, we conclude (**this is the point where this proof fails!**) that

$$\nu^*(lL - K_S) - 2E \text{ is ample on } \tilde{S}.$$

Now, $(*)$ would have been a consequence of the Kodaira vanishing theorem (see Theorem 1-2-13). (Exercise! Figure out why this is a fake proof.)

"Real" Proof: Now we give a real proof, which is slightly tedious, since we try to stick to elememtary methods without using the Kawamata–Viehweg vanishing theorem at the moment.

Note that $L^2 \geq 0$. In fact,

$$L^2 = \lim_{n \to \infty} \frac{1}{n^2}(nL + A)^2,$$

where A is an ample divisor on S. By Kleiman's criterion, $nL + A$ is also ample for $\forall n \in \mathbb{N}$ and hence $(nL + A)^2 > 0$. Thus we have the desired inequality.

> Case: $L^2 > 0$

Suppose $L \cdot G > 0$ for any curve $G \subset S$. Then since $L^2 > 0$, the Nakai–Moishezon criterion implies that L is ample, in which case the assertion is clear.

Thus we may assume that there exists a curve E such that $L \cdot E = 0$ (and thus r must be strictly positive). But then $K_S \cdot E < 0$ and $E^2 < 0$ by the Hodge index theorem. This implies that E is a (-1)-curve. Let $\mu : S \to S'$ be the contraction of E.

Then it is easy to see (Exercise! Why?) that

(i) $L = \mu^* L'$ for some nef divisor L' on S', and
(ii) $L' = \mu_* L = \mu_*(A + rK_S) = \mu_* A + rK_{S'}$ and $\mu_* A$ is ample on S'.

We repeat the same argument as above replacing L on S with L' on S'. We conclude that either L' is ample or there is a contraction of a (-1)-curve E' with $L' \cdot E' = 0$. In the first case the assertion follows immediately, while in the second case we keep repeating the process. The process must come to an end after finitely many steps, since we cannot contract (-1)-curves infinitely many times (recall that the Betti number drops by 1 when we contract a (-1)-curve.), reaching the stage where we are eventually in the first case and hence the assertion follows.

> Case: $L^2 = 0$ and $L \not\equiv 0$, i.e., there is a curve D such that $L \cdot D > 0$

We claim that $L \cdot K_S < 0$. In fact, first take $h \gg 0$ such that $|hA - D| \neq \emptyset$ by Serre's criterion for ampleness (cf. theorem 1-1-7). Then

$$L \cdot hA = L \cdot ((hA - D) + D) > 0.$$

Now $L \cdot A > 0$ and $L^2 = L \cdot (A + rK_S) = 0$ implies (since r must be positive in this case)

$$L \cdot K_S < 0.$$

It also follows from Kleiman's criterion that

$$lL - K_S = \frac{1}{r}A + \frac{lr - 1}{r}L$$

is ample for $l \gg 0$.

The Riemann–Roch theorem yields

$$\chi(\mathcal{O}_S(lL)) = h^0(S, \mathcal{O}_S(lL)) - h^1(S, \mathcal{O}_S(lL)) + h^2(S, \mathcal{O}_S(lL))$$

$$= \frac{1}{2}(lL - K_S) \cdot lL + \chi(\mathcal{O}_S)$$

$$= \frac{1}{2}l(-K_S \cdot L) + \chi(\mathcal{O}_S).$$

Since $lL - K_S$ is ample, the Kodaira vanishing theorem implies that

$$H^i(S, \mathcal{O}_S(lL)) = H^i(S, \mathcal{O}_S(K_S + lL - K_S)) = 0 \text{ for } i > 0.$$

Thus $h^0(S, \mathcal{O}_S(lL))$ grows linearly in l, since $L \cdot K_S < 0$. Therefore, we may take $l \gg 0$ such that

$$|lL| = |M| + F,$$

where $|M|$ is the nonempty movable part of $|lL|$ and F is the fixed part. Observe that M is nef, since it has no fixed component. Thus we conclude that

$$0 \le M^2 \le M \cdot (M + F) = M \cdot lL \le (M + F) \cdot lL = (lL)^2 = 0,$$

which implies

$$M^2 = M \cdot F = F^2 = 0.$$

Thus $|M|$ is base point free, since $|M|$ has no fixed component and $M^2 = 0$. Let $\Phi_{|M|} : S \to V'$ be the associated morphism and $S \overset{\Phi}{\to} V \overset{\Psi}{\to} V'$ its Stein factorization. Then $M \cdot F = 0$ implies that F is contained in fibers of Φ. From Lemma 1-2-10 it follows that $F^2 = 0$ implies that F is proportional to full fibers and hence that $F = \Phi^*(\Sigma a_i P_i)$ for some $a_i \in \mathbb{Q}_{>0}$, where $P_i \in V$. Therefore,

$$|l'lL| = |l'\Phi^*(\Psi^*\mathcal{O}_{V'}(1) + \Sigma a_i P_i)|$$

is base point free for sufficiently divisible $l' \in \mathbb{N}$, since $\Psi^*\mathcal{O}_{V'}(1) + \Sigma a_i P_i$ is ample on V.

Case: $L^2 = 0$ and $L \equiv 0.$

We have only to show that

$$h^0(S, \mathcal{O}_S(lL)) \ne 0 \text{ for sufficiently divisible } l \in \mathbb{N}.$$

Note that

$$lL - K_S \equiv -K_S \equiv \frac{1}{r}A$$

is ample for any $l \in \mathbb{Z}$. Therefore, the Kodaira vanishing theorem and Riemann–Roch theorem yield

$$h^0(S, \mathcal{O}_S(lL)) = \chi(\mathcal{O}_S(lL)) = \frac{1}{2}(lL - K_S) \cdot lL + \chi(\mathcal{O}_S)$$

$$= \frac{1}{2}(0 - K_S) \cdot 0 + \chi(\mathcal{O}_S) = \chi(\mathcal{O}_S) = h^0(S, \mathcal{O}_S) = 1 \ne 0.$$

This completes the proof of the base point freeness theorem in dimension 2.   □

We now discuss a special feature, which will be one of the key ingredients in the proof of the Cone theorem in the next section, of the rational number r in the statement of the rationality theorem.

Corollary 1-2-15 (Boundedness of the Denominator in Dimension 2). *Let S be a nonsingular projective surface whose canonical bundle is not nef, A an ample divisor on S. Let $r := \sup\{t; A + tK_S$ is nef$\}$ be the rational number as in the rationality theorem. Then the denominator of r is bounded. More specifically,*

$$r = \frac{p}{q}, \text{ where } p \in \mathbb{N} \text{ and } q = 1, 2, \text{ or } 3.$$

PROOF.   We use the notation of the proof of the base point freeness theorem.

Case: $L^2 > 0$

In this case there exists a (-1)-curve E such that $L \cdot E = 0$. Thus $r = \frac{A \cdot E}{-K_S \cdot E} = A \cdot E$ is an integer.

Case: $L^2 = 0$ and $L \not\equiv 0$

In this case, $\Phi_{|lL|} : S \to V$ is a morphism onto a curve for sufficiently (divisible and) large $l \in \mathbb{N}$. For a general fiber F of Φ (the Stein factorization of $\phi_{|lL|}$), we have $F^2 = 0$ and $K_S \cdot F < 0$. Thus the arithmetic genus formula $0 \le h^1(\mathcal{O}_F) = \frac{1}{2}(K_S + F) \cdot F + 1$ implies $K_S \cdot F = -2$. Therefore, $r = \frac{A \cdot F}{-K_S \cdot F}$ has denominator equal to either 1 or 2.

Case: $L^2 = 0$ and $L \equiv 0$

If $\rho(S) = \dim_{\mathbb{R}} N_1(S) > 1$, then take another ample divisor A' that is not numerically proportional to A. Then $L' = A' + r'K_S$ must be in one of the previous cases (r' being defined similarly for A'), and thus there is a curve G with $K_S \cdot G = -1$ or -2. Then since $L \equiv 0$, we have $L \cdot G = 0$, which implies that $r = \frac{A \cdot G}{-K_S \cdot G}$ has denominator equal to either 1 or 2.

If $\rho(S) = 1$, then $-K_S$ is ample. Thus the exponential sequence and Kodaira vanishing theorem give the exact cohomology sequence

$$0 = H^1(S, \mathcal{O}_S) \to H^1(S, \mathcal{O}_S^*) \to H^2(S, \mathbb{Z}) \to H^2(S, \mathcal{O}_S) = 0,$$

and thus

$$\text{Pic}(S) \cong H^1(S, \mathcal{O}_S^*) \cong H^2(S, \mathbb{Z}).$$

Therefore, all the cohomology classes arise from the line bundles and hence

$$\text{Pic}(S) \cong \mathbb{Z} \oplus \text{torsion},$$

since $\dim_{\mathbb{R}} H^2(S, \mathbb{Z}) \otimes \mathbb{R} = \dim_{\mathbb{R}} N_1(S) = 1$. Let H be an ample divisor that corresponds to $1 \in \mathbb{Z}$. Set $-K_S \equiv kH$, where $k \in \mathbb{N}$. Suppose $k > 3$. Then for $x = 1, 2, 3$ we have by the Kodaira vanishing theorem

$$\chi(\mathcal{O}_S(K_S + xH)) = h^0(S, \mathcal{O}_S(K_S + xH)) = 0,$$

since $K_S + xH \equiv (x - k)H$ and $x - k < 0$. This would imply that $\chi(\mathcal{O}_S(K_S + xH)) = 0$ as a polynomial in x, a contradiction!

(For $x \gg 0$ we should have $\chi(\mathcal{O}_S(K_S + xH)) = h^0(S, \mathcal{O}_S(K_S + xH)) > 0$ by Serre's ampleness criterion.) Therefore, we conclude that $k = 1, 2,$ or 3 and hence that $r = \frac{A \cdot C}{-K_S \cdot C}$ (C any curve on S) has denominator equal to either 1, 2, or 3. $\qquad\square$

We make explicit at this point a characterization of the Stein factorization of the morphism $\Phi_{|lL|} : S \to W'$ given by a base point free linear system $|lL|$ that we have been using implicitly.

Proposition 1-2-16 (Stein Factorization). *Let L be a line bundle on a normal projective variety such that the linear system $|lL|$ is base point free for all sufficiently large $l \in \mathbb{N}$.*

Let $\Phi_{|lL|} : S \to W'$ be the morphism associated to $|lL|$, and $S \xrightarrow{\Phi} W \xrightarrow{\Psi} W'$ its Stein factorization (cf. Hartshorne [3] Chapter III). Then

(i) *Φ is a morphism with connected fibers onto a normal projective variety W,*
(ii) *$L \cdot C = 0$ for a curve $C \subset S$ iff C is in a fiber of Φ,*
(iii) *$\Phi_* \mathcal{O}_S = \mathcal{O}_W$,*
(iv) *$L = \Phi^* H$ for some ample line bundle H on W,*
(v) *Φ coincides with $\Phi_{|lL|}$ for any sufficiently large $l \in \mathbb{N}$.*

Moreover, Φ is characterized either by conditions (i) and (ii) or (iii) and (ii).

PROOF.   It is immediate that the Stein factorization $\Phi : S \to W$ with W constructed as

$$\Psi : W = Spec\, \Phi_{|lL|_*} \mathcal{O}_S \to W'$$

satisfies conditions (ii) and (iii), Ψ being a finite morphism, since $\Phi_{|lL|_*} \mathcal{O}_S$ is a coherent $\mathcal{O}_{W'}$-module by Hartshorne [3], Corollary 5.20 in Chapter II. The morphism Φ also has connected fibers, thanks to Zariski's Main Theorem below, verifying condition (i).

It is easy to see that conditions (i) and (ii) characterize W with the Zariski topology up to homeomorphism: The points of W are set-theoretically just the equivalence ($\sim$) classes of points of S, where $P \sim Q$ iff there exists a chain of curves $C_1, C_2, \ldots, C_h$ such that $P \in C_1$, $Q \in C_h$, $C_a \cap C_{a+1} \neq \emptyset$ and such that $L \cdot C_a = 0$ for all a. The topology is given by the rule that $U \subset W$ is open iff $\phi^{-1}(U) \subset S$ is open. Since $\Phi_* \mathcal{O}_S$ is a coherent module over $\mathcal{O}_W$, again by Hartshorne [3], and since W is normal, we conclude that $\Phi_* \mathcal{O}_S = \mathcal{O}_W$. This determines the ringed space structure of W, providing the characterization.

If we assume conditions (ii) and (iii) instead, then by Zariski's Main Theorem Φ has connected fibers. As in the previous argument, this connectedness with the condition (ii) determines W as a set together with the (Zariski) topology. Condition (iii) now determines the ringed space structure of W, hence giving the characterization.

In order to see (iv), take sufficiently large and relatively prime $p, q \in \mathbb{N}$ with $rp + sq = 1$ for some $r, s \in \mathbb{Z}$. Then by the characterizations we see that the Stein factorizations of $\Phi_{|pL|} : S \to W_p$ and $\Phi_{|qL|} : S \to S_q$ coincide to give

$$S \xrightarrow{\Phi} W \xrightarrow{\Psi_p} W_p,$$
$$S \xrightarrow{\Phi} W \xrightarrow{\Psi_q} W_q.$$

Since $pL = \Phi_{|pL|}^* H_p$ and $qL = \Phi_{|qL|}^* H_q$ for some ample divisors H_p on W_p and H_q on W_q, we conclude that

$$L = (rp + sq)L$$
$$= r\Phi_{|pL|}^* H_p + s\Phi_{|qL|}^* H_q$$
$$= \Phi^*(r\Psi_p^* H_p + s\Psi_q^* H_q) = \Phi^* H,$$

where $H = r\Psi_p^* H_p + s\Psi_q^* H_q$ is ample on W (Exercise! Why?), establishing condition (iv).

Take $l_0 \in \mathbb{N}$ such that lH is very ample for $l \geq l_0$. Then $\Phi_{|lL|}$ coincides with the Stein factorization Φ for $l \geq l_0$. This proves condition (v). $\square$

Theorem 1-2-17 (Zariski's Main Theorem). *Let* $\Phi : S \to W$ *be a projective morphism of noetherian schemes, and assume that* $\Phi_*(\mathcal{O}_S) = \mathcal{O}_W$. *Then* $\Phi^{-1}(p)$ *is connected for any point* $p \in W$.

PROOF.   For a proof, see Hartshorne [3], Corollary 11.3, page 279. $\square$

We conclude this section by presenting a flowchart for MMP in dimension 2, still in a provisional form, with the key criterion being, "Is K_S nef?"

Flowchart 1-2-18.

Minimal Model Program in Dimension 2 (still a provisional form)

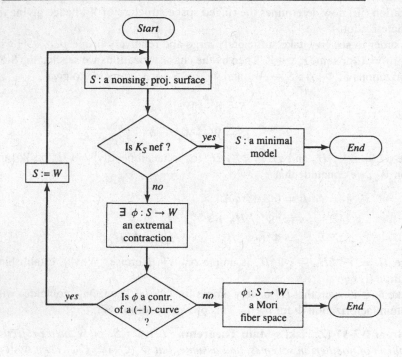

1.3   Surfaces Whose Canonical Bundles Are Not Nef II

In the previous section we presented MMP in dimension 2 in a provisional form.
What is missing is the characterization of the half-line called an "**extremal ray**,"

$$R_l = \mathbb{R}_+[C] \subset \overline{NE}(S) \subset N_1(S),$$

which contains the numerical classes of all the curves C contracted by an extremal contraction $\phi : S \to W$.

The main purpose of this section is to discuss the

cone theorem
contraction theorem,

which give the desired characterization. The cone theorem describes the extremal rays purely in terms of convex geometry of the cone $\overline{NE}(S)$ in regard to the intersection with the canonical bundle K_S, while the contraction theorem guarantees that each of those extremal rays described in the cone theorem is associated to a geometric extremal contraction.

Our main tools of the proof in this section are the rationality and base point freeness theorems and the boundedness of the denominator of the previous section.

Finally, at the end of the section we present the complete form of the minimal model program in dimension 2.

Theorem 1-3-1 (Cone Theorem in Dimension 2). *Let S be a nonsingular projective surface. Then the closure of the cone of effective curves has the description*

$$\overline{NE}(S) = \overline{NE}(S)_{K_S \geq 0} + \Sigma R_l,$$

where

$$\overline{NE}(S)_{K_S \geq 0} := \{z \in \overline{NE}(S); K_S \cdot z \geq 0\}$$

and the R_l are half-lines such that $R_l - \{0\}$ are in $\overline{NE}(S)_{K_S < 0}$ and such that they are of the form

$$R_l = \overline{NE}(S) \cap L^{\perp}$$

for some nef line bundles L.

Moreover, the R_l are discrete in the half-space $N_1(S)_{K_S < 0}$.

Remark-Definition 1-3-2. (i) The half lines R_l are called **extremal rays**. In fact, it is easy to see that the R_l are extremal in the sense of convex geometry: If $z_1 + z_2 \in R_l$ for $z_1, z_2 \in \overline{NE}(S)$, then $z_1, z_2 \in R_l$.

When we want to specify the condition $K_S \cdot z < 0$ for $z \in R_l - \{0\}$, we call these the extremal rays **with respect to K_S**.

(ii) Though the extremal rays R_l are discrete in the half-space $\overline{NE}(S)_{<0}$, they could have accumulation points on the hyperplane $\overline{NE}(S) \cap K_S^{\perp}$.

PROOF. Step 1. For any nef line bundle M with

$$\overline{NE}(S)_{K_S < 0} \cap M^{\perp} \neq \emptyset,$$

there exists an extremal ray R_l as above such that

$$R_l \subset \overline{NE}(S) \cap M^{\perp}.$$

The basic idea to verify the assertion above is to perturb M by some (small portion of) an ample divisor B so that $M + \epsilon B + r K_S$ will cut out a face of smaller dimension from $\overline{NE}(S)$.

Figure 1-3-3.

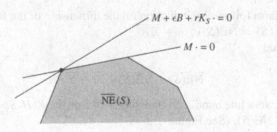

$M + \epsilon B + r K_S \cdot = 0$

$M \cdot = 0$

$\overline{NE}(S)$

What makes this idea legitimate is the boundedness of the denominator (cf. Corollary 1-2-15). In the following, instead of taking $M + \epsilon B$ with ϵ small, we consider $\nu M + B$ with ν big.

Define

$$r_M(\nu, B) := \sup\{t; \nu M + B + t K_S \text{ is nef}\}.$$

Since M is nef, $r_M(\nu, B)$ is a nondecreasing function of ν. Since the denominator of $r_M(\nu, B)$ is bounded and since $r_M(\nu, B)$ is bounded from above uniformly,

$$r_M(\nu, B) \leq \frac{B \cdot z}{-K_S \cdot z} \text{ for some fixed } 0 \neq z \in \overline{NE}(S)_{K_S < 0} \cap M^{\perp},$$

with the right-hand side independent of ν, $r_M(\nu, B)$ stabilizes to a fixed $r_M(B)$ for $\nu \geq \nu_B$. Therefore,

$$0 \neq \overline{NE}(S) \cap ((\nu_B + 1)M + B + r_M(B)K_S)^{\perp} \subset \overline{NE}(S) \cap M^{\perp}.$$

Take $B_1, B_2, \ldots, B_\rho$ to be ample divisors forming a basis of $N^1(S)$ with $\rho = \dim_{\mathbb{R}} N^1(S)$. Then

$$((\nu_{B_i} + 1)M + B_i + r_M(B_i)K_S) \cdot \{\} = 0, \quad i = 1, 2, \ldots, \rho,$$

give at least $\rho - 1$ independent linear conditions (by increasing some ν_{B_i} if necessary). Thus if

$$\dim \overline{NE}(S) \cap M^{\perp} > 1,$$

then there exists i such that

$$0 \neq \overline{NE}(S) \cap ((\nu_{B_i} + 1)M + B_i + r_M(B_i)K_S)^{\perp} \overset{\neq}{\subset} \overline{NE}(S) \cap M^{\perp}.$$

(In fact, $(\nu_{B_i} + 1)M + B_i + r_M(B_i)K_S$ is nef but not ample by the choice of $r_M(B_i)$, and hence by Kleiman's criterion there exists $0 \neq z \in \overline{NE}(S)$ such that $\{(\nu_{B_i} + 1)M + B_i + r_M(B_i)K_S\} \cdot z = 0$.)

Note that

$$\overline{NE}(S)_{K_S < 0} \cap ((\nu_{B_i} + 1)M + B_i + r_M(B_i)K_S)^{\perp} \neq \emptyset.$$

In order to see this, observe that since $(\nu_{B_i} + 1)M + B_i$ is ample and hence $(\nu_{B_i} + 1)M + B_i$ is strictly positive on $\overline{NE}(S) - \{0\}$ by Kleiman's criterion, for $0 \neq z \in ((\nu_{B_i} + 1)M + B_i + r_M(B_i)K_S)^{\perp}$ we have $(\nu_{B_i} + 1)M + B_i \cdot z > 0$, and hence $K_S \cdot z < 0$.

Now the claim follows by induction on the dimension of the face $\overline{NE}(S) \cap M^{\perp}$.

Step 2. $\overline{NE}(S) = \overline{NE}(S)_{K_S \geq 0} + \Sigma R_l$.

Suppose that

$$\overline{NE}(S) \overset{\neq}{\subset} \overline{NE}(S)_{K_S \geq 0} + \Sigma R_l.$$

Then there exists a line bundle D such that $D > 0$ on the R.H.S. $- \{0\}$ but $D \cdot z < 0$ for some $z \in \overline{NE}(S)$. (See Figure 1-3-4)

Figure 1-3-4.

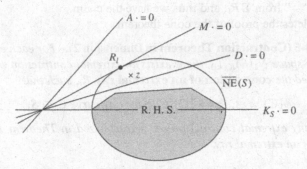

It follows immediately that $A = h(D + aK_S)$ for some negative rational number a and sufficiently divisible $h \in \mathbb{N}$ is an ample integral divisor. By the rationality theorem we obtain a nef line bundle $M = l(A + rK_S)$ ($r \in \mathbb{Q}_{>0}$ and some $l \in \mathbb{N}$) such that $\overline{\mathrm{NE}}(S)_{K_S<0} \cap M^{\perp} \neq \{0\}$ and that R.H.S. $\cap M^{\perp} = \{0\}$. But by Step 1, there exists $R_l \subset \overline{\mathrm{NE}}(S) \cap M^{\perp}$ and $R_l \subset$ R.H.S., a contradiction!

Step 3. The R_l are discrete in the half-space $N_1(S)_{K_S<0}$.

If K_S is numerically trivial, then there is no extremal ray (with respect to K_S) and nothing to prove. So we may assume that K_S is not numerically trivial.

Choose ample divisors $A_1, A_2, \ldots, A_{\rho-1}$ such that

$$K_S, A_1, A_2, \ldots, A_{\rho-1}$$

form a basis of $N^1(S)$. For each R_l and an associated nef line bundle L such that

$$R_l = \overline{\mathrm{NE}}(S) \cap L^{\perp},$$

take $r_L(A_i)$ and ν_{A_i} as in Step 1. Then

$$0 \neq \overline{\mathrm{NE}}(S) \cap ((\nu_{A_i} + 1)L + A_i + r_L(A_i)K_S)^{\perp} = \overline{\mathrm{NE}}(S) \cap L^{\perp} = R_l.$$

Therefore,

$$\{A_i + r_L(A_i)K_S\} \cdot \xi_l = 0 \text{ for } 0 \neq \xi_l \in R_l,$$

which implies

(∗)
$$\frac{A_i \cdot \xi_l}{-K_S \cdot \xi_l} = r_L(A_i).$$

Since we can take

$$\frac{A_i \cdot \{\}}{-K_S \cdot \{\}} \text{ for } i = 1, 2, \ldots \rho - 1$$

as coordinates of the hyperplane $\{z \in N_1(S); -K_S \cdot z = 1\}$ and since the denominators of the $r_L(A_i)$ are bounded by Corollary 1-2-15, we conclude from (∗) that the intersections of the R_l and the hyperplane are discrete, which is exactly the claim.

Step 4. $\overline{\mathrm{NE}}(S) = \overline{\mathrm{NE}}(S)_{K_S \geq 0} + \Sigma R_l$.

Since the R_l are discrete in the half-space $N_1(S)_{K_S<0}$, we can get rid of the closure sign "—" from ΣR_l, and thus we have the claim.

This completes the proof of the cone theorem.   □

Theorem 1-3-5 (Contraction Theorem in Dimension 2). *For each extremal ray R_l in the half-space $N_1(S)_{K_S<0}$, there exists an extremal contraction $\phi = \text{cont}_{R_l} :$ $S \to W$, called the* **contraction of an extremal ray R_l,** *such that*

$$\phi(C) = \text{pt.} \iff [C] \in R_l \text{ for any curve } C \subset S.$$

Conversely, any extremal contraction as characterized in Theorem 1-2-1 is the contraction of an extremal ray.

PROOF.   For each extremal ray R_l, there exists by definition a nef line bundle L such that

$$R_l = \overline{NE}(S) \cap L^\perp.$$

Then $A = L - rK_S$ is ample for some small $r \in \mathbb{Q}_{>0}$ by Kleiman's criterion, and thus $L = A + rK_S$ has the property that $|lL|$ is an integral and base point free linear system for sufficiently divisible and large $l \in \mathbb{N}$ by the base point freeness theorem. It follows from Proposition 1-2-16 that $\phi = \Phi_{|lL|} : S \to W$ is an extremal contraction with the required property.

Conversely, suppose $\phi : S \to W$ is an extremal contraction as defined in Theorem 1-2-1. Take an ample divisor H on W. Then $M = \phi^* H$ satisfies the conditions in Step 1 of the cone theorem. Therefore, there exists an extremal ray R_l such that

$$R_l \subset \overline{NE}(S) \cap M^\perp.$$

Now it is easy to see (Exercise! Why?) that the contraction cont_{R_l} of the extremal ray R_l coincides with ϕ.   □

With the cone theorem and contraction theorem at hand, we can now present the complete form of the minimal model program in dimension 2, which from now on we simply refer to as MMP in dimension 2.

Flowchart 1-3-6.

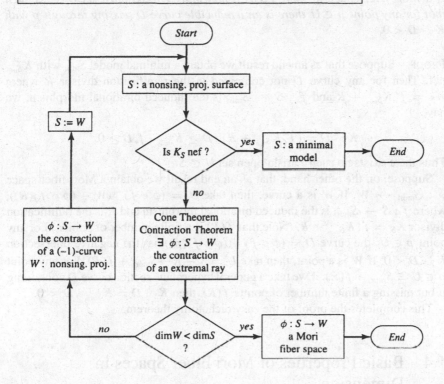

Minimal Model Program in Dimension 2 (the complete form)

Remark 1-3-7. (i) It is easy to see that an extremal contraction $\phi : S \to W$ with dim W = dim S is the contraction of a (-1)-curve $\mu : S \to T$. (See Theorem 1-4-8.) An extremal contraction $\phi : S \to W$ with dim W < dim S is by definition a Mori fiber space. We refer the reader to the next section for the detailed study of the structures of Mori fiber spaces in dimension 2.

(ii) The complete form of MMP in dimension 2, presented as above, is slightly different from the one described in Mori's paper [5]. His flowchart asks whether there exists an extremal contraction $\phi : S \to W$ with dim W = dim S when K_S is not nef, and then takes such an extremal contraction (i.e., a contraction of a (-1)-curve) whenever there is one. Accordingly, his flowchart produces end results without any (-1)-curves. For example, the surface $\mathbb{F}_1$ (one-point blowup of $\mathbb{P}^2$) cannot be an end result of his flowchart. But this is too restrictive from our point of view, where $\mathbb{F}_1$ should be a Mori fiber space (an end result of MMP in dimension 2 other than minimal models) and hence should be included in the category to run the Sarkisov program (cf. Section 1-8 and Chapter 13).

We end this section by giving a simple criterion for a nonsingular surface S to produce a Mori fiber space after going through MMP in dimension 2.

Theorem 1-3-9 (Easy Dichotomy Theorem of MMP in Dimension 2). *Let S be a nonsingular projective surface. Then an end result of MMP starting from S is a Mori fiber space iff there exists a nonempty Zariski open set $U \subset S$ such that for any point $p \in U$ there is an irreducible curve D passing through p with $K_S \cdot D < 0$.*

PROOF.   Suppose that as an end result we obtain a minimal model $S_{\min}$ with $K_{S_{\min}}$ nef. Then for any curve D not contained in the ramification divisor R where $K_S = f^* K_{S_{\min}} + R$ and $f : S \to S_{\min}$ is the induced birational morphism, we have

$$K_S \cdot D = (f^* K_{S_{\min}} + R) \cdot D \geq K_{S_{\min}} \cdot f_* D \geq 0.$$

Thus there exists no such Zariski open set $U \subset S$.

Suppose, on the other hand, that as an end result we obtain a Mori fiber space $\phi : S_{\text{mori}} \to W$. If W is a curve, then take $U = (\phi \circ f)^{-1}(W - (\phi \circ f)(R))$, where $f : S \to S_{\text{mori}}$ is the induced birational morphism and R is the ramification divisor $K_S = f^* K_{S_{\text{mori}}} + R$. (Note that $f(R)$ is a finite number of points.) For any point $p \in U$, the curve $D = (\phi \circ f)^{-1}((\phi \circ f)(p))$ has the negative intersection $K_S \cdot D < 0$. If W is a point, then take $U = S - R \overset{\sim}{\to} S_{\text{mori}} - f(R)$. For any point $p \in U \cong S_{\text{mori}} - f(R)$, if we take a general hyperplane section curve D containing p but missing a finite number of points $f(R)$, then $K_S \cdot D = K_{S_{\text{mori}}} \cdot D < 0$.

This completes the proof of the easy dichotomy theorem.   □

1.4   Basic Properties of Mori Fiber Spaces in Dimension 2

The purpose of this section is to study **Mori fiber spaces in dimension 2**, one half of the end results of MMP in dimension 2, the other half minimal models in dimension 2, to be studied in Section 1-5. Note that the easy dichotomy theorem of the last section can be considered as a (birational) property of Mori fiber spaces.

Definition 1-4-1 (Mori Fiber Space in Dimension 2). *A nonsingular projective surface S with a morphism $\phi : S \to W$ is a* **Mori fiber space** *if it is an extremal contraction $\phi : S \to W$ with $\dim W < \dim S$, or equivalently if*

(i) *ϕ is a morphism with connected fibers onto a normal projective variety W of $\dim W < \dim S$ (i.e., in dimension 2, W is either a nonsingular projective curve or a point), and*

(ii) *all the curves C in fibers of ϕ are numerically proportional with $K_S \cdot C < 0$.*

Remark 1-4-2.   (i) Recall as a consequence of the cone theorem and contraction theorem that $\phi : S \to W$ is a Mori fiber space if and only if it is the contraction of an extremal ray $\phi = \text{cont}_{R_l} : S \to W$ with $\dim W < \dim S$.

(ii) The complete form of MMP in dimension 2 produces an end result that is either a minimal model S_{min}, a surface with nef canonical bundle $K_{S_{min}}$ by definition, or a Mori fiber space S_{mori} with an extremal contraction $\phi : S_{mori} \to W$ with dim W < dim S. Conversely, a minimal model or a Mori fiber space is an end result of MMP in dimension 2.

The following feature of Mori fiber spaces is almost trivial to prove but turns out to give an important birational characterization.

Theorem 1-4-3. *Let $\phi : S \to W$ be a Mori fiber space in dimension 2. Then*

$$H^0(S, \mathcal{O}_S(mK_S)) = 0 \quad \forall m \in \mathbb{N}.$$

(That is to say, $\kappa(S) = -\infty$. See Section 1-5 for the definition of the Kodaira dimension κ.)

PROOF.   Suppose that $H^0(S, \mathcal{O}_S(mK_S)) \neq 0$ for some $m \in \mathbb{N}$. Take an effective member $D \in |mK_S| \neq \emptyset$. We can take a curve C that is contained in a fiber of ϕ but not in D. Then

$$0 > mK_S \cdot C = D \cdot C \geq 0,$$

a contradiction! ☐

The following is the main structure theorem for Mori fiber spaces in dimension 2.

Theorem 1-4-4 (Characterization of Mori Fiber Spaces in Dimension 2). *Let $\phi : S \to C$ be a Mori fiber space in dimension 2. Then one of the following two cases occurs:*

Case: dim $W = 1$

In this case, every fiber of ϕ is isomorphic to $\mathbb{P}^1$, i.e.,

$$\phi^{-1}(p) \cong \mathbb{P}^1 \quad \forall p \in W.$$

More precisely, $\phi : S \to W$ is a $\mathbb{P}^1$-bundle over a nonsingular projective curve W in the algebraic category, i.e., for each point $p \in W$ there exists a Zariski open neighborhood $p \in U \subset W$ such that we have the following commutative diagram:

$$
\begin{array}{ccc}
\phi^{-1}(U) & \xrightarrow{\sim} & \mathbb{P}^1 \times U \\
\phi \downarrow & & p_2 \downarrow \\
p \in U & \xrightarrow{\sim} & U
\end{array}
$$

Case: dim $W = 0$

$S \cong \mathbb{P}^2$ *and* $\phi : S \cong \mathbb{P}^2 \to W \cong$ pt. *is the obvious morphism.*

PROOF.

$\boxed{\text{Case: } \dim W = 1}$

First we claim that every fiber of ϕ is irreducible. If not, then we can take two distinct irreducible components G_1 and G_2 in a fiber $\phi^{-1}(p)$ for some $p \in W$. Since $\phi^{-1}(p)$ is connected, we may choose G_1 and G_2 such that $G_1 \cdot G_2 > 0$. But then for a general fiber F of ϕ, we have on the one hand that

$$F \cdot G_2 > 0,$$

since F must be numerically proportional to G_1, and on the other that

$$F \cdot G_2 = 0,$$

since the general fiber F is disjoint from G_2, a contradiction!

Now set

$$\phi^{-1}(p) = \mu_p F_p \text{ as divisors on } S,$$

where $F_p = \phi^{-1}(p)_{\text{red}}$ and $\mu_p \in \mathbb{N}$ is the multiplicity of $\Phi^{-1}(p)$ as a divisor. We have the numerical data via the arithmetic genus formula

$$F_p^2 = 0, F_p \cdot K_S < 0,$$

$$\frac{1}{2}(F_p + K_S) \cdot F_p + 1 = h^1(F_p, \mathcal{O}_{F_p}) \geq 0,$$

which imply

$$F_p^2 = 0, F_p \cdot K_S = -2 \text{ and } h^1(F_p, \mathcal{O}_{F_p}) = 0$$

and thus

$$K_S \cdot \phi^{-1}(p) = (-2)\mu_p.$$

Observe also that for a general fiber F we have the multiplicity $\mu = 1$ and

$$K_S \cdot F = -2.$$

Since all the fibers are algebraically and hence numerically equivalent,

$$K_S \cdot F = K_S \cdot \phi^{-1}(p).$$

Therefore, we conclude that

$$\mu_p = 1 \quad \forall p \in W$$

and hence

$$\phi^{-1}(p) \text{ is reduced } \forall p \in C.$$

Finally,

$$\phi^{-1}(p) = \phi^{-1}(p)_{\text{red}} = F_p \quad \text{and} \quad h^1(F_p, \mathcal{O}_{F_p}) = 0$$

imply (cf. Exercise 1-1-5)

$$\phi^{-1}(p) \cong \mathbb{P}^1.$$

To show that $\phi : S \to C$ is actually an algebraic $\mathbb{P}^1$-bundle, it suffices to prove the proposition below, following the argument in Beauville [1]. We also refer the reader to Tsen's lemma (cf. Theorem 3-2-10).

Proposition 1-4-5. *Let $\phi : S \to W$ be a morphism from a nonsingular projective surface onto a nonsingular projective curve, and $p \in W$ a point on W such that*

$$\phi^{-1}(p) = F \cong \mathbb{P}^1.$$

Then there exists a Zariski open neighborhood U of p that makes the following diagram commute:

$$
\begin{array}{ccc}
\phi^{-1}(U) & \xrightarrow{\sim} & \mathbb{P}^1 \times U \\
\phi\downarrow & & p_2\downarrow \\
p \in U & \xrightarrow{\sim} & U
\end{array}
$$

PROOF. First note that $H^2(S, \mathcal{O}_S) = 0$, since otherwise $H^2(S, \mathcal{O}_S) \cong H^0(S, \mathcal{O}_S(K_S)) \neq 0$ by Serre duality (cf. Theorem 1-4-10), which would contradict

$$F^2 = 0 \text{ and } F \cdot K_S = -2.$$

The key of the proof is to show that there exists a divisor D such that

$$D \cdot F = 1.$$

(The essential point is to show the existence of a section $s : W \to S$ for the morphism ϕ, though we do not explicitly formulate the assertion this way. Another way to show the existence of a section is via the use of Tsen's lemma. See Section 3-2.)

As a consequence of $H^2(S, \mathcal{O}_S) = 0$ and the exponential sequence, we have the surjection

$$\mathrm{Pic}(S) \cong H^1(S, \mathcal{O}_S^*) \to H^2(S, \mathbb{Z}) \to 0.$$

Therefore, if

$$\{z \cdot F; z \in \mathrm{Pic}(S)\} = d\mathbb{Z} \text{ for some } d \in \mathbb{N},$$

then Poincaré duality implies that there exists a class in $H^2(S, \mathbb{Z})$ representing the linear form $\cdot \frac{1}{d} F$ and hence a divisor F' such that

$$z \cdot F' = z \cdot \frac{1}{d} F \quad \text{for all } z \in \mathrm{Pic}(S),$$

i.e.,

$$F \equiv dF'.$$

Therefore,

$$F'^2 = 0 \text{ and } K_S \cdot F' = \frac{-2}{d} \quad \text{(an integer)}.$$

On the other hand,

$$F'2 + K_S \cdot F' \equiv 0 \mod (2)$$

(Exercise! Why?), which implies $d = 1$, i.e., that there exists D such that $D \cdot F = 1$.

Now for $r \in \mathbb{N}$ consider the exact sequence

$$0 \to \mathcal{O}_S(D + (r - 1)F) \to \mathcal{O}_S(D + rF) \to \mathcal{O}_F(D) \cong \mathcal{O}_{\mathbb{P}^1}(1) \to 0,$$

which induces the cohomology sequence

$$H^0(S, \mathcal{O}_S(D + rF)) \to H^0(F, \mathcal{O}_F(D)) \cong H^0(\mathbb{P}^1, \mathcal{O}_{\mathbb{P}^1}(1))$$
$$\to H^1(S, \mathcal{O}_S(D + (r - 1)F)) \to H^1(S, \mathcal{O}_S(D + rF)) \to 0.$$

Thus for $r \gg 0$, $H^1(S, \mathcal{O}_S(D + rF))$ becomes stationary, i.e.,

$$H^1(S, \mathcal{O}_S(D + (r - 1)F)) \to H^1(S, \mathcal{O}_S(D + rF))$$

is an isomorphism, and hence we have a surjection

$$H^0(S, \mathcal{O}_S(D + rF)) \to H^0(\mathbb{P}^1, \mathcal{O}_{\mathbb{P}^1}(1)).$$

Take a 2-dimensional subspace $V \subset H^0(\mathcal{O}_S(D + rF))$ that maps onto $H^0(\mathbb{P}^1, \mathcal{O}_{\mathbb{P}^1}(1))$. Since the associated pencil $|V|$ has no base points on F, there exists a neighborhood U of $p \in W$ such that $|V|$ has no base points on $\phi^{-1}(U)$. Then every member $D_t \in |V|$ is a section over U (we may assume that D_t has no vertical components in $\phi^{-1}(U)$), and no two have an intersection (since V is 2-dimensional). We have the commutative diagram

$$
\begin{array}{ccc}
 & \Phi_{|V|} \times \phi & \\
\phi^{-1}(U) & \xrightarrow{\hspace{1cm}} & \mathbb{P}^1 \times U \\
\phi\downarrow & & p_2\downarrow \\
p \in U & \xrightarrow{\hspace{1cm}} & U
\end{array}
$$

where $\phi \times \Phi_{|V|}$ is an isomorphism, since it is 1-to-1 and proper over $\mathbb{P}^1 \times U$ (Exercise! Why?) onto a normal variety $\mathbb{P}^1 \times U$. This completes the proof of the proposition and hence the analysis of the case dim $W = 1$.

Case: dim $W = 0$

The analysis of this case is quite delicate, giving the characterization of $\mathbb{P}^2$. We use the unimodularity of Poincaré duality and Hodge theory, taking full advantage of the fact that we are working over $\mathbb{C}$. (We warn the reader that a Mori fiber space S in dimension 2 with $\rho(S) = 1$ over some nonalgebraically closed field k of characteristic zero may not be isomorphic to $\mathbb{P}^2_k$. See Section 3-2 for an example.)

Consider the exponential sequence

$$0 \to \mathbb{Z} \to \mathcal{O}_S \to \mathcal{O}_S^* \to 0$$

and the cohomology sequence

$$H^1(S, \mathcal{O}_S) \to H^1(S, \mathcal{O}_S^*) \cong \operatorname{Pic}(S)$$
$$\to H^2(S, \mathbb{Z}) \to H^2(S, \mathcal{O}_S).$$

Since $-K_S$ is ample in this case, we have by the Kodaira vanishing theorem that

$$H^1(S, \mathcal{O}_S) = H^2(S, \mathcal{O}_S) = 0.$$

Thus we conclude that

$$H^2(S, \mathbb{Z}) \cong \mathrm{Pic}(S)$$

and hence that numerical equivalence among the divisors coincides with (co)homo-logical equivalence.

Condition (ii) of the definition of a Mori fiber space in this case implies that all the curves in S are numerically proportional, and thus

$$1 = \rho(S) = \dim_{\mathbb{R}} N_1(S) = \dim_{\mathbb{R}} N^1(S) = \dim_{\mathbb{R}} H^2(S, \mathbb{R}).$$

Therefore,

$$\mathrm{Pic}(S) \cong \mathbb{Z} \oplus \text{torsion}.$$

Take an ample divisor H that gives rise to $1 \in \mathbb{Z}$ and set

$$-K_S \equiv rH \text{ for } r \in \mathbb{N}.$$

The Kodaira vanishing theorem implies

$$\chi(\mathcal{O}_S) = h^0(S, \mathcal{O}_S) = 1,$$

and by the Riemann–Roch theorem we have

$$\chi(\mathcal{O}_S(H)) = h^0(S, \mathcal{O}_S(H))$$
$$= \frac{1}{2}(H - K_S) \cdot H + \chi(\mathcal{O}_S) = \frac{1+r}{2}H^2 + 1 \geq 2.$$

Therefore, there exists a member $D \in |H|$, which must be irreducible and reduced, since H is an ample generator.

Subcase: $r > 1$.

The arithmetic genus formula yields

$$h^1(D, \mathcal{O}_D) = \frac{1}{2}(D + K_S) \cdot D + 1 = \frac{1}{2}(1 - r)H^2 + 1 < 1,$$

since $r > 1$ and H is ample. This implies

$$h^1(D, \mathcal{O}_D) = 0 \text{ and } D \cong \mathbb{P}^1$$

and

$$\text{either} \left\{ \begin{array}{c} r = 3 \\ H^2 = 1 \end{array} \right\} \text{ or } \left\{ \begin{array}{c} r = 2 \\ H^2 = 2 \end{array} \right\}.$$

We further conclude that the second case is impossible, since $H^2 = 1$ by the unimodularity of the intersection form in Poincaré duality.

We claim that $|H|$ is base point free: The base locus of $|H|$, if any, must be on D. But the exact sequence

$$H^0(S, \mathcal{O}_S(D)) \to H^0(D, \mathcal{O}_D(D)) \to H^1(S, \mathcal{O}_S) = 0,$$

where $H^0(D, \mathcal{O}_D(D)) \cong H^0(\mathbb{P}^1, \mathcal{O}_{\mathbb{P}^1}(1))$, implies that $|H|$ does not have a base point on D. Thus $|H|$ is base point free.

Now we consider the morphism

$$\Phi_{|H|} : S \to \mathbb{P}^{h^0(H)-1} = \mathbb{P}^2.$$

Since $H = \Phi_{|H|}^* \mathcal{O}(1)$ is ample on S, $\Phi_{|H|}$ is a finite surjective morphism. But then

$$1 = H^2 = \deg \Phi_{|H|} \cdot \mathcal{O}(1)^2$$

implies that $\Phi_{|H|}$ is an isomorphism, since it is a 1-to-1 proper morphism onto a normal variety. Thus $S \cong \mathbb{P}^2$ in this subcase.

Subcase: $r = 1$.

Since $-K_S$ gives the ample generator in this subcase and since the intersection pairing

$$H^2(S, \mathbb{Z}) \times H^2(S, \mathbb{Z}) \to \mathbb{Z}$$

is nondegenerate and unimodular by Poincaré duality, we have

$$(-K_S)^2 = 1.$$

Moreover, looking at the Betti numbers and via Hodge theory obtaining

$$b_0(S) = b_4(S) = 1,$$
$$b_1(S) = b_3(S) = h^1(S, \mathbb{C}) = h^{1,0}(S) + h^{0,1}(S) = 2h^1(S, \mathcal{O}_S) = 0,$$
$$b_2(S) = h^2(S, \mathbb{C}) = \dim_{\mathbb{C}} H^2(S, \mathbb{Z}) \otimes \mathbb{C} = 1,$$

we conclude that

$$e(S) := \text{the topological Euler characteristic} = \Sigma(-1)^i b_i(S) = 3.$$

On the other hand, the Hirzebruch–Riemann–Roch theorem says that

$$\chi(\mathcal{O}_S)(\text{an integer}) = \frac{1}{12}((-K_S)^2 + e(S)) = \frac{1}{3},$$

a contradiction! Thus this subcase cannot occur.

This completes the proof of Theorem 1-4-4.   □

We state the Hirzebruch–Riemann–Roch theorem in dimension 2, known as the Noether formula.

Theorem 1-4-6 (Hirzebruch–Riemann–Roch Theorem in Dimension 2). *If D is a divisor on a nonsingular projective surface S, then*

$$\chi(\mathcal{O}_S(D)) = \frac{1}{2}(D - K_S) \cdot D + \frac{1}{12}((-K_S)^2 + c_2),$$

where $c_2 = e(S)$ is the topological Euler characteristic.

We also recall the basic fact that we used from Hodge theory.

Theorem 1-4-7 (Hodge Decomposition). *Let S be a nonsingular projective variety over $\mathbb{C}$. Then we have the Hodge decomposition*

$$H^i(S, \mathbb{C}) = \oplus_{p+q=i} H^{p,q}(S),$$

where

$$H^{p,q}(S) \cong H^q(S, \Omega_S^p)$$

and

$$H^{p,q}(S) = \overline{H^{q,p}(S)},$$

i.e., they are complex conjugate to each other when represented as harmonic forms. In particular,

$$h^{p,q}(S) = h^{q,p}(S).$$

We can rephrase Theorem 1-4-4 as the classification theorem for extremal rays in dimension 2.

Theorem 1-4-8 (Classification of Extremal Rays in Dimension 2). *Let S be a nonsingular projective surface whose canonical bundle K_S is not nef. Then an extremal ray R_l with $R_l - \{0\}$ in the half-space $N_1(S)_{K_S < 0}$ is one of the following:*

(i) *$R_l = \mathbb{R}_+[l]$, where l is a (-1)-curve with $K_S \cdot l = -1$, and $\mathrm{cont}_{R_l} : S \to W$ is the contraction of the (-1)-curve l,*

(ii) *$R_l = \mathbb{R}_+[l]$, where l is a fiber with $K_S \cdot l = -2$ of the algebraic $\mathbb{P}^1$-bundle $\mathrm{cont}_{R_l} : S \to W$ over a nonsingular projective curve W, or*

(iii) *$R_l = \mathbb{R}_+[l]$, where l is a line in $S \cong \mathbb{P}^2$ with $K_S \cdot l = -3$ and $\mathrm{cont}_{R_l} : S \to W$ is the structure morphism from $\mathbb{P}^2$ to $W = \mathrm{Spec}\,\mathbb{C}$.*

Conversely, any curve as described in (i), (ii), or (iii) spans an extremal ray $R_l = \mathbb{R}_+[l]$ with respect to K_S, and it has the minimum intersection with $-K_S$ among the curves in its numerical class.

PROOF.   We prove that when the dimension dim W of the image of the contraction morphism $\phi = \mathrm{cont}_{R_l} : S \to W$ equals 2, 1, or 0, we have case (i), (ii) or (iii), respectively. We have already proved in Theorem 1-4-4 that when dim $W = 1$ or 0 we have case (ii) or (iii), respectively.

Suppose dim $W = 2$. Then there exists a curve l such that l contracts to a point on W. Then $l^2 < 0$ by Theorem 1-2-9 and $K_S \cdot l < 0$, since $[l] \in R_l$. Thus l is a (-1)-curve. Suppose there is another curve l' contracting to a point on W. Then on the one hand, we would have $l \cdot l' \geq 0$, l being a different curve from l'. On the other hand, since $[l'] \in R_l$ is numerically proportional to l, we would have $l \cdot l' < 0$, a contradiction. Thus $\mathrm{cont}_{R_l} : S \to W$ is the contraction of the (-1)-curve l (and no other curve) described by Theorem 1-1-6.

Conversely, suppose l is any curve described in (i), (ii), or (iii) as above. Observe in each case that the morphism $\phi : S \to W$ induces an injective homomorphism of vector spaces

$$\phi^* : N^1(W) \to N^1(S) \text{ with } \dim_{\mathbb{R}} N^1(W) = \dim_{\mathbb{R}} N^1(S) - 1.$$

Take an ample divisor H on W. It suffices to prove

$$\overline{NE}(S) \cap \phi^* H^\perp = R_l = \mathbb{R}_+[l].$$

Suppose not, i.e., suppose that there is another half line R (R may not be extremal) different from R_l such that

$$\overline{NE}(S) \cap \phi^* H^\perp \supset R.$$

On the one hand, the vector subspace

$$\{L \in N^1(S); L \cdot z \geq 0, z \in R_l + R\}$$

could only be of dimension $\dim_{\mathbb{R}} N^1(S) - 2$. On the other hand, since $\phi^* H \cdot z = H \cdot \phi_* z = 0$ for $0 \neq z \in R \subset \overline{NE}(S)$ and hence $\phi_* z = 0$ in $N_1(S)$, the vector subspace has to contain the pullbacks of all the ample divisors on W, which span a space of dimension $\dim_{\mathbb{R}} N^1(S) - 1$. A contradiction! $\square$

Finally, as a corollary to the analysis of Mori fiber spaces in dimension 2, we obtain Castelnuovo's criterion for rationality.

Corollary 1-4-9 (Castelnuovo's Criterion for Rationality). *Let S be a nonsingular projective surface. Then*

$$S \text{ is birational to } \mathbb{P}^2 \Leftrightarrow h^1(S, \mathcal{O}_S) = h^0(S, \mathcal{O}_S(2K_S)) = 0.$$

PROOF. Note that both $h^1(S, \mathcal{O}_S)$ and $h^0(S, \mathcal{O}_S(2K_S))$ are birational invariants (cf. Section 1-8).

($\Rightarrow$) The numerical data $h^1(\mathbb{P}^2, \mathcal{O}_{\mathbb{P}^2}) = h^0(\mathbb{P}^2, \mathcal{O}_{\mathbb{P}^2}(2K_{\mathbb{P}^2})) = 0$ and the above note give the implication.

($\Leftarrow$) To show this implication, we first prove the following lemma.

Lemma 1-4-10. *If $h^1(S, \mathcal{O}_S) = h^0(S, \mathcal{O}_S(2K_S)) = 0$, then K_S is not nef.*

PROOF. Proof of the lemma Suppose $h^1(S, \mathcal{O}_S) = h^0(S, \mathcal{O}_S(2K_S)) = 0$. Then the Riemann–Roch theorem yields

$$\chi(\mathcal{O}_S) = h^0(S, \mathcal{O}_S) - h^1(S, \mathcal{O}_S) + h^2(S, \mathcal{O}_S) = h^0(S, \mathcal{O}_S) = 1,$$

since

$$h^2(S, \mathcal{O}_S) = h^0(S, \mathcal{O}_S(K_S)) = 0$$

by Serre duality and the assumption. Thus the Riemann–Roch theorem again yields

$$h^0(S, \mathcal{O}_S(-K_S)) = h^0(S, \mathcal{O}_S(-K_S)) + h^2(S, \mathcal{O}_S(-K_S))$$
$$\geq \chi(\mathcal{O}_S(-K_S)) = K_S^2 + \chi(\mathcal{O}_S) = K_S^2 + 1,$$

since

$$h^2(S, \mathcal{O}_S(-K_S)) = h^0(S, \mathcal{O}_S(2K_S)) = 0$$

by Serre duality and the assumption.

If K_S is nef, then $K_S{}^2 \geq 0$ (cf. the proof of Theorem 1-2-14), which would imply $h^0(S, \mathcal{O}_S(-K_S)) \geq 1$. Thus there would be an effective member $D \in |-K_S|$. Then for a very ample divisor A we would have

$$0 \leq A \cdot D = A \cdot (-K_S) \leq 0,$$

which would imply that $D = 0$ as a divisor, i.e., $-K_S \sim 0$. But then

$$h^0(S, \mathcal{O}_S(2K_S)) = h^0(S, \mathcal{O}_S) = 1, \text{ a contradiction!}$$

Now we finish the proof of the implication ($\Leftarrow$). Starting from S with $h^1(S, \mathcal{O}_S) = h^0(S, \mathcal{O}_S(2K_S)) = 0$, we apply MMP in dimension 2. Then since both invariants remain zero throughout the birational transformations of MMP, the end result cannot be a minimal model with a nef canonical divisor by the lemma. Thus we obtain a Mori fiber space $\phi : S_{\text{mori}} \to W$ birational to S. If $\dim W = 0$, then $S_{\text{mori}} \cong \mathbb{P}^2$, and we are done. If $\dim W = 1$, then it is birational to $\mathbb{P}^1 \times W$ by Proposition 1-4-5, and thus by the Künneth formula we obtain

$$0 = h^1(S, \mathcal{O}_S) = h^1(\mathbb{P}^1 \times W, \mathcal{O}_{\mathbb{P}^1 \times W}) = h^1(W, \mathcal{O}_W).$$

Thus $W \cong \mathbb{P}^1$ and S is birational to $\mathbb{P}^1 \times \mathbb{P}^1$, which is of course birational to $\mathbb{P}^2$ and rational. This completes the proof of Castelnuovo's criterion for rationality.   □

We make explicit the statement of Serre duality that we used in the middle of the proof of Lemma 1-4-10.

Theorem 1-4-11 (Serre Duality). *Let D be a divisor on a nonsingular projective variety S of dimension n. Then there is a natural isomorphism (of vector spaces)*

$$H^i(S, \mathcal{O}_S(D)) \cong Ext^i(\mathcal{O}_S(K_S - D), \mathcal{O}_S(K_S)) \cong H^{n-i}(S, \mathcal{O}_S(K_S - D))^*,$$

where $*$ denotes the dual, and in particular,

$$h^i(S, \mathcal{O}_S(D)) = h^{n-i}(S, \mathcal{O}_S(K_S - D)).$$

1.5   Basic Properties of Minimal Models in Dimension 2

In the previous section we studied Mori fiber spaces in dimension 2. The purpose of this section is to study **minimal models in dimension 2**, the "other" half of the end results of MMP in dimension 2. The two most important results of this section are the

hard dichotomy theorem, and the
abundance theorem.

The hard dichotomy theorem gives the characterization of minimal models in terms of the Kodaira dimension, in contrast to the easy dichotomy theorem of the previous section. The abundance theorem, which asserts that a minimal model S has base point free (pluri)canonical linear systems $|mK_S|$ for m sufficiently divisible and large, paves the way to make a fusion between the minimal model program and the Iitaka program, by providing the Iitaka fibration $\Phi = \Phi_{|mK_S|} : S \to S_{can}$. (The study of the canonical model S_{can} is the subject of the next section.)

To the untrained eye, these two theorems do not seem to provide much information about the structure of minimal models. We will demonstrate their power in Section 1-7 by deriving the Enriques classification of algebraic surfaces as an easy corollary.

Definition 1-5-1 (Minimal Model in Dimension 2). *A nonsingular projective surface S is a **minimal model** iff the canonical bundle K_S is nef.*

We remark that obviously a minimal model S as defined above is an end result of MMP in dimension 2.

The Kodaira dimension, a naive but fundamental birational invariant, plays a crucial role in our analysis.

Definition 1-5-2 (Kodaira Dimension). *Let S be a nonsingular projective variety. The Kodaira dimension $\kappa(S)$ of S is defined to be*

$$\kappa(S) = \kappa(S, K_S)$$

$$:= \begin{cases} -\infty & \\ & \text{if } H^0(S, \mathcal{O}_S(mK_S)) = 0 \quad \forall m \in \mathbb{N}, \\ \text{transdeg.}_{\mathbb{C}} \oplus_{m \geq 0} H^0(S, \mathcal{O}_S(mK_S)) - 1 & \\ & \text{if } H^0(S, \mathcal{O}_S(mK_S)) \neq 0 \text{ for some } m \in \mathbb{N}. \end{cases}$$

We remark that $\kappa(S, D)$ can be defined similarly for any divisor D on S and that for a singular variety T the Kodaira dimension is defined to be that of its desingularization $\kappa(T) = \kappa(S)$, where $v : S \to T$ is a birational morphism from a nonsingular projective variety S.

Proposition 1-5-3 (Basic Properties of the Kodaira Dimension).

(i) *The Kodaira dimension is a birational invariant, i.e., if S' is another nonsingular projective variety birational to S, then*

$$\kappa(S) = \kappa(S').$$

(ii)

$$\kappa(S) := \begin{cases} -\infty & \text{if } H^0(S, \mathcal{O}_S(mK_S)) = 0 \quad \forall m \in \mathbb{N}, \\ \max\{\dim \Phi_{|mK_S|}(S)\} & \text{if } H^0(S, \mathcal{O}_S(mK_S)) \neq 0 \text{ for some } m \in \mathbb{N}. \end{cases}$$

(iii) *There exist $m_0 \in \mathbb{N}$ and $\alpha, \beta > 0$ such that*

$$\alpha m^{\kappa(S)} \leq h^0(S, \mathcal{O}_S(mm_0 K_S)) \leq \beta m^{\kappa(S)} \quad \forall m \in \mathbb{N}$$

and

$$\kappa(S) = 0 \text{ iff } h^0(S, \mathcal{O}_S(mK_S)) = 0 \text{ or } 1 \quad \forall m \in \mathbb{N} \text{ and}$$
$$h^0(S, \mathcal{O}_S(mK_S)) \neq 0 \text{ for some } m \in \mathbb{N}.$$

PROOF.   We note that for two birationally equivalent varieties S and S',

$$H^0(S, \mathcal{O}_S(mK_S)) \cong H^0(S', \mathcal{O}_{S'}(mK_{S'}))$$

canonically (cf. Section 1-8), which implies

$$\kappa(S) = \kappa(S').$$

We refer the reader to Iitaka [5], Chapter 10, for the proofs of (ii) and (iii), which give characterizations of $\kappa(S)$ and each of which can be used as an alternative definition of the Kodaira dimension.   □

The most important property of a minimal model (in Dimension 2) is the following:

Theorem 1-5-4 (Existence of an Effective Pluricanonical Divisor for a Minimal Model in Dimension 2).   *Let S be a minimal model in dimension 2. Then*

$$\kappa(S) \geq 0,$$

i.e.,

$$H^0(S, \mathcal{O}_S(mK_S)) \neq 0 \text{ for some } m \in \mathbb{N}.$$

Note that this theorem, together with Theorem 1-4-3, leads to the following.

Theorem 1-5-5 (Hard Dichotomy Theorem of MMP in Dimension 2).   *Let S be a nonsingular projective surface. Then an end result of MMP in dimension 2 starting from S is a minimal model (respectively a Mori fiber space) iff $\kappa(S) \geq 0$ (respectively $\kappa(S) = -\infty$).*

Once Theorem 1-5-4 is established, the so-called abundance theorem for minimal models in dimension 2 follows rather easily.

Theorem 1-5-6 (Abundance Theorem in Dimension 2).   *Let S be a minimal model in dimension 2. Then $|mK_S|$ is base point free for sufficiently divisible and large $m \in \mathbb{N}$.*

We will prove Theorem 1-5-6 first assuming Theorem 1-5-4 and then complete the proof by verifying Theorem 1-5-4. The key methods for our rather subtle proof of these two theorems are:

the nonnegativity of the second Chern class c_2 for minimal models;
the Albanese map;
the canonical bundle formula for elliptic fibrations.

Proof of Abundance Assuming $\kappa(S) \geq 0$

Let S be a minimal model in dimension 2. Then by Theorem 1-5-4, $\kappa(S) \geq 0$. We prove the Abundance Theorem, differentiating our arguments according to the value of the Kodaira dimension $\kappa(S) = 2, 1,$ or 0.

Case: $\kappa(S) = 2$

The abundance theorem for this case can be derived as an easy corollary to the very powerful base point freeness theorem in a more general setting via the Kawamata–Viehweg vanishing (cf. Section 6-2.) Here we present an elementary proof, based upon the specific construction of the canonical model in the next section via the contraction of (-2)-curves.

Lemma 1-5-7 (Kodaira's Lemma). *Let S be a minimal model in dimension 2. Then*

$$\kappa(S) = \dim S = 2 \Leftrightarrow K_S^2 > 0.$$

PROOF.   Suppose $\kappa(S) = 2$. Take $m_0 \in \mathbb{N}$ and $\alpha, \beta > 0$ as in Proposition 1-5-3 (iii). Take a very ample divisor A on S and consider the exact sequence

$$0 \to \mathcal{O}_S(mm_0K_S - A) \to \mathcal{O}_S(mm_0K_S) \to \mathcal{O}_A(mm_0K_S) \to 0,$$

which induces the cohomology sequence

$$0 \to H^0(S, \mathcal{O}_S(mm_0K_S - A)) \to H^0(S, \mathcal{O}_S(mm_0K_S)) \overset{\eta}{\to} H^0(A, \mathcal{O}_A(mm_0K_S)).$$

Since

$$h^0(S, \mathcal{O}_S(mm_0K_S)) \geq \alpha m^{\kappa(S)}$$

and since

$$h^0(A, \mathcal{O}_A(mm_0K_S|_A)) \approx m \cdot \deg_A(m_0K_S|_A),$$

(Exercise! Why?) we conclude that

$$0 \neq \ker(\eta) \subset H^0(S, \mathcal{O}_S(mm_0K_S - A)) \text{ for some } m \in \mathbb{N}$$

and thus

$$mm_0K_S \sim D + A,$$

where $D \in |mm_0K_S - A|$ is an effective divisor. Then

$$mm_0K_S^2 = K_S \cdot (D + A) \geq K_S \cdot A = \frac{1}{mm_0}(D + A) \cdot A > 0.$$

On the other hand, if $K_S^2 > 0$, then first the Riemann–Roch theorem yields

$$\chi(\mathcal{O}_S(mK_S)) = h^0(S, \mathcal{O}_S(mK_S)) - h^1(S, \mathcal{O}_S(mK_S)) + h^2(S, \mathcal{O}_S(mK_S))$$

$$= \frac{1}{2}(mK_S - K_S) \cdot mK_S + \chi(\mathcal{O}_S)$$

$$\leq h^0(S, \mathcal{O}_S(mK_S)) + h^2(S, \mathcal{O}_S(mK_S)),$$

whereas for $m \gg 0$

$$h^2(S, \mathcal{O}_S(mK_S)) = h^0(S, \mathcal{O}_S(K_S - mK_S)) = 0$$

by Serre duality. (If the second term is not zero, then there should be a member $G \in |K_S - mK_S|$, which would imply

$$0 > (1-m)A \cdot K_S = A \cdot (K_S - mK_S) = A \cdot G \geq 0, \text{ a contradiction!})$$

Thus

$$h^0(S, \mathcal{O}_S(mK_S)) \approx \frac{1}{2}K_S{}^2 m^2 \text{ for } m \gg 0,$$

and by Proposition 1-5-3 (iii) we conclude that

$$\kappa(S) = 2. \qquad \square$$

We go back to the proof of the abundance theorem in the case $\kappa(S) = 2$. The key to the analysis in this case is the morphism $\Phi : S \to S_{\text{can}}$ onto the canonical model, whose existence will be proved in Section 1-6.

Proposition 1-5-8. *Let S be a minimal model in dimension 2 with $\kappa(S) = 2$. Then there exists a birational morphism*

$$\Phi : S \to S_{\text{can}}$$

onto a normal projective surface S_{can} such that

(i) *for any curve $C \subset S$*

$$\Phi(C) = \text{pt.} \Leftrightarrow K_S \cdot C = 0,$$

    and such that

(ii)

$$\mathcal{O}_S(K_S) = \Phi^* \mathcal{K} \text{ for some line bundle } \mathcal{K} \text{ on } S_{\text{can}}.$$

Now we look at the line bundle $\mathcal{K}$ on S_{can}. By Lemma 1-5-7 we have

$$\mathcal{K}^2 = (\Phi^*\mathcal{K})^2 = K_S{}^2 > 0.$$

For any curve $D \subset S_{\text{can}}$, take an irreducible component C of $\Phi^{-1}(D)$ that maps onto D. Then

$$\mathcal{K} \cdot D = \mathcal{K} \cdot \Phi_* C = \Phi^*\mathcal{K} \cdot C = K_S \cdot C > 0, \text{ since } \Phi(C) \neq \text{pt.}$$

Therefore, $\mathcal{K}$ is ample on S_{can} by the Nakai–Moishezon criterion, and thus $|mK_S| = |\Phi^* m\mathcal{K}|$ is base point free for sufficiently large $m \in \mathbb{N}$.

Case: $\kappa(S) = 1$

By Proposition 1-5-3 there exists $m \in \mathbb{N}$ such that the image of the rational map $\Phi_{|mK_S|}$ has dimension $\kappa(S) = 1$. Let

$$|mK_S| = M + F,$$

where M is the movable part and F the fixed part. Then by Lemma 1-5-7

$$0 = K_S{}^2 = (M+F)^2 \geq M \cdot (M+F) \geq M^2 \geq 0,$$

which implies

$$0 = M^2 = M \cdot F = F^2.$$

Using the same argument as in the proof of the Base Point Freeness Theorem in dimension 2 (Theorem 1-2-14) in the case $L^2 = 0$ and $L \not\equiv 0$, we conclude that $|mK_S|$ is base point free for sufficiently large $m \in \mathbb{N}$.

Case: $\kappa(S) = 0$

Proposition 1-5-9 (Nonnegativity of c_2 for Minimal Models in Dimension 2).
Let S be a minimal model in dimension 2 with $\kappa(S) \leq 0$. Then

$$\chi(\mathcal{O}_S) \geq 0.$$

Remark 1-5-10. (i) Though Proposition 1-5-9 looks technical at first sight, it has a nice generalization to higher dimensions in the following sense: Since S is a minimal model in dimension 2 with $\kappa(S) < \dim S$, the Kodaira lemma implies $K_S^2 = 0$. Thus claiming (cf. Theorem 1-4-6 (Noether formula))

$$\chi(\mathcal{O}_S) = \frac{1}{12}((-K_S)^2 + c_2) = \frac{1}{12}c_2 \geq 0$$

is equivalent to claiming the nonnegativity of the second Chern class c_2. Miyaoka [1] generalizes this to higher-dimensions, claiming that for a minimal model X in dimension 3 the second Chern class c_2 is pseudoeffective, i.e.,

$$H \cdot c_2 \geq 0 \text{ for any ample divisor } H \text{ on } X.$$

(To be precise, we have to take a resolution $f : Y \to X$ and claim that $f_*c_2(Y)$ is pseudoeffective.) The proof goes beyond the scope of this book, using Bogomolov's inequality blended with Mori's bend and break technique. (See also the argument by Shepherd–Barron in Kollár et al. [1].) We will discuss briefly Mori's bend and break technique later in Chapter 10.

(ii) In order to make the proof simple and make the lemma suitable for verifying Theorem 1-5-4 later, we put the extra assumtion $\kappa(S) \leq 0$, which leads to the numerical condition $K_S^2 = 0$.

But in fact, we have the **Miyaoka–Yau inequality**

$$3c_2 \geq K_S^2$$

for any minimal model in dimension 2, including the one with $K_S^2 > 0$. Thus for a minimal model in dimension 2, without any extra assumption, we always have

$$c_2 \geq 0 \text{ and } \chi(\mathcal{O}_S) \geq 0$$

(with both inequalities being strict in the case $K_S^2 > 0$). It is actually this form that Miyaoka generalizes to higher-dimensional minimal models, claiming that

$$3c_2 - c_1^2 \text{ is pseudoeffective.}$$

(Again, to be precise, we have to take a resolution $f : Y \to X$ and claim that $f_*\{3c_2(Y) - c_1(Y)^2\}$ is pseudoeffective.)

PROOF. By Noether's formula (Theorem 1-4-6) and Lemma 1-5-7, which implies $K_S^2 = 0$ under the assumption, we compute

$$12\chi(\mathcal{O}_S) = K_S^2 + c_2(S) = c_2(S) = e(S) = 2(b_0 - b_1) + b_2 = 2(1 - b_1) + b_2.$$

On the other hand, Serre duality and Hodge theory imply

$$\chi(\mathcal{O}_S) = \Sigma(-1)^i h^i(S, \mathcal{O}_S) = 1 - h^1(S, \mathcal{O}_S) + h^0(S, \mathcal{O}_S(K_S)),$$
$$b_1 = h^0(S, \mathbb{C}) = h^1(S, \mathcal{O}_S) + h^0(S, \Omega_S^1) = 2h^1(S, \mathcal{O}_S).$$

Therefore, we conclude that

$$8\chi(\mathcal{O}_S) = -2 - 4h^0(S, \mathcal{O}_S(K_S)) + b_2 \geq -6,$$

since $b_2 \geq 0$ and since $h^0(S, \mathcal{O}_S(K_S)) \leq 1$ by the assumption and Proposition 1-5-3 (iii). Thus

$$\chi(\mathcal{O}_S) \geq 0.$$

This completes the proof of Proposition 1-5-9.   □

The nonnegativity of $\frac{1}{12}c_2 = \chi(\mathcal{O}_S)$ by Proposition 1-5-9,

$$\chi(\mathcal{O}_S) = 1 - h^1(S, \mathcal{O}_S) + h^0(S, \mathcal{O}_S(K_S)) \geq 0,$$

puts a very strong restriction on the numerical possibilities for the minimal models in dimension 2 with $\kappa(S) = 0$. As one can easily see in the following, there are only 5 possibilities for $h^1(S, \mathcal{O}_S)$ and $h^0(S, \mathcal{O}_S(K_S))$. (It turns out that subcase 5 cannot occur and we have only four subcases that actually occur. Inside of parentheses we put the names commonly used for the corresponding surfaces.)

Subcase 1: $h^1(S, \mathcal{O}_S) = 0$ and $h^0(S, \mathcal{O}_S(K_S)) = 1$	(a K3 surface).
Subcase 2: $h^1(S, \mathcal{O}_S) = 0$ and $h^0(S, \mathcal{O}_S(K_S)) = 0$	(an Enriques surface).
Subcase 3: $h^1(S, \mathcal{O}_S) = 2$ and $h^0(S, \mathcal{O}_S(K_S)) = 1$	(an Abelian surface).
Subcase 4: $h^1(S, \mathcal{O}_S) = 1$ and $h^0(S, \mathcal{O}_S(K_S)) = 0$	(a bielliptic surface).
Subcase 5: $h^1(S, \mathcal{O}_S) = 1$ and $h^0(S, \mathcal{O}_S(K_S)) = 1$	(no such surfaces).

Note that the assertion of the abundance theorem in the case $\kappa(S) = 0$ is equivalent to

$$mK_S \sim 0 \text{ for some } m \in \mathbb{N}.$$

We analyze each individual subcase.

Subcase 1: $h^1(S, \mathcal{O}_S) = 0$ and $h^0(S, \mathcal{O}_S(K_S)) = 1$

In this subcase

$$\chi(\mathcal{O}_S) = 1 - h^1(S, \mathcal{O}_S) + h^0(S, \mathcal{O}_S(K_S)) = 2,$$

and thus we compute by the Riemann–Roch theorem and Serre duality

$$h^0(S, \mathcal{O}_S(-K_S)) + h^0(S, \mathcal{O}_S(2K_S))$$
$$\geq h^0(S, \mathcal{O}_S(-K_S)) - h^1(S, \mathcal{O}_S(-K_S)) + h^2(S, \mathcal{O}_S(-K_S))$$
$$= \chi(\mathcal{O}_S(-K_S)) = \frac{1}{2}(-K_S - K_S)(-K_S) + \chi(\mathcal{O}_S) = 2.$$

But then since $\kappa(S) = 0$ implies $h^0(S, \mathcal{O}_S(2K_S)) \leq 1$, we conclude that

$$h^0(S, \mathcal{O}_S(-K_S)) \geq 1.$$

Now that

$$h^0(S, \mathcal{O}_S(K_S)) = 1 \& h^0(S, \mathcal{O}_S(-K_S)) \geq 1,$$

we conclude that

$$K_S \sim 0.$$

Subcase 2: $h^1(S, \mathcal{O}_S) = 0$ and $h^0(S, \mathcal{O}_S(K_S)) = 0$

Since $h^1(S, \mathcal{O}_S) = 0$ and K_S is nef, Lemma 1-4-10 implies $h^0(S, \mathcal{O}_S(2K_S)) \neq 0$. On the other hand, since in this subcase

$$\chi(\mathcal{O}_S) = 1 - h^1(S, \mathcal{O}_S) + h^0(S, \mathcal{O}_S(K_S)) = 1,$$

we compute by the Riemann–Roch theorem and Serre duality

$$h^0(S, \mathcal{O}_S(-2K_S)) + h^0(S, \mathcal{O}_S(3K_S))$$
$$\geq h^0(S, \mathcal{O}_S(-2K_S)) - h^1(S, \mathcal{O}_S(-2K_S)) + h^2(S, \mathcal{O}_S(-2K_S))$$
$$= \chi(\mathcal{O}_S(-2K_S)) = \frac{1}{2}(-2K_S - K_S)(-2K_S) + \chi(\mathcal{O}_S) = 1.$$

If $h^0(S, \mathcal{O}_S(3K_S)) \neq 0$ (i.e., $= 1$ cf. Proposition 1-5-3 (iii)), then $h^0(S, \mathcal{O}_S(2K_S)) \neq 0$ (i.e., $= 1$) would imply that $h^0(S, \mathcal{O}_S(K_S)) \neq 0$ (i.e., $= 1$). (In fact, if $D_a = \Sigma a_i D_i$ is the unique member of $|3K_S|$ and $D_b = \Sigma b_i D_i$ that of $|2K_S|$, then $2D_a = \Sigma 2a_i D_i = 3D_b = \Sigma 3b_i D_i$ is the unique member of $|6K_S|$. But this would imply that $D_a - D_b$ is an effective member of $|K_S|$.) This is a contradiction! Thus $h^0(S, \mathcal{O}_S(3K_S)) = 0$, and hence $h^0(S, \mathcal{O}_S(-2K_S)) \geq 1$. Again, now that

$$h^0(S, \mathcal{O}_S(-2K_S)) \neq 0 \text{ and } h^0(S, \mathcal{O}_S(2K_S)) \neq 0,$$

we conclude that

$$2K_S \sim 0.$$

Subcase 3: $h^1(S, \mathcal{O}_S) = 2$ and $h^0(S, \mathcal{O}_S(K_S)) = 1$

Since $h^1(S, \mathcal{O}_S) = h^0(S, \Omega_S^1) \neq 0$ by Hodge theory, we utilize the Albanese map, which provides a nontrivial morphism to an Abelian variety.

Theorem 1-5-11 (Albanese Map). *Let S be a nonsingular projective variety. Then there exists an Abelian variety A and a morphism $\alpha : S \to A$ called the **Albanese map**, characterized by the following universal property: For any morphism $f : S \to T$ to an Abelian variety T, there exists a unique morphism $\tilde{f} : A \to T$ such that $\tilde{f} \circ \alpha = f$.*

Moreover, we have a canonical isomorphism

$$\alpha^* : H^0(A, \Omega_A^1) \overset{\sim}{\to} H^0(S, \Omega_S^1).$$

PROOF.   For a proof, see, e.g., Beauville [1], Theorem V.13, page 63.   □

Let $\alpha : S \to A$ be the Albanese map as described in Theorem 1-5-11.
First we claim that $\dim \alpha(S) \neq 0$ or 1.
We leave it to the reader as an exercise to see why $\dim \alpha(S) \neq 0$.
Suppose $\dim \alpha(S) = 1$. Let $S \to C \to \alpha(S)$ be the Stein factorization. Then $g(C) = h^0(A, \Omega_A^1) = h^0(S, \Omega_S^1) = 2$. Take an étale cover $\tilde{A} \to A$ of degree $d > 1$ (Exercise! Find such a map!) and "pull back" the entire setting by a base change:

$$
\begin{array}{ccccccc}
\tilde{S} & \longrightarrow & \tilde{C} & \longrightarrow & \tilde{\alpha}(S) & \hookrightarrow & \tilde{A} \\
\downarrow & & \downarrow & & \downarrow & & \downarrow \\
S & \longrightarrow & C & \longrightarrow & \alpha(S) & \hookrightarrow & A.
\end{array}
$$

Since $\tilde{S} \to S$ is étale, $\tilde{S}$ is again a minimal model with $\kappa(\tilde{S}) = 0$. (Exercise! Why?) But then

$$h^0(\tilde{S}, \mathcal{O}_{\tilde{S}}(K_{\tilde{S}})) \geq h^0(S, \mathcal{O}_S(K_S)) = 1,$$

$$h^1(\tilde{S}, \mathcal{O}_{\tilde{S}}) \geq h^1(C, \mathcal{O}_C) = g(\tilde{C}) = d\{g(C) - 1\} + 1 > g(C) = 2.$$

(Note that the Hurwitz formula implies

$$2g(\tilde{C}) - 2 = d \cdot \{2g(C) - 2\},$$

since there is no ramification.) But this is impossible, since there is no such numerical possibility for the invariants of $\tilde{S}$.

Thus we have to consider only the case $\dim \alpha(S) = 2$, i.e., $\alpha(S) = A$. Let $D \in |K_S|$ be the unique effective member. We claim that

$$D = 0 \sim K_S.$$

We will derive a contradiction assuming $D = \Sigma d_i D_i \neq 0$. Under the assumption there exists an index o such that D_o is a nonsingular elliptic curve. In fact, the numerical data

(*) $0 = K_S^2 = K_S \cdot D \geq K_S \cdot D_i = D \cdot D_i = (\Sigma_{j \neq i} d_j D_j + d_i D_i) \cdot D_i \geq d_i D_i^2$

imply by the arithmetic genus formula that

(**) $$h^1(D_i, \mathcal{O}_{D_i}) = \frac{1}{2}(D_i + K_S) \cdot D_i + 1 \leq 1,$$

and hence that D_i is either a nonsingular elliptic curve or a (possibly singular) rational curve (Exercise! Why?). If all the D_i were (possibly singular) rational curves, then their images $\alpha(D_i)$ would all be points on an Abelian variety A and thus $\alpha(D) = $ pts. on A. But then Theorem 1-2-9 (negativity of self-intersection for contractible curves) would imply

$$0 > D^2 = K_S{}^2 = 0,$$

a contradiction! (Exercise! Show that the negativity of self-intersection for contractible curves holds as long as the morphism is generically finite (and does not have to be birational) and hence the theorem can be applied to α.) Thus there exists a nonsingular elliptic curve D_o. For this nonsingular elliptic curve D_o we have by (∗) and (∗∗)

$$D_o^2 = 0,$$

which implies that $\alpha(D_o)$ is not a point and hence an elliptic curve (Exercise! Why?). Then by Poincaré's complete reducibility theorem (See, e.g., Mumford [5], page 173) there exists a morphism $\beta : A \to B$ from A to another Abelian variety B such that

$$\alpha(D_o) = \beta^{-1}(p) \text{ for some } p \in B.$$

Take a composite

$$\gamma := \beta \circ \alpha : S \to B.$$

Then D_o is contained in a fiber of γ. Since $D_o^2 = 0$, Lemma 1-2-10 implies that D_o is proportional to the full fiber. But this would imply $\kappa(S) = \kappa(S, K_S) \geq \kappa(S, D_o) = 1$, a contradiction!

Subcase 4: $h^1(S, \mathcal{O}_S) = 1$ and $h^0(S, \mathcal{O}_S(K_S)) = 0$

Let $\alpha : S \to A$ be the Albanese map, where in this subcase it is a surjective morphism onto an elliptic curve A. Take the Stein factorization $S \to C \to A$, where C is also an elliptic curve. (In fact, it can be proved that $C \cong A$, and by abuse of notation we write $\alpha : S \to C$ for the Stein factorization.) Let F be a general fiber of α.

If $g(F) = 0$, then

$$0 = h^1(F, \mathcal{O}_F) = \frac{1}{2}(K_S + F) \cdot F + 1 \text{ and } F^2 = 0,$$

which would imply $K_S \cdot F = -2$, contradicting the assumption that K_S is nef.

If $g(F) = 1$, then

$$0 = F^2 \text{ and } 0 = K_S \cdot F = mK_S \cdot F = D \cdot F,$$

where D is the unique member of $|mK_S|$ for some $m \in \mathbb{N}$. This implies that D is contained in fibers of α. Moreover,

$$D^2 = m^2 K_S{}^2 = 0$$

implies that D is proportional to the full fibers by Lemma 1-2-10. Thus we conclude that

$$0 = D \sim mK_S.$$

(If $D \neq 0$, then we would have $\kappa(S) = \kappa(S, K_S) = \kappa(S, D) = 1$, a contradiction!)

If $g(F) \geq 2$, then by Proposition 1-5-12 there exists an elliptic fibration $\beta : S \to B$ onto a nonsingular projective curve B. Now carry the argument of the previous case with β instead of α to conclude that $0 = D \sim mK_S$. But then $F^2 = 0$ and $mK_S \cdot F = 0$ would imply $g(F) = 1$, a contradiction! Thus $g(F)$ cannot be greater than or equal to 2.

Subcase 5: $h^1(S, \mathcal{O}_S) = 1$ and $h^0(S, \mathcal{O}_S(K_S)) = 1$

In this subcase we look at the exponential sequence

$$0 \to \mathbb{Z} \to \mathcal{O}_S \to \mathcal{O}_S^* \to 0,$$

which induces the cohomology sequence

$$H^0(S, \mathcal{O}_S) \tilde{\to} H^0(S, \mathcal{O}_S^*)$$
$$\to H^1(S, \mathbb{Z}) \overset{i}{\to} H^1(S, \mathcal{O}_S) \overset{exp}{\to} H^1(S, \mathcal{O}_S^*)$$
$$\to H^2(S, \mathbb{Z}).$$

Via the canonical isomorphism $\text{Pic}(S) \cong H^1(S, \mathcal{O}_S^*)$, the trivial line bundle $\mathcal{O}_S$ belongs to the image under the map exp, which is isomorphic to $H^1(S, \mathcal{O}_S)/i(H^1(S, \mathbb{Z}))$. Since $H^1(S, \mathbb{Z}) \neq 0$ by the assumption $h^1(S, \mathcal{O}_S) = 1$, $i(H^1(S, \mathbb{Z}))$ is a nonzero subgroup finitely generated over $\mathbb{Z}$ inside of the vector space $H^1(\mathcal{O}_S)$. In particular, we can choose a divisor M such that $M \not\sim 0$ but $2M \sim 0$. But then the Riemann–Roch theorem and Serre duality yield

$$h^0(S, \mathcal{O}_S(M)) + h^0(S, \mathcal{O}_S(K_S - M))$$
$$= h^0(S, \mathcal{O}_S(M)) + h^2(S, \mathcal{O}_S(M))$$
$$\geq h^0(S, \mathcal{O}_S(M)) - h^1(S, \mathcal{O}_S(M)) + h^2(S, \mathcal{O}_S(M))$$
$$= \chi(\mathcal{O}_S(M)) = \frac{1}{2}(M - K_S) \cdot M + \chi(\mathcal{O}_S) = 1.$$

Since $h^0(S, \mathcal{O}_S(M)) = 0$, this implies $h^0(S, \mathcal{O}_S(K_S - M)) \neq 0$, and thus there exists a member $G \in |K_S - M|$, which gives rise to $2G \in |2(K_S - M)| = |2K_S|$. On the other hand, since $h^0(S, \mathcal{O}_S(K_S)) = h^0(S, \mathcal{O}_S(2K_S)) = 1$ by the assumption, we also have a member $D \in |K_S|$, which gives rise to a unique member $2D \in |2K_S|$. Therefore, we have $2D = 2G$, i.e., $D = G$, which implies

$$K_S \sim D = G \sim K_S - M.$$

Thus we would have $M \sim 0$, contradicting the choice of M! Therefore, this subcase cannot happen.

This completes the proof of Theorem 1-5-6 assuming Theorem 1-5-4.   □

Proof of $\kappa(S) \geq 0$ for a Minimal Model S in Dimension 2

We give a proof of Theorem 1-5-4.

Let S be a minimal model in dimension 2. We derive a contradiction assuming $\kappa(S) = -\infty$.

Note first that since K_S is nef, Lemma 1-4-10 (cf. Castelnuovo's criterion for rationality) implies either $h^1(S, \mathcal{O}_S) \neq 0$ or $h^0(S, \mathcal{O}_S(2K_S)) \neq 0$. Since $\kappa(S) = -\infty$ by assumption and hence $h^0(S, \mathcal{O}_S(2K_S)) = 0$, we have $h^1(S, \mathcal{O}_S) = h^{0,1}(S) = h^{1,0}(S) = h^0(S, \Omega^1_S) \neq 0$. Note second that by Proposition 1-5-9 and the Serre duality

$$\chi(\mathcal{O}_S) = 1 - h^1(S, \mathcal{O}_S) + h^0(S, \mathcal{O}_S(K_S)) \geq 0.$$

Since $h^0(S, \mathcal{O}_S(K_S)) = 0$ and $h^1(S, \mathcal{O}_S) \neq 0$, we conclude that

$$h^1(S, \mathcal{O}_S) = 1.$$

Let $\alpha : S \to A$ be the Albanese map, and let $S \to C \to A$ be the Stein factorization, where C as well as A is necessarily an elliptic curve. (We write $\alpha : S \to C$ as before by abuse of notation.) Let F be a general fiber of α. The following proposition is the most subtle part of the proof.

Proposition 1-5-12. *Let S be a nonsingular projective surface whose canonical bundle K_S is nef and with*

$$\kappa(S) \leq 0, h^1(S, \mathcal{O}_S) = 1 \text{ and } h^0(S, \mathcal{O}_S(K_S)) = 0.$$

Let $\alpha : S \to C$ be (the Stein factorization of) the Albanese map onto an elliptic curve. Then there exists an elliptic fibration $\beta : S \to B$ (a morphism with connected fibers onto a nonsingular projective curve B whose general fiber is an elliptic curve) such that $\beta(F) = B$ for a general fiber F of α.

PROOF.   First note that

$$g(F) \geq 1.$$

In fact, if $g(F) = 0$, then

$$0 = h^1(F, \mathcal{O}_F) = \frac{1}{2}(K_S + F) \cdot F + 1 \text{ and } F^2 = 0$$

would imply $K_S \cdot F = -2$, contradicting the assumption that K_S is nef. Note also that by Lemma 1-5-7

$$K_S^2 = 0.$$

Step 1. The Albanese map $\alpha : S \to C$ is a smooth morphism.

First we claim that all the fibers of α are irreducible.

Suppose there is a reducible fiber containing two irreducible components F_1 and F_2. Take an ample divisor H. Then F_1, F_2, and H would give rise to linearly independent classes in $H^2(S, \mathbb{R})$ and thus the second Betti number

$$b_2 \geq 3.$$

(Suppose $hH + f_1F_1 + f_2F_2 = 0$ in $H^2(S, \mathbb{R})$, where $h, f_1, f_2 \in \mathbb{R}$. First, $0 = (hH + f_1F_1 + f_2F_2) \cdot F = hH \cdot F$ implies $h = 0$. Second, $f_1F_1 + f_2F_2 \equiv 0$ iff $f_1 = f_2 = 0$ by Lemma 1-2-10. Thus $h = f_1 = f_2 = 0$, proving that H, F_1, and F_2 are linearly independent in $H^2(S, \mathbb{R})$.) On the other hand, by Noether's formula and Hodge theory

$$e(S) = 12\chi(\mathcal{O}_S) - K_S^2 = 0$$
$$= 2(b_0 - b_1) + b_2 = 2(1 - 2h^1(S, \mathcal{O}_S)) + b_2,$$

which implies

$$b_2 = 2,$$

a contradiction! Thus all the fibers are irreducible.

Set

$$F_p := \{\alpha^{-1}(p)\}_{\text{red}}$$

and

$$\alpha^{-1}(p) = m_p F_p.$$

Lemma 1-5-13. *Let F_p be a projective curve that is irreducible and reduced (but may be singular). Then*

$$e(F_p) \geq 2\chi(\mathcal{O}_{F_p}),$$

and equality holds iff F_p is nonsingular.

PROOF.   Note first that if F_p is nonsingular, then

$$e(F_p) = b_0 - b_1 + b_2 = 2 - b_1 = 2\{1 - h^{0,1}\}$$
$$= 2\{h^0(F_p, \mathcal{O}_{F_p}) -^1 (F_p, \mathcal{O}_{F_p})\} = 2\chi(\mathcal{O}_{F_p})$$

is an easy consequence of Hodge theory.

Let $\mu : \tilde{F}_p \to F_p$ be the normalization. Consider the diagram as below where the rows are exact, the two left vertical arrows are given by the natural inclusions, and the third arrow ϕ is such that the diagram commutes:

$$
\begin{array}{ccccccccc}
0 & \longrightarrow & \mathbb{C}_{F_p} & \overset{\sigma}{\longrightarrow} & \mu_*\mathbb{C}_{\tilde{F}_p} & \longrightarrow & \text{Coker}(\sigma) & \longrightarrow & 0 \\
 & & \downarrow & & \downarrow & & \downarrow\phi & & \\
0 & \longrightarrow & \mathcal{O}_{F_p} & \overset{\eta}{\longrightarrow} & \mu_*\mathcal{O}_{\tilde{F}_p} & \longrightarrow & \text{Coker}(\eta) & \longrightarrow & 0.
\end{array}
$$

Since

$$\mu_*\mathbb{C}_{\tilde{F}_p} \cap \mathcal{O}_{F_p} = \mathbb{C}_{F_p},$$

ϕ is injective. Therefore,

$$h^0(\text{Coker}(\sigma)) \leq h^0(\text{Coker}(\eta)).$$

Since both $\text{Coker}(\sigma)$ and $\text{Coker}(\eta)$ are skyscraper sheaves, we get

$$e(\tilde{F}_p) = e(F_p) + h^0(\text{Coker}(\sigma)),$$
$$\chi(\mathcal{O}_{\tilde{F}_p}) = \chi(\mathcal{O}_{F_p}) + h^0(\text{Coker}(\eta)).$$

These together with

$$e(\tilde{F}_p) = 2\chi(\mathcal{O}_{\tilde{F}_p})$$

imply

$$e(F_p) = 2\chi(\mathcal{O}_{F_p}) + 2h^0(\text{Coker}(\eta)) - h^0(\text{Coker}(\sigma)) \geq 2\chi(\mathcal{O}_{F_p}),$$

equality holding iff $h^0(\text{Coker } \eta) = 0$, i.e., $\tilde{F}_p \cong F_p$ is nonsingular.

The above lemma implies

$$e(F_p) \geq 2\chi(\mathcal{O}_{F_p}),$$

and by the computation using the Riemann–Roch theorem as in Lemma 1-1-4,

$$e(F_p) \geq 2\chi(\mathcal{O}_{F_p}) = -(F_p + K_S) \cdot F_p$$
$$= -\left(\frac{1}{m_p}F + K_S\right) \cdot \frac{1}{m_p}F = -\frac{1}{m_p}K_S \cdot F$$
$$= -\frac{1}{m_p}(F + K_S) \cdot F = \frac{1}{m_p}2\chi(\mathcal{O}_F) = \frac{1}{m_p}e(F).$$

Since $g(F) \geq 1$, we have

$$e(F) \leq 0.$$

Therefore,

$$e(F_p) \geq e(F) \quad \forall p \in C$$

and

$$e(F_p) = e(F) \text{ iff } e(F_p) = 2\chi(\mathcal{O}_{F_p}) \text{ and } \frac{1}{m_p}e(F) = e(F).$$

On the other hand, let $\Theta \subset C$ be the locus where $\alpha^{-1}(p)(p \in \Theta)$ fails to be smooth, i.e., where it is either singular or multiple. We have

$$0 = e(S) = e(S - \alpha^{-1}\Theta) + e(\alpha^{-1}\Theta)$$
$$= e(F)e(C - \Theta) + \Sigma_{p \in \Theta}e(\alpha^{-1}(p))$$
$$= e(F)e(C) + \Sigma_{p \in \Theta}(e(F_p) - e(F))$$
$$= \Sigma_{p \in \Theta}(e(F_p) - e(F)) \quad (\text{Recall that } e(C) = 0, \text{ since } C \text{ is an elliptic curve.})$$

Thus

$$e(F_p) = e(F) \quad \forall p \in \Theta.$$

Therefore, we conclude by Lemma 1-5-13 that

$$F_p \text{ is nonsingular } \forall p \in \Theta, \text{ since } e(F_p) = 2\chi(\mathcal{O}_{F_p})$$

and that

either $\{g(F) \geq 2$ and $m_p = 1 \quad \forall p \in \Theta\}$ or $\{g(F) = 1$ and m_p arbitrary$\}$,

since $\frac{1}{m_p} e(F) = e(F)$.

But if $g(F) = 1$ and $m_p \neq 1$ for some $p \in \Theta$, then the canonical bundle formula for the elliptic fibration $\alpha : S \to C$ (cf. Theorem 1-5-17) would give

$$mK_S = \alpha^* \mathcal{K} \text{ for some } m \in \mathbb{N},$$

where $\mathcal{K}$ is a line bundle on C of positive degree and thus $\kappa(S) > 0$, a contradiction!

Thus $\alpha^{-1}(p)$ is irreducible, reduced, and smooth for all $p \in C$, i.e., $\alpha : S \to C$ is a smooth morphism.

Step 2. There exists an elliptic curve $G \subset S$ such that $G^2 = K_S \cdot G = 0$ and $\alpha(G) = C$.

Case: $g(F) = 1$.

Since $\alpha : S \to C$ is a smooth morphism, in this case every fiber $F = \alpha^{-1}(p)p \in C$ is a smooth elliptic curve. Take a very ample divisor H such that $H|_F$ is an effective divisor of degree $m = H \cdot F$. Then since $F \cong \text{Pic}^0(F)$, there exist exactly m^2 distinct points $p_1^F, \ldots, p_{m^2}^F$ such that

$$mp_i^F \sim H|_F.$$

It is easy to see (Exercise!) that

$$G' = \{p_i^F ; i = 1, \ldots, m^2; F \text{ all the fibers}\} \to C$$
$$\cup \hspace{5.5cm} \cup$$
$$p_i^F \hspace{4.5cm} \mapsto p$$

is an étale morphism from G' to an elliptic curve C, and thus an irreducible component G of G' is also an elliptic curve with $\alpha(G) = C$.

Figure 1-5-14.

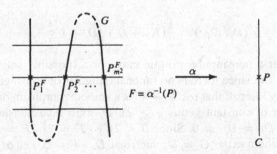

Note that by the canonical bundle formula for the elliptic fibration $\alpha : S \to C$ we have

$$K_S = \alpha^* \{K_C \otimes (R^1 \alpha_* \mathcal{O}_S)^\vee\},$$

which must be numerically trivial by the assumption $\kappa(S) \leq 0$ and K_S being nef. Thus $K_S \cdot G = 0$. Since G is an elliptic curve,

$$1 = h^1(G, \mathcal{O}_G) = \frac{1}{2}(K_S + G) \cdot G + 1$$

implies $G^2 = 0$.

Case: $g(F) \geq 2$.

(Note first that this case should not happen, since the conclusion of the abundance theorem would eventually tell us that

$$g(F) = h^1(F, \mathcal{O}_F) = \frac{1}{2}(K_S + F) \cdot F + 1 = 1, \quad \text{a contradiction!}$$

Note second that the analysis of this case, extracted from Griffiths–Harris [1], is classically called the "Enriques trick" and constitutes the most difficult part of the proof.)

Since $\alpha : S \to C$ is a smooth morphism, in this case every fiber $F = \alpha^{-1}(p)$ for $p \in C$ is a smooth curve of constant genus $g \geq 2$.

Fix a fiber F_o of α. It is sufficient to show that there is a member $D = \Sigma d_i D_i \in |2K_S + F_p - F_o|$ for some fiber F_p of α.

In fact, such a member D cannot be a zero divisor. (If it were a zero divisor, then $2K_S + F_p - F_o \equiv 0$ and thus K_S would be numerically trivial. But then

$$h^1(F, \mathcal{O}_F) = \frac{1}{2}(K_S + F) \cdot F + 1 = 1,$$

contradicting the assumption $g(F) \geq 2$.) Thus $D = \Sigma d_i D_i \neq 0$ and

(*) $\qquad 0 = (2K_S)^2 = (2K_S + F_p - F_o)^2 = 2K_S \cdot D$
$$\geq 2K_S \cdot D_i = D \cdot D_i = (\Sigma_{j \neq i} d_j D_j + d_i D_i) \cdot D_i \geq d_i D_i^2,$$

which implies

(**) $\qquad h^1(D_i, \mathcal{O}_{D_i}) = \frac{1}{2}(K_S + D_i) \cdot D_i + 1 \leq 1.$

Thus D_i is either a nonsingular elliptic curve or a (possibly singular) rational curve for each i. But since there is no rational (singular or nonsingular) curve on S (Exercise! Why? Recall that $\alpha : S \to C$ is a smooth morphsim onto an elliptic curve with fibers of constant genus $g \geq 2$), D_i must be a nonsingular elliptic curve with $K_S \cdot D_i = D_i^2 = 0$. Since $0 < 2K_S \cdot F = D \cdot F = (\Sigma d_i D_i) \cdot F$, since $g(F) \geq 2$, there exists $G = D_o$ such that $D_o \cdot F > 0$, i.e., $\alpha(G) = C$ with $G^2 = K_S \cdot G = 0$.

We now prove that there is a member $D = \Sigma d_i D_i \in |2K_S + F_p - F_o|$ for some fiber F_p of α.

Note that since

$$h^2(S, \mathcal{O}_S(2K_S + F_p)) = h^0(S, \mathcal{O}_S(-K_S - F_p)) = 0$$

by Serre duality, the Riemann–Roch theorem yields

$$h^0(S, \mathcal{O}_S(2K_S + F_p)) \geq \chi(\mathcal{O}_S(2K_S + F_p))$$

$$= \frac{1}{2}(2K_S + F_p - K_S) \cdot (2K_S + F_p) + \chi(\mathcal{O}_S) = 3g - 3,$$

where we note that $K_S \cdot F_p = 2g - 2$ and $\chi(\mathcal{O}_S) = 0$. On the other hand,

$$h^0(F_o, \mathcal{O}_{F_o}(2K_S + F_p|_{F_o})) = h^0(F_o, \mathcal{O}_{F_o}(2K_{F_o})) = 3g - 3.$$

Now, if for some $p \in C$ the restriction map

$$r_p : H^0(S, \mathcal{O}_S(2K_S + F_p)) \to H^0(F_o, \mathcal{O}_{F_o} 2K_{F_o}))$$

is not injective, then the kernel

$$H^0(S, 2K_S + F_p - F_o)$$

is nonzero, and we are done. So we may assume that r_p is injective and thus an isomorphism for all $p \in C$. We fix a divisor $D_{F_o} = P_1 + \cdots + P_{4g-4} \in |2K_{F_o}|$. Consider the incidence relation

$$I := \{(p, q); q \in D_p\} \subset C \times S,$$

where $D_p \in |2K_S + F_p|$ is the unique lift of D_{F_o}. (Exercise! Show that I is a closed algebraic subset of $C \times S$.) Since F_p and $F_{p'}$ are not linearly equivalent (unless $p = p'$), D_p and $D_{p'}$ are not linearly equivalent (unless $p = p'$) and thus the second projection $I \to S$ is surjective (Exercise! Why?). Thus for any point $q \in F_o, q \neq P_1, \ldots, P_{4g-4}$, there exists D_p that contains q. But this D_p, containing $4g - 3$ points $q, P_1, \ldots, P_{4g-4}$, has to contain the entire curve F_o. Therefore,

$$D_p - F_o \in |2K_S + F_p - F_o| \neq \emptyset,$$

and we are done!

Step 3. There exists another elliptic curve $G' \subset S$ such that $G'^2 = K_S \cdot G' = 0$, $\alpha(G') = C$, and G' is disjoint from G.

Consider the exact sequence for $n \in \mathbb{N}$

$$0 \to \mathcal{O}_S(nK_S + (n - 1)G) \to \mathcal{O}_S(nK_S + nG)$$
$$\to \mathcal{O}_S(nK_S + nG)|_G = nK_G \cong \mathcal{O}_G \to 0,$$

which induces the cohomology sequence

$$H^1(S, \mathcal{O}_S(nK_S + nG)) \to H^1(G, \mathcal{O}_G) \to H^2(S, \mathcal{O}_S(nK_S + (n - 1)G)).$$

By Serre duality and since K_S is nef, we have for $n \geq 2$

$$H^2(S, \mathcal{O}_S(nK_S + (n - 1)G)) \cong H^0(S, \mathcal{O}_S(-(n - 1)K_S - (n-)G)) = 0,$$
$$H^2(S, \mathcal{O}_S(nK_S + nG)) \cong H^0(S, \mathcal{O}_S(-(n - 1)K_S - nG)) = 0.$$

Thus $h^1(G, \mathcal{O}_G) = 1$ implies

$$h^1(S, \mathcal{O}_S(nK_S + nG)) \geq 1,$$

and hence the Riemann–Roch theorem yields

$$h^0(S, \mathcal{O}_S(nK_S + nG)) = \frac{1}{2}(nK_S + nG - K_S) \cdot (nK_S + nG) + \chi(\mathcal{O}_S)$$
$$+ h^1(S, \mathcal{O}_S(nK_S + nG)) \geq 1.$$

Thus there exists a member $0 \neq G_n \in |nK_S + nG|$. Note that the G_n cannot be a multiples of G for all $n \geq 2$. (Suppose they are. Then

$$G_n = mG \sim nK_S + nG,$$
$$G_{n+1} = m'G \sim (n+1)K_S + (n+1)G.$$

It is easy to see that for an ample divisor A on S

$$\{(n+1)K_S + (n+1)G\} \cdot A > \{nK_S + nG\} \cdot A$$

and hence

$$m' > m.$$

But then

$$K_S \sim G_{n+1} - G_n - G \sim (m' - m - 1)G,$$

and hence we would have $h^0(S, \mathcal{O}_S(K_S)) > 0$, a contradiction!) Take n such that G_n is not a multiple of G, i.e.,

$$G_n = mG + \Sigma d_i D_i \text{ with } \Sigma d_i D_i \neq 0, \ D_i \neq G.$$

Since

$$0 = G \cdot (nK_S + nG) = G \cdot (mG + \Sigma d_i D_i),$$

we have

$$G \cdot D_i = 0 \quad \forall i.$$

This implies by the Hodge index theorem that

$$D_i^2 \leq 0 \text{ as } G^2 = 0.$$

Since

$$0 = K_S \cdot (nK_S + nG) = K_S \cdot (mG + \Sigma d_i D_i) \geq K_S \cdot D_i,$$

the arithmetic genus formula implies that D_i is either a nonsingular elliptic curve or a (possibly singular) rational curve. But since $\alpha : S \to C$ is a smooth morphism onto an elliptic curve having fibers F with $g(F) \geq 2$, there is no rational curve on S. Therefore, all the D_i are nonsingular elliptic curves with

$$K_S \cdot D_i = D_i^2 = 0.$$

Now we only have to set one of the D_i to be G'. In fact, then $G \cdot G' = 0$ implies that G' is disjoint from G and that G' cannot be a fiber F of α, since $g(F) \geq 2$. Thus $\alpha(G') = C$.

Step 4. There exists an elliptic fibration $\beta : S \to B$ such that $\beta(F) = B$.

Consider the exact sequence

$$0 \to \mathcal{O}_S(2K_S + G + G') \to \mathcal{O}_S(2K_S + 2G + 2G')$$
$$\to \mathcal{O}_S(2K_S + 2G + 2G')|_{G \coprod G'} \cong \mathcal{O}_G \oplus \mathcal{O}_{G'} \to 0,$$

which induces the cohomology sequence

$$H^1(S, \mathcal{O}_S(2K_S + 2G + 2G')) \to H^1(G \coprod G', \mathcal{O}_G \oplus \mathcal{O}_{G'})$$
$$\to H^2(S, \mathcal{O}_S(2K_S + G + G')).$$

Since by Serre duality we have

$$H^2(S, \mathcal{O}_S(2K_S + G + G')) \cong H^0(S, \mathcal{O}_S(-K_S - G - G')) = 0,$$
$$H^2(S, \mathcal{O}_S(2K_S + 2G + 2G')) \cong H^0(S, \mathcal{O}_S(-K_S - 2G - 2G')) = 0,$$

the equality $h^1(G \coprod G', \mathcal{O}_G \oplus \mathcal{O}_{G'}) = h^1(G, \mathcal{O}_G) + h^1(G', \mathcal{O}_{G'}) = 2$ implies

$$h^1(S, \mathcal{O}_S(2K_S + 2G + 2G')) \geq 2.$$

Therefore, the Riemann–Roch theorem yields

$$h^0(S, \mathcal{O}_S(2K_S + 2G + 2G'))$$
$$= \frac{1}{2}(2K_S + 2G + 2G' - K_S) \cdot (2K_S + 2G + 2G')$$
$$+ \chi(\mathcal{O}_S) + h^1(S, \mathcal{O}_S(2K_S + 2G + 2G'))$$
$$= h^1(S, \mathcal{O}_S(2K_S + 2G + 2G')) \geq 2.$$

Set

$$|2K_S + 2G + 2G'| = |M| + \text{Fix},$$

where $|M|$ is the movable part and *Fix* the fixed part. Note that the movable part $|M|$ is not empty, since $h^0(S, \mathcal{O}_S(2K_S + 2G + 2G')) \geq 2$. Moreover, since

$$0 = (2K_S + 2G + 2G')^2 \geq M^2 \geq 0,$$
$$0 = (2K_S + 2G + 2G')^2 \geq 2K_S \cdot (M + \text{Fix}) \geq 2K_S \cdot M \geq 0,$$

we conclude that $|M|$ is base point free and by Bertini's theorem that for any general member $D = \Sigma d_i D_i \in |M|$ each component D_i is an elliptic curve with $K_S \cdot D_i = D_i^2 = 0$, D_i and D_j being disjoint if $i \neq j$. Thus the Stein factorization $\beta : S \to B$ of the morphism $\Phi_{|M|} : S \to \Phi_{|M|}(S)$ is an elliptic fibration with general fibers consisting of these D_i's. Moreover, this elliptic fibration cannot coincide with $\alpha : S \to C$, since otherwise we would have

$$0 < G \cdot F = G \cdot D_i \leq G \cdot (M + \text{Fix}) = G \cdot (2K_S + 2G + 2G') = 0,$$

a contradiction! Therefore, $\beta(F) = B$.

This completes the proof of Proposition 1-5-12.

Now we finish the proof of Theorem 1-5-4 based upon Proposition 1-5-12.

Looking at the elliptic fibration $\beta : S \to B$ we conclude that

$$mK_S = \beta^*\mathcal{K}$$

for some $m \in \mathbb{N}$ and some line bundle $\mathcal{K}$ on B by the canonical bundle formula for the elliptic fibration $\beta : S \to B$. The adjunction formula implies that for a fiber F of $\alpha : S \to C$,

$$mK_F = (mK_S + mF)|_F = \beta^*\mathcal{K}|_F$$

is either trivial or of positive degree, depending on whether $g(F) = 1$ or $g(F) > 1$. This would imply $h^0(S, \mathcal{O}_S(dmK_S)) \geq 1$, for some $d \in \mathbb{N}$, contradicting our assumption $\kappa(S) = -\infty$!

This completes the proof of Theorem 1-5-4. $\qquad\qquad\qquad\qquad\qquad\qquad\square$

Remark 1-5-15. There is no conceptual proof of the hard dichotomy theorem or the abundance theorem so far, and any of the known (to the author) proofs involves Proposition 1-5-12 sitting at the technical heart. The essential part is to show, under the assumption of Proposition 1-5-12, that the Albanese map $\alpha : S \to C$ is an elliptic fibration, i.e., $g(F) = 1$.

Beauville [1] utilizes the fact, after showing that α is smooth, that there is no nonconstant map from a projective elliptic curve C to the "moduli" space of nonsingular projective curves of genus $g \geq 2$. (A higher-dimensional analogue of this fact can be observed in the recent research literature; see Kovács [2] and Migliorini [1].) Then he concludes that by taking a fiber product over an étale base change $C' \to C$,

$$\begin{array}{ccc} F \times C' \cong \; S \times_C C' & \longrightarrow & S \\ \downarrow & & \downarrow \\ C' & \longrightarrow & C \end{array}$$

if $g(F) \geq 2$, one would have

$$\kappa(S) = \kappa(S \times_C C') = \kappa(F \times C') = 1,$$

a contradiction! Bart–Peters–Van de Ven [1] overcome this point by proving the Iitaka conjecture $C_{2,1}$,

$$\kappa(S) \geq \kappa(C) + \kappa(F),$$

which would again imply $\kappa(S) \geq 1$ if we assume $g(F) \geq 2$, a contradiction.

We state the adjunction formula that we used several times in our proof of the hard dichotomy theorem and abundance theorem. We also analyze the canonical bundle formula for elliptic fibrations that played a key role in the proof.

Proposition 1-5-16 (Adjunction Formula). *Let F be a nonsingular curve on a nonsingular surface S. Then*

$$\mathcal{O}_F(K_F) = \mathcal{O}_S(K_S + F)|_F.$$

PROOF.   For a proof, see Hartshorne [3], Chapter II, Proposition 8.20 (cf. Lemma 1-1-4). $\qquad\qquad\qquad\qquad\qquad\qquad\qquad\qquad\qquad\qquad\qquad\qquad\qquad\square$

Theorem 1-5-17 (Canonical Bundle Formula for Elliptic Fibrations). *Let $\alpha : S \to C$ be a morphism with connected fibers from a nonsingular projective surface S onto a nonsingular projective curve C such that the general fiber F is an elliptic curve and there is no (-1)-curve in any fiber. Then*

$$\mathcal{O}_S(K_S) = \alpha^*\{\mathcal{O}_C(K_C) \otimes (R^1\alpha_*\mathcal{O}_S)^\vee\} \otimes \mathcal{O}_S(\Sigma(m_i - 1)F_i),$$

where $(R^1\alpha_*\mathcal{O}_S)^\vee$ is a line bundle on C of degree $\chi(\mathcal{O}_S)$ and where $\alpha^{-1}(p_i) = m_i F_i$ are multiple fibers (not necessarily irreducible) with $F_i = \{\alpha^{-1}(p_i)\}_{\text{red}}$. In particular,

$$\mathcal{O}_S(mK_S) = \alpha^*\mathcal{K}$$

for some $m \in \mathbb{N}$ and some line bundle $\mathcal{K}$ on C.

PROOF.   The following arguments are extracted from Barth–Peters–Van de Ven [1]. We start the proof by verifying the following lemma.   □

Lemma 1-5-18. *Let $\alpha : S \to C$ be an elliptic fibration as in Theorem 1-5-17. Then*

(i) *if $\alpha^{-1}(p_i) = m_i F_i$ is a multiple fiber, then $\mathcal{O}_{F_i}(F_i)$ is torsion of order exactly m_i and $\omega_{F_i} = \mathcal{O}_{F_i}$,*

(ii) $h^0(\alpha^{-1}(p), \mathcal{O}_{\alpha^{-1}(p)}) = h^1(\alpha^{-1}(p), \mathcal{O}_{\alpha^{-1}(p)}) = 1 \quad \forall p \in C.$

PROOF.   (i) We take a small analytic neighborhood Δ of p_i so that F_i is the deformation retract of $S_\Delta := \alpha^{-1}(\Delta)$. First note that the order of $\mathcal{O}_{S_\Delta}(F_i)$ is exactly m_i, since

$$\mathcal{O}_{S_\Delta}(m_i F_i) \cong \mathcal{O}_{S_\Delta}(\alpha^* p_i) \cong \mathcal{O}_{S_\Delta},$$

and thus it is of order less than or equal to m_i, and since if it is of order $n_i < m_i$, then $n_i F_i = \text{div}(f)$ for some entire holomorphic function $f \in \Gamma(S_\Delta, \mathcal{O}_{S_\Delta}) = \alpha^*\Gamma(\Delta, \mathcal{O}_\Delta)$, which is impossible! Second, consider the exact sequence associated to the exponential sequence

$$
\begin{array}{ccccccc}
H^1(S_\Delta, \mathbb{Z}) & \longrightarrow & H^1(S_\Delta, \mathcal{O}_{S_\Delta}) & \longrightarrow & H^1(S_\Delta, \mathcal{O}_{S_\Delta}^*) & \longrightarrow & H^2(S_\Delta, \mathbb{Z}) \\
\| & & \downarrow & & \downarrow & & \| \\
H^1(F_i, \mathbb{Z}) & \longrightarrow & H^1(F_i, \mathcal{O}_{F_i}) & \longrightarrow & H^1(F_i, \mathcal{O}_{F_i}^*) & \longrightarrow & H^2(F_i, \mathbb{Z})
\end{array}
$$

Since $\mathcal{O}_{S_\Delta}(F_i) \in H^1(S_\Delta, \mathcal{O}_{S_\Delta}^*)$ is of finite order, its image in torsion-free $H^2(S_\Delta, \mathbb{Z}) \cong H^2(F_i, \mathbb{Z})$ is zero. Therefore, it is the image of an element $\xi \in H^1(S_\Delta, \mathcal{O}_{S_\Delta})$. The element ξ maps to $\xi|_{F_i} \in H^1(F_i, \mathcal{O}_{F_i})$, that in turn maps to $\mathcal{O}_{F_i}(F_i) \in H^1(F_i, \mathcal{O}_{F_i}^*)$. Suppose $\mathcal{O}_{F_i}(F_i)$ is of order m. Then m divides m_i. On the other hand, there is an element $c \in H^1(F_i, \mathbb{Z})$ that maps to $m(\xi|_{F_i})$. Note that

$$H^1(F_i, \mathbb{Z}) \hookrightarrow H^1(F_i, \mathcal{O}_{F_i})$$

is injective and hence also injective are

$$H^1(S_\Delta, \mathbb{Z}) \to V \text{ and } V \to H^1(F_i, \mathcal{O}_{F_i}),$$

where V is the $\mathbb{Q}$-linear span of the image of $H^1(S_\Delta, \mathbb{Z})$ inside of $H^1(S_\Delta, \mathcal{O}_{S_\Delta})$. Since

$$m_i \xi \in V \text{ and thus } m\xi \in V,$$

we conclude that

$$c = m\xi \text{ in } V,$$

which implies

$$\mathcal{O}_{S_\Delta}(mF_i) \cong \mathcal{O}_{S_\Delta}.$$

Thus m_i must divide m. Therefore, we finally have

$$m = m_i.$$

It is also easy to see (Exercise! Why?) that if $H^1(F_i, \mathbb{Z}) = 0$, then there is no nontrivial torsion line bundle on S_Δ. Thus a multiple fiber F_i cannot be simply connected. Since F_i is either a nonsingular elliptic curve, a nodal or cuspidal rational curve, or consists of (-2)-curves, and since the intersection form of F_i is negative semidefinite (cf. Lemma 1-2-10), it follows easily that a nonsimply connected F_i is either a nonsingular elliptic curve, a nodal rational curve, or a cycle of (-2)-curves. (Exercise! Check this assertion. For the detailed answer to this exercise, we refer the reader to Kodaira's classification of degenerations of elliptic curves.)

Figure 1-5-19.

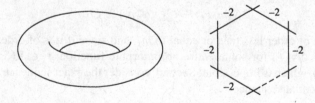

In each case, we have

$$\omega_{F_i} = \mathcal{O}_{F_i}.$$

(ii) Suppose $\alpha^{-1}(p) = m_p F_p$ is nonsingular, i.e., F_p is a nonsingular elliptic curve and $m_p = 1$. Then clearly,

$$h^0(\alpha^{-1}(p), \mathcal{O}_{\alpha^{-1}(p)}) = h^1(\alpha^{-1}(p), \mathcal{O}_{\alpha^{-1}(p)}) = 1.$$

Suppose $\alpha^{-1}(p) = m_p F_p$ is multiple, i.e., $m_p > 1$. Consider the exact sequence

$$0 \to \mathcal{O}_{F_p}(-\nu F_p) \to \mathcal{O}_{(\nu+1)F_p} \to \mathcal{O}_{\nu F_p} \to 0 \text{ for } 1 \le \nu \le m_p - 1.$$

Since $\mathcal{O}_{F_p}(-\nu F_p)$ is a nontrivial torsion bundle and since F_p is either a nonsingular elliptic curve, a nodal rational curve, or a cycle of (-2)-curves, we have

$$H^0(F_p, \mathcal{O}_{F_p}(-\nu F_p)) = 0 \text{ for } 1 \le \nu \le m_p - 1.$$

Thus by induction on ν we have

$$1 \le h^0(\alpha^{-1}(p), \mathcal{O}_{\alpha^{-1}(p)}) = h^0(m_p F_p, \mathcal{O}_{m_p F_p}) \le h^0(F_p, \mathcal{O}_{F_p}) = 1,$$

that is,

$$h^0(\alpha^{-1}(p), \mathcal{O}_{\alpha^{-1}(p)}) = 1.$$

Suppose $\alpha^{-1}(p) = F_p$ is singular and not multiple, i.e., F_p is not a nonsingular elliptic curve but $m_p = 1$. If F_p is reduced, then since F_p is also connected, we have obviously

$$h^0(\alpha^{-1}(p), \mathcal{O}_{\alpha^{-1}(p)}) = 1.$$

If F_p is not reduced, then consider the exact sequence

$$0 \to \mathcal{O}_{F_p - \{F_p\}_{\text{red}}}(-\{F_p\}_{\text{red}}) \to \mathcal{O}_{F_p} \to \mathcal{O}_{\{F_p\}_{\text{red}}} \to 0.$$

If $h^0(\alpha^{-1}(p), \mathcal{O}_{\alpha^{-1}(p)}) = h^0(F_p, \mathcal{O}_{F_p}) > 1$, then there exists $0 \ne h \in H^0(F_p, \mathcal{O}_{F_p})$ such that $h|_{\{F_p\}_{\text{red}}} = 0$. Take a maximal divisor $F_p \ge D \ge \{F_p\}_{\text{red}}$ such that $h|_D = 0$. Then $F_p > D$, and by setting $G = F_p - D$ we have the exact sequence

$$0 \to \mathcal{O}_G \xrightarrow{h} \mathcal{O}_G(-D) \to \text{Coker} \to 0.$$

It follows from the maximality of D that the map obtained by the multiplication of h is injective and that Coker has a finite support. Now,

$$G \cdot D = \chi(\mathcal{O}_S) - \chi(\mathcal{O}_S(-G)) - \chi(\mathcal{O}_S(-D)) - \chi(\mathcal{O}_S(-G - D))$$

$$\text{(cf. Exercise 1-2-2)}$$

$$= -\{\chi(\mathcal{O}_G(-D)) - \chi(\mathcal{O}_G)\} = -h^0(\text{Coker}) \le 0.$$

On the other hand,

$$0 = F_p^2 = (G + D)^2 = G^2 + 2G \cdot D + D^2, \quad G^2 < 0, \quad \text{and } D^2 < 0$$

$$\text{(cf. Lemma 1-2-9)}$$

would imply

$$G \cdot D > 0,$$

a contradiction! Thus we have

$$h^0(\alpha^{-1}(p), \mathcal{O}_{\alpha^{-1}(p)}) \le 1,$$

that is,

$$h^0(\alpha^{-1}(p), \mathcal{O}_{\alpha^{-1}(p)}) = 1.$$

In conclusion, we have

$$h^0(\alpha^{-1}(p), \mathcal{O}_{\alpha^{-1}(p)}) = 1 \quad \forall p \in C.$$

Now, since α is flat, the Euler characteristic $\chi(\mathcal{O}_{\alpha^{-1}(p)})$ is constant (cf. Hartshorne [3], Chapter III, Corollary 9.10). Thus we also have

$$h^1(\alpha^{-1}(p), \mathcal{O}_{\alpha^{-1}(p)}) = 1 \quad \forall p \in C.$$

This completes the proof of Lemma 1-5-18.

Now we go back to the proof of Proposition 1-5-17.

First note that $\alpha_* \mathcal{O}_S(K_S)$ is locally free of rank one, i.e., a line bundle, since

$$h^0(\alpha^{-1}(p), \mathcal{O}_S(K_S)|_{\alpha^{-1}(p)}) = h^0(\alpha^{-1}(p), \omega_{\alpha^{-1}(p)})$$
$$= h^1(\alpha^{-1}(p), \mathcal{O}_{\alpha^{-1}(p)}) = 1 \quad \forall p \in C,$$

where the second equality is given by Serre duality for a projective Cohen–Macaulay scheme $\alpha^{-1}(p)$ (cf. Hartshorne [3], Chapter III, Theorem 7.6) and since we have Grauert's theorem (Hartshorne [3], Chapter III, Corollary 12.9). By considering the canonical homomorphism

$$\lambda : \alpha^* \alpha_* \mathcal{O}_S(K_S) \to \mathcal{O}_S(K_S),$$

we obtain

$$\mathcal{O}_S(K_S) = \alpha^* \alpha_* \mathcal{O}_S(K_S) \otimes \mathcal{O}_S(D),$$

where D is an effective divisor corresponding to the vanishing locus of λ.

Note that

$$D \not\geq \alpha^{-1}(p) \quad \forall p \in C.$$

In fact, suppose

$$D \geq \alpha^{-1}(p) \text{ for some } p \in C.$$

Take a local generator ω for the line bundle $\alpha_* \mathcal{O}_S(K_S)$ in a neighborhood of p. The section ω can be considered as a 2-form in a neighborhood of $\alpha^{-1}(p)$. The 2-form ω vanishes up to the order described by D, and thus $\omega / \alpha^* f$ is still holomorphic under the assumption, where $\{ f = 0 \}$ gives the divisor p on C. But then

$$\omega / f \in \alpha_* \mathcal{O}_S(K_S) \text{ and } \omega / f \notin \mathcal{O}_{C,p} \cdot \omega,$$

a contradiction! In particular, D does not contain any nonsingular fiber. Moreover, D does not contain any horizontal component, since λ is an isomorphism over the general fibers.

Since $K_S \cdot C = 0$ for any curve contained in a fiber (Exercise! Why? recall the assumption that there is no (-1)-curve in any fiber), we have

$$K_S^2 = (\alpha^* \alpha_* \mathcal{O}_S(K_S) \otimes \mathcal{O}_S(D))^2 = K_S \cdot (\alpha^* \alpha_* \mathcal{O}_S(K_S) \otimes \mathcal{O}_S(D)) = D^2 = 0.$$

Thus $D = \Sigma n_p F_p$ is proportional to the full fibers. The argument above implies $D = \Sigma n_i F_i$ with $n_i < m_i$, where the $\alpha^{-1}(p_i) = m_i F_i$ are multiple fibers. Now the adjunction formula tells us that

$$\omega_{F_i} = \mathcal{O}_S(K_S + F_i)|_{F_i} = \mathcal{O}_{F_i}((n_i + 1)F_i).$$

Since $\omega_{F_i} = \mathcal{O}_{F_i}$ and $\mathcal{O}_{F_i}(F_i)$ is torsion of order m_i and since $n_i < m_i$ from the above argument, we have $n_i = m_i - 1$.

Finally, by relative duality we have

$$\alpha_* \omega_{S/C} \cong (R^1 \alpha_* \mathcal{O}_S)^\vee$$

and hence

$$\alpha_* \mathcal{O}_S(K_S) = \omega_C \otimes \alpha_* \omega_{S/C} = \mathcal{O}_C(K_C) \otimes (R^1 \alpha_* \mathcal{O}_S)^\vee.$$

Moreover, by Leray's spectral sequence we have isomorphisms

$$H^0(S, \mathcal{O}_S) \cong H^0(S, \alpha_* \mathcal{O}_S) \text{ and } H^2(S, \mathcal{O}_S) \cong H^1(C, R^1 \alpha_* \mathcal{O}_S)$$

and the exact sequence

$$0 \to H^1(C, \alpha_* \mathcal{O}_S) \to H^1(S, \mathcal{O}_S) \to H^0(C, R^1 \alpha_* \mathcal{O}_S) \to 0.$$

Therefore, the Riemann–Roch theorem on C yields

$$\chi(\mathcal{O}_S) = \chi(\alpha_* \mathcal{O}_S) - \chi(R^1 \alpha_* \mathcal{O}_S) = -\deg(R^1 \alpha_* \mathcal{O}_S).$$

This completes the proof of the canonical bundle formula for elliptic fibrations. □

1.6 Basic Properties of Canonical Models in Dimension 2

The purpose of this section is to study **canonical models in dimension 2**.

Definition-Proposition 1-6-1 (Iitaka Fibration in Dimension 2). *Let S be a minimal model in dimension 2. Then* $|mK_S|$ *is base point free for sufficiently divisible and large* $m \in \mathbb{N}$ *by the abundance theorem. Thus*

$$\Phi = \Phi_{|mK_S|} : S \to S_{\text{can}} \text{ for sufficiently divisible and large } m \in \mathbb{N}$$

satisfies

(i) $\Phi : S \to S_{\text{can}}$ *is a morphism with connected fibers onto a normal projective variety* S_{can},

(ii) *for any curve* $C \subset S$

$$\Phi(C) = \text{pt.} \Leftrightarrow K_S \cdot C = 0,$$

(iii) $\mathcal{O}_S(mK_S) = \Phi^* \mathcal{K}$ *for some ample line bundle* $\mathcal{K}$ *on* S_{can},

(iv) $\kappa(S) = \dim S_{\text{can}}$.

Moreover, properties (i) and (ii) characterize $\Phi : S \to S_{\text{can}}$, *which we call the* **Iitaka fibration** *of S onto its* **canonical model**.

PROOF.   The assertions follow immediately from the abundance theorem and Proposition 1-2-16.   □

Exercise 1-6-2.

$$R := \oplus_{m \geq 0} H^0(S, \mathcal{O}_S(mK_S))$$

has the natural structure of a graded $\mathbb{C}$*-algebra and is called the* **canonical ring** *of a nonsingular projective surface S. Show that R is finitely generated as a* $\mathbb{C}$*-algebra and that*

$$S_{\text{can}} = \operatorname{Proj} R,$$

where S_{can} *is constructed first by taking a minimal model of S and then by the method described above. Thus R being a birational invariant (cf. Corollary 1-8-5),* S_{can} *is canonically determined only by the birational equivalence class of the surface.*

In the following, we will study the properties of the Iitaka fibration onto the canonical model $\Phi : S \to S_{\text{can}}$ in more detail according to the Kodaira dimension $\kappa(S) = \dim S_{\text{can}}$.

Case: $\kappa(S) = \dim S_{\text{can}} = 2$

We start the analysis of this case by giving a proof to Proposition 1-5-8 = Theorem 1-6-3 (which completes the proof of the abundance theorem in this case to guarantee the existence of the canonical model). Our strategy for the proof consists of two major parts. The first is the analysis of the configurations of curves that are supposed to be contracted by Φ. An easy combinatorial consideration leads to the so-called $A - D - E$ types of the dual graphs of the curves. The second is the contractibility of these curves, which we show by an argument similar to the one used for Castelnuovo's contractibility criterion for (-1)-curves.

Theorem 1-6-3.   *Let S be a minimal model in dimension 2 with* $\kappa(S) = 2$. *Then there exists a birational morphism*

$$\Phi : S \to S_{\text{can}}$$

onto a normal projective surface such that

(i) *for any curve* $C \subset S$

$$\Phi(C) = \text{pt.} \Leftrightarrow K_S \cdot C = 0,$$

   and that

(ii) $\mathcal{O}_S(K_S) = \Phi^* \mathcal{K}$ *for some line bundle* $\mathcal{K}$ *on* S_{can}.

PROOF.   Step 1. Analysis of the configurations of curves to be contracted.

Theorem 1-6-4. *Let S be a minimal model in dimension 2 with $\kappa(S) = 2$. Then*

(i) *there exists only a finite number of curves C with $K_S \cdot C = 0$,*

(ii) *any (irreducible reduced) curve C with $K_S \cdot C = 0$ is a (-2)-curve, i.e.,*

$$C \cong \mathbb{P}^1 \text{ and } C^2 = -2,$$

and

(iii) *the dual graph of any connected component of $\cup_{C:K_S \cdot C=0} C$ is one of the following types:*

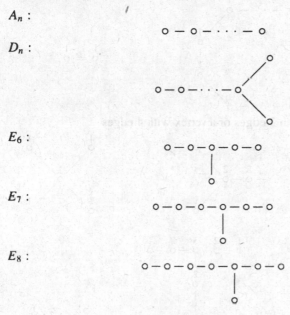

A_n :

D_n :

E_6 :

E_7 :

E_8 :

PROOF.   (i) Since $\kappa(S) = 2$, there exists $m \in \mathbb{N}$ such that

$$|mK_S| = |M| + F,$$

where $|M|$ is the movable part and F the fixed part, and such that

$$\Phi_{|M|} : S - Z \to T$$

is a birational morphism defined everywhere except possibly finitely many points Z, where Z is the fixed locus of $|M|$. If $K_S \cdot C = 0$, then either $C \subset F$ or $C \cap Z = \emptyset$ and $\Phi_{|M|}(C) = \mathrm{pt}$. Thus there are only finitely many such curves.

(ii) Since $K_S^2 > 0$ by Lemma 1-5-7, the Hodge index theorem implies $C^2 < 0$ if $K_S \cdot C = 0$. Thus the arithmetic genus formula

$$0 \le h^1(C, \mathcal{O}_C) = \frac{1}{2}(K_S + C) \cdot C + 1 < 1$$

implies $C \cong \mathbb{P}^1$ with $C^2 = -2$ (cf. Exercise 1-1-5).

(iii) Fix a connected component $E = \Sigma E_i$ of $\cup_{C:K_S \cdot C=0} C$. Let $\Sigma e_i E_i$ with $e_i \in \mathbb{N} \cup \{0\}$ be any effective divisor whose support lies in this connected component.

Since $K_S \cdot \Sigma e_i E_i = 0$, the Hodge index theorem implies $(\Sigma e_i E_i)^2 < 0$. Since this holds for any effective divisor, i.e., for any combination $\{e_i \in \mathbb{N} \cup \{0\}\}$, and since all the components are (-2)-curves isomorphic to $\mathbb{P}^1$, the following is the list of impossible subgraphs of the dual graph of E. (The numbers in the picture indicate a combination $\{e_i\}$ that makes $(\Sigma e_i E_i)^2 = 0$. The coefficients of the components not appearing in the subgraph are taken to be 0.)

(a) $\tilde{A}_n$: a cycle or 2 vertices connected by 2 edges

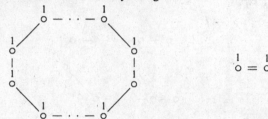

(b) $\tilde{D}_n$: 2 vertices, each with 3 edges or a vertex with 4 edges

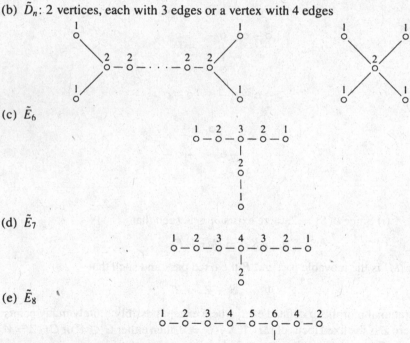

(c) $\tilde{E}_6$

(d) $\tilde{E}_7$

(e) $\tilde{E}_8$

Now it is just a combinatorial exercise that the only dual graphs possible are the ones listed in the theorem.

Step 2. Contraction of a connected component $E = \Sigma E_i$ of $\cup_{C; K_S \cdot C = 0} C$.

We claim that there exists a birational morphism

$$\Psi : S \to T$$

onto a normal projective surface such that

$$\Psi(E) = \text{pt.}(= p) \text{ and } \Psi : S - E \tilde{\to} T - p.$$

In order to establish the claim we follow the line of argument used to prove Castelnuovo's contractibility criterion for (-1)-curves in Theorem 1-1-6.

Choose a very ample divisor A on S. We would like to find

$$M := A + \Sigma a_i E_i, a_i \in \mathbb{N},$$

such that

$$M \cdot E_j = 0 \forall j.$$

M will then satisfy the conditions

(a) M is nef,
(b) for any curve $C \subset S$,

$$M \cdot C = 0 \Leftrightarrow C = e_j \text{ for some } j.$$

Lemma 1-6-5. *For the configuration of (-2)-curves of type A_n, D_n, E_6, E_7, or E_8, the intersection form $I = (I_{ij}) = (E_i \cdot E_j)$ is negative definite.*

PROOF.   Exercise! (See, e.g., Humphreys [1].)   □

Now, since I is negative definite with

$$I_{ii} < 0 \quad \forall i,$$
$$I_{ij} \geq 0 \text{ for } i \neq j,$$

and hence I^{-1} consists of nonzero columns whose entries are all nonpositive (Exercise! Why? actually, in the case of A_n, D_n, E_6, E_7, or E_8, all the entries of I^{-1} are strictly negative, since their dual graphs are connected), the system of simultaneous linear equations

$$\{(A + \Sigma a_i E_i) \cdot E_j = 0\}$$

has a unique solution

$$\{a_i\}, \text{ where } a_i \in \mathbb{Q} \quad \text{and } a_i > 0 \quad \forall i.$$

Replacing A by its sufficiently divisible multiple, we may assume

$$a_i \in \mathbb{N} \quad \forall i \text{ and } H^1(S, \mathcal{O}_S(A)) = 0.$$

Next we claim that $|M|$ is base point free. Since

$$|M| \supset |A| + \Sigma a_i E_i \text{ with } a_i \in \mathbb{N},$$

where A is very ample, the base points of M, if any, could only be on E. Thus we have only to show that $|M|$ has no base point on any irreducible component E_j of E.

Consider the exact sequence

$$0 \to \mathcal{O}_S(M - E_j) \to \mathcal{O}_S(M) \to \mathcal{O}_{E_j}(M) \to 0,$$

which induces the cohomology sequence

$$H^0(S, \mathcal{O}_S(M)) \to H^0(E_j, \mathcal{O}_{E_j}(M)) \to H^1(S, \mathcal{O}_S(M - E_j)).$$

Since

$$H^0(E_j, \mathcal{O}_{E_j}(M)) \cong H^0(\mathbb{P}^1, \mathcal{O}_{\mathbb{P}^1}),$$

we have only to show that

$$H^1(S, \mathcal{O}_S(M - E_j)) = 0,$$

which follows from the following claim.

Claim 1-6-6. *For each E_j there exists a sequence of curves*

$$E_{j0} = e_j, E_{j1}, \ldots, E_{jl} \text{ with } \Sigma_{t=1}^{l} E_{jt} = \Sigma a_i E_i$$

such that if we set

$$B_0 = M - E_{j0}, B_1 = B_0 - E_{j1}, B_2 = B_1 - E_{j2},$$

$$\ldots$$

$$B_l = B_{l-1} - E_{jl} = M - \Sigma a_i E_i = A,$$

then

$$\mathcal{O}_S(B_{r-1})|_{E_{j_r}} \cong \mathcal{O}_{\mathbb{P}^1}(k_r) \text{ with } k_r \geq -1$$

for all $r = 1, 2, \ldots, l$.

PROOF.   Just consider the columns of height a_i over the E_i, where the base is the corresponding dual graph consisting of the E_i. Then choose a sequence to fill in the columns starting with E_j so that at each stage you are not filling in any column that is adjacent to a column higher by more than one story. We illustrate one example of how we build up a sequence in the case where the dual graph is E_6 with the given $\{a_i\}$ and starting E_j below. Details are left to the reader as an exercise.

Figure 1-6-7.

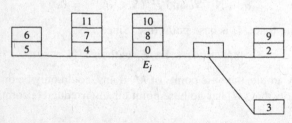

In fact, by the claim we have the exact sequence

$$0 \to \mathcal{O}_S(B_r) \to \mathcal{O}_S(B_{r-1}) \to \mathcal{O}_{E_{j_r}}(B_{r-1}) \cong \mathcal{O}_{\mathbb{P}^1}(k_r) \to 0,$$

which induces the cohomology sequence

$$H^1(S, \mathcal{O}_S(B_r)) \to H^1(S, \mathcal{O}_S(B_{r-1})) \to H^1(E_{j_r}, \mathcal{O}_{E_{j_r}}(B_{r-1})) \cong H^1(\mathbb{P}^1, \mathcal{O}_{\mathbb{P}^1}(k_r)).$$

Since

$$H^1(\mathbb{P}^1, \mathcal{O}_{\mathbb{P}^1}(k_r)) = 0 (k_r \geq -1)$$

and since

$$H^1(S, \mathcal{O}_S(B_l)) = H^1(S, \mathcal{O}_S(A)) = 0$$

by the choice of A, the descending induction on r gives the desired vanishing of

$$H^1(\mathcal{O}_S(M - E_j)) = 0.$$

Now that $|M|$ is base point free, the conditions

(a) M is nef,
(b) for any curve $C \subset S$,

$$M \cdot C = 0 \Leftrightarrow C = e_j \text{ for some } j,$$

and Proposition 1-2-16 imply that we have only to set $\Psi = \Phi_{|mM|} : S \to T$ for some sufficiently large $m \in \mathbb{N}$ to obtain the desired contraction morphism.

Step 3. Contraction of all the connected components of $\cup_{C; K_S \cdot C = 0} C$.

Note that all the arguments in Step 2 are local in a neighborhood of a connected component of $\cup_{C; K_S \cdot C = 0} C$. Then just by taking a sufficiently divisible multiple of A (if necessary) so that the systems of simultaneous equations corresponding to all the connected components should have integral solutions, we can construct the birational morphism $\Phi : S \to S_{\mathrm{can}}$ such that

(i) $\Phi : S \to S_{\mathrm{can}}$ is a morphism with connected fibers onto a normal projective variety S_{can},
(ii) for any curve $C \subset S$,

$$\Phi(C) = \mathrm{pt.} \Leftrightarrow K_S \cdot C = 0.$$

Finally, note that the entire argument above goes through if we replace A with

$$A' := K_S + A$$

with the same solutions for the systems of simultaneous equations. (We take A sufficiently very ample so that $H^1(S, \mathcal{O}_S(A')) = 0$.)

By Proposition 1-2-16

$$\mathcal{O}_S(M') = \mathcal{O}_S(A' + \Sigma a_i E_i) = \Phi^* \mathcal{L}' \text{ for some line bundle } \mathcal{L}' \text{ on } S_{\mathrm{can}}$$
$$\mathcal{O}_S(M) = \mathcal{O}_S(A + \Sigma a_i E_i) = \Phi^* \mathcal{L} \text{ for some line bundle } \mathcal{L} \text{ on } S_{\mathrm{can}}.$$

Therefore,

$$\mathcal{O}_S(K_S) = \mathcal{O}_S(M' - M) = \Phi^* \mathcal{K},$$

where

$$K = \mathcal{L}' \otimes \mathcal{L}^{-1}.$$

This completes the proof of Theorem 1-6-3.   □

Remark 1-6-8. Once we establish the more general base point freeness theorem via the Kawamata–Viehweg vanishing theorem, as will be discussed in Chapters 5 and 6, the assertions of Theorem 1-6-3 follow immediately as Corollary 6-2-5.

Note that K coincides with $\mathcal{O}_{S_{\mathrm{can}}}(K_{S_{\mathrm{can}}})$, where $K_{S_{\mathrm{can}}}$ is the canonical divisor of the canonical model S_{can}. (See Section 4-2 for the precise definition.) Thus

$$K_S = \Phi^* K_{S_{\mathrm{can}}},$$

where K_{can} is the ample canonical divisor on the canonical model S_{can}.

Case: $\kappa(S) = \dim S_{\mathrm{can}} = 1$

Note that in this case $\Phi : S \to S_{\mathrm{can}} = C$ is an elliptic fibration over a nonsingular projective curve C, since $K_S \cdot F = 0$ for a general fiber implies via the arithmetic genus formula

$$h^1(F, \mathcal{O}_F) = \frac{1}{2}(K_S + F) \cdot F + 1 = 1.$$

Then the canonical bundle formula gives

$$\mathcal{O}_S(K_S) = \Phi^* \left\{ \mathcal{O}_C(K_C) \otimes (R^1 \Phi_* \mathcal{O}_S)^\vee \otimes \mathcal{O}_C \left(\Sigma \frac{m_i - 1}{m_i} p_i \right) \right\}.$$

It can be shown that $(R^1 \Phi_* \mathcal{O}_S)^\vee$ is either ample or torsion on C (cf. Barth–Peters–Van de Ven [1], Chapter III, Theorems (18.2), (18.3)). Thus we can finally write

$$K_S = \Phi^*(K_{S_{\mathrm{can}}} + A + B) \text{ as } \mathbb{Q}\text{-divisors on,}$$

where $K_{S_{\mathrm{can}}} + A + B$ is the ample (log) canonical divisor on the canonical model S_{can} with

$A := (R^1 \Phi_* \mathcal{O}_S)^\vee$ is ample or torsion,

$B := \Sigma \dfrac{m_i - 1}{m_i} p_i$ is the boundary divisor arising from multiple fibers.

Case: $\kappa(S) = \dim S_{\mathrm{can}} = 0$

In this case, $\Phi : S \to S_{\mathrm{can}}$ is the trivial morphism from S to a point S_{can}. Of course, we have

$$K_S = \Phi^* K_{S_{\mathrm{can}}} \text{ as } \mathbb{Q}\text{-divisors on } S,$$

where $K_{S_{\mathrm{can}}}$ is the ample canonical divisor of S_{can}.

Summary 1-6-9. *Let* $\Phi : S \to S_{\text{can}}$ *be a morphism from a minimal model onto its canonical model. Then*

$$K_S = \Phi^* K_{S_{\text{can}}} (\, or \; K_{S_{\text{can}}} + A + B),$$

i.e., the canonical divisor K_S of S is the pullback of the ample (log) canonical divisor $K_{S_{\text{can}}}$ (or $K_{S_{\text{can}}} + A + B$) of the canonical model S_{can} by the Iitaka fibration. In particular,

$$\kappa(S) = \kappa(S_{\text{can}}, K_{S_{\text{can}}}) (\, or \; = \kappa(S_{\text{can}}, K_{S_{\text{can}}} + A + B)) = \dim S_{\text{can}}.$$

Observe that there is a minimal model S in dimension 2 with $\kappa(S) = 1$ such that $\Phi : S \to S_{\text{can}}$ maps onto $S_{\text{can}} \cong \mathbb{P}^1$ or $S_{\text{can}} \cong$ an elliptic curve. Thus

$$\kappa(S) \neq \kappa(S_{\text{can}}, K_{S_{\text{can}}}) = \kappa(S_{\text{can}}).$$

In general, it is necessary to take an appropriate *log* canonical divisor $K_{S_{\text{can}}} + A + B$ instead of a mere canonical divisor $K_{S_{\text{can}}}$ in order to obtain the equality between the Kodaira dimension upstairs $\kappa(S) = \kappa(S, K_S)$ and the one downstairs, $\kappa(S_{\text{can}}, K_{S_{\text{can}}} + A + B)$.

1.7   The Enriques Classification of Surfaces

The purpose of this section is to present the **Enriques classification of surfaces**, which is now a straightforward consequence of what we have accomplished through Sections 1-1–1-6.

Theorem 1-7-1 (The Enriques Classification of Surfaces). *Let S be a nonsingular projective surface. Then the birational classification of S is given by the following table:*

$\kappa(S)$	p_g	q	a biratioal representative
$-\infty$	0	0	$\mathbb{P}^2$
$-\infty$	0	> 0	$\mathbb{P}^1 \times C$ C a smooth curve of genus q
0	1	0	a K3 surface
0	0	0	an Enriques surface, i.e., a surface that admits a double cover by a K3 surface
0	1	2	an Abelian surface
0	0	1	a bielliptic surface, i.e., a surface that admits a finite étale cover by a product of two elliptic curves
1	≥ 0	≥ 0	an elliptic surface, i.e., a surface with an elliptic fibration over a smooth curve
2	≥ 0	≥ 0	a surface of general type

$$p_g = p_g(S) = h^0(S, \mathcal{O}_S(K_S)) : \text{the geometric genus of } S,$$
$$q = q(S) = h^1(S, \mathcal{O}_S) = h^0(S, \Omega_S^1) : \text{the irregularity of } S$$

We should emphasize that the Mori program, via the execution of MMP, reduces the classification to that of Mori fiber spaces and minimal models. The description of Mori fiber spaces was already given through their rigid characterization (cf. Theorem 1-4-4). The description of minimal models will be obtained through the abundance theorem using the method of (global) canonical covers.

It is this unifying power of the Mori program that allows us to give a rather short and self-contained proof of the (otherwise formidable) Enriques classification of surfaces, using only the following:

intersection theory
Hirzebruch–Riemann–Roch theorem
Serre duality
Hodge theory
Kodaira vanishing theorem
Albanese map
Canonical bundle formula for elliptic fibrations

PROOF.   We start with a nonsingular projective surface S.

We input S into **MMP in dimension 2**, so that it will produce as an end result a **Mori fiber space** $\phi : S_{\text{mori}} \to W$ when $\kappa(S) = -\infty$ or a **minimal model** S_{min} when $\kappa(S) \geq 0$, according to the **hard dichotomy theorem**.

Case: $\kappa(S) = -\infty$

The description of the Mori fiber space $\phi : S_{\text{mori}} \to W$ is given according to the dimension of W by Theorem 1-4-4.

When dim $W = 1$, S_{mori} is a $\mathbb{P}^1$-bundle in the Zariski topology over a nonsingular projective curve $W = C$, and hence birational to $\mathbb{P}^1 \times C$. By the Künneth formula, we have $q(S) = q(C) = g(C)$.

When dim $W = 0$, $S_{\text{mori}} \cong \mathbb{P}^2$ with $q(S) = q(\mathbb{P}^2) = 0$.

Case: $\kappa(S) \geq 0$

Thanks to the **abundance theorem**, the **Iitaka fibration** provides a morphism $\Phi : S_{min} \to S_{\text{can}}$ from the minimal model onto its **canonical model**.

When $\kappa(S) = \dim S_{\text{can}} = 2$, S is by definition called a surface of general type. The properties of the canonical model were briefly studied in Section 1-6. Their singularities, called canonical singularities, will be analyzed and characterized as rational double poits in Section 4-6.

When $\kappa(S) = \dim S_{\text{can}} = 1$, the arithmetic genus formula gives

$$g(F) = (K_S + F) \cdot F + 1 = 1$$

for the general fiber F of Φ. Thus Φ is an elliptic fibration and S is an elliptic surface by definition. Its structure can be analyzed further in relation to the study of the degenerations of elliptic curves by Kodaira.

When $\kappa(S) = \dim S_{can} = 0$, the Iitaka fibration is trivial and does not yield much information. First, by the **nonnegativity of** c_2 we have

$$0 \le c_2(S) = 12\chi(\mathcal{O}_S) = 12\{1 - q(S) + p_g(S)\},$$

which as in the proof of Theorem 1-5-4 and Theorem 1-5-6 restricts the numerical possibilities to the following four:

Subcase 1: $q(S) = 0$ and $p_g(S) = 1$ (a K3 surface).
Subcase 2: $q(S) = 0$ and $p_g(S) = 0$ (an Enriques surface).
Subcase 3: $q(S) = 2$ and $p_g(S) = 1$ (an Abelian surface).
Subcase 4: $q(S) = 1$ and $p_g(S) = 0$ (a bielliptic surface).

(We omit here the mere numerical possibility $q(S) = p_g(S) = 1$, which was shown not to occur geometrically in Section 1-5.)

| Subcase 1: $q(S) = 0$ and $p_g(S) = 1$ |

In this subcase, S is called a K3 surface by definition.

| Subcase 2: $q(S) = 0$ and $p_g(S) = 0$ |

In this subcase, since K_S is nef and $q(S) = 0$, Lemma 1-4-10 (cf. **Castelnuovo's criterion for rationality**) implies $h^0(S, \mathcal{O}_S(2K_S)) \ne 0$. Thus $\mathcal{O}_S(2K_S) \cong \mathcal{O}_S$ as a consequence of the abundance theorem, while $\mathcal{O}_S(K_S) \not\cong \mathcal{O}_S$, since $p_g(S) = h^0(S, \mathcal{O}_S(K_S)) = 0$. $\qquad\qquad \square$

Lemma 1-7-2 (Global Canonical Cover). *Let S be a nonsingular projective surface whose canonical bundle has the property*

$$\mathcal{O}_S(iK_S) \not\cong \mathcal{O}_S \text{ for } i = 1, 2, \ldots, r - 1,$$
$$\mathcal{O}_S(rK_S) \cong \mathcal{O}_S, r \in \mathbb{N}.$$

Then there exists a degree-r étale cover

$$c : \tilde{S} \to S,$$

where $\tilde{S}$ is a nonsingular projective surface with

$$K_{\tilde{S}} = c^*K_S \cong \mathcal{O}_{\tilde{S}}.$$

PROOF.   Consider the total space $\mathcal{L}(K_S)$ (respectively $\mathcal{L}(rK_S)$) of the canonical bundle K_S (respectively rK_S) over S regarded as vector bundles of rank 1 and the natural map induced by taking the r-th power,

$$\mathcal{L}(K_S) = \mathrm{Spec} \oplus_{m \ge 0} \mathcal{O}_S(mK_S^\vee) \xrightarrow[r]{r\text{-th power}} \mathcal{L}(rK_S) = \mathrm{Spec} \oplus_{m \ge 0} \mathcal{O}_S(mrK_S^\vee)$$

$$\pi_1 \searrow \qquad\qquad \swarrow \pi_r$$

$$S.$$

Since

$$rK_S \cong \mathcal{O}_S,$$

there is a nowhere-vanishing section

$$s \in H^0(S, rK_S).$$

Regard the section s as a map $s : S \to \mathcal{L}(rK_S)$ such that $\pi_r \circ s = \mathrm{id}$ on S. Then

$$\tilde{S} = r^{-1}(s(S))(\xrightarrow{r} s(S) \cong) \to S$$

is an étale cover of degree r. (The space $\tilde{S}$ is nonsingular projective since it is an etále cover of a nonsingular projective variety $s(S)$, but has yet to be shown to be irreducible.)

Choose an affine cover $\{U_\alpha\}$ of S so that locally

$U_\alpha = \mathrm{Spec}\, A_\alpha$

$\mathcal{O}_S(K_S)|_{U_\alpha} = \mathcal{O}_{U_\alpha} \cdot t_\alpha$ as an $\mathcal{O}_{U_\alpha}$-module where t_α is a generator

$\mathcal{L}(K_S)|_{U_\alpha} = \mathrm{Spec}\, A_\alpha[x_\alpha]$ as the total space of the line bundle

$\mathcal{O}(rK_S)|_{U_\alpha} = \mathcal{O}_{U_\alpha} \cdot t_\alpha^r$ as an $\mathcal{O}_{U_\alpha}$-module

$\mathcal{L}(rK_S)|_{U_\alpha} = \mathrm{Spec}\, A_\alpha[x_\alpha^r]$ as the total space of the line bundle

$s|_{U_\alpha} = f_\alpha \cdot t_\alpha^r$ as a section of an $\mathcal{O}_{U_\alpha}$-module $K_S|_{U_\alpha}$.

$s(U_\alpha) = \mathrm{Spec}\, A_\alpha[x_\alpha^r]/(x_\alpha^r - f_\alpha)$ as a subspace

$t_\alpha = g_{\alpha\beta} \cdot t_\beta$, where $\{g_{\alpha\beta}\}$ are the transition functions,

and hence the coordinate variables satisfy the relations

$x_\alpha = g_{\alpha\beta}^{-1} \cdot x_\beta$

Then $\tilde{S}$ over U_α is

$$r^{-1}(s(U_\alpha)) = \mathrm{Spec}\, A_\alpha[x_\alpha]/(x_\alpha{}^r - f_\alpha),$$

and

$$\{x_\alpha \in \Gamma(r^{-1}(U_\alpha), \mathcal{O}_{\tilde{S}})\}$$

gives a nonzero global section $\tilde{s}$ of

$$K_{\tilde{S}} = c^* K_S$$

as

$$t_\alpha = c^* g_{\alpha\beta} \cdot t_\beta \text{ and } x_\alpha = c^* g_{\alpha\beta}^{-1} \cdot x_\beta.$$

The line bundle $\mathcal{O}_{\tilde{S}}(K_{\tilde{S}})$, having a nonzero global section, is trivial:

$$\mathcal{O}_{\tilde{S}}(K_{\tilde{S}}) \cong \mathcal{O}_{\tilde{S}}.$$

If $\tilde{S} = r^{-1}(s(S))$ were reducible, then a connected component

$$\tilde{S} \supset \tilde{S}' \xrightarrow{c'} S$$

would give an étale cover of degree

$$r' < r \text{ with } \mathcal{O}_{\tilde{S}'}(K_{\tilde{S}'}) \cong \mathcal{O}_{\tilde{S}'}.$$

But then

$$r'K_S = c'_* K_{\tilde{S}'} \sim 0, \text{ i.e., } \mathcal{O}_S(r'K_S) \cong \mathcal{O}_S,$$

contradicting the assumption. Thus $\tilde{S} := r^{-1}(s(S))$ is irreducible. $\square$

Remark-Exercise 1-7-3. The key point of the construction above is that starting from a section

$$s \in H^0(S, rK_S) \text{ given by } \{f_\alpha\} \text{ with } f_\alpha = (g_{\alpha\beta}^{-1})^r f_\beta$$

we construct a section by taking the rth root

$$\tilde{s} \in H^0(\tilde{S}, K_{\tilde{S}}) \text{ given by } \{\sqrt[r]{f_\alpha}\} \text{ with } \sqrt[r]{f_\alpha} = (g_{\alpha\beta}^{-1})\sqrt[r]{f_\beta}.$$

An equivalent and yet another way of constructing the global canonical cover is to set

$$c : \tilde{S} := Spec \oplus_{i=0}^{r-1} \mathcal{O}_S(-iK_S) \to S,$$

where the $\mathcal{O}_S$-algebra structure on $\oplus_{i=0}^{r-1}\mathcal{O}_S(-iK_S)$ is given by

$$\mathcal{O}_S(-iK_S) \times \mathcal{O}_S(-jK_S) \to \mathcal{O}_S(-(i+j-r)K_S) \text{ when } i+j \geq r,$$
$$a \times b \mapsto a \cdot b \cdot s.$$

Then it is immediate (Exercise! Why?) that $\tilde{S}$ is nonsingular projective and that we have isomorphisms of $c_*\mathcal{O}_{\tilde{S}}$-modules

$$\begin{aligned}
c_*\mathcal{O}_{\tilde{S}}(K_{\tilde{S}}) = c_*c^*K_S &\cong \{\oplus_{i=0}^{r-1}\mathcal{O}_S(-iK_S)\} \otimes \mathcal{O}_S(K_S) \\
&\cong \oplus_{i=0}^{r-2}\mathcal{O}_S(-iK_S) \oplus \mathcal{O}_S(K_S) \\
&\cong \oplus_{i=0}^{r-2}\mathcal{O}_S(-iK_S) \oplus \mathcal{O}_S(-(r-1)K_S) \cdot s \cong \oplus_{i=0}^{r-1}\mathcal{O}_S(-iK_S) \\
&\cong c_*\mathcal{O}_{\tilde{S}}.
\end{aligned}$$

We leave it to the reader to verify that this new way of construction coincides with the old one in the proof and that if $\hat{c} : \hat{S} \to S$ is any étale Galois cover of degree r with Galois group isomorphic to $\mathbb{Z}/r$ such that $K_{\hat{S}} = \hat{c}^*K_S \cong \mathcal{O}_{\hat{S}}$, then we can describe $\hat{S}$ as

$$\hat{S} := Spec \oplus_{i=0}^{r-1} \mathcal{M}^{\otimes i} \otimes \mathcal{O}_S(-iK_S)$$

for some line bundle $\mathcal{M}$ on S such that $\mathcal{M} \otimes \mathcal{O}_S(-K_S)$ is torsion of order r.

We go back to the analysis of Subcase 2.

Let $c : \tilde{S} \to S$ be an étale morphism of degree 2 obtained as the global canonical cover. Then since $\mathcal{O}_{\tilde{S}}(K_{\tilde{S}}) \cong \mathcal{O}_{\tilde{S}}$, we have the following two numerical possibilities:

$$\begin{cases} h^1(\tilde{S}, \mathcal{O}_{\tilde{S}}) = 0, \\ h^0(\tilde{S}, \mathcal{O}_{\tilde{S}}(K_{\tilde{S}})) = 1, \end{cases} \text{ or } \begin{cases} h^1(\tilde{S}, \mathcal{O}_{\tilde{S}}) > 0, \\ h^0(\tilde{S}, \mathcal{O}_{\tilde{S}}(K_{\tilde{S}})) = 1. \end{cases}$$

The second case cannot happen. In fact, on the one hand we would have

$$\chi(\mathcal{O}_{\tilde{S}}) = 2\chi(\mathcal{O}_S) = 2,$$

since $\tilde{S}$ is a degree-2 étale cover of S. (Exercise! Verify the equality above.) But on the other hand,

$$\chi(\mathcal{O}_{\tilde{S}}) = 1 - h^1(\tilde{S}, \mathcal{O}_{\tilde{S}}) + h^0(\tilde{S}, \mathcal{O}_{\tilde{S}}(K_{\tilde{S}})) \le 1,$$

a contradiction!

Thus only the first case happens, where S admits a degree-2 étale cover by a K3 surface $\tilde{S}$.

> Subcase 3: $q(S) = 2$ and $p_g(S) = 1$

Take the Albanese mapping $\alpha : S \to A$. Then by the same argument as in Subcase 3 of Theorem 1-5-6, we conclude that dim $\alpha(S) \ne 0$ or 1. Thus dim $\alpha(S) = 2$, i.e., the Albanese mapping is surjective. Let

$$K_S = \alpha^* K_A + R$$

be the ramification formula. Then since $K_A \sim 0$, we conclude that the ramification divisor $R \sim K_S$ is zero by the abundance theorem. This implies that α is étale and hence S is also an Abelian surface.

> Subcase 4: $q(S) = 1$ and $p_g(S) = 0$

Thanks to the abundance theorem, we take the global canonical cover $c : \tilde{S} \to S$. Then the numerical conditions

$$h^0(\tilde{S}, \mathcal{O}_{\tilde{S}}(K_{\tilde{S}})) = 1,$$
$$h^1(\tilde{S}, \mathcal{O}_{\tilde{S}}) = h^0(\tilde{S}, \Omega^1_{\tilde{S}}) \ge h^0(S, \Omega^1_S) \ge 1,$$

imply that $\tilde{S}$ is an Abelian surface by the analysis of Subcase 3. (Note that the merely numerical possibility Subcase 5 cannot happen geometrically.) Let $\alpha : S \to A$ be the Albanese mapping. Since A is an elliptic curve, we conclude by Poincaré's complete reducibility theorem that there exists an étale cover $A \leftarrow A'$ such that

$$
\begin{array}{ccc}
\tilde{S} & \longleftarrow & S' \times_A A' \cong \mathcal{E} \times A' \\
\downarrow & & \downarrow \\
A & \longleftarrow & A'
\end{array}
$$

where $\mathcal{E}$ is an elliptic curve that is the (connected component including the identity of the) kernel of the homomorphism $\tilde{S} \to A$ between Abelian varieties. Thus S admits an étale cover by a product of two elliptic curves $\mathcal{E} \times A'$.

This completes the proof of Theorem 1-7-1.   □

1.8   Birational Relation Among Surfaces

The purpose of this section is to study the **birational relation** among surfaces. First, we factor any given birational map into a composite of blowups and blowdowns.

Second, we discuss the special features of the birational relation among the good representatives, namely minimal models when $\kappa \geq 0$ and Mori fiber spaces when $\kappa = -\infty$.

Uniqueness of the minimal model in a fixed birational equivalence class is the special feature of the birational relation among the good representatives with $\kappa \geq 0$ in dimension 2. The **Castelnuovo–Noether theorem** decomposing any birational map between two birational Mori fiber spaces into a composite of elementary transformations called "**links**" is the special feature of the birational relation among the good representatives with $\kappa = -\infty$ in dimension 2. Our presentation is given in the framework of the **Sarkisov program**, which has been recently developed to give factorization of birational maps among higher-dimensional Mori fiber spaces.

Lemma 1-8-1. *Let*

$$\psi : V \to W,$$
$$\phi : V \to S,$$

be surjective morphisms between normal projective varieties such that

$$\psi_* \mathcal{O}_V = \mathcal{O}_W,$$
$$\phi_* \mathcal{O}_V = \mathcal{O}_S.$$

(Note that thanks to Zariski's Main Theorem the condition holds if and only if ψ and ϕ have connected fibers (cf. Proposition 1-2-16).) Suppose

$$\phi(\psi^{-1}(p)) = \text{pt.} \quad \forall p \in W.$$

Then there exists a unique morphism $\tau : W \to S$ (with connected fibers) such that

$$\phi = \tau \circ \psi.$$

PROOF.   Set-theoretically, we define the map $\tau : W \to S$ by

$$\tau(p) = \phi(\psi^{-1}(p)) \text{ for } p \in W,$$

which is well-defined by the assumption. Since both ψ and ϕ are proper, τ is a continuous map in the Zariski topology. Observe that for any open subset $U_S \subset S$ if we set $U_W = \tau^-(U_S)$, then

$$\phi^{-1}(U_S) = \psi^{-1}(U_W).$$

By the conditions

$$\psi_* \mathcal{O}_V = \mathcal{O}_W,$$
$$\phi_* \mathcal{O}_V = \mathcal{O}_S,$$

we have

$$\Gamma(U_S, \mathcal{O}_S) \tilde{\to} \Gamma(\phi^{-1}(U_S), \mathcal{O}_V) = \Gamma(\psi^{-1}(U_W), \mathcal{O}_V) \tilde{\to} \Gamma(U_W, \mathcal{O}_W),$$

which defines a map (which is actually an isomorphism as demonstrated above)

$$\mathcal{O}_S \to \tau_* \mathcal{O}_W.$$

Now it is easy to see that this, together with the topological description of τ, gives a morphism from W to S as local ringed spaces, i.e., a morphism as varieties. Uniqueness is also immediate.   □

Theorem 1-8-2 (Factorization of Birational Morphisms in Dimension 2). *Let*

$$f : S \to T$$

be a birational morphism between nonsingular projective surfaces S and T. Then f is a composite of blowdowns (or a composite of blowups from the viewpoint of starting from T), i.e., there is a sequence of contractions of (-1)-curves starting from $S_0 = S$,

$$\mu_i : S_i \to S_{i+1} \text{ a contraction of a } (-1)\text{-curve on } S_i, \quad i = 0, 1, \dots, l,$$

ending with $S_{l+1} = T$.

PROOF.   We give two proofs, the first of which goes *top down*, depending on Castelnuovo's contractibility criterion, while the second one goes *bottom up*, using the specific structure of the blowup of a point.

(*top down*) If $f : S \to T$ is an isomorphism, there is nothing to prove. Suppose $f : S \to T$ is not an isomorphism. Then in the ramification formula

$$K_S = f^* K_T + R,$$

R is a nonzero effective divisor (Exercise! Why?). Since f^{-1} is a morphism defined over some Zariski open subset U with $\text{codim}_T(T - U) \geq 2$, the image of the ramification divisor $f(R) \subset T - U$ is a finite number of points. Therefore, by Theorem 1-2-9 we have $R^2 < 0$, and hence there exists a curve $E \subset R$ such that

$$0 > E \cdot R = e \cdot (f^* K_T + R) = e \cdot K_S \quad \text{and} \quad 0 > E^2.$$

Thus E is a (-1)-curve on S (cf. Lemma 1-1-4) such that $f(E) = \text{pt}$. Take $\mu_0 : S = S_0 \to S_1$ to be the contraction of E. Then f factors through μ_0, i.e., there exists a birational morphism $f_1 : S_1 \to T$ such that $f = f_1 \circ \mu_0$ by Lemma 1-8-1. (Exercise! Check the condition of Lemma 1-8-1 by showing that for any birational morphism $f : S \to T$ between normal projective varieties, we have $f_* \mathcal{O}_S = \mathcal{O}_T$.) If $f_1 : S_1 \to T$ is an isomorphism, then we are done. If not, we repeat the argument with $f : S \to T$ replaced by $f_1 : S_1 \to T$. This procedure has to come to an end after finitely many repetitions, since at each repetition $\dim_{\mathbb{R}} H^2(S_i, \mathbb{R})$ drops by 1, providing the desired sequence at the end.

(*bottom up*) As in the first proof, we may assume that $f : S \to T$ is not an isomorphism. Then there is a point $p \in f(R) \subset T$ where f^{-1} is not defined as a morphism. Let $\mu : T_1 \to T$ be the blowup of T with the center at p. We claim that $f : S \to T$ factors through T_1, i.e., $f_1 := \mu^{-1} \circ f$ is a morphism. If we take a local coordinate system (x, y) in a neighborhood U of p,

$$
\begin{array}{ccc}
f^{-1}(U) & & \\
\downarrow & \searrow & f_1 \\
U & \leftarrow & \mu^{-1}(U) = \{xt - ys = 0\} \\
& & \cap \\
& & U \times \mathbb{P}^1,
\end{array}
$$

then the rational map f_1 is given by

$$f_1{}^*t = f^*y \text{ and } f_1{}^*s = f^*x.$$

Observe that f_1 is a morphism at $p' \in S$ ($f(p') = p$) if either f_1^*s/f_1^*t or f_1^*t/f_1^*s is regular at p', i.e., if $f^*x/f^*y \in \mathcal{O}_{S,p'}$ or $f^*y/f^*x \in \mathcal{O}_{S,p'}$.

Let

$$\operatorname{div}(f^*x) = C_x + \Sigma_{f(E_i)=p}\nu_{E_i}(f^*x)E_i,$$
$$\operatorname{div}(f^*y) = D_y + \Sigma_{f(E_i)=p}\nu_{E_i}(f^*y)E_i,$$

where C_x (respectively D_y) is the strict transform of $C = \operatorname{div}(x)$ (respectively $D = \operatorname{div}(y)$). Observe that the valuation ν_{E_i} is upper semicontinuous as a function $\nu_{E_i} : m_{T,p} \to \mathbb{Z}$. Therefore, we see that for a general choice of x and y we may assume

$$1 \leq \nu_{E_i}(x) = \nu_{E_i}(y) \quad \forall E_i \text{ with } f(E_i) = p.$$

Observe also that with x and y the local coordinates, the divisors C and D are smooth at p and hence $f : C_x \to C$ and $f : D_y \to D$ are both isomorphisms. Therefore,

$$
\begin{aligned}
C_x \cap f^{-1}(p) &= \{q_x\}, \\
D_y \cap f^{-1}(p) &= \{q_y\},
\end{aligned}
$$

are both isolated points.

We claim that $q_x \neq q_y$. In fact, suppose $q_x = q_y$. Then (in a neighborhood U of p)

$$
\begin{aligned}
1 &= \deg \operatorname{div}(y|_C) \\
&= \deg \operatorname{div}(f^*y|_{C_x}) \\
&= \deg(D_y|_{C_x}) + \deg(\Sigma_{f(E_i)=p}\nu_{E_i}(f^*y)E_i|_{C_x}) \geq 1 + 1 = 2,
\end{aligned}
$$

a contradiction!

Now for any $q \in f^{-1}(p)$, if $q \neq q_y$, then

$$\operatorname{div}\left(\frac{f^*x}{f^*y}\right) = C_x - D_y, \text{ and hence } \frac{f^*x}{f^*y} \in \mathcal{O}_{S,q}.$$

If $q \neq q_x$, then similarly

$$\operatorname{div}\left(\frac{f^*y}{f^*x}\right) = D_y - C_x, \text{ and hence } \frac{f^*y}{f^*x} \in \mathcal{O}_{S,q}.$$

This completes the proof of the claim.

Thus $f : S \to T = T_0$ factors through T_1 to provide $f_1 : S \to T_1$. If $f_1 : S \to T_1$ is an isomorphism, then we are done. If not, we argue as above to obtain a morphism $f_2 : S \to T_2$, where T_2 is a blowup of T_1 at a point with the exceptional divisor E_2.

This process must come to an end after finitely many repetitions, since (the discrete valuations corresponding to) $E_1, E_2, \ldots$ are (the discrete valuations corresponding to) the distinct prime divisors supporting the ramification divisor R for $f : S \to T$, and hence they are finite.

This completes the proof.   □

Remark 1-8-3. Hartshorne [3] and Mumford [6] give their own proofs along the lines of (*bottom up*). It is crucial to observe that the above theorem is a feature unique to dimension 2 and that both (*top down*) and (*bottom up*) would fail in dimension 3 or higher. In fact, there are examples of birational morphisms between nonsingular projective varieties of dimension greater than or equal to 3 that cannot be factored into a composite of blowups with smooth centers. We refer the reader to Danilov [2] for an example due to Hironaka, and to Shannon [1] (cf. Sally [1]) for another, both of which can be interpreted using toric geometry. (Exercise! Find your own example!)

We also refer the reader to Włodarczyk [2] and Abramovich–Karu–Matsuki–Włodarczyk [1] for the recent solution of the (weak) factorization conjecture of birational maps, based upon the (weak) factorization of toric (or toroidal) birational maps provided by Morelli [1][2], Abramovich–Matsuki–Rashid [1][2] (cf. Oda [1], Danilov [2], Ewald [1], Włodarczyk [1]).

Corollary 1-8-4 (Factorization of Birational Maps in Dimension 2). *Let*

$$\phi : S_1 \dashrightarrow S_2$$

be a birational map between two nonsingular projective surfaces. Then there exists a sequence of blowups $\psi_1 : V \to S_1$ followed by a sequence of blowdowns $\psi_2 : V \to S_2$ that factorizes ϕ.

PROOF.   Take the graph

$$\Gamma \subset S_1 \times S_2$$

(which is defined to be the closure of the graph of the morphism $\phi : U_1 \to S_2$, where the rational map ϕ is well-defined on the open subset U_1 and is independent of the choice of U_1) and its desingularization

$$\tau : V \to \Gamma.$$

Then by Theorem 1-8-2

$$\psi_1 = p_1 \circ \tau$$

is a sequence of blowups, and

$$\psi_2 = p_2 \circ \tau$$

is a sequence of blowdowns. Thus we have the desired factorization. □

Corollary 1-8-5 (Birational Invariants). *Let*

$$\phi : S_1 \dashrightarrow S_2$$

be a birational map between two nonsingular projective surfaces. Then there exist isomorphisms

$$H^i(S_1, \mathcal{O}_{S_1}) \cong H^i(S_2, \mathcal{O}_{S_2}),$$
$$H^0(S_1, \mathcal{O}_{S_1}(mK_{S_1})) \cong H^0(S_2, \mathcal{O}_{S_2}(mK_{S_2})) \text{ for any } m \in \mathbb{N}$$

induced by ϕ.

In particular,

$$h^i(S, \mathcal{O}_S) \text{ and } h^0(S, \mathcal{O}_S(mK_S)),$$

where $m \in \mathbb{N}$, are birational invariants.

PROOF. By Corollary 1-8-4 we may assume that $\phi : S_1 \rightarrow S_2$ is a birational morphism obtained by contracting a (-1)-curve E. Then

$$\phi_* \mathcal{O}_{S_1} \cong \mathcal{O}_{S_2} \text{ and } R^i \phi_* \mathcal{O}_{S_1} = 0 \text{ for } i > 0$$

(cf. Theorem 5-2-8 or Hartshorne [3], Chapter V, Proposition 3.4, page 387), which via the Leray spectral sequence induce isomorphisms

$$H^i(S_1, \mathcal{O}_{S_1}) \cong H^i(S_2, \mathcal{O}_{S_2}).$$

The ramification formula implies

$$H^0(S_1, \mathcal{O}_{S_1}(mK_{S_1})) = H^0(S_1, \mathcal{O}_{S_1}(m(\phi^* K_{S_2} + E))) \cong H^0(S_2, \mathcal{O}_{S_2}(mK_{S_2})). \ \square$$

We remark, however, that we don't need a detailed structure theorem of birational maps to reach the conclusion. In fact, as long as S has rational singularities, $h^i(S, \mathcal{O}_S)$ is a birational invariant and so is $h^i(S, \mathcal{O}_S(mK_S))$, $m \in \mathbb{N}$, as long as we have the ramification formula $K_V = f^* K_S + R$ (where R is some exceptional effective divisor) for a resolution of singularities $f : V \rightarrow S$.

Birational Relation Among Minimal Models

A similar argument to that of Corollary 1-8-4 leads to the following feature of minimal models in dimension 2.

Theorem 1-8-6 (Absolute Minimality of Minimal Models in Dimension 2).
Let S be a minimal model in dimension 2. Then any birational map $\phi : T \dashrightarrow S$ from a nonsingular projective surface T is a morphism.

PROOF.   Take the graph of the birational map ϕ,

$$\Gamma \subset T \times S,$$

and its desingularization $\tau : V \to \Gamma$ as in the proof of Corollary 1-8-4.
    Set

$$\psi_1 = p_1 \circ \tau \text{ and } \psi_2 = p_2 \circ \tau$$

and write down the ramification formulas

$$K_V = \psi_1^* K_T + R_{\psi_1},$$
$$K_V = \psi_2^* K_S + R_{\psi_2}.$$

If ψ_1 is not an isomorphism, then as in the proof of Theorem 1-8-2 there exists a
(-1)-curve $E \subset R_{\psi_1}$ such that

$$0 > E \cdot K_V = e \cdot (\psi_2^* K_S + R_{\psi_2}).$$

Since K_S is nef, this implies that $E \cdot R_{\psi_2} < 0$ and hence

$$E \subset R_{\psi_2}$$

and that by Lemma 1-8-1 the morphisms ψ_1 and ψ_2 factors through the contraction
$\mu : V = V_0 \to V_1$ of E,

$$\psi_1 = \psi_{10} = \psi_{11} \circ \mu,$$
$$\psi_2 = \psi_{20} = \psi_{21} \circ \mu.$$

If $\psi_{11} : V_1 \to T$ is an isomorphism, then $\phi = \psi_{21} \circ \psi_{11}^{-1} : T \to S$ is a
morphism. If not, then we repeat the argument replacing V_0 with V_1. This procedure
has to come to an end, since $\dim_{\mathbb{R}} H^2(V_i, \mathbb{R})$ drops by 1 each time, reaching the
stage where ψ_{1i} is an isomorphism, verifying that $\phi = \psi_{2i} \circ \psi_{1i}^{-1}$ is actually a
morphism. $\square$

Corollary 1-8-7 (Uniqueness of the Minimal Model in Dimension 2).   *There
exists a unique minimal model in a fixed birational equivalence class in dimension
2. More strongly, if*

$$\phi : S_1 \dashrightarrow S_2$$

*is a birational map between two minimal models in dimension 2, then ϕ is an
isomorphism.*

PROOF.   Apply Theorem 1-8-6 to ϕ and ϕ^{-1}. $\square$

Birational Relation Among Mori Fiber Spaces

We now focus our attention on the birational relation among Mori fiber spaces in
dimension 2. We will describe the classical Castelnuovo–Noether theorem in the
framework of the Sarkisov program. The grand picture of the Sarkisov program
will be revealed in Chapter 13 after we study the minimal model program in higher
dimensions and the logarithmic catogory.

Theorem 1-8-8 (Castelnuovo–Noether Theorem = Sarkisov Program in Dimension 2). *Let*

$$
\begin{array}{ccc}
S & \overset{\Phi}{\underset{\text{birat}}{\dashrightarrow}} & S' \\
\downarrow \phi & & \downarrow \phi' \\
W & & W'
\end{array}
$$

be a birational map between two Mori fiber spaces in dimension 2:

$$\phi : S \to W,$$
$$\phi' : S' \to W'.$$

Then there is an algorithm, called the **Sarkisov program** *in dimension 2, to decompose Φ into a composite of the following four types of "links" (elementary transformations):*

Type (I)

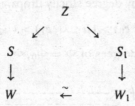

$$
\begin{array}{ccc}
& & \mathbb{F}_1 \\
& \swarrow & \\
\mathbb{P}^2 & & \downarrow \\
\downarrow & & \\
\text{pt.} & \leftarrow & \mathbb{P}^1
\end{array}
$$

Type (II)

$$
\begin{array}{ccc}
& Z & \\
\swarrow & & \searrow \\
S & & S_1 \\
\downarrow & & \downarrow \\
W & \overset{\sim}{\leftarrow} & W_1
\end{array}
$$

where $S \to W$ is a $\mathbb{P}^1$ bundle over a nonsingular projective curve W, $Z \to S$ is a blowup of a point in one ruling, $Z \to S_1$ is the contraction of the strict transform of that ruling to obtain another $\mathbb{P}^1$-bundle $S_1 \to W_1 = W$.

Type (III) (*Inverse of* Type (I))

$$
\begin{array}{ccc}
& \mathbb{F}_1 & \\
& \searrow & \\
\downarrow & & \mathbb{P}^2 \\
& & \downarrow \\
\mathbb{P}^1 & \to & \text{pt.}
\end{array}
$$

Type (IV)

$$
\begin{array}{ccc}
\mathbb{P}_1^1 \times \mathbb{P}_2^1 & = & \mathbb{P}_1^1 \times \mathbb{P}_2^1 \\
\downarrow p_1 & & \downarrow p_2 \\
\mathbb{P}_1^1 & & \mathbb{P}_2^1
\end{array}
$$

$$\searrow \qquad \swarrow$$

$$\mathrm{pt.}$$

PROOF.   The rest of this section will be spent on the proof of the Sarkisov program in dimension 2.

Strategy for "Untwisting" Φ

The strategy for decomposing Φ into a composite of links, the operation that we call **"untwisting"** of Φ, is to set up a good invariant of Φ with reference to the fixed Mori fiber space $\phi' : S' \to W'$ (called the **Sarkisov degree**, the triplets of numbers (μ, λ, e) in lexicographical order, as will be defined below), which should tell us how far ϕ is from being an isomorphism of Mori fiber spaces.

It is the **Noether–Fano–Iskovskikh criterion** that allows us to judge precisely when Φ is an isomorphism of Mori fiber spaces in terms of the Sarkisov degree and the canonical divisor of S.

Starting with a given birational map Φ between two Mori fiber spaces, we ask whether Φ is actually an isomorphism of Mori fiber spaces via the NFI criterion. If the answer is *yes*, then there is nothing more to do, and we stop right there. If the answer is *no*, then we "untwist" Φ by an appropriate link of Type (I), (II), (III), or (IV) to obtain a new birational map Φ_1. We repeat the process with Φ_1. Each time we "untwist," the Sarkisov degree strictly drops, i.e.,

$$
(\mu, \lambda, e) = (\mu_0, \lambda_0, e_0) > (\mu_1, \lambda_1, e_1) > (\mu_2, \lambda_2, e_2) > \dots,
$$

where these are the Sarkisov degrees of $\Phi = \Phi_0, \Phi_1, \Phi_2, \dots$ with respect to the fixed reference $\phi' : S' \to W'$:

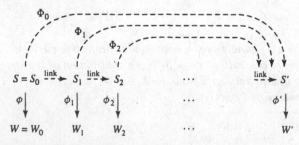

Finally, by observing that the set of the Sarkisov degrees satisfies the descending chain condition, we conclude that the process must come to an end after finitely many steps, reaching the Mori fiber space $\phi' : S' \to W'$, expressing Φ as a composite of links.

Sarkisov degree. First we choose and fix $\mu' \in \mathbb{N}$ and an ample divisor A' on W' so that

$$H' = -\mu' K_{S'} + \phi'^* A'$$

is a very ample divisor on S'. (Note that $-K_{S'}$ is relatively ample. We refer the reader to Hartshorne [3], Chapter II, Proposition 7.10, or Iitaka [5], Theorem 7.11.)

We take a nonsingular projective surface V that dominates both S and S' by birational morphisms (which are compatible with Φ), i.e.,

$$S \xleftarrow{\sigma} V \xrightarrow{\sigma'} S' \text{ with } \sigma' \circ \sigma^{-1} = \Phi.$$

For a member

$$\mathcal{H}' \in |H'|$$

we define the **"homaloidal" transform** $\mathcal{H}$ on S of $\mathcal{H}'$ to be

$$\mathcal{H} = \sigma_* \sigma'^* \mathcal{H}'.$$

We note that the homaloidal transform does not depend on the choice of V. We are ready to define the Sarkisov degee.

(I) μ: the quasi-effective threshold.

The first of the triplet, the quasi-effective threshold μ, is defined to be a rational number (necessarily positive) such that

$$\mu K_S + \mathcal{H} \equiv_W 0,$$

that is to say,

$$(\mu K_S + \mathcal{H}) \cdot F = 0 \text{ for any curve } F \text{ in a fiber of } \phi : S \to W.$$

Observe that μ is independent of the choice of a member $\mathcal{H}$ and that since all the curves contracted by ϕ are numerically proportional, we have to check $(\mu K_S + \mathcal{H}) \cdot F = 0$ for only one curve F in a fiber of ϕ.

Note that since for a rational curve l as specified in Theorem 1-4-8, which generates the extremal ray corresponding to the Mori fiber space $\phi : S \to W$,

$$K_S \cdot l = -2 \text{ or } -3$$

(depending on whether l is of type (ii) or (iii) in the classification of extremal rays in dimension 2 in Theorem 1-4-8), we conclude by setting $F = l$ that

$$\mu \in \frac{1}{3!} \mathbb{N}.$$

(ii) λ: the maximal multiplicity.

In order to define the second of the triplet, the maximal multiplicity λ, we consider the linear system consisting of the homaloidal transforms $\mathcal{H}$ for $\mathcal{H}' \in |H'|$, which we denote by $\Phi_{\text{homaloidal}}^{-1} |H'|$. We note that this linear system may be smaller than the complete linear system $|\mathcal{H}|$.

If $\Phi_{\text{homaloidal}}^{-1}|H'|$ does not have any base point, then $\lambda = 0$ by definition. If $\Phi_{\text{homaloidal}}^{-1}|H'|$ has some base points, call them $p_1, p_2, \ldots, p_l$. (Observe that $\Phi_{\text{homaloidal}}^{-1}|H'|$ does not have any base components.) Then

$$\lambda = \max_i \{\min_{\mathcal{H} \in \Phi_{\text{homaloidal}}^{-1}|H'|} \{\text{mult}_{p_i} \mathcal{H}\}\}$$

$$= \max_i \{\text{mult}_{p_i} \mathcal{H}\} \text{ where } \mathcal{H} \in \Phi_{\text{homaloidal}}^{-1}|H'| \text{ is a general member.}$$

In short, λ is the maximum multiplicity of a general member $\mathcal{H} \in \Phi_{\text{homaloidal}}^{-1}|H'|$ at the base points.

Obviously,

$$\lambda \in \mathbb{Z}_{\geq 0}.$$

(iii) e: the number of the crepant exceptional divisors.

In order to define the third of the triplet, the number of the crepant exceptional divisors (with respect to $K_S + \frac{1}{\lambda}\mathcal{H}$), write

$$K_V = \sigma^* K_S + \Sigma a_k E_k,$$
$$\mathcal{H}_V = \sigma'^* \mathcal{H}' = \sigma^* \mathcal{H} - \Sigma b_k E_k,$$

for a general member $\mathcal{H}' \in |H'|$.

First note that

$$\lambda = \max\left\{\frac{b_k}{a_k}\right\},$$

which follows from the fact that the multiplicity of a divisor at infinitely near points cannot exceed the multiplicity at the original point (Exercise! Confirm this fact and prove the equality above.)

Therefore, when $\lambda > 0$, we have

$$K_V + \frac{1}{\lambda}\mathcal{H}_V = \sigma^*\left(K_S + \frac{1}{\lambda}\mathcal{H}\right) + \Sigma\left(a_k - \frac{1}{\lambda}b_k\right)E_k$$

with

$$a_k - \frac{1}{\lambda}b_k \geq 0 \quad \forall E_k \text{ and } a_{k_o} - \frac{1}{\lambda}b_{k_o} = 0 \text{ for some } E_{k_o}.$$

We define

$$e := \#\left\{E_k; a_k - \frac{1}{\lambda}b_k = 0 \text{ or equivalently } \frac{b_k}{a_k} = \lambda\right\}.$$

Note that this does not depend on the choice of V (Exercise! Why?) and that clearly,

$$e \in \mathbb{N}.$$

(When $\lambda = 0$, i.e., when $\Phi_{\text{homaloidal}}^{-1}|H'|$ has no base points, all the exceptional divisors E_k have $b_k/a_k = 0 = \lambda$, and e is not well-defined. But when $\lambda = 0$, we will always be in the case $\lambda \leq \mu$ in the algorithm of the Sarkisov program. The untwisting in this case decreases the quasi-effective threshold strictly. Therefore, the invariant e plays no role, and we leave e undefined in this case.)

This completes the definition of the triplet (μ, λ, e) associated to Φ with reference to $\phi' : S' \to W'$ after fixing μ' and A'. We consider these triplets in lexicographical order. The next criterion tells us when Φ is an isomorphism of Mori fiber spaces, in terms of the Sarkisov degree (μ, λ, e) and the canonical divisor of S.

In the following $\mathcal{H}'$ always refers to a *general* member of $|H'|$, and $\mathcal{H}$ to its homaloidal transform.

Proposition 1-8-9 (Noether–Fano–Iskovskikh Criterion). *Suppose*

$$\lambda \leq \mu \quad and \quad K_S + \frac{1}{\mu}\mathcal{H} \text{ is nef on } S.$$

Then Φ is an isomorphism of Mori fiber spaces, i.e., there exists a commutative diagram

$$
\begin{array}{ccc}
S & \xrightarrow{\sim}{\Phi} & S' \\
\downarrow{\phi} & & \downarrow{\phi'} \\
W & \xrightarrow{\sim} & W'
\end{array}
$$

PROOF. Step 1. Show that $\mu' = \mu$.

In fact,

$$K_{S'} + \frac{1}{t}\mathcal{H}' \text{ is ample on } S', \text{ for any } t \text{ with } \quad 0 < t < \mu',$$

and thus its homaloidal transform has a positive intersection with the general fiber F of ϕ,

$$\left(K_S + \frac{1}{t}\mathcal{H}\right) \cdot F > 0, \text{ for any } t \text{ with } \quad 0 < t < \mu'.$$

Therefore,

$$\mu' \leq \mu.$$

(Note that the inequality $\mu' \leq \mu$ always holds without the extra assumption that $K_S + \frac{1}{\mu}\mathcal{H}$ is nef.)

On the other hand, since

$$K_{S'} + \frac{1}{t}\mathcal{H}' \text{ is negative over } W' \quad \forall t > \mu',$$

its homaloidal transform

$$K_S + \frac{1}{t}\mathcal{H}$$

can never be nef for all $t > \mu'$. Therefore,

$$\mu' \geq \mu.$$

(Note that we slightly cheated in the above argument, since as K_S is not exactly the homaloidal transform of $K_{S'}$. We leave it as an exercise for the reader to make this part of the argument rigorous. The condition $\lambda \leq \mu$ should also be used, though it did not show up in the above argument. For details, we refer the reader to the proof of Proposition 13-1-3.)

Take a nonsingular projective surface V that dominates both S and S' by birational morphisms as in the definition of the Sarkisov degree,

$$S \xleftarrow{\sigma} V \xrightarrow{\sigma'} S'.$$

We write down the ramification formulas

$$K_V = \sigma^* K_S + R,$$
$$K_V = \sigma'^* K_{S'} + R',$$
$$K_V + \frac{1}{\mu}\mathcal{H}_V = \sigma^*\left(K_S + \frac{1}{\mu}\mathcal{H}\right) + \Sigma r_k E_k,$$
$$K_V + \frac{1}{\mu}\mathcal{H}_V = \sigma'^*\left(K_{S'} + \frac{1}{\mu}\mathcal{H}'\right) + \Sigma r_j' E_j'.$$

Note that

$$R \geq \Sigma r_k E_k \geq 0, \text{ since } \lambda \leq \mu,$$

and that

$$R' = \Sigma r_j' E_j', \text{ since } \mathcal{H}_V = \sigma'^* \mathcal{H}'.$$

Step 2. In the obvious equality

$$\sigma^*\left(K_S + \frac{1}{\mu}\mathcal{H}\right) + \Sigma r_k E_k = \sigma'^*\left(K_{S'} + \frac{1}{\mu}\mathcal{H}'\right) + \Sigma r_j' E_j',$$

show that

$$\sigma^*\left(K_S + \frac{1}{\mu}\mathcal{H}\right) = \sigma'^*\left(K_{S'} + \frac{1}{\mu}\mathcal{H}'\right),$$

that is,

$$\Sigma r_k E_k = \Sigma r_j' E_j'.$$

We introduce the notation

$$(\cup E_k) \cup (\cup E_j') = \cup F_l$$

and

$$\Sigma r_k E_k = \Sigma_{l \in \mathcal{S}_\sigma} r_l F_l + \Sigma_{l \in \mathcal{S}_{\sigma,\sigma'}} r_l F_l,$$
$$\Sigma r_j' E_j' = \Sigma_{l \in \mathcal{S}_{\sigma,\sigma'}} r_l' F_l + \Sigma_{l \in \mathcal{S}_{\sigma'}} r_l' F_l,$$

where

$$
\text{the divisor } F_l \text{ is } \begin{cases} \sigma\text{-exceptional but } not\ \sigma'\text{-exceptional,} \\ \sigma\text{-exceptional and } \sigma'\text{-exceptional,} \\ not\ \sigma\text{-exceptional but } \sigma'\text{-exceptional,} \end{cases} \text{ iff } \begin{cases} l \in S_\sigma, \\ l \in S_{\sigma,\sigma'}, \\ l \in S_{\sigma'}. \end{cases}
$$

Then by the two equalities

$$
\Sigma_{l \in S_\sigma} r_l F_l + \Sigma_{l \in S_{\sigma,\sigma'}} (r_l - r_l') F_l
$$

$$
\equiv -\sigma^* \left(K_S + \frac{1}{\mu} \mathcal{H} \right) + \sigma'^* \left(K_{S'} + \frac{1}{\mu} \mathcal{H}' \right) + \Sigma_{l \in S_{\sigma'}} r_l' F_l,
$$

$$
\Sigma_{l \in S_{\sigma,\sigma'}} (r_l' - r_l) F_l + \Sigma_{l \in S_{\sigma'}} r_l' F_l
$$

$$
\equiv -\sigma'^* \left(K_{S'} + \frac{1}{\mu} \mathcal{H}' \right) + \sigma^* \left(K_S + \frac{1}{\mu} \mathcal{H} \right) + \Sigma_{l \in S_\sigma} r_l F_l
$$

and by the negativity lemma below applied to σ over W and to σ' over W', we conclude that

$$
r_l \le 0 \text{ if } l \in S_\sigma,
$$
$$
r_l - r_l' \le 0 \text{ if } l \in S_{\sigma,\sigma'},
$$
$$
r_l' - r_l \le 0 \text{ if } l \in S_{\sigma,\sigma'},
$$
$$
r_l' \le 0 \text{ if } l \in S_{\sigma'},
$$

and hence

$$
r_l = 0 \text{ if } l \in S_\sigma,
$$
$$
r_l = r_l' \text{ if } l \in S_{\sigma,\sigma'},
$$
$$
r_l' = 0 \text{ if } l \in S_{\sigma'}.
$$

Therefore, we have

$$
\Sigma r_k E_k = \Sigma_{l \in S_{\sigma,\sigma'}} r_l F_l = \Sigma_{l \in S_{\sigma,\sigma'}} r_l' F_l = \Sigma r_j' E_j'.
$$

Lemma 1-8-10 (Negativity Lemma in Dimension 2). *Let $f : V \to T$ be a birational morphism from a nonsingular projective surface V to another surface T. Suppose that a divisor $\Sigma \alpha_i E_i$ consisting of the exceptional divisors for f is relatively nef over T, i.e.,*

$$
(\Sigma \alpha_i E_i) \cdot E_j \ge 0 \quad \forall E_j.
$$

Then we have

$$
\alpha_i \le 0 \quad \forall i.
$$

PROOF. Proof of Lemma 1-8-10 Suppose that some α_i is greater than 0. Then on one hand, we have by Theorem 1-2-9

$$
(\Sigma \alpha_i E_i) \cdot (\Sigma_{\alpha_i > 0} \alpha_i E_i) = (\Sigma_{\alpha_i \le 0} \alpha_i E_i) \cdot (\Sigma_{\alpha_i > 0} \alpha_i E_i) + (\Sigma_{\alpha_i > 0} \alpha_i E_i)^2 < 0.
$$

On the other hand, we have by the assumption of the divisor being relatively nef over T that

$$(\Sigma \alpha_i E_i) \cdot (\Sigma_{\alpha_i > 0} \alpha_i E_i) \geq 0,$$

a contradiction!   □

Step 3. Show that Φ induces an isomorphism of the Mori fiber spaces.

Observe that the inequality and equality obtained by Step 2

$$R \geq \Sigma r_k E_k = \Sigma r'_j E'_j = R'$$

by Step 2 imply that all the exceptional divisors for σ' are also exceptional for σ. Therefore, by Lemma 1-8-1, we conclude that $\Phi^{-1} = \sigma \circ \sigma'^{-1}$ is a morphism

$$\Phi^{-1} : S \leftarrow S'.$$

On the other hand, the other equality of Step 2,

$$\sigma^* \left(K_S + \frac{1}{\mu} \mathcal{H} \right) = \sigma'^* \left(K_{S'} + \frac{1}{\mu} \mathcal{H}' \right),$$

implies that $\forall p \in W$ we have

$$
\begin{aligned}
0 &= \sigma^*(K_S + \frac{1}{\mu}\mathcal{H}) \cdot (\phi \circ \sigma)^{-1}(p) \\
&= \sigma'^*(K_{S'} + \frac{1}{\mu}\mathcal{H}') \cdot (\phi \circ \sigma)^{-1}(p) \\
&= \frac{1}{\mu}(\phi' \circ \sigma')^* A' \cdot (\phi \circ \sigma)^{-1}(p) \\
&= A' \cdot (\phi' \circ \sigma')_*(\phi \circ \sigma)^{-1}(p).
\end{aligned}
$$

Since A' is ample and since $(\phi \circ \sigma)^{-1}(p)$ is connected (Exercise! Why?),

$$(\phi' \circ \sigma')((\phi \circ \sigma)^{-1}(p)) \text{ is a point.}$$

Therefore, using by Lemma 1-8-1 again, we conclude that there exists a morphism $\tau : W \to W'$ that makes the following diagram commute:

$$
\begin{array}{ccc}
S & \overset{\Phi^{-1}}{\longleftarrow} & S' \\
\downarrow \phi & & \downarrow \phi' \\
W & \overset{\tau}{\longrightarrow} & W'
\end{array}
$$

Since $\phi' : S' \to W'$ is the contraction of an extremal ray and thus all the curves in fibers are numerically proportional, σ and τ must be isomorphisms giving the desired isomorphism of the two Mori fiber spaces.

This completes the proof of Proposition 1-8-9.   □

Flowchart for Sarkisov Program in Dimension 2

Now we present the flowchart to untwist a given birational map

$$
\begin{array}{ccc}
S & \xrightarrow[\text{birat}]{\Phi} & S' \\
\downarrow{\phi} & & \downarrow{\phi'V} \\
W & & W'
\end{array}
$$

between two Mori fiber spaces in dimension 2.

Start.

The first question to ask is

$$\lambda \leq \mu?$$

According to whether the answer to this question is *yes* or *no*, we proceed separately into the case $\lambda \leq \mu$ or into the case $\lambda > \mu$.

Case: $\lambda \leq \mu$

If $\lambda \leq \mu$, then the next question to ask is

$$K_S + \frac{1}{\mu}\mathcal{H} \text{ nef?}$$

If the answer to this question is *yes*, then $K_S + \frac{1}{\mu}\mathcal{H}$ is nef and $\lambda \leq \mu$ by the case assumption. Thus the Noether–Fano–Iskovskikh criterion applies to conclude that Φ is an isomorphism of Mori fiber spaces. This leads to an

End.

Suppose that $K_S + \frac{1}{\mu}\mathcal{H}$ is not nef. Then $\phi : S \to W$ cannot be the morphism of $\mathbb{P}^2$ mapping to a point, since if it were, then $K_S + \frac{1}{\mu}\mathcal{H}$ would be numerically trivial and hence nef. Thus $\phi : S \to W$ must be a $\mathbb{P}^1$-bundle over a nonsingular projective curve W (cf. Theorem 1-4-8) and hence $\dim_{\mathbb{R}} N_1(S) = 2$ (Exercise! Why?). One edge of the 2-dimensional cone $\overline{\text{NE}}(S)$ is the extremal ray R whose contraction gives the very Mori fiber space structure $\text{cont}_R = \phi : S \to W$ and whose intersection with $K_S + \frac{1}{\mu}\mathcal{H}$ is 0. Since $K_S + \frac{1}{\mu}\mathcal{H}$ is not nef, the other edge has negative intersection with $K_S + \frac{1}{\mu}\mathcal{H}$ and hence has also negative intersection with K_S (We note that $\Phi_{\text{homaloidal}}^{-1}|H'|$ has no base components, and thus $\mathcal{H}$ is nef.). Therefore, by the cone theorem this edge gives rise to another extremal ray R' with its contraction $\text{cont}_{R'} : S \to T$. (See Figure 1-8-11.)

Figure 1-8-11.

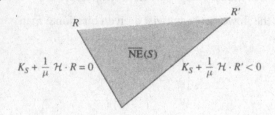

$$S = \mathbb{P}^2 \quad \overset{\Phi}{\dashrightarrow} \qquad\qquad S' = \mathbb{P}^2$$

$$(X : Y : Z) \quad \mapsto \quad \left(\frac{1}{X} : \frac{1}{Y} : \frac{1}{Z}\right) = (YZ : ZX : XY)$$

$$\phi\downarrow \qquad\qquad\qquad\qquad \phi'\downarrow$$

$$W = \mathrm{Spec}\,\mathbb{C} \qquad\qquad\qquad Y' = \mathrm{Spec}\,\mathbb{C}$$

Again the contraction $\mathrm{cont}_{R'} : S \to T$ cannot be the morphism from $\mathbb{P}^2$ to a point. Thus by the classification of extremal rays in dimension 2 (Theorem 1-4-8) $\mathrm{cont}_{R'}$ is either a contraction of a (-1)-curve $\mathrm{cont}_{R'} : S \to S_1$ with $\phi_1 : S_1 \to W_1 = $ pt. or a $\mathbb{P}^1$-bundle over a nonsingular projective curve $\mathrm{cont}_{R'} = \phi_1 : S = S_1 \to W_1$. In the first case, $\phi_1 : S_1 \to W_1$ is a Mori fiber space with $\dim_{\mathbb{R}} N_1(S_1) = 1$, i.e., the morphism from $\mathbb{P}^2$ to a point. Thus this untwisting is of Type (III). In the second case S must be isomorphic to $\mathbb{P}^1 \times \mathbb{P}^1$ (Exercise! Why?), and the two contractions of extremal rays

$$W \overset{\mathrm{cont}_R}{\leftarrow} S = S_1 \overset{\mathrm{cont}_{R'}}{\to} W_1$$

correspond to the two projections

$$\mathbb{P}^1_1 \overset{p_1}{\leftarrow} \mathbb{P}^1_1 \times \mathbb{P}^1_2 \overset{p_2}{\to} \mathbb{P}^1_2.$$

Thus this untwisting is of Type (IV).

In both cases, we have (Exercise! Why?)

$$\left(K_{S_1} + \frac{1}{\mu}\mathcal{H}_{S_1}\right) \cdot F < 0$$

for a curve F in a fiber of $\phi_1 : S_1 \to W_1$, where $\mathcal{H}_{S_1}$ is the homaloidal transform of $\mathcal{H}'$ on S_1. Thus the quasi-effective threshold strictly drops:

$$\mu_1 < \mu.$$

Therefore, in this case of $\lambda \le \mu$, after untwisting Φ by a link of Type (III) or Type (IV), we go back to *Start* with strictly decreased Sarkisov degree.

Case: $\lambda > \mu$

In this case we blowup the base point $P \in S$ where $\mathcal{H}'$ has the maximal multiplicity λ to obtain $p : Z \to S$. If $S \cong \mathbb{P}^2$ and hence $Z \cong \mathbb{F}_1$, then we take

$\phi_1 : S_1 = Z \cong \mathbb{F}_1 \to W_1 \cong \mathbb{P}^1$ to be the unique $\mathbb{P}^1$-bundle structure on $\mathbb{F}_1$ over $\mathbb{P}^1$. If $\phi : S \to W$ is a $\mathbb{P}^1$-bundle over W, then the strict transform of the fiber containing P becomes a new (-1)-curve E_1 (Exercise! Why?) and we contract E_1 to obtain $q : Z \to S_1$ and a new $\mathbb{P}^1$-bundle $\phi_1 : S_1 \to W_1 = W$. Thus in the first case the untwisting is of Type (I), in the second of Type (II).

In the first case, we have

$$K_{S_1} + \frac{1}{\mu}\mathcal{H}_{S_1} = p^* \left(K_S + \frac{1}{\mu}\mathcal{H} \right) - bE \equiv -bE$$

with $b > 0$, since $\lambda > \mu$, where E is the exceptional curve for p, and hence

$$\left(K_{S_1} + \frac{1}{\mu}\mathcal{H}_{S_1} \right) \cdot F < 0$$

for a general fiber F of $\phi_1 : S_1 \to W_1$. Therefore,

$$\mu_1 < \mu.$$

In the second case, since the two Mori fiber spaces $\phi : S \to W$ and $\phi_1 : S_1 \to W_1$ are isomorphic over a dense open subset of $W = W_1$, we have clearly

$$\mu_1 = \mu.$$

Moreover, since

$$K_Z + \frac{1}{\lambda}\mathcal{H}_Z = p^* \left(K_S + \frac{1}{\lambda}\mathcal{H} \right),$$

we conclude that, since $\lambda > \mu$,

$$\left(K_Z + \frac{1}{\lambda}\mathcal{H}_Z \right) \cdot E_1 = \left(K_S + \frac{1}{\lambda}\mathcal{H} \right) \cdot p_* E_1 < 0.$$

This implies

$$K_Z + \frac{1}{\lambda}\mathcal{H}_Z = q^* \left(K_{S_1} + \frac{1}{\lambda}\mathcal{H}_{S_1} \right) + b_1 E_1 \text{ with } b_1 > 0,$$

which allows us to conclude that

$$\lambda_1 \leq \lambda$$

and that if $\lambda_1 = \lambda$, then E is not a (crepant) exceptional divisor any more and E_1 is not a crepant exceptional divisor either. Therefore,

$$\lambda_1 \leq \lambda, \text{ and if } \lambda_1 = \lambda, \text{ then } e_1 \leq e - 1 < e.$$

Therefore, in this case of $\lambda > \mu$, after untwisting Φ by a link of Type (I) or Type (II), we also go back to *Start* with strictly decreased Sarkisov degree.

Since

$$\mu \in \frac{1}{3!}\mathbb{N}, \qquad \lambda \in \mathbb{Z}_{\geq 0}, \qquad e \in \mathbb{N},$$

the set of the Sarkisov degrees satisfies the descending chain condition. Thus the process must come to an end after finitely many steps, providing the required untwisting of Φ by a composite of links of Type (I), (II), (III), or (IV).

This completes the proof of Theorem 1-8-8. $\square$

Flowchart 1-8-12.

Sarkisov Program in Dimension 2

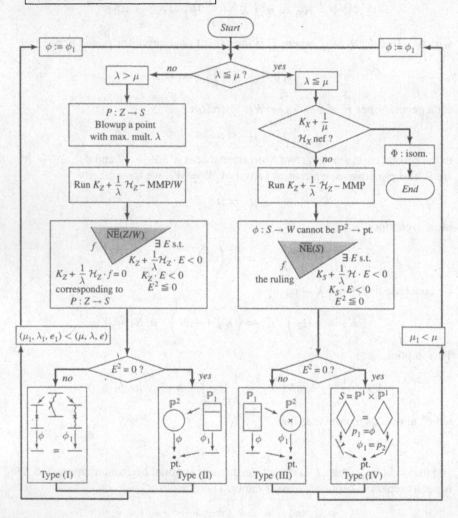

Example 1-8-13. We demonstrate how to untwist the quadratic transformation of $\mathbb{P}^2$ according to the Sarkisov program.

Figure 1-8-14.

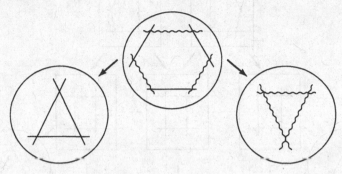

Take the quadratic transformation

$$S = \mathbb{P}^2 \quad \overset{\Phi}{\dashrightarrow} \quad S' = \mathbb{P}^2$$

$$(X : Y : Z) \quad \mapsto \quad \left(\frac{1}{X} : \frac{1}{Y} : \frac{1}{Z}\right) = (YZ : ZX : XY)$$

$$\phi\downarrow \qquad\qquad\qquad \phi'\downarrow$$

$$W = \operatorname{Spec} \mathbb{C} \qquad\qquad Y' = \operatorname{Spec} \mathbb{C}$$

We take a very ample divisor

$$H' = -\mu' K_{S'} + \phi'^* A' = -\frac{1}{3}(-3H'),$$

where H' is the hyperplane of $\mathbb{P}^2$ with $\mu' = \frac{1}{3}$. (Though we do not follow the rule in the presentation that $\mu' \in \mathbb{N}$, the Sarkisov program works as long as H' is a very ample divisor on S'.) Note that for a general member $\mathcal{H}' \in |H'|$, its homaloidal transform $\mathcal{H}$ is a conic passing through the three points

$$(1:0:0), \quad (0:1:0), \quad (0:0:1).$$

We illustrate the untwisting of the quadratic transformation in Figure 1-8-15 keeping the track of the Sarkisov degree. We leave the verification of the details to the reader.

Figure 1-8-15.

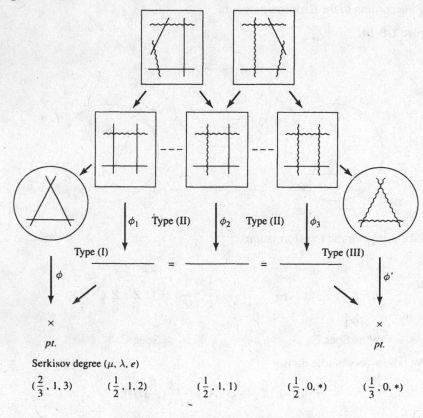

Serkisov degree (μ, λ, e)

$(\frac{2}{3}, 1, 3)$ $(\frac{1}{2}, 1, 2)$ $(\frac{1}{2}, 1, 1)$ $(\frac{1}{2}, 0, *)$ $(\frac{1}{3}, 0, *)$

Exercise 1-8-16. As an application of the Sarkisov program, show that the group of the Cremona transformations, birational maps of $\mathbb{P}^2$ to itself, is generated by the quadratic transformations and projective linear transformations (i.e., the automorphisms of $\mathbb{P}^2$).

CHAPTER 2

Logarithmic Category

The purpose of this chapter is to give an introductory account of the **logarithmic category** introduced by Iitaka and inspired by the works of Grothendieck [2] and Deligne [1][2][3].

Working with the logarithmic category has the following three great advantages:

(i) **Iitaka's Philosophy**: It allows us to deal with open varieties U by considering **logarithmic pairs** (X, D) (called **log pairs** for short) consisting of the compactifications X of U and the boundaries $D = X - U$, which we choose to be of pure codimension one and hence called the **boundary divisor**. Iitaka's philosophy states that a theory or theorem about complete varieties X dictated by the canonical divisor K_X should find its counterpart in the logarithmic category as a theory or theorem about log pairs (X, D) dictated by the **log canonical divisor** $K_X + D$, and vice versa.

(ii) **Generalized Adjunction**: The log canonical divisor $K_X + D$ of a log pair (X, D) naturally restricts to the log canonical divisor on the boundary through the (generalized) adjunction formula

$$K_X + D|_D = K_D + \text{Diff}_X D.$$

It may give an inductional structure to the scheme of arguments when one tries to prove a property of the log canonical divisor $K_X + D$, if the property "propagates" through the adjunction to the log canonical divisor of the boundary $K_D + \text{Diff}_X D$.

In the presence of singularities on X, the usual adjunction formula $K_X + D|_D = K_D$ needs a correction term $\text{Diff}_X D$ called "**different**" to hold as above. Thus the logarithmic category provides a more natural stage for the (generalized) adjunction to work.

(iii) **Logarithmic Ramification Formula**: As we have the ramification formula for a morphism between (smooth) varieties $f : Y \to X$

$$K_Y = f^* K_X + R,$$

we have a similar ramification formula called the logarithmic ramification formula for a morphism between logarithmic pairs (smooth with only normal crossings) $f : (Y, D_Y) \to (X, D_X)$,

$$K_Y + D_Y = f^*(K_X + D_X) + R_{\log}.$$

The logarithmic ramification formula behaves more naturally than the usual one under some circumstances.

In Section 2-1 we demonstrate principle (i) by studying the analogy between the number of rational (or integral) points on a curve and the number of nonconstant holomorphic maps from the entire complex plane to a curve. We observe, for example, that Dirichlet's unit theorem is a theorem in the logarithmic category corresponding to the Mordell–Weil theorem in the usual category. We also observe an easy case of the generalized adjunction (ii) in the presence of singularities, finishing the section by verifying the logarithmic ramification formula mentioned in (iii).

In Section 2-2, according to Iitaka's philosophy of Section 2-1, we formulate the **log birational geometry of surfaces**.

We show how the line of arguments in Chapter 1 can be modified to the one in the logarithmic category. This is a prelude to the later chapters, where Iitaka's philosophy intricately blends in with the minimal model program, and together they form the indispensable parts of the Mori program.

2.1   Iitaka's Philosophy

(i) Iitaka's Philosophy: First we demonstrate principle (i) on the two subjects of

- o the number of nonconstant holomorphic maps from $\mathbb{C}$ to a (punctured) Riemann surface;
- o the number of rational (or integral) points on a nonsingular projective (or open) curve.

Compare the following two tables, one for the number of nonconstant holomorphic maps from $\mathbb{C}$ to a compact Riemann surface C and the other for the number of nonconstant holomorphic maps from $\mathbb{C}$ to a punctured Riemann surface.

Table 2.1.1.

C	$\mathbb{P}^1$	elliptic	hyperbolic
$g(C) = h^0(\mathcal{O}_C(K_C))$	0	1	≥ 2
# of nonconst. hol. maps	∞		0
$\deg K_C = 2g(C) - 2$	≤ 0		> 0

Table 2.1.2.

$C - \cup_{i=1}^n p_i$	$\mathbb{P}^1 - \cup_{i=1}^n p_i$ $n < 2$	$\mathbb{P}^1 - \cup_{i=1}^n p_i$ $n = 2$ E elliptic	$\mathbb{P}^1 - \cup_{i=1}^n p_i$ $n > 2$ $E - \cup_{i=1}^n p_i$ $n > 0$ $C - \cup_{i=1}^n p_i$ $n \geq 0,\ g(C) \geq 2$
$h^0(\mathcal{O}_C(K_C + \Sigma_{i=1}^n p_i))$	< 0	1	≥ 2
# of nonconst. holomorphic maps	∞		0
$\deg(K_C + \Sigma_{i=1}^n p_i)$ $= 2g(C) - 2 + n$	≤ 0		> 0

It is remarkable that the number of nonconstant holomorphic maps from $\mathbb{C}$ to a compact Riemann surface C is determined by the signature of the number $2g(C) - 2$, whereas the number of nonconstant holomorphic maps from $\mathbb{C}$ to a punctured Riemann surface $C - \cup_{i=1}^n p_i$ is determined by the signature of the number $2g(C) - 2 + n$.

The number $2g(C) - 2$ is the degree of the line bundle corresponding to the sheaf Ω_C^1 of regular 1-forms on C. Thus it is this line bundle $\mathcal{O}_C(K_C) = \Omega_C^1$ (called the canonical bundle of C associated to the canonical divisor K_C) that controls the number of holomorphic maps from $\mathbb{C}$ in the case of a compact Riemann surface.

Then it is a natural and important question to ask what object gives rise to the crucial number $2g(C) - 2 + n$ and thus controls the number of holomorphic maps from $\mathbb{C}$ to a punctured Riemann surface. The notion of the logarithmic 1-forms (the logarithmic canonical bundle associated to the logarithmic canonical divisor) given below answers this question.

Definition 2-1-3 (Logarithmic Forms). Let X be a nonsingular variety of dimension d and D a boundary divisor with only normal crossings. Then the sheaf of logarithmic 1-forms along D is defined to be

$$\Omega_X^1(\log D)_p := \Sigma_{i=1}^s \mathcal{O}_{X,p} \cdot \frac{dz_i}{z_i} + \Sigma_{i=s+1}^d \mathcal{O}_{X,p} \cdot dz_i \subset (\Omega_X^1 \otimes k(X))_p$$

where at a point $p \in X$ a local coordinate system $(z_1, \ldots, z_s, z_{s+1}, \ldots, z_d)$ is given such that

$$D = \left\{ \prod_{i=1}^s z_i = 0 \right\}.$$

The sheaf of logarithmic m-forms is defined to be

$$\Omega_X^m(\log D) := \wedge^m \Omega_X^1(\log D),$$

and in particular, the sheaf of logarithmic d-forms

$$\Omega_X^d(\log D) \cong \mathcal{O}_X(K_X + D)$$

is a line bundle called the **logarithmic canonical bundle** for the pair (X, D). We call $K_X + D$ the **logarithmic canonical divisor** (or **log canonical divisor** for short).

In the case of a punctured Riemann surface $C - \cup_{i=1}^n p_i$, the number $2g(C) - 2 + n$ is the degree of the corresponding logarithmic canonical bundle for the pair $(X = C, D = \Sigma p_i)$.

In summary, we have observed a simple example of the following grand principle, the so-called Iitaka philosophy.

Principle 2-1-4 (Iitaka's Philosophy). *Whenever we have a theorem about nonsingular complete varieties whose statement is dictated by the behavior of the regular differential forms (the canonical bundles), there should exist a corresponding theorem about logarithmic pairs (pairs consisting of nonsingular complete varieties and boundary divisors with only normal crossings) whose statement is dictated by the behavior of the logarithmic forms (the logarithmic canonical bundles), and vice versa.*

To demonstrate one more example of this principle, we now turn our attention to the number of rational (respectively, integral) points on a nonsingular projective curve C_K defined over a number field K (respectively, over the ring $\mathcal{O}_K$ of algebraic integers in the number field K). (We assume that $C_K \times \overline{K}$, where $\overline{K}$ is the algebraic closure of K, is irreducible and nonsingular and that C_K has a section $\operatorname{Spec} K \to C_K$ when $g(C_K) = 0$ or 1.)

Table 2.1.5.

C_K	$\mathbb{P}^1_K$	elliptic	hyperbolic
$g(C) = h^0(\mathcal{O}_C(K_C))$	0	1	≥ 2
$\deg K_C = 2g(C) - 2$	< 0	0	> 0
K-rat. pts.	$K \cup \{\infty\}$	f.g. abelian group	$< \infty$

Though the analogy for the number of rational points for a punctured algebraic curve may not be observed, the analogy for the number of integral points for a punctured algebraic curve defined over a ring of algebraic integers is again striking.

Table 2.1.6.

$C - \cup_{i=1}^n p_i$	(i) $\mathbb{P}^1_K$ (ii) $\mathbb{P}^1 - \{\infty\}$	(iii) $\mathbb{P}^1 - \{0, \infty\}$ (iv) E elliptic	(v) $\mathbb{P}^1 - \cup_{i=1}^{n>2} p_i$ (vi) $E - \cup_{i=1}^{n>0} p_i$ (vii) $C - \cup_{i=1}^{n\geq 0} p_i$ $g(C) \geq 2$
$h^0(K_C + \Sigma_{i=1}^n p_i)$	0	1	≥ 2
$\deg(K_C + \Sigma_{i=1}^n p_i)$ $= 2g(C) - 2 + n$	< 0	0	> 0
K-rat. pts.	(i) $K \cup \{\infty\}$ (ii) K	(iii) K^* (iv) MW grp	(v) $K \cup \{\infty\}$ $- \cup_{i=1}^n p_i$ (vi) MW grp $- \cup_{i=1}^n p_i$ (vii) $< \infty$
integral pts.	infinite (i) $K \cup \{\infty\}$ (ii) $\mathcal{O}_K$	finitely generated (iii) $\mathcal{O}_K^*$ (iv) MW grp	$< \infty$

It is noteworthy that the set of rational points on an elliptic curve C defined over a number field, by the Mordell–Weil theorem, is a finitely generated abelian group. The canonical divisor K_C, the controlling factor, has degree $2g(C) - 2 = 0$. According to Iitaka's philosophy, we look for a corresponding theorem for a punctured algebraic curve whose logarithmic canonical divisor $K_C + \Sigma_{i=1}^n p_i$, the controlling factor, has degree 0. For example, we take $\mathbb{P}^1_{\mathcal{O}_K} - \{0, \infty\} = \operatorname{Spec} \mathcal{O}_K[x, \frac{1}{x}]$, where the corresponding logarithmic canonical divisor $K_{\mathbb{P}^1} + D$ with $D = 0 + \infty$ has the degree $2g(\mathbb{P}^1) - 2 + 2 = 0$. The integral points are nothing but the sections of the morphism

$$\mathbb{P}^1_{\mathcal{O}_K} - \{0, \infty\} = \operatorname{Spec} \mathcal{O}_K \left[x, \frac{1}{x} \right] \to \operatorname{Spec} \mathcal{O}_K,$$

and the sections

$$\operatorname{Spec} \mathcal{O}_K \to \operatorname{Spec} \mathcal{O}_K [x, \frac{1}{x}] = \mathbb{P}^1_{\mathcal{O}_K} - \{0, \infty\}$$

are in 1-to-1 correspondence with the homomorphisms

$$h : \mathcal{O}_K \left[x, \frac{1}{x} \right] \to \mathcal{O}_K,$$

which in turn are equivalent to choosing a unit

$$h(x) \in \mathcal{O}^*.$$

Thus Dirichlet's unit theorem, which states that the set of units in the algebraic integers $\mathcal{O}_K$ forms a finitely generated abelian group, can be regarded as the logarithmic counterpart of the Mordell–Weil theorem.

The celebrated theorem of Faltings (the Mordell conjecture) states that the number of rational points on a nonsingular projective curve C defined over a number field is finite if $g(C) \geq 2$, i.e., the canonical divisor K_C has a positive degree $2g(C) - 2 > 0$. If we look for a corresponding theorem in the logarithmic category, then we should encounter the finiteness statement for the integral points on a punctured curve whose logarithmic canonical divisor $K_C + D$ has a positive degree. In fact, on a curve C with $g(C) \geq 2$, there are only finitely many integral points whether it is punctured or not, since $\deg K_C > 0$ and we have only finitely many rational points. As for curves with $g(C) = 1$, we have the Siegel's theorem stating that there are only finitely many integral points on an affine elliptic curve $C - \{\infty\} \cong \{f = 0\} \subset \mathbb{A}^2$ with an integral Weierstrass equation f, having the logarithmic canonical divisor of positive degree $\deg(K_C + (\infty)) = 1 > 0$. It is left to the reader as an exercise to check another theorem of Siegel stating that there are only finitely many integral points on $\mathbb{P}^1_{\mathcal{O}_K} - \cup_{i=1}^n p_i$ if $n > 2$, and thus the logarithmic canonical divisor has positive degree

$$\deg(K_{\mathbb{P}^1} + \Sigma_{i=1}^n p_i) = -2 + n > 0.$$

(ii) Generalized Adjunction: Next we present an easy example of the generalized adjunction (ii).

We take a singular quadric

$$Q = \{XY - Z^2 = 0\} \subset \mathbb{P}^3_{\mathbb{C}} = \operatorname{Proj} \mathbb{C}[X, Y, Z, W].$$

Now, for the dualizing sheaf for the quadric, which is a Cartier divisor on a nonsingular variety $\mathbb{P}^3_{\mathbb{C}}$, the ordinary adjunction formula works. We leave it to the reader to check the following isomorphisms (cf. Hartshorne [3], Section III.7)

$$\omega_Q = Ext^1(\mathcal{O}_Q, \omega_{\mathbb{P}^3_{\mathbb{C}}}) \cong \mathcal{O}_{\mathbb{P}^3}(K_{\mathbb{P}^3_{\mathbb{C}}} + Q)|_Q \cong \mathcal{O}_Q(-2).$$

We take a line $l \subset Q$ passing through the singular point P of the quadric. Note that if we take a hyperplane $H \subset \mathbb{P}^3_{\mathbb{C}}$ containing l but not tangent to Q, then

$$H|_Q = l + l'$$

for some line $l' \neq l$, and that if we take a hyperplane $G \subset \mathbb{P}^3_{\mathbb{C}}$ containing l' and tangent to Q, then

$$G|_Q = 2l'.$$

Figure 2.1.7.

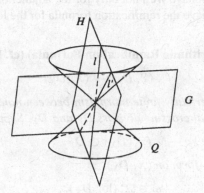

Thus

$$2l = 2(l + l') - 2l' \sim 2H|_Q - G|_Q \sim H|_Q.$$

While the dualizing sheaf for the line is given by

$$\mathcal{O}_l(K_l) \cong \omega_l \cong \mathcal{O}_{\mathbb{P}^1}(-2),$$

a naive application of adjunction would give

$$\mathcal{O}_Q(K_Q + l)|_l = \mathcal{O}\left(K_Q + \frac{1}{2}(H|_Q)\right)\Big|_l = \mathcal{O}_{\mathbb{P}^1}\left(-2 + \frac{1}{2}\right).$$

Therefore, in order to have the equality relating these two, we have to introduce a correcting term

$$\mathrm{Diff}_Q l = \mathcal{O}_{\mathbb{P}^1}\left(\frac{1}{2}\right) = \mathcal{O}_{\mathbb{P}^1}\left(\frac{1}{2}P\right)$$

and hence observe the generalized adjunction

$$K_Q + l|_l = K_l + \mathrm{Diff}_Q l.$$

It should be emphasized that the correcting term appears with the rational coefficient $\frac{1}{2}$.

Observation 2-1-8. *In the natural course of working with the logarithmic category, not only do we consider the pairs (X, D_X) consisting of a nonsingular projective variety X and a normal crossing divisor $D_X = \Sigma D_i = \Sigma 1 \cdot D_i$ with all* *the coefficients equal to* 1 *but also the pairs* (Y, D_Y), *where Y is a nonsingular (or mildly singular) projective variety and* $D_Y = \Sigma d_i D_i$ *is a $\mathbb{Q}$-divisor whose support is a normal crossing divisor (or a divisor that satisfies some mild conditions) but with rational coefficients*

$$0 \le d_i \le 1, d_i \in \mathbb{Q}.$$

(iii) Logarithmic Ramification Formula: We end this section by verifying that the logarithmic forms behave well not only for the adjunctions but also under the pullbacks and that we have the ramification formula for the logarithmic canonical bundles.

Theorem 2-1-9 (Logarithmic Ramification Formula) (cf. Iitaka [5]). *Let*

$$f : (Y, D_Y) \to (X, D_X)$$

be a birational or generically finite morphism between nonsingular varieties of dimension d with normal crossing divisors D_Y and D_X. Suppose

$$f^{-1}(D_X) \subset D_Y.$$

Then any logarithmic 1-*form on* (X, D_X),

$$\omega \in \Omega_X^1(\log D_X)_p,$$

pulls back to a logarithmic 1-*form on* (Y, D_Y),

$$f^*(\omega) \in \Omega_Y^1(\log D_Y)_q, \, f(q) = p.$$

In particular, we have the ramification formula for the logarithmic canonical divisors

$$K_Y + D_Y = f^*(K_X + D_X) + R_{\log},$$

where $R_{\log}$ is some effective divisor, called the logarithmic ramification divisor.

PROOF.   Take a local coordinate system around $p \in X$,

$$(z_1, \dots, z_s, z_{s+1}, \dots, z_d),$$

as in the definition of logarithmic forms so that

$$D_X = \{\Pi_{i=1}^s z_i = 0\}$$

and

$$\Omega_X^1(\log D_X)_p = \Sigma_{i=1}^s \mathcal{O}_X \cdot \frac{dz_i}{z_i} + \Sigma_{i=s+1}^d \mathcal{O}_X \cdot dz_i \subset (\Omega_X^1 \otimes k(X))_p,$$

and a local coordinate system around $q \in Y$,

$$(w_1, \dots, w_t, w_{t+1}, \dots, w_d),$$

so that

$$D_Y = \{\Pi_{j=1}^t w_j = 0\}$$

and

$$\Omega_Y^1(\log D_Y)_q = \Sigma_{j=1}^t \mathcal{O}_Y \cdot \frac{dw_j}{w_j} + \Sigma_{j=t+1}^d \mathcal{O}_Y \cdot dw_j \subset (\Omega_Y^1 \otimes k(Y))_q.$$

Since

$$f^{-1}(D_X) \subset D_Y,$$

we have

$$f^*z_i = u_i \Pi_{j=1}^t w_j^{n_{ij}}, u_i \in \mathcal{O}_Y^* \text{ and } n_{ij} \geq 0 \text{ for } 1 \leq i \leq s.$$

Thus, for $1 \leq i \leq s$,

$$f^* \frac{dz_i}{z_i} = f^* d\log z_i = d\log f^*z_i = \Sigma_{j=1}^t n_{ij} \frac{dw_j}{w_j} + \frac{du_i}{u_i} \in \Omega_Y^1(\log D_Y)_q.$$

On the other hand, for $s + 1 \leq i \leq d$, obviously

$$f^* dz_i = df^*z_i \in \Omega_{Y_q}^1 \subset \Omega_Y^1(\log D_Y)_q.$$

Therefore, the pullback of a logarithmic 1-form along D_X is a logarithmic 1-form along D_Y.

Moreover, since

$$f^* \left(\frac{dz_1}{z_1} \wedge \ldots \wedge \frac{dz_s}{z_s} \wedge dz_{s+1} \wedge \ldots \wedge dz_d \right)$$

$$= \wedge_{i=1}^s \left(\Sigma_{j=1}^t n_{ij} \frac{dw_j}{w_j} + \frac{du_i}{u_i} \right) \wedge_{i=s+1}^d f^* dz_i$$

$$= r \left(\frac{dw_1}{w_1} \wedge \ldots \wedge \frac{dw_t}{w_t} \wedge dw_{t+1} \wedge \ldots \wedge dw_d \right),$$

where $r \in \mathcal{O}_{Y_q}$, and since

$$f^* \left(\frac{dz_1}{z_1} \wedge \ldots \wedge \frac{dz_s}{z_s} \wedge dz_{s+1} \wedge \ldots \wedge dz_d \right) \neq 0$$

which follows from the fact that f^* is injective, i.e.,

$$(\Omega_X^d \otimes k(X))_p \overset{f^*}{\underset{injective}{\hookrightarrow}} (\Omega_Y^d \otimes k(Y))_q,$$

the logarithmic ramification formula follows.

This completes the proof of the logarithmic ramification formula.   □

Exercise 2-1-10. *In the setting for the logarithmic ramification formula, assume further that $f(D_Y) \subset D_X$, i.e., $D_Y = f^{-1}(D_X)$, and that $f_*(R) \subset D_X$, where R is the usual ramification divisor $K_Y = f^*K_X + R$ and where $f_*(R)$ is the codimension-one part of $f(R)$. Show that then the log ramification divisor $R_{\log}$ is exceptional for f.*

In particular, if $\dim Y = \dim X = 1$ with $D_Y = f^{-1}(D_X)$ and if $f : Y - G \to X - D$ is unramified (i.e., étale), then

$$K_Y + D_Y = f^*(K_X + D_X).$$

2.2   Log Birational Geometry of Surfaces

In this section we apply Iitaka's philosophy of Section 2-1 to the Mori program in dimension 2 to discuss the **log birational geometry of surfaces**, following its main strategies MP 1 through MP 3.

$\boxed{\text{MP 1}}$ Log Minimal Model Program in Dimension 2.

Recall that MMP in dimension 2, given an input of a nonsingular projective surface S, produces a minimal model or a Mori fiber space as output. The decisive criterion of the program is whether or not the canonical bundle (divisor) is nef, and thus "dictated" by the behavior of the canonical bundle:

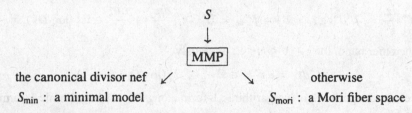

the canonical divisor nef $\swarrow$ $\searrow$ otherwise

$S_{\min}$: a minimal model S_{mori} : a Mori fiber space

In short, the aim of this section is to establish the program that takes as input a logarithmic pair (S, B) consisting of a nonsingular projective surface and a boundary $\mathbb{Q}$-divisor $B = \Sigma b_i B_i$ with only normal crossings and rational coefficients $0 \le b_i \le 1$, and then produces a log minimal model or log Mori fiber space as output. The key criterion of log MMP in dimension 2, according to Iitaka's philosophy, should be whether the log canonical divisor is nef and thus dictated by its behavior:

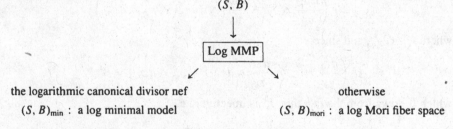

the logarithmic canonical divisor nef   otherwise

$(S, B)_{\min}$: a log minimal model $(S, B)_{\text{mori}}$: a log Mori fiber space

Recall that in order to establish MMP in dimension 2 we need only:

cone theorem,
contraction theorem,

based upon

rationality theorem,
boundedness of denominator,
base point freeness theorem.

Thus in order to establish the logarithmic minimal model program in dimension 2 we need only to verify the logarithmic version of these theorems:

log cone theorem,

log contraction theorem,

based upon

log rationality theorem,
log boundedness of denominator,
log base point freeness theorem.

Actually, these theorems can be shown to hold in the category $\mathfrak{E}$ of log pairs (S, B) consisting of nonsingular projective surfaces and boundary $\mathbb{Q}$-divisors $B = \Sigma b_i B_i$ with only normal crossings and rational coefficients $0 \leq b_i \leq 1$.

So have we established log MMP in dimension 2?

Trying to run log MMP in the category $\mathfrak{E}$, we immediately face the following fundamental obstacle:

Obstacle 2-2-1. *There is a logarithmic pair (S, B) in the category $\mathfrak{E}$ such that the logarithmic canonical divisor $K_S + B$ is not nef, and such that* any *contraction of an extremal ray with respect to $K_S + B$ results in a projective but* singular *surface.*

Therefore, we inevitably go out of the category $\mathfrak{E}$ in the process of running log MMP.

Example 2-2-2. Take a singular quadric

$$Q = \{XY - Z^2 = 0\} \subset \mathbb{P}^3_{\mathbb{C}} = \mathrm{Proj}\mathbb{C}[X, Y, Z, W].$$

We take a general member

$$H \in |\mathcal{O}_Q(4)| \text{ associated to } s \in H^0(Q, \mathcal{O}_Q(4))$$

such that H is irreducible nonsingular and does not pass through the vertex p of the quadric. We take the double cover

$$d : T = Spec(\mathcal{O}_Q \oplus \mathcal{O}_Q(-2)) \to Q,$$

where the $\mathcal{O}_Q$-algebra structure on $\mathcal{O}_Q \oplus \mathcal{O}_Q(-2)$ is given by

$$\mathcal{O}_Q(-2) \times \mathcal{O}_Q(-2) \to \mathcal{O}_Q,$$
$$a \times b \mapsto a \cdot b \cdot s.$$

Then d is ramified along H, and T is nonsingular except at two points q_1 and q_2 that are the inverse images of p. (Exercise! Verify the statements about T.) Blow up T at q_1 and q_2 to obtain $f : S \to T$. Locally, each blowup is isomorphic to the blowup of Q at p. Therefore, S is nonsingular, and the exceptional locus consists of two (-2)-curves D_1 and D_2.

Figure 2-2-3.

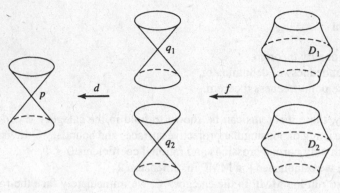

Moreover,

$$\mathcal{O}_S(K_S) \equiv f^*\omega_T \equiv d^*\omega_Q \otimes d^*\mathcal{O}_Q\left(\frac{1}{2}H\right) \equiv d^*\mathcal{O}_Q\left(-2 + \frac{1}{2}\cdot 4\right) \equiv 0.$$

Set

$$B = d_1 D_1 + d_2 D_2,$$

where

$$0 < d_1, d_2 \leq 1 \text{ with } d_1, d_2 \in \mathbb{Q}.$$

Then the only extremal rays with respect to $K_S + B$ are (Exercise ! Why ?)

$$R_{l_1} = \mathbb{R}_+[D_1] \text{ and } R_{l_2} = \mathbb{R}_+[D_2],$$

and the contraction of each extremal ray R_{l_1} or R_{l_2} introduces a birational morphism onto a normal projective surface with a singular point q_1 or q_2, which are analytically isomorphic to $p \in Q$.

Exercise 2-2-4. *Find an example of a logarithmic pair (S, B) in the category $\mathfrak{E}$ such that the logarithmic canonical divisor $K_S + B$ is not nef, and such that any contraction $\phi : (S, B) \to (T, B_T)$ of an extremal ray with respect to $K_S + B$ results in a nonsingular projective surface T but $B_T = \phi_*(B)$ is no longer a divisor with only normal crossings.*

In general, even starting from $(S, B) \in \mathfrak{E}$, we could obtain a singular surface T and/or a boundary that is not a divisor with only normal crossings as the result of the contraction of an extremal ray $\phi : (S, B) \to (T, B_T)$.

Thus Obstacle 2-2-1 (together with Example 2-2-2 and Exercise 2-2-4) initiates the search for a new and right category $\mathfrak{D}$, in which all the key theorems for log MMP should work and in which we should be able to stay after any birational extremal contraction.

This naturally leads us to the following notion of log terminal singularities.

Definition 2-2-5 (Log Terminal Singularities in Dimension 2). *A log pair* (S, B) *consisting of a normal surface and a boundary* $\mathbb{Q}$-*divisor* $B = \Sigma b_i B_i$ *with rational coefficients* $0 \leq b_i \leq 1$ *has only log terminal singularities if*

(1) $K_S + B$ *is* $\mathbb{Q}$-*Cartier, i.e., there exists* $e \in \mathbb{N}$ *such that* $e(K_S + B)$ *is a Cartier divisor,*

(2) *there exists a projective birational morphism* $f : V \to S$ *from a nonsingular surface such that in the logarithmic ramification formula*

$$K_V + B_V = f^*(K_S + B) + \Sigma a_j E_j$$

all the coefficients for the exceptional divisors are strictly positive

$$a_j > 0 \quad \forall E_j \text{ exceptional,}$$

where

$$B_V = f_*^{-1} B + \Sigma E_j, \; f_*^{-1} B \text{ being the strict transform of } B,$$

is a divisor with only normal crossings.

For the details we refer the reader to Chapter 4, where we discuss the basic properties of several classes of singularities, including log terminal singularities.

Now all the key theorems hold in the category $\mathfrak{D}$ of log pairs (S, B) consisting of normal projective surfaces S and boundary divisors B with only log terminal singularities. We leave the precise statements and proofs in the category $\mathfrak{D}$ of

log cone theorem,
log contraction theorem,

and

log rationality theorem,
log boundedness of denominator,
log base point freeness theorem.

to the reader as an exercise for the moment, as we give the precise statements and proofs of the more general theorems in arbitrary dimension in later chapters (cf. Chapters 6 and 7).

After establishing the key theorems, it is the following basic observation that justifies our choice of the new category $\mathfrak{D}$.

Theorem 2-2-6. *The category* $\mathfrak{D}$ *of log pairs* (S, B) *consisting of normal projective surfaces* S *and boundary divisors* B *with only log terminal singularities is the smallest category in which log MMP in dimension 2 works and that contains the category* $\mathfrak{E}$. *That is to say, the category* $\mathfrak{D}$ *satisfies the following three conditions:*

(i) $\mathfrak{D}$ *contains* $\mathfrak{E}$.

(ii) $\mathfrak{D}$ *is closed under the operations of log MMP in dimension 2, i.e., for any birational extremal contraction on* $(S, B) \in \mathfrak{D}$,

$$\phi : (S, B) \to (T, B_T = \phi_*(B)),$$

the resulting log pair again belongs to $\mathfrak{D}$, *i.e.,* $(T, B_T) \in \mathfrak{D}$.

(iii) *Any object* $(S, B) \in \mathfrak{D}$ *can be obtained from some object in* $\mathfrak{E}$ *through the process of log MMP in dimension* 2.

PROOF.   Let $\mathfrak{D}'$ be the category of log pairs (S, B) consisting of normal projective surfaces and boundary divisors B with only $\mathbb{Q}$-**factorial** and log terminal singularities. (We say that S is $\mathbb{Q}$-factorial if any Weil divisor $D \subset S$ is $\mathbb{Q}$-Cartier, i.e., there exists $d \in \mathbb{N}$ such that dD is a Cartier divisor.) First we show that the category $\mathfrak{D}'$ satisfies conditions (i), (ii), and (iii), and second we show that $\mathfrak{D}'$ actually coincides with the original category $\mathfrak{D}$.

Condition (i) is obvious.

In order to see condition (ii), let

$$\phi : (S, B) \to (T, B_T = \phi_*(B))$$

be a birational extremal contraction on $(S, B) \in \mathfrak{D}'$.

Note first that the exceptional locus of ϕ consists of a single curve E. In fact, the $\mathbb{Q}$-Cartier divisor E and the curve E that belongs to the extremal ray have intersection number

$$E \cdot E = E^2 < 0$$

by Theorem 1-2-9, while any other curve $C \neq E$ has nonnegative intersection with E. Thus C cannot belong to the extremal ray and thus cannot be in the exceptional locus.

$\mathbb{Q}$-factoriality of T can be checked in the following way. For any Weil divisor D on T, take a rational number $a \in \mathbb{Q}$ such that

$$\{\phi_*^{-1}D + aE\} \cdot E = 0,$$

where $\phi_*^{-1}D$ is the strict transform of D on S. It is then a consequence of the contraction theorem (Exercise! Why?) that there exists a $\mathbb{Q}$-Cartier divisor H on T such that

$$\phi_*^{-1}D + aE = \phi^*H.$$

It now follows that

$$D = \phi_*\{\phi_*^{-1}D + aE\} = \phi_*\phi^*H = H$$

is $\mathbb{Q}$-Cartier. Thus T is $\mathbb{Q}$-factorial.

Now we check that (T, B_T) has only log terminal singularities. Condition (0) is immediate, since T is $\mathbb{Q}$-factorial and hence $K_T + B_T$ is $\mathbb{Q}$-Cartier. Condition (1) can be checked as follows.

First consider the case where $E = B_o \subset B$. Since (S, B) has only log terminal singularities, we can take a projective birational morphism $f : V \to S$, as in the definition of log terminal singularities, such that

$$K_V + B_V = f^*(K_S + B) + \Sigma a_j E_j, a_j > 0 \quad \forall E_j \text{ exceptional for } f,$$

where

$$B_V = f_*^{-1}B + \Sigma E_j$$

is a divisor with only normal crossings. Since $(K_S + B) \cdot E < 0$, in the ramification formula

$$K_S + B = \phi^*(K_T + B_T) + aE$$

the coefficient a is greater than 0. Moreover, by definition and by the assumption $E = B_o \subset B$ we have

$$(B_T)_V = (\phi \circ f)_*^{-1}B_T + \Sigma E_j + f_*^{-1}E = B_V + f_*^{-1}bE,$$

where

$$b = 1 - d_o \geq 0.$$

Therefore,

$$
\begin{aligned}
K_V + (B_T)_V &= K_V + B_V + f_*^{-1}bE \\
&= f^*(K_S + B) + \Sigma a_j E_j + f_*^{-1}bE \\
&= f^*(\phi^*(K_T + B_T) + aE) + \Sigma a_j E_j + f_*^{-1}bE \\
&= (\phi \circ f)^*(K_T + B_T) + \Sigma a_j E_j + f_*^{-1}bE + f^*aE \\
&= (\phi \circ f)^*(K_T + B_T) + \Sigma c_k E_k, \quad c_k > 0 \\
&\qquad\qquad\qquad\qquad\qquad \forall E_k \text{ exceptional for } \phi \circ f
\end{aligned}
$$

and

$$\mathrm{Supp}\, B_V = \mathrm{Supp}(B_T)_V$$

is a divisor with only normal crossings. Thus (T, B_T) has only log terminal singularities.

Second, consider the case where $E \not\subset B$. Every argument goes parallel to the one in the case $E \subset B$, except that

$$\mathrm{Supp}(B_T)_V = \mathrm{Supp}(B_V + f_*^{-1}E)$$

may not be a divisor with only normal crossings. Note that the locus where $\mathrm{Supp}(B_T)_V$ may not have normal crossings has to be over the locus $f_*^{-1}(E)$. We take a further blowup $g : W \to V$ with centers on $f_*^{-1}(E)$ such that

$$\mathrm{Supp}((B_T)_V)_W = \mathrm{Supp}(B_T)_W$$

is a divisor with only normal crossings. By the logarithmic ramification formula, we have

$$K_W + (B_V)_W = g^*(K_V + B_V) + R,$$

where R is some effective (maybe zero) divisor exceptional for g. Since

$$K_V + B_V = f^*(K_S + B) + \Sigma a_j E_j \quad a_j > 0 \quad \forall E_j \text{ exceptional for } f,$$

$$K_S + B = \phi^*(K_T + B_T) + aE \text{ with } a > 0,$$

and since

$$(B_T)_W = (B_V)_W + (\phi \circ f \circ g)_*^{-1}(E),$$

we conclude that

$$
\begin{aligned}
K_W + (B_T)_W &= K_W + (B_V)_W + (\phi \circ f \circ g)_*^{-1}(E) \\
&= g^*(K_V + B_V) + R + (\phi \circ f \circ g)_*^{-1}(E) \\
&= g^*(f^*(K_S + B) + \Sigma a_j E_j) + R + (\phi \circ f \circ g)_*^{-1}(E) \\
&= g^*(f^*(\phi^*(K_T + B_T) + aE) + \Sigma a_j E_j) + R + (\phi \circ f \circ g)_*^{-1}(E) \\
&= (\phi \circ f \circ g)^*(K_T + B_T) + \Sigma e_l E_l, \, e_l > 0
\end{aligned}
$$

$$\forall E_l \text{ exceptional for } \phi \circ f \circ g.$$

Thus again (T, B_T) has only log terminal singularities.

Finally, we prove condition (iii) and also show that the categories $\mathfrak{D}$ and $\mathfrak{D}'$ coincide.

Let (S, B) be an object in $\mathfrak{D}$. Since (S, B) has only log terminal singularities, we can take a birational morphism $f : V \to S$ from a nonsingular projective surface V such that

$$K_V + B_V = f^*(K_S + B) + \Sigma a_j E_j, a_j > 0 \quad \forall E_j \text{ exceptional for } f,$$

where

$$B_V = f_*^{-1}(B) + \Sigma E_j$$

is a divisor with only normal crossings.

Since by Theorem 1-2-9

$$0 > (\Sigma a_j E_j)^2 = (K_V + B_V) \cdot (\Sigma a_j E_j),$$

we conclude that there is E_{j_o} such that

$$(K_V + B_V) \cdot E_{j_o} < 0 \text{ and } (E_{j_o})^2 < 0.$$

It is easy to see (Exercise! Why?) that $R_l = \mathbb{R}_+[E_{j_o}]$ is an extremal ray with respect to $K_V + B_V$, which can be contracted by the log contraction theorem to produce a birational morphism $\mu : (V, B_V) \to (V_1, B_{V_1} = \mu_* B_V)$ contracting E_{j_o} to a point.

We know that $(V_1, B_{V_1}) \in \mathfrak{D}'$ by condition (i) and that f factors through μ by Lemma 1-8-1 to give $f_1 : (V_1, B_{V_1}) \to (S, B)$. The ramification formula now reads

$$K_{V_1} + B_{V_1} = f_1^*(K_S + B) + \Sigma_{j \neq j_o} a_j E_j.$$

We repeat the process with $f_1 : (V_1, B_{V_1}) \to (S, B)$ instead of $f : (V, B_V) \to (S, B)$. It has to come to an end after finitely many iterations, finally to yield $f_l : (V_l, B_{V_l}) \to (S, B)$ with $(V_l, B_{V_l}) \in \mathfrak{D}'$ and no exceptional divisor. Since f_l is a projective birational morphism and since S is normal, we conclude from the fact that f_l has no exceptional divisor that

$$(S, B) \underset{f_l^{-1}}{\tilde{\rightarrow}} (V_l, B_{V_l}) \in \mathfrak{D}'.$$

Since $(S, B) \in \mathfrak{D}$ is arbitrary, this shows that the category $\mathfrak{D}$ coincides with $\mathfrak{D}'$.

Moreover, since the process is a repetition of extremal contractions, operations in log MMP starting from $(V, B_V) \in \mathfrak{E}$, we verify condition (iii).

This completes the proof of Theorem 2-2-6. □

Remark 2-2-7. It is a feature unique to dimension 2 that the category of projective log pairs with only log terminal singularities and that of projective log pairs with only $\mathbb{Q}$-factorial and log terminal singularities coincide. In higher dimensions, due to the existence of (log crepant) small contractions, the former is strictly larger than the latter. It is also worthwhile noting that it is the latter category that is the smallest category that contains all the log pairs consisting of nonsingualr projective varieties with boundary divisors with only normal crossings and in which log MMP works. Thus the notion of "$\mathbb{Q}$-**factoriality**" is indispensable and essential in higher dimensions. Here in dimension 2, $\mathbb{Q}$-factoriality is implicitly swallowed into the notion of log terminal singularities. For more details, we refer the reader to Chapter 4, where we characterize log terminal singularities in dimension 2 as quotient singularities (and hence obviously $\mathbb{Q}$-factorial).

We present the flowchart for log MMP in dimension 2, which goes completely parallel to MMP in dimension 2 of Section 1-3.

Flowchart 2-2-8.

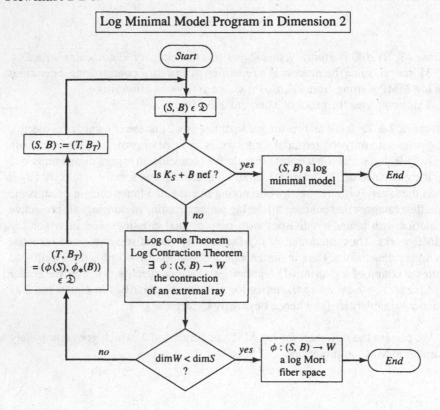

| Log Minimal Model Program in Dimension 2 |

Start

$(S, B) \in \mathfrak{D}$

$(S, B) := (T, B_T)$

Is $K_S + B$ nef ? — yes → (S, B) a log minimal model → End

no

(T, B_T)
$= (\phi(S), \phi_*(B))$
$\in \mathfrak{D}$

Log Cone Theorem
Log Contraction Theorem
$\exists \ \phi : (S, B) \to W$
the contraction
of an extremal ray

no ← dim W < dim S ? — yes → $\phi : (S, B) \to W$ a log Mori fiber space → End

---

MP 2 Basic properties of log minimal models and log Mori fiber spaces in dimension 2.

Once log MMP in dimension 2 with log terminal singularities is established, the easy dichotomy theorem of MMP in dimension 2 and its proof can be easily generalized to

easy dichotomy theorem of log MMP in dimension 2,

whose precise statement and proof are left to the reader as an excercise.

hard dichotomy theorem of log MMP in dimension 2, and
log abundance theorem for log minimal models in dimension 2

also hold. Here we only outline rather subtle proofs of these two theorems.

In order to prove the hard dichotomy theorem of log MMP in dimension 2, thanks to the easy dichotomy of log MMP in dimension 2, we only have to show the logarithmic version of Theorem 1-5-4, i.e.,

existence of an effective pluri-log canonical divisor of a log minimal model in dimension 2 which states the following:

Let (S, B) a log minimal model in dimension 2 with only log terminal singularities. Then $\kappa(K_S + B) \geq 0$.

For the proof of this theorem, first we show that we may assume that S is nonsingular by taking the so-called minimal resolution (cf. Section 4-5). Now if $0 \leq \kappa(K_S)$ $(\leq \kappa(K_S + B))$, then there is nothing more to prove. If $-\infty = \kappa(K_S)$, then S is birationally equivalent to a $\mathbb{P}^1$-bundle over a nonsingular projective curve, and thus its structure is well understood (cf. Section 1-4). We apply this understanding to the analysis. (For details see for example, Lemma 11.2.1 in Kollár et al. [1].)

Once $\kappa(K_S + B) \geq 0$ for a log minimal model (S, B) has been proved, a proof of the log abundance theorem for log minimal models in dimension 2 can be carried out almost parallel to that of the abundance theorem for minimal models in dimension 2 (cf. Theorem 1-5-6). First we show that we may assume that S is nonsingular. When $\kappa(K_S + B) = 2$, which is equivalent to saying that $(K_S + B)^2 > 0$, we may show that the curves having zero intersection with $K_S + B$ can be contracted to obtain a normal projective surface T with only simple elliptic, cusp, or quotient singularities (cf. Section 4-6) in a similar (but more tedious) manner to the argument in Section 1-6. Then by the Nakai–Moishezon criterion $K_S + B$ is the pullback of an ample divisor on T, and thus the log abundance theorem holds. (For a different proof, see Theorem 8.4 in Kollár et al. [1] and modify the 3-dimensional argument there to the 2-dimensional one here.) When $\kappa(K_S + B) = 1$, the proof is straightforward and as easy as that of Theorem 1-5-6. The only remaining case is the one where $\kappa(K_S + B) = 0$. The Kodaira dimension of the underlining surface $\kappa(K_S)$ must be either 0 or $-\infty$. If $\kappa(K_S) = 0$, then by taking some étale cover if necessary, we may assume that S is birationally equivalent to an Abelian surface or to a K3 surface (cf. Section 1-7). If S is birationally equivalent to an Abelian surface, the argument in the proof of Theorem 1-5-6 goes through with little change. If S is birationally equivalent to a K3 surface, then first consider the case $B \neq 0$. The Riemann–Roch theorem and the vanishing of $h^2(S, \mathcal{O}_S(m(K_S + B))) = h^0(S, \mathcal{O}_S(K_S - m(K_S + B)))$ (since $B \neq 0$) gives $h^0(S, \mathcal{O}_S(m(K_S + B))) \geq 2$ for m sufficiently divisible, which is against the assumption $\kappa(K_S + B) = 0$. The case where $B = 0$ is the ordinary abundance theorem. If $\kappa(K_S) = -\infty$, then S is birationally equivalent to a $\mathbb{P}^1$-bundle, and we exploit the understanding of the structure again. (For a different proof, see Theorem 11.3.1 in Kollár et al. [1].)

We have described only the outline of the proofs for the logarithmic versions of the hard dichotomy and abundance theorems in dimension 2. Details are left to the reader as an exercise.

MP 3 Birational relation among log minimal models and log Mori fiber spaces.

In Section 1-8 we considered the relation among good representatives, i.e., minimal models or Mori fiber spaces, in a fixed birational equivalence class. When we deal with the log pairs and the relation among good representatives obtained through log MMP, the mere notion of birational equivalence does not work well, and one is naturally led to the notion of the log MMP relation:

Definition 2-2-9 (Log MMP-Relation). *Two projective log pairs (S_1, B_1) and (S_2, B_2) of dimension 2 with only log terminal singularities are said to be log MMP related if there exists a log pair (V, B_V) consisting of a nonsingular projective surface V and a boundary divisor B_V with only normal crossings and rational coefficients (between 0 and 1) such that both (S_1, B_1) and (S_2, B_2) are obtained from (V, B_V) via some processes of log MMP in dimension 2.*

Exercise 2-2-10. *Define the notion of the MMP relation for nonsingular projective surfaces as an analogy to Definition 2-2-9 and prove that they are MMP related if they are birational.*

Now, the

Uniqueness of the log minimal model in dimension 2, and
log Sarkisov program in dimension 2

hold, replacing the notion of "birational" (or equivalently "MMP related" as in Exercise 2-2-10) with that of "log MMP related".

Again the details are left to the reader as an exercise (cf. Chapters 11, 12, and 13).

CHAPTER 3

Overview of the Mori Program

The purpose of this chapter is to outline the key points of the **Mori program in dimension 3 (or higher)**, extending the analogy from dimension 2 to higher dimensions as much as possible. At the same time we try to distinguish the remarkable features, such as flips, that are unique to higher dimensions. We will not give any detailed proofs in this chapter. The emphasis is on presenting the global picture at an early stage by taking the reader on a quick roller-coaster ride of the Mori program.

We review the Mori program in dimension 2 demonstrated in the (log) birational geometry of surfaces in Chapetrs 1 and 2 via the main strategies MP 1 through MP 4.

  MP 1 Find good representatives in a given birational equivalence class, a minimal model, or a Mori fiber space via the minimal model program (called MMP for short).

The two key ingredients of MMP in dimension 2 are

    cone theorem,
    contraction theorem,

based upon

    rationality theorem,
    base point freeness theorem,
    boundedness of the denominator.

  MP 2 Study the properties of the good representatives.
The basic birational properties of Mori fiber spaces are

    easy dichotomy theorem,
    ruledness,

while those of minimal models are

> hard dichotomy theorem,
> abundance theorem.

MP 3 Study the (birational) relation among the good representatives. Uniqueness of the minimal model (in a given birational equivalence class) is the special feature of the relation among minimal models in dimension 2, while

> Castelnuovo–Noether theorem

describes the relation among Mori fiber spaces by decomposing birational maps into elementary operations called links.

MP 4 Construct the moduli space of the good representatives, fixing some discrete invariants but varying birational equivalence classes.

In Section 3-1 we discuss MP 1 focused on MMP in dimension 3 or higher. Though the basic five theorems as reviewed still hold in higher dimensions, MMP works only with the introduction of the two new features:

> **singularities**,
> **flips**.

In Section 3-2 we briefly discuss the basic properties of Mori fiber spaces in dimension 3 or higher. While the easy dichotomy theorem still provides a coarse description of Mori fiber spaces, in dimension 3 or higher we have to replace the notion of ruledness with that of

> **uniruledness**.

In Section 3-3 we mention the basic properties of minimal models, which are completely parallel to those in dimension 2.

In Section 3-4 we discuss the birational relation among the good representatives. The relation among minimal models and birational maps among them can be best described by

> **finiteness of the minimal models (in a given birational equivalence class) up to isomorphism (still conjectural)**,
> **flops**,

while the relation among Mori fiber spaces is described by

> **Sarkisov program**,

which provides an algorithm to decompose any birational map between two Mori fiber spaces into elementary operations called "**links**." We observe that the notion of "**flops**" and the "Sarkiov program" are most natural when put into the framework of the log category discussed in Chapter 2.

3.1   Minimal Model Program in Dimension 3 or Higher

According to the first major strategy MP 1 of the Mori program, we would like to establish the minimal model program in dimension 3 or higher (or rather a minimal model program that should work in arbitrary dimension), which, given an input of a nonsingular projective variety X, produces a minimal model or a Mori fiber space:

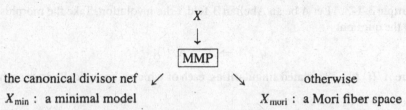

The basic five theorems for MP 1, as presented in dimension 2, hold in the category $\mathfrak{S}$ of *nonsingular* projective varieties. Thus we might hope naively that MMP in dimension 3 or higher would work in the form shown in Flowchart 3-1-1:

Flowchart (a naive hope) 3-1-1.

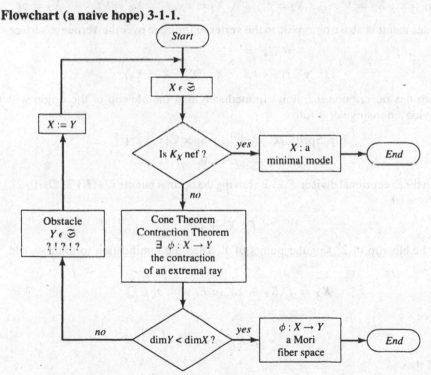

But we face the following fundamental obstacle (cf. Section 2-2).

Obstacle 3-1-2. *Some of the birational contractions of extremal rays from nonsingular projective varieties result in* singular *varieties. There is even an example* *of a nonsingular projective 3-fold X (respectively n-fold) such that the canonical divisor K_X is not nef, and any contraction of an extremal ray $\phi : X \to Y$ with respect to K_X is birational and results in a projective but singular 3-fold (respectively n-fold) Y.*

Therefore, we *inevitably go out of the category $\mathfrak{S}$ in the process of running MMP through the cone theorem and contraction theorem.*

Example 3-1-3. Let A be an Abelian 3-fold, i the involution. Take the morphism onto the quotient

$$q : A \to Y = A/\langle i \rangle,$$

where $A/\langle i \rangle$ has 2^6 isolated singularities, each of which is analytically isomorphic to

$$0 \in \operatorname{Spec} \mathbb{C}[x, y, z]^{(i)}, \text{ where } i : (x, y, z) \to (-x, -y, -z).$$

By setting

$$X_0 = x^2, X_1 = y^2, \qquad X_2 = z^2, \qquad X_3 = xy, \qquad X_4 = yz, \qquad X_5 = zx$$

we see that it is also isomorphic to the vertex of the cone over the Veronese surface

$$\mathbb{P}^2 \hookrightarrow \mathbb{P}^5,$$

$$(x : y : z) \mapsto (X_0 : X_1 : X_2 : X_3 : X_4).$$

From this description it follows immediately that the blowup of the origin will provide a nonsingular 3-fold

$$\operatorname{Blp}_0 \operatorname{Spec} \mathbb{C}[x, y, z]^{(i)} \to \operatorname{Spec} \mathbb{C}[x, y, z]^{(i)},$$

$$E \mapsto 0,$$

with the exceptional divisor $E \cong \mathbb{P}^2$ having the normal bundle $\mathcal{O}_E(E) \cong \mathcal{O}_{\mathbb{P}^2}(-2)$.
    Now let

$$f : X \to Y$$

be the blowup of 2^6 singular points of Y. Then the ramification formula should read

$$K_X = f^* K_Y + \Sigma_{i=1}^{64} a_i E_i \text{ with } a_i \in \mathbb{Q}.$$

Since

$$q^* K_Y = K_A$$

and thus

$$2K_Y = q_* q^* K_Y = q_* K_A \sim 0$$

and since

$$\mathcal{O}_{E_i}(E_i) \cong \mathcal{O}_{\mathbb{P}^2}(-2)$$

and

$$\mathcal{O}_X(K_X + E_i)|_{E_i} \cong \mathcal{O}_{E_i}(K_{E_i}) \cong \mathcal{O}_{\mathbb{P}^2}(-3),$$

we conclude (Exercise! Why?) that

$$K_X = f^* K_Y + \Sigma_{i=1}^{64} \frac{1}{2} E_i = \Sigma_{i=1}^{64} \frac{1}{2} E_i.$$

If a curve l generates an extremal ray, then since l has a negative intersection with $K_X = \Sigma_{i=1}^{64} \frac{1}{2} E_i$ it has to be contained in one of the E_i. On the other hand, for a curve $l_i \subset E_i$ we see that

$$\mathrm{NE}(X) = \mathbb{R}_+[l_i] + B_i,$$

where B_i is the convex cone generated by curves having nonnegative intersections with E_i, and that

$$\overline{\mathrm{NE}}(X) = \mathbb{R}_+[l_i] + \overline{B_i}.$$

This shows that $\mathbb{R}_+[l_i]$ is an extremal ray with respect to K_X. Therefore, the extremal rays on X with respect to K_X are the 2^6 half-lines

$$\mathbb{R}_+[l_i] \subset \overline{\mathrm{NE}}(X), \text{ where } l_i \subset E_i,$$

and the contraction of each results in a 3-fold with a singular point, analytically isomorphic to the vertex of a cone over the Veronese surface.

Solution 3-1-4 to Obstacle 3-1-2 (Inevitable Introduction of Singularities). *As demonstrated in Obstacle 3-1-2 and Example 3-1-3, it is impossible to stay in the category $\mathfrak{S}$ of nonsingular projective varieties for MMP in higher dimensions (≥ 3) to work. Therefore, we introduce the new category $\mathfrak{C}$ of normal projective varieties with only $\mathbb{Q}$ factorial and terminal singularities.*

Definition 3-1-5 (Terminal Singularities). *A normal variety X of dimension n has only* **terminal singularities** *if*

(1) *K_X is $\mathbb{Q}$-Cartier, i.e., there exists $e \in \mathbb{N}$ such that eK_X is a Cartier divisor,*
(2) *there exists a projective birational morphism $f : V \to X$ from a nonsingular variety such that in the ramification formula*

$$K_V = f^* K_X + \Sigma a_j E_j$$

all the coefficients for the exceptional divisors are strictly positive:

$$a_j > 0 \forall E_j \text{ exceptional.}$$

We also need the notion of $\mathbb{Q}$-factoriality.

Definition 3-1-6 ($\mathbb{Q}$-Factoriality). *A normal variety X has only* $\mathbb{Q}$**-factorial singularities** *if every Weil divisor D on X is $\mathbb{Q}$-Cartier, i.e., there exists $d \in \mathbb{N}$ such that dD is Cartier.*

At this point, the definition of terminal singularities or $\mathbb{Q}$-factoriality may look heuristic. The following observation supports our choice of the category $\mathfrak{C}$. (A further justification of why this category $\mathfrak{C}$ is the "right" choice for our MMP will be given in Theorem 3-1-16.)

Observation 3-1-7.   *The category $\mathfrak{C}$ enjoys the following three properties:*

(i) *The basic five theorems for MMP,*

> *cone theorem,*
> *contraction theorem,*
> *rationality theorem,*
> *base point freeness theorem,*
> *boundedness of the denominator,*

hold in the category $\mathfrak{C}$.

(ii) *The key criterion "Is K_X nef?" for MMP makes perfect sense for any object $X \in \mathfrak{C}$, since the canonical divisor K_X is $\mathbb{Q}$-Cartier, and hence the intersection number $K_X \cdot C$ is well-defined for any curve $C \subset X$.*

(iii) *The category $\mathfrak{C}$ behaves well under the extremal birational contraction: Let*

$$\phi : X \to Y$$

be the birational contraction of an extremal ray with respect to K_X from

$$X \in \mathfrak{C}.$$

If

$$\operatorname{codim}_X \operatorname{Exc}(\phi) = 1$$

*(in this case we call ϕ a "**divisorial**" contraction (or a contraction **of divisorial type**), since the exceptional locus turns out to consist of a prime divisor), then*

$$Y \in \mathfrak{C}.$$

If

$$\operatorname{codim}_X \operatorname{Exc}(\phi) \geq 2$$

*(in this case we call ϕ a "**small**" contraction, since the exceptional locus $\operatorname{Exc}(\phi)$ of ϕ has higher codimension and hence is small), then the canonical divisor K_Y is not $\mathbb{Q}$-Cartier, and hence*

$$Y \notin \mathfrak{C}.$$

The detailed verification of the assertions in (i), (ii), and (iii) above is one of the main themes of the later chapters. Here we only present the easy reason why $Y \notin \mathfrak{C}$ for a small extremal contraction $\phi : X \to Y$: Suppose K_Y is $\mathbb{Q}$-Cartier. Then $K_X = \phi^* K_Y$, since ϕ is small and hence induces an isomorphism of X and Y

outside of a codimension-2 (or more) locus Exc (ϕ). But then $K_X \cdot l = \phi^* K_Y \cdot l = 0$ for a curve l contracted by ϕ, whereas $K_X \cdot l < 0$, since ϕ is the contraction of an extremal ray with respect to K_X, a contradiction!

So we replace our old category $\mathfrak{S}$ with a larger one $\mathfrak{C}$. In this new category $\mathfrak{C}$, as long as our extremal contraction is birational and divisorial, the resulting variety will stay in the category $\mathfrak{C}$, and hence we can proceed with the program. Recall that as soon as the extremal contraction is not birational, we obtain a Mori fiber space by definition and come to an end of the program. But we still have to face one more obstacle as below.

Obstacle 3-1-8 (Small Contractions). *Some of the birational contractions ϕ : $X \to Y$ of extremal rays from the objects $X \in \mathfrak{C}$ are* small *and hence result in the objects $Y \notin \mathfrak{C}$. There is even an example of $X \in \mathfrak{C}$ such that the canonical divisor K_X is not nef, and any contraction $\psi : X \to Y$ of an extremal ray with respect to K_X is birational and* small.

(We remark that there is no small contraction of an extremal ray with respect to K_X on a nonsingular projective 3-fold and that hence if a small contraction occurs, it has to be on a singular 3-fold.)

Example 3-1-9 (Francia [1], Hironaka [1]) (cf. Toric Examples in Example-Claim 14-2-5). Take a nonsingular projective 3-fold Z such that K_Z is nef and such that it contains three distinct smooth divisors D_1, D_2, D_3 meeting transversally with each other, so that

$$C_1 = D_1 \cap D_2,$$
$$C_2 = D_2 \cap D_3,$$
$$C_3 = D_3 \cap D_1,$$

are curves of genus $g(C_1), g(C_2), g(C_3) > 0$ meeting at a point $p \in Z$. (Exercise! Find such Z and three divisors D_1, D_2, D_3 on it.) Blow up $p \in Z$. Let $L_1', L_2',$ and L_3' be the lines joining two of the three intersection points of the C_i' with P', where the C_i' are the strict transforms of the curves C_i and $P' \cong \mathbb{P}^2$ is the exceptional divisor. Then blow up $C_1', C_2',$ and C_3' to obtain the exceptional divisors $\hat{E}_1, \hat{E}_2,$ and $\hat{E}_3$, respectively. Now the strict transforms $\hat{L}_i \cong \mathbb{P}^1$ of the lines L_i' have the normal bundle $\mathcal{O}_{\mathbb{P}^1}(-1) \oplus \mathcal{O}_{\mathbb{P}^1}(-1)$. Thus if we blow them up, the exceptional divisors $G_i \cong \mathbb{P}^1 \times \mathbb{P}^1$ have the normal bundle $\mathcal{O}_{G_i}(G_i) \cong \mathcal{O}_{\mathbb{P}^1 \times \mathbb{P}^1}(-1, -1)$ and can be contracted in the "other" direction. Now the strict transform $P \cong \mathbb{P}^2$ of the first exceptional divisor has the normal bundle $\mathcal{O}_E(E) \cong \mathcal{O}_{\mathbb{P}^2}(-2)$ to be contracted to obtain

$$f : X \to Z.$$

Note that X has only $\mathbb{Q}$-factorial and terminal singularities and hence $X \in \mathfrak{C}$. (Exercise! Why?)

We illustrate the operations to obtain X from Z in Figure 3-1-10.

Figure 3-1-10.

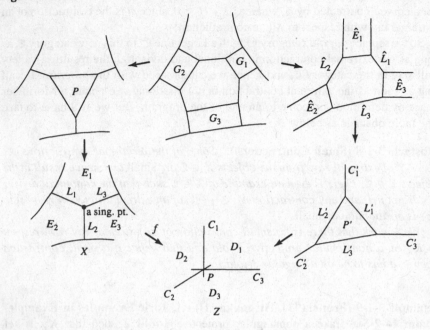

The ramification formula reads

$$K_X = f^* K_Z + E_1 + E_2 + E_3,$$

where E_1, E_2, E_3 are the strict transforms of $\hat{E}_1$, $\hat{E}_2$, $\hat{E}_3$. If a curve l on X generates an extremal ray R_l, then since, l has a negative intersection with K_X, it has to be contained in E_1, E_2, or E_3. Now, R_l being an extremal ray of $\overline{NE}(X)$, it is the image of an extremal ray of $\overline{NE}(E_i)$ under the natural map $i_* : \overline{NE}(E_i) \to \overline{NE}(X)$. It is easy to see that the extremal rays for E_i are generated by the rational curves l_{i-1} and l_i ($l_0 = l_3$). On the other hand, a direct computation shows that

$$K_X \cdot l_1 = E_1 \cdot l_1 = E_2 \cdot l_1 = -\frac{1}{2}, \qquad E_3 \cdot l_1 = \frac{1}{2},$$
$$K_X \cdot l_2 = E_2 \cdot l_2 = E_3 \cdot l_2 = -\frac{1}{2}, \qquad E_1 \cdot l_1 = \frac{1}{2},$$
$$K_X \cdot l_3 = E_3 \cdot l_3 = E_1 \cdot l_3 = -\frac{1}{2}, \qquad E_2 \cdot l_1 = \frac{1}{2}.$$

Therefore,

$$NE(X) = \mathbb{R}_+[l_i] + B_i,$$

where B_i is the convex cone generated by curves having nonnegative intersections either with E_i or E_{i+1} (We use the convention that $E_{3+1} = E_1$), and that

$$\overline{NE}(X) = \mathbb{R}_+[l_i] + \overline{B_i}.$$

This shows that $\mathbb{R}_+[l_1]$, $\mathbb{R}_+[l_2]$, $\mathbb{R}_+[l_3]$ are indeed extremal rays of $\overline{NE}(X)$ with respect to K_X.

Since the rational curve l_i is the only curve whose class belongs to $\mathbb{R}_+[l_i]$ ($i = 1, 2, 3$), the contraction of each extremal ray is birational and small.

At first sight, Obstacle 3-1-8, as demonstrated by Example 3-1-9, seems fatal to our strategy of replacing the category $\mathfrak{S}$ with the new category $\mathfrak{C}$ in order to establish MMP in higher dimensions, since it seems that we would still be forced out of the category! whenever we face a small extremal contraction.

Mori's brilliant solution to rescue us from this possibly fatal situation is the introduction of a new (and still conjectural in dimension ≥ 4) operation called "**flip**," which brings us right back into the category $\mathfrak{C}$ even when we are thrown out of it by a small contraction.

Solution 3-1-11 to Obstacle 3-1-8 (Existence of "Flip"). *Let*

$$\phi : X \to Y$$

be a small *contraction of an extremal ray with espect to K_X from an object $X \in \mathfrak{C}$. Then there exists a commutative diagram*

$$
\begin{array}{ccc}
X & \overset{\not\cong}{\dashrightarrow} & X^+ \\
\phi \searrow & & \swarrow \phi^+ \\
& Y &
\end{array}
$$

where ϕ^+ is another small morphism from

$$X^+ \in \mathfrak{C}.$$

*The morphism $\phi^+ : X^+ \to Y$ is called the "**flip**" of ϕ. (The word "flip" also refers to the commutative diagram above or to the object X^+ by abuse of language.)*

Whenever we have a small contraction $\phi : X \to Y$ of an extremal ray, we proceed in the program with X^+, which is in our category $\mathfrak{C}$, instead of Y, which is not in the category $\mathfrak{C}$.

It is worth studying the first few basic properties of a flip (assuming its existence). It is a consequence of the contraction theorem that

$$\mathrm{Div}(X) \otimes \mathbb{Q} = \phi^* \mathrm{Div}(Y) \otimes \mathbb{Q} \oplus \mathbb{Q} \cdot K_X,$$

where $\mathrm{Div}(X)$ refers to the group of Cartier divisors on X.

Then since ϕ^+ is small, i.e., $\mathrm{codim}(\phi^+) \geq 2$, we also have

$$\mathrm{Div}(X^+) \otimes \mathbb{Q} = \phi^{+*} \mathrm{Div}(Y) \otimes \mathbb{Q} \oplus \mathbb{Q} \cdot K_{X^+}.$$

Since $X^+ \in \mathfrak{C}$, $Y \notin \mathfrak{C}$ and hence ϕ^+ is projective and not an isomorphism, there has to be a ϕ^+-ample divisor on X^+ that is not ϕ^+-trivial. It has to be the divisor K_{X^+}! (Exercise! Show why it is not $-K_{X^+}$ that is supposed to be ϕ^+-ample.)

Therefore, we conclude that

$$X^+ = Proj \oplus_{m \geq 0} \phi^+_* \mathcal{O}_{X^+}(mK_{X^+})$$
$$= Proj \oplus_{m \geq 0} \phi_* \mathcal{O}_X(mK_X),$$

obtaining the description of $\phi^+ : X^+ \to Y$ by the relative canonical ring of X with respect to ϕ and hence showing the uniqueness of the flip (if it exists).

Example 3-1-12. We demonstrate how to "flip" a small contraction in the example of Francia and Hironaka. Say, we take the small contraction $\phi = cont_{l_1}$ of the extremal ray generated by l_1. First we blow up the singular point of X to obtain again the exceptional divisor $E \cong \mathbb{P}^2$. Then blow up the strict transform $l_1' \cong \mathbb{P}^1$ of l_1 with normal bundle $\mathcal{O}_{\mathbb{P}^1}(-1) \oplus \mathcal{O}_{\mathbb{P}^1}(-1)$. The exceptional locus $D_1 \cong \mathbb{P}^1 \times \mathbb{P}^1$ has normal bundle $\mathcal{O}_{D_1}(D_1) \cong \mathcal{O}_{\mathbb{P}^1 \times \mathbb{P}^1}(-1, -1)$, so that it can be contracted in the other direction. Now, the strict transform of E is isomorphic to $\mathbb{F}_1$, which can be contracted in the direction of the ruling to obtain $\phi^+ : X^+ \to Y$.

We illustrate the construction of the flip in Figure 3-1-13.

Figure 3-1-13.

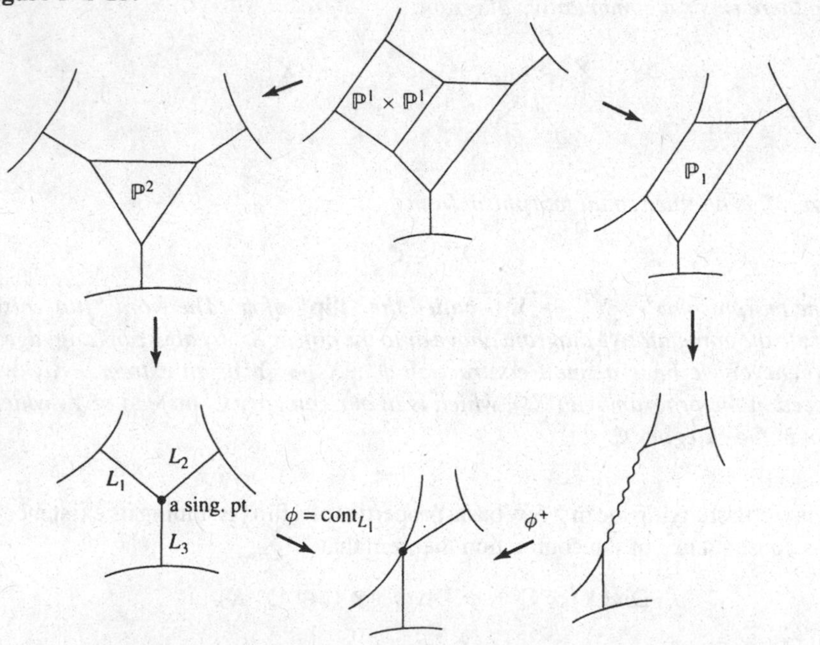

Now we have gathered all the necessary ingredients to present the flowchart for MMP in higher dimensions in Flowchart 3-1-15, except for the last step, which guarantees that this flowchart does not contain an infinite loop. This can be achieved via the study of the behavior of the Picard number $\rho(X)$ (cf. Observation 3-1-7) and by the following.

Solution 3-1-14 (Termination of Flips). *There is* no *infinite sequence of flips*

$$X = X_0 \dashrightarrow X_0^+ = X_1 \dashrightarrow X_1^+ = X_2 \dashrightarrow \quad \cdots \quad X_{i-1}^+ = X_i \dashrightarrow X_i^+ = X_{i+1} \cdots$$

$$Y = Y_0 \qquad\qquad Y_1 \qquad\qquad Y_2 \qquad \cdots \qquad Y_i \quad \cdots$$

Flowchart 3-1-15.

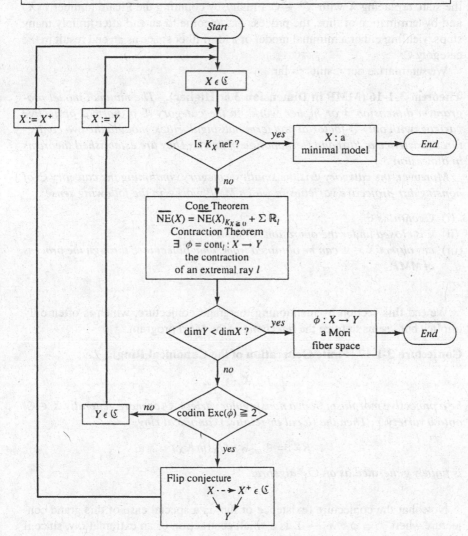

We start with $X \in \mathfrak{C}$. The first question to ask is whether the canonical divisor K_X is nef, which makes sense, since K_X is $\mathbb{Q}$-Cartier. If the answer is *yes*, then

X is a minimal model in the category $\mathfrak{C}$ by definition and we come to an end. If the answer is *no*, then we apply the cone theorem to find an extremal ray and by the contraction theorem its contraction $\phi : X \to Y$. Then the next question to ask is whether $\dim Y < \dim X$. If the answer is *yes*, then $\phi : X \to Y$ is a Mori fiber space in the category $\mathfrak{C}$ by definition and we come to another end. If the answer is *no*, then $\phi : X \to Y$ is necessarily a birational morphism, and we proceed to the next question, asking whether $\mathrm{codimExc}(\phi) = 1$. If the answer is *yes*, then ϕ is divisorial and $Y \in \mathfrak{C}$. We go back to the start replacing X with Y. If the answer is *no*, then ϕ is small and the flip $\phi^+ : X^+ \to Y$ exists. We go back to the start, replacing X with $X^+ \in \mathfrak{C}$. Finally, by counting the Picard number $\rho(X)$ and by termination of flips, the process has to come to an end after finitely many steps, yielding either a minimal model or a Mori fiber space as an end result in the category $\mathfrak{C}$.

We summarize our results so far.

Theorem 3-1-16 (MMP in Dimension 3 or Higher). *The minimal model program in dimension 3 or higher works in the category $\mathfrak{C}$ of normal projective varieties with only $\mathbb{Q}$-factorial and terminal singularities, modulo the two conjectures (existence of flip) and (termination of flips). (They are established theorems in dimension 3.)*

Moreover, the category $\mathfrak{C}$ is the smallest category containing the category $\mathfrak{S}$ of nonsingular projective varieties in which MMP works, in the following sense:

(i) *$\mathfrak{C}$ contains $\mathfrak{S}$,*
(ii) *$\mathfrak{C}$ is closed under the operations of MMP,*
(iii) *any object $X \in \mathfrak{C}$ can be obtained from some object in $\mathfrak{C}$ through the process of MMP.*

We end this section by mentioning the grand conjecture, which is often only implicit but seems to form the *backbone* of the Mori program.

Conjecture 3-1-17 (Finite Generation of the Canonical Ring). *Let*

$$f : X \to Y$$

be a projective morphism from a nonsingular variety X (or more generally $X \in \mathfrak{C}$) onto a variety Y. Then the (local or relative) canonical ring

$$R_Y := \oplus_{m \geq 0} f_* \mathcal{O}_X(m K_X)$$

is finitely generated as an $\mathcal{O}_Y$-algebra.

Note that the conjecture (existence of flip) is a special case of this grand conjecture when $f = \phi : X \to Y$ is a small contraction of an extremal ray, since if Conjecture 3-1-17 holds, then we can set

$$X^+ = \mathrm{Proj} R_Y.$$

(The verification that $X^+ = \operatorname{Proj} R_Y \in \mathfrak{C}$ and that $\phi^+ : \operatorname{Proj} R_Y \to Y$ is small will be given in Chapter 9.)

Note also that the conjecture of the finite generation of the (global or usual) canonical ring

$$R = \oplus_{m \geq 0} H^0(X, \mathcal{O}_X(mK_X))$$

is also a special case of the grand conjecture when $f : X \to Y = \operatorname{Spec} \mathbb{C}$ is the structure morphism over the base field $\mathbb{C}$. The latter leads to the existence of the canonical model (cf. Section 1-6)

$$X_{\mathrm{can}} = \operatorname{Proj} R.$$

(Singularities of the canonical models will be discussed in Section 4-2.)

Remark 3-1-18. The minimal model program can be regarded as a method to prove the finite generation of the canonical ring of a variety X (of general type): If we could find a minimal model X_{min} through MMP, then the canonical divisor $K_{X_{\mathrm{min}}}$ is nef (and big) by construction, from which the finite generation follows immediately via the use of the base point freeness theorem (cf. Section 6-2). Since the key of the mechanism of MMP lies in the existence of flip, which is equivalent to the finite generation of $\oplus_{m \geq 0} \phi_*(mK_X)$ for a small contraction $\phi : X \to Y$, it can be said that we reduce the problem of showing the finite generation of the *global* (or usual) canonical ring to that of the *local* (or relative) canonical ring.

3.2 Basic Properties of Mori Fiber Spaces in Dimension 3 or Higher

Once the minimal model program is established via (existence of flip) and (termination of flips) along with the basic five theorems, we have minimal models or Mori fiber spaces as its end results. The purpose of this section is to browse through the basic properties of **Mori fiber spaces**.

As in the previous section, $\mathfrak{C}$ denotes the category of normal projective varieties with only $\mathbb{Q}$-factorial and terminal singularities.

Definition 3-2-1 (Mori Fiber Space). *A normal projective variety X with only $\mathbb{Q}$-factorial and terminal singularities, i.e., $X \in \mathfrak{C}$, with a morphism $\phi : X \to Y$ is a **Mori fiber space** if it is an extremal contraction with $\dim Y < \dim X$, or equivalently if*

(i) *ϕ is a morphism with connected fibers onto a normal projective variety Y of $\dim Y < \dim X$, and*

(ii) *all the curves C in fibers of ϕ are numerically proportional and $K_X \cdot C < 0$.*

Remark 3-2-2. (i) Recall as a consequence of the cone theorem and contraction theorem that $\phi : X \to Y$ is a Mori fiber space iff it is the contraction of an extremal ray (with respect to K_X) $\phi = \operatorname{cont}_{R_l} : X \to Y$ on $X \in \mathfrak{C}$ with $\dim Y < \dim X$.

(ii) Assume that MMP in dimension n holds via (existence of flip) and (termination of flips). Then it produces either a minimal model X_{min}, an object $X_{min} \in \mathfrak{C}$ with nef canonical divisor $K_{X_{min}}$ by definition, or a Mori fiber space X_{mori} with an extremal contraction $\phi : X_{mori} \to Y$ with dim $Y <$ dim X. Conversely, a minimal model or a Mori fiber space is an end result of MMP in dimension n.

First note the following easy but fundamental property of Mori fiber spaces.

Theorem 3-2-3. *Let $\phi : X \to Y$ be a Mori fiber space. Then for any $p \in X$ there exists an irreducible curve D passing through p with $K_X \cdot D < 0$, and thus*

$$H^0(X, \mathcal{O}_X(mK_X)) = 0 \quad \forall m \in \mathbb{N} \quad (i.e., \kappa(X) = -\infty).$$

The easy dichotomy theorem also holds in arbitrary dimension.

Theorem 3-2-4 (Easy Dichotomy Theorem of MMP). *Assume that MMP in dimension n holds via (existence of flip) and (termination of flips). Then an end result of MMP starting from $X \in \mathfrak{C}$ of dimension n is a Mori fiber space iff there exists a nonempty Zariski open set $U \subset X$ such that for any point $p \in U$ there is an irreducible curve D passing through p and $K_X \cdot D < 0$.*

A more rigid description of the structure of a Mori fiber space in arbitrary dimension can be given by producing rational curves through Mori's "bend and break" technique (cf. Chapter 10): A Mori fiber space is covered by rational curves. This leads to the description of Mori fiber spaces in dimension 2 as "ruled" surfaces, i.e., surfaces birational to $C \times \mathbb{P}^1$, but in dimension 3 as "uniruled" varieties, i.e., varieties dominated by $W \times \mathbb{P}^1$ by a generically finite map. Though the difference between the two notions of "ruled" and "uniruled" may seem minor at first, especially when there is no surface that is uniruled but not ruled, it plays a crucial and subtle role in the classification of Mori fiber spaces, surrounding the question of rationality.

Definition 3-2-5 (Ruled and Uniruled). *A variety X of dimension n is "**ruled**" (respectively "**uniruled**") if there exists a birational (respectively generically finite) map*

$$W \times \mathbb{P}^1 \dashrightarrow X,$$

where W is a variety of dimension $n - 1$.

Theorem 3-2-6 (Uniruledness of Mori Fiber Spaces) (cf. Miyaoka–Mori [1]). *Let $X \in \mathfrak{C}$ be a Mori fiber space of dimension n. Then X is covered by rational curves, i.e., for any point $p \in X$ there exists a rational curve C passing through p. (For a general point $p \in X$ a rational curve C can be taken such that $K_X \cdot C < 0$.) In particular, X is uniruled.*

Exercise 3-2-7. Show that if X is uniruled, then for any point $p \in X$ there exists a rational curve C passing through p, and for a general point $p \in X$ a rational curve C can be taken such that $K_X \cdot C < 0$.

Ruled vs. Uniruled

Though we will postpone the proof of Theorem 3-2-6 until the introduction of Mori's "bend and break" technique in Chapter 10, here we discuss briefly the subtle but essential difference between the two notions "ruled" and "uniruled," and why we need to introduce the latter in dimension 3 or higher.

This part is intended *only* for the highly motivated reader who dares the technical detail of the subtle arguments. We advise the general reader to skip this part for his first reading of this section.

For simplicity, we restrict ourselves to dimension 3 in the following discussion. Let $\phi : X \to Y$ be a Mori fiber space in dimension 3. We analyze the situation case by case according to the dimension of Y.

Case: $\dim Y = \dim X - 1 = 2$

We take a Zariski open set $U \subset Y$ over which ϕ is smooth and thus

$$\phi : X_U = \phi^{-1}(U) \to U$$

is a $\mathbb{P}^1$-bundle.

Question 3-2-8. *When is $\phi : X_U \to U$ a trivial $\mathbb{P}^1$-bundle (by shrinking U if necessary), i.e., we have a commutative diagram as below*

$$
\begin{array}{ccc}
X_U & \xrightarrow{\sim} & U \times \mathbb{P}^1 \\
\phi \downarrow & & \downarrow p_1 \\
U & = & U,
\end{array}
$$

or equivalently when is $\phi : X \to Y$ birationally trivial, i.e., we have a commutative diagram of the form

$$
\begin{array}{ccc}
X & \xdashrightarrow{\sim}_{\text{birational}} & Y \times \mathbb{P}^1 \\
\phi \downarrow & & \downarrow p_1 \\
Y & = & Y
\end{array}
$$

It is easy to see that $\phi : X_U \to U$ is a conic bundle

$$X_U = Proj \oplus_{m \geq 0} \phi_*(-mK_X) \hookrightarrow Proj \oplus_{m \geq 0} Sym^m \phi_*(-K_X),$$

since fiberwise this gives an embedding of $\mathbb{P}^1$ as a conic in $\mathbb{P}^2$:

$$\mathbb{P}^1 \hookrightarrow \mathbb{P}^2 = \mathbb{P}(H^0(\mathbb{P}^1, \mathcal{O}_{\mathbb{P}^1}(-K_{\mathbb{P}^1}))).$$

By shrinking U again if necessary, we can describe X_U as the zero locus of a nondegenerate quadratic form $f(X_0, X_1, X_2)$ with coefficients in $\Gamma(U, \mathcal{O}_U) \in \mathbb{C}(Y)$:

$$X_U = \{f(X_0, X_1, X_2) = 0\} \hookrightarrow Proj\mathcal{O}_U[X_0, X_1, X_2].$$

It is also easy to see that now the birational triviality of $\phi : X \to Y$ is equivalent to whether $f(X_0, X_1, X_2) = 0$ has a solution in $\Gamma(U, \mathcal{O}_U)$, i.e., whether this conic bundle has a section (again shrinking U if necessary). In fact, if ϕ is birationally trivial, then it obviously has a section over some open set $U \subset Y$. Conversely, suppose it has a section over some $U \subset Y$. Then regarding $X_U \subset U \times \mathbb{P}^2$ as a conic bundle, take a projection from the section to $U \times$ (a hyperplane in $\mathbb{P}^2$). This becomes a morphism (as long as ϕ is smooth over U) and actually is an isomorphism. Thus $\phi : X_U \to U$ is a trivial $\mathbb{P}^1$-bundle.

Conclusion 3-2-9. $\phi : X \to Y$ *is birationally trivial if and only if the associated nondegenerate quadratic form*

$$f(X_0, X_1, X_2) = 0$$

with coefficients in $\mathbb{C}(Y)$ has a solution in $\mathbb{C}(Y)$.

We recall that in the case where $\dim X = 2$ and thus $\dim Y = \dim X - 1 = 1$ any nondegenerate quadratic form has a solution and thus has a section (**Tsen's lemma**). Therefore, every $\mathbb{P}^1$-bundle is locally trivial. The point of Beauville's proof for Proposition 1-4-5 is to find a section by a topological argument without directly referring to Tsen's lemma.

Here, however, we present an elegant proof of Tsen's lemma that I learned from Reid [9].

Theorem 3-2-10 (Tsen's Lemma). *Let*

$$f(X_0, X_1, X_2) = \Sigma_{i+j+k=2} f_{ijk} X_0^i X_1^j X_2^k$$

be a quadratic form with coefficients in the function field $\mathbb{C}(Y)$ of a (nonsingular projective) curve Y over $\mathbb{C}$. Then $f = 0$ has a (nontrivial) solution for X_0, X_1, X_2 in $\mathbb{C}(Y)$.

PROOF.   We take a very ample divisor H on Y and let

$$\mathbb{C}[Y] = \oplus_{m \geq 0} H^0(Y, \mathcal{O}_Y(mH)) = \oplus_{m \geq 0} \mathbb{C}[Y]_m$$

be the homogeneous coordinate ring of Y with the natural grading. Since

$$\mathbb{C}(Y) = \left\{ \frac{a}{b}; a, b \in \mathbb{C}[Y]_m \text{ for some } m \in \mathbb{N}, b \neq 0 \right\},$$

we may assume that all the coefficients of the quadratic form f belong to $\mathbb{C}[Y]_l$ for some $l \in \mathbb{N}$. We want to find elements (not all zero)

$$X_0, X_1, X_2 \in \mathbb{C}[Y]_m$$

such that

$$f(X_0, X_1, X_2) = 0 \in \mathbb{C}[Y]_{2m+l}.$$

By the Riemann–Roch theorem the number of free variables for X_0, which is the dimension of $\mathbb{C}[Y]_m$, can be computed ($m \gg 0$) as

$$h^0(Y, \mathcal{O}_Y(mH)) = \chi(\mathcal{O}_Y(mH)) = 1 - g(Y) + m \cdot \deg H.$$

As we have two more variables X_1, X_2, the total number of free variables is

$$3 \cdot \{1 - g(Y) + m \cdot \deg H\}.$$

On the other hand, when we plug these variables into $f(X_0, X_1, X_2) = 0$ in $\mathbb{C}[Y]_{2m+l}$, the number of homogeneous polynomials we get as the condition is the dimension of $\mathbb{C}[Y]_{2m+l}$, which can again be computed by the Riemann–Roch theorem:

$$h^0(Y, \mathcal{O}_Y((2m+l)H)) = 1 - g(Y) + (2m+l) \cdot \deg H.$$

For $m \gg 0$, since $3 > 2$, we have

$$3m \cdot \deg H \sim 3 \cdot (1 - g(Y) + m \cdot \deg H) > 1 - g(Y) + (2m+l) \cdot \deg H \sim 2m \cdot \deg H,$$

that is to say, the number of free variables exceeds the number of conditions (homogeneous polynomials) imposed on them. Since $\mathbb{C}$ is algebraically closed, this implies that we have a nontrivial solution for the variables.

This completes the proof of Tsen's lemma.   □

It is worthwhile to note that the above proof fails when $\dim Y \geq 2$ (since the assertion is not true), since, e.g., the number of free variables is of order $3m^2 H^2$, whereas that of the conditions is of order $2^2 m^2 H^2$ in the case $\dim Y = 2$.

Now we go back to the case when $\dim X = 3$ and thus $\dim Y = \dim X - 1 = 2$, and the landscape changes drastically. We no longer have Tsen's lemma, and actually, we can create a nondegenerate quadatic form $f(X_0, X_1, X_2) = 0$ with coefficients in $\mathbb{C}(Y)$ of a function field of transcendental degree 2 over $\mathbb{C}$ such that it has no solution in $\mathbb{C}(Y)$. (See Artin–Mumford [1] and the definition of a **Brauer group**.) The corresponding conic bundle is not birationally trivial.

For our purpose, we choose a nonsingular projective surface A with no rational curves, e.g., an Abelian surface, and a nondegenerate quadratic form f over $\mathbb{C}(A)$ that has no solution in $\mathbb{C}(A)$. (Exercise! Show that there exists such A with a nondegenerate quadratic form f as described.) We also take a corresponding Mori fiber space $\phi : X \to A$, which does not give a trivial $\mathbb{P}^1$-bundle no matter how much we shrink A. (In order to find such a Mori fiber space, first take the closure W

$$W = \overline{\{f(X_0, X_1, X_0) = 0\}} \subset A \times \mathbb{P}^2 = Proj \mathcal{O}_A[X_0, X_1, X_2].$$

Then take its desingularization $V \to W \to A$. Finally, apply MMP in dimension 3 (over A) to obtain a desired Mori fiber space $\phi : X \to A$.)

We claim that X is not "ruled." In fact, suppose there exists a birational map

$$W \times \mathbb{P}^1 \dashrightarrow X,$$

where W is a nonsingular projective surface. Observe that for a general point $p \in W$, since A has no rational curve, $\{p\} \times \mathbb{P}^1$ maps isomorphically onto a fiber of ϕ by the birational map. Therefore, the image of $W \times \{q\}$ for some fixed $q \in \mathbb{P}^1$ gives a (birational) section of ϕ (as a consequence, we conclude that W must be birational to A), which violates our choice at the beginning.

Note that X is "uniruled," since it is a conic bundle over some base $U \subset A$. In fact, to create a section over U (again shrinking U if necessary) we have only to take an appropriate degree-2 cover $d : \tilde{A} \to A$ such that the corresponding quadratic from has a solution in $\mathbb{C}(\tilde{A})$. Then

$$
\begin{array}{ccccc}
X & \longleftarrow & X \times_A \tilde{A} & \overset{\text{birational}}{\dashleftarrow} & \tilde{A} \times \mathbb{P}^1 \\
\phi\downarrow & & \downarrow & & \downarrow \\
A & \overset{d}{\longleftarrow} & \tilde{A} & = & \tilde{A}
\end{array}
$$

Another example of a nonruled but uniruled 3-fold can be given using a general cubic hypersurface $W_3 \subset \mathbb{P}^4$. Blow up W_3 along a line $L \subset W_3 \subset \mathbb{P}^4$:

$$
\begin{array}{ccccc}
X & \subset & \mathrm{Blp}_L \mathbb{P}^4 & \longrightarrow & \mathbb{P}^2 \\
\downarrow & & \downarrow & & \\
W_3 & \subset & \mathbb{P}^4 & &
\end{array}
$$

(Exercise! Show that such a line L exists.) Then the induced map $\phi : X \to \mathbb{P}^2 = Y$ is a Mori fiber space. It is the celebrated result of Clemens–Griffiths [1] that W_3 is not rational (i.e., not birational to $\mathbb{P}^3$) and hence neither is X. Therefore, ϕ cannot be birationally trivial. Since X is a conic bundle over Y, it is uniruled. However, X cannot be ruled. In fact, if there exists a birational map $W \times \mathbb{P}^1 \dashrightarrow X$, then W must be birational to $\mathbb{P}^2$, since X dominates it (Exercise! Why?), which would imply that X is rational, a contradiction. (See the arguments for Subcase $q(X) = 0$ in Case $\dim Y = \dim X - 2 = 1$.)

> Case: $\dim Y = \dim X - 2 = 1$

In this case, a general fiber S of ϕ is a nonsingular projective surface with an ample $-K_S$, and hence called a Del Pezzo surface. As $H^1(S, \mathcal{O}_S) = 0$ by the Kodaira vanishing theorem and obviously $H^0(S, \mathcal{O}_S(2K_S)) = 0$, Castelnuovo's criterion for rationality (cf. Corollary 1-4-9) tells us that S is birational to $\mathbb{P}^2$.

Question 3-2-11. *When is a Mori fiber space* $\phi : X \to Y$ *birationally trivial, i.e., there exists a commutative diagram of birational maps*

$$
\begin{array}{ccc}
X & \overset{\sim}{\underset{\text{birational}}{\dashrightarrow}} & Y \times \mathbb{P}^2 \\
\phi\downarrow & & \downarrow{p_1} \quad ? \\
Y & = & Y
\end{array}
$$

Note that

$$R^i \phi_* \mathcal{O}_X = R^i \phi_* \mathcal{O}_X(-K_X + K_X) = 0 \text{ for } i > 0,$$

by the (relative version of the) Kodaira vanishing theorem, and hence

$$q(Y) = h^1(Y, \mathcal{O}_Y) = h^1(Y, \phi_* \mathcal{O}_X) = h^1(X, \mathcal{O}_X) = q(X).$$

Subcase: $q(X) > 0$.

Exercise 3-2-12. Show in this subcase of $q(X) > 0$ that $\phi : X \to Y$ with $\dim Y = \dim X - 2 = 1$ is birationally trivial if and only if X is ruled.

Take a nonsingular projective curve Y with $g(Y) = q(Y) > 0$ with a very ample divisor A. Consider the product with the projective space $\mathbb{P}^3 \times Y$ and a general member

$$X_a \in |p_1^* \mathcal{O}_{\mathbb{P}^3}(a) \otimes p_2^* \mathcal{O}_Y(A)|, \quad a = 2, 3,$$

where p_1 and p_2 are the projections onto the first and second factors, respectively.
  We claim that

$$\phi = p_2|_{X_a} : X_a \to Y$$

is a Mori fiber space.
  In fact, X_a is a nonsingular projective 3-fold by Bertini's theorem.
  Secondly,

$$-K_{X_a}|_{\phi^{-1}(p)} \equiv \mathcal{O}_{\mathbb{P}^3}(-a + 4)|_{\phi^{-1}(p)} \text{ is ample } \forall p \in Y,$$

and hence $-K_{X_a}$ is ϕ-ample.
  Thirdly, we have

$$H^1(\mathbb{P}^3 \times Y, \mathcal{O}_{\mathbb{P}^3 \times Y}) \cong H^1(Y, \mathcal{O}_Y) \cong H^1(X_a, \mathcal{O}_{X_a})$$

and

$$H^2(\mathbb{P}^3 \times Y, \mathbb{Z}) \cong H^2(X_a, \mathbb{Z})$$

by the Lefschetz hyperplane section theorem. (Note that though the Lefschetz hyperplane section theorem is usually stated for the cohomology groups with $\mathbb{Q}$-coefficients (cf., e.g., Griffiths–Harris [1]), it also holds with $\mathbb{Z}$-coefficients.)
  By chasing the cohomology sequences associated to the exponential sequences on $\mathbb{P}^3 \times Y$ and on Y, we conclude that

$$\mathrm{Pic}(\mathbb{P}^3 \times Y) \cong \mathrm{Pic}(X_a).$$

Therefore, since $\rho(\mathbb{P}^3 \times Y/Y) = 1$, we also conclude that

$$\rho(X_a/Y) = 1.$$

This shows that $\phi : X_a \to Y$ is a Mori fiber space.
  When $a = 2$, the Mori fiber space $\phi : X_2 \to Y$ is a quadric bundle over a nonsingular projective curve Y. Then by (the argument in the proof of) Tsen's

lemma, we have a section for ϕ. We see that the projection from the section to the hyperplane $H \times Y \cong \mathbb{P}^2 \times Y$ is a birational map. Therefore, $\phi : X_2 \to Y$ is birationally trivial, and hence X_2 is ruled.

When $a = 3$, applying the argument via the Sarksiov program (See Remark 3-2-15), we see that $\phi : X_3 \to Y$ is not birationally trivial, and hence X_3 is not ruled by the exercise above, though it is uniruled by Theorem 3-2-6.

$\boxed{\text{Subcase: } q(X) = 0}$

The implication

$$\phi \text{ is birationally trivial} \Longrightarrow X \text{ is ruled}$$

is obvious. If X is ruled, i.e., birational to $W \times \mathbb{P}^1$, then it is easy to see that W itself is covered by rational curves (Exercise! Why?) and hence W is birational to $C \times \mathbb{P}^1$. Since $q(X) = q(C) = 0$, we conclude that

$$X \text{ is ruled} \Longleftrightarrow X \text{ is rational.}$$

For an example, we take a general cubic hypersurface $W_3 \subset \mathbb{P}^4$. Blow up W_3 along the center $C = W_3 \cap P$, where P is a general plane $\mathbb{P}^2 \cong P \subset \mathbb{P}^4$:

$$
\begin{array}{ccc}
X & \subset & \text{Blp}_P \mathbb{P}^4 \longrightarrow \mathbb{P}^1 \\
\downarrow & & \downarrow \\
W_3 & \subset & \mathbb{P}^4.
\end{array}
$$

The induced map $\phi : X \to \mathbb{P}^1 = Y$ is a Mori fiber space, whose general fiber is a cubic surface. Again by the reslult of Clemens–Griffiths [1], W_3 is not rational, and hence neither is X. Thus by the above argument, we conclude that X is not ruled and that ϕ is not birationally trivial.

How about the implication

$$X \text{ is rational} \Longrightarrow \phi \text{ is birationally trivial} \quad ??$$

Unfortunately, or rather to say, very fortunately to make the subject matter much more interesting, this implication does not hold, as the following example shows.

Example 3-2-13 (A Birationally Nontrivial Mori Fiber Space Structure, with a Dense Set of Sections, on a Rational 3-Fold). This example, due classically to Castelnuovo and Iskovskikh (cf. Manin [1]), is a beautiful application of the Sarkisov program in dimension 2 (over an algebraically *non*-closed field) to the geometry of 3-folds by Corti–Pukhlikov–Reid [1]. It was communicated and taught to the author by Professors. Reid and Mukai during his pleasant stay at the University of Warwick.

We take two cubic surfaces S_3, $S_3' \subset \mathbb{P}^3$ such that the intersection $C = S_3 \cap S_3'$ is a smooth curve of genus 10. We blow up $\mathbb{P}^3$ along C to obtain

$$\phi : X = \text{Blp}_C \mathbb{P}^3 \to \mathbb{P}^1,$$

which parametrizes the linear pencil of cubic surfaces containing S_3 and S_3'. We claim that $\phi : X \to \mathbb{P}^1$ is not birationally trivial.

We have only to show that the generic fiber of ϕ is not birational to $\mathbb{P}^2_{\mathbb{C}(t)}$ over $\operatorname{Spec}\mathbb{C}(t)$, where $\mathbb{C}(t)$ is the function field of $\mathbb{P}^1$. We denote the generic fiber by $\phi : X \to \operatorname{Spec}\mathbb{C}(t)$ by abuse of notation. Observe that the generic fiber can be considered as a cubic surface $X_3 \subset \mathbb{P}^3_{\mathbb{C}(t)}$.

Suppose that there exists a birational map

$$
\begin{array}{ccc}
X & \overset{\Phi}{\underset{\text{birat}}{\dashrightarrow}} & \mathbb{P}^2_{\mathbb{C}(t)} \\
\downarrow{\phi} & & \downarrow{\phi'} \\
\operatorname{Spec}\mathbb{C}(t) & = & \operatorname{Spec}\mathbb{C}(t)
\end{array}
$$

Since Φ is a birational map between two Mori fiber spaces over $\operatorname{Spec}\mathbb{C}(t)$, the Sarkisov program in dimension 2 of Section 1-8 tells us that we should be able to untwist Φ by some links. (We discussed the Sarkisov program over $\mathbb{C}$ in Section 1-8. We leave it to the reader as an exercise to make the necessary modifications when one wants to apply the program over $\mathbb{C}(t)$.) We show that any link starting from ϕ would result in the same ϕ (up to isomorphism). Thus no matter how many times we untwist Φ, the resulting Mori fiber space would be isomorphic to $\phi : X \to \operatorname{Spec}\mathbb{C}(t)$. (Note that the isomorphism may be given through a different birational map than the link itself.) But on the other hand, $\phi : X \to \operatorname{Spec}\mathbb{C}(t)$ is not isomorphic to $\phi' : \mathbb{P}^2_{\mathbb{C}(t)} \to \operatorname{Spec}\mathbb{C}(t)$, a contradiction!

We analyze the untwisting by the Sarkisov program.

Note first that the first untwisting link must be in the case $\lambda > \mu$. (If $\lambda \leq \mu$, then since $K_X + \frac{1}{\mu}\mathcal{H}$ must necessarily be nef, the NFI criterion shows that Φ is an isomorphism of Mori fiber spaces, a contradiction!) Thus we have to blow up the base point $P \in X$ with the maximal multiplicity λ of $\mathcal{H}$ to obtain $p : Z \to X$. Observe that the extension degree satisfies

$$
e = [k(P) : \mathbb{C}(t)] \leq 2.
$$

In fact, $P_1, \ldots, P_e$ being the $\overline{\mathbb{C}(t)}$-rational points conjugate to each other over $\mathbb{C}(t)$, we have

$$
3\mu^2 = \mathcal{H}^2 \geq \Sigma_{i=1}^{e}(\operatorname{mult}_{P_i}\mathcal{H})^2 = \lambda^2 \cdot e.
$$

Subcase: $e = [k(P) : \mathbb{C}(t)] = 1$.

In this case, P is a $\mathbb{C}(t)$-rational point. Let $T_{X,P} \subset \mathbb{P}^3_{\mathbb{C}(t)}$ be the tangent space to X at P. Observe that $T_{X,P} \cap X$ is either a rational curve C with a node or a cusp or a union of three distinct lines meeting at P, i.e., P is an Eckardt point

(A priori there are several other possibilities for $T_{X,P} \cap X$, including a union of a conic and a line or a union of three lines not meeting at P, since $T_{X,P} \cap X$ is a cubic curve on the plane $T_{X,P}$. But in each case there would be a line L defined over $\mathbb{C}(t)$ on X with $L^2 = -1$, contradicting the condition that $\phi : X \to \operatorname{Spec}\mathbb{C}(t)$ is a Mori fiber space and hence that the Picard number $\rho(X/\operatorname{Spec}\mathbb{C}(t))$ is equal to 1.)

Geiser Involution

In the first case, the strict transfrom C' of C becomes a (-1)-curve on Z. We contract C' by $\mu : Z \to X_1$ to obtain another Mori fiber space $\phi_1 : X_1 \to$ Spec $\mathbb{C}(t)$. This is the link.

In order to describe the link, we introduce the following specific birational selfmap of X, called the Geiser involution. For any point $Q \in X(Q \neq P)$ we take a plane H_Q containing P and Q. The intersection $H_Q \cap X$ is an irreducible (Exercise! Why?) cubic, and hence the line joining P and Q meets $H_Q \cap X$ at another point Q'. The Geiser involution i_G is set-theoretically the map $i_G(Q) = Q'$. It is easy to see (Exercise!) that i_G is a birational map defined everywhere but P on X and that i_G induces an automorphism of order 2 on Z such that $i_G(E_P) = C'$, where E_P is the exceptional divisor for $p : Z \to X$. Now observe that the link going from ϕ to ϕ_1 is nothing but going from ϕ to its "mirror" image by the Geiser involution i_G:

$$
\left\{
\begin{array}{cc}
 & Z \\
 \swarrow & \searrow \\
X & \quad X_1 \\
\downarrow \phi & \downarrow \phi_1 \\
\text{Spec } \mathbb{C}(t) & \text{Spec } \mathbb{C}(t)
\end{array}
\right\}
=
\left\{
\begin{array}{cc}
 & Z \overset{i_G}{\to} Z \\
 \swarrow & \searrow \\
X & \quad X_1 \\
\downarrow & \downarrow \\
\text{Spec } \mathbb{C}(t) & \text{Spec } \mathbb{C}(t)
\end{array}
\right\}
$$

Therefore, $\phi_1 : X_1 \to$ Spec $\mathbb{C}(t)$ is isomorphic to $\phi : X \to$ Spec $\mathbb{C}(t)$ (not through Φ but through $\Phi \circ i_G^{-1}$).

Bad Link

In the second case where $T_{X,P} \cap X$ is a union of three lines meeting at P, the strict transforms of these three lines become (-2)-curves disjoint from each other, conjugate over Spec $\mathbb{C}(t)$. Thus they give rise to the other extremal edge (other than the one corresponding to the original blowup) of $\overline{NE}(Z/\text{Spec } \mathbb{C}(t))$ to be contracted to produce $Z \to X_1$. But X_1 is not nonsingular any more, and hence this link is against the conclusion of the Sarkisov program that X_1 stays in the category of nonsingular projective surfaces. It is a bad link! Thus this case cannot happen.

Subcase: $e = [k(P) : \mathbb{C}(t)] = 2$.

In this case, since $e = 2$, there are two $\overline{\mathbb{C}(t)}$-rational points P_1, P_2 over P. Blowing up P is equivalent to blowing up P_1 and P_2 simultaneously. Let E_{P_1} and E_{P_2} be the exceptional divisors for p over P_1 and P_2, respectively.

Bertini Involution

We take a line L joining P_1 and P_2 in $\mathbb{P}^3$. Note that the line L is not contained in X. In fact, if it were, then L would be a (-1)-curve defined over $\mathbb{C}(t)$ on X, contradicting the condition $\rho(X/\text{Spec } \mathbb{C}(t)) = 1$. Therefore, L meets X at the third point $R(\neq P_1, P_2)$, which is necessarily a $\mathbb{C}(t)$-rational point. For any point

$Q \in X(Q \neq P_1, P_2, R)$, take the plane H_Q containing Q, P_1, P_2 (and hence R). As before, by the condition $\rho(X/\mathrm{Spec}\,\mathbb{C}(t)) = 1$, we see that $H_Q \cap X$ is an irreducible cubic with R being a nonsingular point on it. Thus there is an additive group structure on $H_Q \cap X$ with R the zero element. The Bertini involution i_B is set-theoretically the map $i_B(Q) = Q' = -Q$, where -1 is the involution for the additive group. It is easy to see (Exercise!) that i_B is a birational map on X defined everywhere but P_1, P_2. After blowing up P_1 and P_2, the linear pencil $\{H_Q \cap X\}$ still has the unique base point R. We blow up R to obtain $q : W \to Z$. The space W can be considered as the parametrization of the linear pencil $g : W \to \mathbb{P}^1$ with the section E_R, where E_R is the exceptional divisor for q. Thus $g : W \to \mathbb{P}^1$ has the additive group (scheme) structure over $\mathbb{P}^1$ with E_R the zero element. The Bertini involution on W is nothing but the involution -1 for this additive group. Since $i_B(E_R) = E_R$, the Bertini involution descends to an automorphism i_B of order 2 on Z. Observe that for a general $Q \in X$, the point R is not the inflection point of $H_Q \cap X$, and hence $P_1 \neq i_B(P_2), P_2 \neq i_B(P_1)$. Therefore, the curves $\{i_B(E_{P_1}), i_B(E_{P_2})\}$ give rise to the other edge (other than the one coresponding to $\{E_{P_1}, E_{P_2}\}$ for the contraction p) of $\overline{NE}(Z/\mathrm{Spec}\,\mathbb{C}(t))$. The contraction of this extremal ray $Z \to X_1$ gives us another Mori fiber space $\phi_1 : X_1 \to \mathrm{Spec}\,\mathbb{C}(t)$ and thus a link. Again from the description it is clear that $\phi_1 : X_1 \to \mathrm{Spec}\,\mathbb{C}(t)$ is obtained as the mirror image of $\phi : X \to \mathrm{Spec}\,\mathbb{C}(t)$ through the Bertini involution and hence is isomorphic to $\phi : X \to \mathrm{Spec}\,\mathbb{C}(t)$:

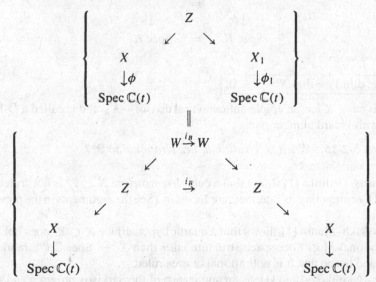

Thus we conclude that $\phi : X = \mathrm{Blp}_C \mathbb{P}^3 \to \mathbb{P}^1$ is not birationally trivial.

It should be observed, as pointed out to us by Mukai, that in contrast to the previous examples there are **infinitely many sections** for this Mori fiber space $\phi : X = \mathrm{Blp}_C \mathbb{P}^3 \to \mathbb{P}^1$, though it is not birationally trivial: Any fiber $\mathrm{Blp}_C^{-1}(p)$ for $p \in C$ is a section for ϕ, and also, any secant line of C in $\mathbb{P}^3$, except for the lines contained in the members of the linear pencil of the cubics, gives rise to a section of ϕ. Actually, these sections are even **dense** in X. (Exercise! Check this last assertion.)

Exercise 3-2-14. The Mori fiber space $\phi : X \to \mathbb{P}^1$ over the generic point $\operatorname{Spec} \mathbb{C}(t)$ can be considered a cubic surface

$$\{c(X_0, X_1, X_2, X_3) = \Sigma_{i+j+k+l=3} c_{ijkl} X_0^i X_1^j X_2^k X_3^l = 0\} \subset \mathbb{P}^3_{\mathbb{C}(t)}, \quad c_{ijkl} \in \mathbb{C}(t).$$

The sections of ϕ are nothing but the solutions of the cubic form $c = 0$ in $\mathbb{C}(t)$. Show by the same argument as in the proof of Tsen's lemma that this cubic form has a solution and hence the Mori fiber space has a section.

Remark 3-2-15. The above argument actually proves the following statement: Let

$$X_3 \subset \mathbb{P}^3_K$$

be a nonsingular cubic surface over a field K (of characteristic zero) such that

$$\rho(X/\operatorname{Spec} K) = 1.$$

Then X is not birationally trivial over K, i.e., there is no commutative diagram of the form

$$
\begin{array}{ccc}
X & \overset{\text{birat}}{\dashrightarrow} & \mathbb{P}^2_K \\
\downarrow{\phi} & & \downarrow{\phi'} \\
\operatorname{Spec} K & = & \operatorname{Spec} K.
\end{array}
$$

Case: $\dim Y = \dim X - 3 = 0.$

In this case, X has an ample anticanonical divisor $-K_X$ and is called a $\mathbb{Q}$-Fano 3-fold with Picard number one.

Question 3-2-16. *When is X rational, i.e., birational to $\mathbb{P}^3$?*

Clemens–Griffiths [1] shows that a cubic hypersurface $X \subset \mathbb{P}^4$ is not "ruled" or rational by computing its intermediate Jacobian. (See the arguments in the previous cases.)

Iskovskikh–Manin [1] shows that a quartic hypersurface $X \subset \mathbb{P}^4$ does not have any birational Mori fiber space structure other than $X \to \operatorname{Spec} \mathbb{C}$ ("birationally rigid") and hence that it is not rational or even ruled.

Corti–Pukhlikov–Reid [1], as an application of the Sarkisov program and as an extension of the method of Iskovskikh–Manin, shows that a certain kind of $\mathbb{Q}$-Fano hypersurfaces in some weighted projective spaces are also birationally rigid and hence never ruled or rational.

In general, it is a very, very hard question to judge whether a given variety is rational or not. We expect that the Sarkisov program will shed more light on this subject and on the structure of Mori fiber spaces in higher dimensions in general.

3.3 Basic Properties of Minimal Models in Dimension 3 or Higher

In the previous section we studied the basic properties of Mori fiber spaces in dimension 3 or higher. The purpose of this section is to describe those of **minimal models**, the "other half" of the end results of MMP in dimension 3 or higher. In dimension 3, thanks to the work of Kawamata [12], Miyaoka [1][2][3], and others, these properties are firmly established theorems. In higher dimensions, however, most of them remain as important conjectures.

As in the previous section, $\mathfrak{C}$ denotes the category of normal projective varieties with only $\mathbb{Q}$-factorial and terminal singularities.

Definition 3-3-1 (Minimal Model). *An object $X \in \mathfrak{C}$ is a minimal model if the canonical divisor K_X, which is $\mathbb{Q}$-Cartier since X has terminal singularities, is nef.*

We remark that obviously a minimal model $X \in \mathfrak{C}$ of dimension n as defined above is an end result of MMP in dimension n.

Conjecture 3-3-2 (Existence of an Effective Pluricanonical Divisor). *Let $X \in \mathfrak{C}$ be a minimal model. Then*

$$\kappa(X) \geq 0,$$

i.e.,

$$H^0(X, \mathcal{O}_X(mK_X)) \neq 0 \text{ for some } m \in \mathbb{N}.$$

Together with Theorem 3-2-3, this leads to the following.

Conjecture 3-3-3 (Hard Dichotomy Conjecture of MMP). *Let X be an object in the category $\mathfrak{C}$ of dimension n. Assume that MMP in dimension n holds via (existence of flip) and (termination of flips). Then an end result of MMP in dimension n starting from X is a minimal model (respectively a Mori fiber space) iff $\kappa(X) \geq 0$ (respectively $\kappa(X) = -\infty$).*

As in the Enriques classification of surfaces demonstrated in Section 1-7, the abundance conjecture constitutes the core of our study of the structure of the minimal models.

Conjecture 3-3-4 (Abundance Conjecture). *Let $X \in \mathfrak{C}$ be a minimal model. Then $|mK_X|$ is base point free for sufficiently divisible and large $m \in \mathbb{N}$.*

Recall that one of the key ingredients to prove the hard dichotomy theorem and abundance theorem in dimension 2 is to show the nonnegativity of c_2 for minimal models in dimension 2 (cf. Proposition 1-5-9). Miyaoka [1] (see also the argument by Shepherd–Barron in Kollár et al. [1]) generalizes this to the following theorem in dimension 3 or higher.

Theorem 3-3-5 (Miyaoka Inequality). *Let* $X \in \mathfrak{C}$ *be a minimal model in dimension* n. *Then*

$$3c_2(X) - c_1(X)^2 \geq 0$$

in the following sense: Let $f : Y \rightarrow X$ *be a birational morphism from a nonsingular projective variety* Y *onto* X. *Then*

$$3c_2(X) - c_1(X)^2 := f_*\{3c_2(Y) - c_1(Y)^2\}$$

is independent of the choice of the resolution f *and*

$$\{3c_2(X) - c_1(X)^2\} \cdot H_1 \cdots H_{n-2} \geq 0$$

for any ample divisors $H_1, \ldots, H_{n-2}$ *on* X.

The abundance conjecture guarantees that the Iitaka fibration provides a morphism from a minimal model onto its canonical model.

Theorem 3-3-6 (Iitaka Fibration). *Let* $X \in \mathfrak{C}$ *be a minimal model. Assume that the abundance conjecture holds for* X. *Then for sufficiently divisible and large* $m \in \mathbb{N}$,

$$\Phi = \Phi_{|mK_X|} : X \rightarrow X_{\mathrm{can}}$$

satisfies the following properties:

(i) $\Phi : X \rightarrow X_{\mathrm{can}}$ *is a morphism with connected fibers onto a normal projective variety* X_{can},

(ii) *for any curve* $C \subset X$

$$\Phi(C) = \mathrm{pt.} \iff K_X \cdot C = 0,$$

(iii) $\mathcal{O}_X(mK_X) = \Phi^* \mathcal{K}$ *for some ample divisor* $\mathcal{K}$ *on* X_{can},

(iv) $\kappa(X) = \dim X_{\mathrm{can}}$,

(v) *there exists a nonempty Zariski open subset* $U \subset X_{\mathrm{can}}$ *such that* $F_p = \Phi^{-1}(p)$ *is a normal projective variety (of dimension* $\dim X - \dim X_{\mathrm{can}}$) *with only terminal singularities and* $mK_{F_p} \sim 0$ *for all* $p \in U$.

Moreover, properties (i) and (ii) characterize $\Phi : X \rightarrow X_{\mathrm{can}}$, *which we call the* **Iitaka fibration** *of* X *onto its* **canonical model** X_{can}.

Remark 3-3-7. (i) Let X be an object in the category $\mathfrak{C}$. Then the ring

$$R := \oplus_{m \geq 0} H^0(X, \mathcal{O}_X(mK_X))$$

has a natural structure of a graded $\mathbb{C}$-algebra and is called the **canonical ring** of X. The minimal model program, and the hard dichotomy and abundance conjectures imply that R is finitely generated as a $\mathbb{C}$-algebra (when $\kappa(X) = -\infty$, R is nothing but $\mathbb{C}$ itself, all the summands in the positive grades being zero) and that we have the description of the canonical model

$$X_{\mathrm{can}} = \mathrm{Proj}\, R,$$

where X_{can} is constructed first by taking a minimal model X_{min} through MMP and then by the hard dichotomy and abundance conjectures as above. Since R is a birational invariant, X_{can} is "canonically" determined by the birational equivalence class of the variety X.

(ii) It can be shown in dimension 3 that there exists a boundary $D_{X_{can}}$ such that $(X_{can}, D_{X_{can}})$ has only log terminal singularities such that

$$(*) \qquad \Phi^*(K_{X_{can}} + D_{X_{can}}) = K_X \text{ as } \mathbb{Q}\text{-Cartier divisors.}$$

(When $\kappa = 0$, $K_X = 0$ as a $\mathbb{Q}$-Cartier divisor as a consequence of the abundance theorem in dimension 3. When $\kappa = 1 = \dim X_{can}$, the sheaf $\Phi_*(\omega_{X/X_{can}}^{\otimes m})$ is a line bundle that is either torsion or of positive degree (see, e.g., Kawamata [8]). Formula $(*)$ follows from this for a suitable choice of $D_{X_{can}}$. When $\kappa = 2$, we have the canonical bundle formula for the elliptic fibration over the surface X_{can}. (see, e.g., Grassi [1]), and formula $(*)$ follows from this. When $\kappa = 3$, we have $K_X = \Phi^* K_{X_{can}}$.) In higher dimension, Nakayama [3] shows that there exists a boundary divisor $B_{X_{can}}$ such that $(X_{can}, B_{X_{can}})$ has only log terminal singularities, but the existence of a boundary divisor as above is still conjectural.

With property (v) of the Iitaka fibration one is led to the following.

Classification Principle 3-3-8. The study of the structure of an algebraic variety X with $\kappa(X) \geq 0$, in principle via MMP, the abundance conjecture, and the Iitaka fibration, is reduced to that of

the structure of normal projective varieties F with only terminal singularities such that $mK_F \sim 0$ for some $m \in \mathbb{N}$ and hence $\kappa(F) = 0$ (if $q(F) > 0$, then we can utilize the Albanese map to study the structure of F, thus the remaining and most difficult is the study of the structure of those with $q(F) = 0$, called the **Calabi–Yau varieties**) and their moduli, and

the structure of the log pairs of general type $(X_{can}, D_{X_{can}})$ with only log terminal singularities such that the log canonical divisor $K_{X_{can}} + D_{X_{can}}$ is ample and hence $\kappa(X_{can}, K_{X_{can}} + D_{X_{can}}) = \dim X_{can}$.

3.4 Birational Relations Among Minimal Models and Mori Fiber Spaces in Dimension 3 or Higher

In this section we discuss the **birational relations** among minimal models and Mori fiber spaces in dimension 3 or higher according to the main strategy MP 3. For detailed proofs of the assertions, we refer the reader to Chapters 12 and 13.

Birational Relation Among Minimal Models

In Section 1-8 we observed that in dimension 2 we have a unique minimal model in a fixed birational equivalence class. This is no longer true in dimension 3 or higher: There may be many minimal models even in a fixed birational equivalence class. This gives rise to a need to study the relations among them. The first basic feature of the relation among minimal models in a fixed birational equivalence class is that they are isomorphic in codimension one.

Proposition 3-4-1 (Minimal Models Are Isomorphic in Codimension One).
Let $X_1, X_2 \in \mathfrak{C}$ be two minimal models, birational to each other. Then X_1 and X_2 are isomorphic in codimension one. More precisely, if $\phi : X_1 \dashrightarrow X_2$ is a birational map between them, then there exist closed subsets $B_1 \subset X_1$ and $B_2 \subset X_2$ such that ϕ induces an isomorphism

$$X_1 - B_1 \overset{\phi}{\underset{\sim}{\to}} X_2 - B_2$$

and such that

$$\operatorname{codim}_{X_1} B_1 \geq 2 \quad \text{and} \quad \operatorname{codim}_{X_2} B_2 \geq 2.$$

Now the question is how the differences in codimension 2 among minimal models arise and how to describe them. A special codimension 2 operation called "flop," a brother of "flip," answers the question.

Conjecture 3-4-2 (Existence of Flop). *Let*

$$\phi : X \to Y$$

be a small *contraction of an extremal ray R_l (with respect to $K_X + \epsilon D$, where ϵ is a very small positive rational number and D is an effective divisor) from $X \in \mathfrak{C}$ such that*

$$K_X \cdot l = 0, \quad \text{where } R_l = \mathbb{R}_+[l],$$

and such that

$$-D \text{ is } \phi\text{-ample.}$$

Then there exists a commutative diagram

$$
\begin{array}{ccc}
D \subset X & \overset{\Phi}{\dashrightarrow} & X^+ \supset D^+ \\
\phi \searrow & & \swarrow \phi^+ \\
& Y &
\end{array}
$$

where ϕ^+ is another small morphism from $X^+ \in \mathfrak{C}$ such that

$$K_{X^+} \cdot l^+ = 0 \text{ for any curve } l^+ \text{ contracted by } \phi^+$$

and such that

$$D^+ \text{ is } \phi^+\text{-ample, where } D^+ \text{ is the strict transform of } D.$$

(Notice the sign change between $-D$ and D^+ for these divisors to be relatively ample with respect to ϕ and ϕ^+, respectively.) The morphism $\phi^+ : X^+ \to Y$

is called the "**D-flop**" of ϕ. The word "D-flop" also refers to the commutative diagram above or to X^+ by abuse of language.

Once it exists, $\phi^+ : X^+ \to Y$ has necessarily to be unique, since D^+ is ϕ^+-ample and hence

$$X^+ = Proj \oplus_{m \geq 0} \phi^+{}_*(mD^+) = Proj \oplus_{m \geq 0} \phi_*(mD).$$

Actually, it does not even depend on D but only on ϕ.

Observe also that a flop is symmetric in the following sense:

$$\phi^+ : X^+ \to Y$$

is also a small contraction of an extremal ray R_{l+} (with respect to $K_{X+} + \epsilon'D'$, where ϵ' is a very small positive rational number and D' is some effective divisor on X^+) from $X^+ \in \mathfrak{C}$ such that

$$K_{X+} \cdot l^+ = 0, \quad \text{where } R_{l+} = \mathbb{R}_+[l^+],$$

and such that

$$-D' \text{ is } \phi^+\text{-ample.}$$

For example, we can take D' to be an effective member of the linear system $|-D^+ + \phi^{+*}A|$, where A is a sufficiently ample divisor on Y. Then the reverse Φ^{-1} of the commutative diagram above is a D'-flop.

Example 3-4-3 (Atiyah's Flop (cf. Atiyah [1])). Take the cone over a quadric

$$0 \in Y = \{xy - zw = 0\} \subset \mathbb{A}^4.$$

If we blow up the origin, then we get a resolution $f : V \to Y$ whose exceptional locus E is isomorphic to the quadric

$$E \cong \mathbb{P}^1 \times \mathbb{P}^1 \text{ with normal bundle } \mathcal{O}_E(E) \cong \mathcal{O}_{\mathbb{P}^1 \times \mathbb{P}^1}(-1, -1).$$

Therefore, the exceptional locus E can be contracted (Exercise! Check this!) in two different ways corresponding to the two different rulings of $\mathbb{P}^1 \times \mathbb{P}^1$:

$$f_1 : V \to X_1 \quad \text{and} \quad f_2 : V \to X_2.$$

It is easy to see that the exceptional loci l_1 and l_2 of

$$\phi_1 : X_1 \to Y \quad \text{and} \quad \phi_2 : X_2 \to Y$$

are both isomorphic to $\mathbb{P}^1$, and thus ϕ_1 and ϕ_2 are small. Since K_Y is a Cartier divisor, we observe that

$$K_{X_1} = \phi_1^*K_Y \quad \text{and} \quad K_{X_2} = \phi_2^*K_Y$$

and hence

$$K_{X_1} \cdot l_1 = 0 \quad \text{and} \quad K_{X_2} \cdot l_2 = 0.$$

We take D_1 and D_2 to be the divisors on X_1 and X_2 corresponding to the two distinct rulings on $E \cong \mathbb{P}^1 \times \mathbb{P}^1$ not contracted by f_1 and f_2, respectively. Then

$\Phi : X_1 \dashrightarrow X_2$ is a D_1-flop with $\phi_2 = \phi_1^+$, while $\Phi^{-1} : X_2 \dashrightarrow X_1$ is a D_2-flop with $\phi_1 = \phi_2^+$.

Figure 3-4-4.

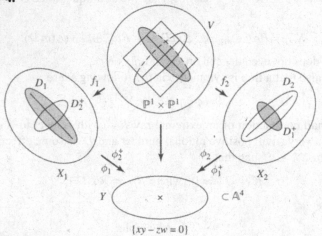

We also need the statement of termination.

Conjecture 3-4-5 (Termination of Flops). *There is* no *infinite sequence of D-flops for a fixed effective divisor D.*

As a consequence of (existence of flop) and (termination of flops) we can now describe the relation of any two minimal models in a fixed birational equivalence class.

Theorem 3-4-6 (Minimal Models Are Connected by Flops.). *Minimal models (in a birational equivalence class) are connected by sequences of flops, assuming (existence of flop) and (termination of flops).*

We can give a fairly explicit description of flops in dimension 3 due to Reid, Kawamata, and Kollár, and thus we may say we understand the local modifications via flops among the minimal models in dimension 3. As for the global nature of the birational relation among minimal models we have the following results.

Corollary 3-4-7 (Finiteness of Minimal Models of General Type). *There are only finitely many minimal models in a birational equivalence class of a variety of general type, assuming (existence of flop) and (termination of flops).*

It is a consequence of the study of the so-called chamber structure formed by the ample cones of the minimal models. For a variety of general type, we can show that the number of the chambers corresponding to the ample cones of the minimal models (in a fixed birational equivalence class) is finite, and hence so is the number of the minimal models itself.

Though for a variety of nongeneral type the number of the chambers may be infinite, it is conjectured that the number of minimal models is finite *up to automorphisms*.

Conjecture 3-4-8 (Finiteness of Minimal Models (!?)). *There are only finitely many minimal models in a fixed birational equivalence class (up to automorphisms).*

Again, for the subtleties of counting the number of minimal models, we refer the reader to Chapter 12.

We remark that all the conjectures mentioned above are proven theorems in dimension 3, except that the finiteness of minimal models, not necessarily of general type and up to automorphisms, is established only in the case $\kappa > 0$ by a recent result of Kawamata [17].

Birational Relation Among Mori Fiber Spaces

We now shift our attention to the birational relation among Mori fiber spaces. The most important subject here is the Sarkisov program, which gives an algorithm to factor a given birational map between Mori fiber spaces into certain elementary operations called "links." It can be considered a higher-dimensional analogue of the classical Castelnuovo–Noether theorem decomposing birational maps among ruled or rational surfaces, which we presented in Section 1-8 as the Sarkisov program in dimension 2.

Conjecture 3-4-9 (Sarkisov Program in Dimension 3 or Higher). *Let*

$$
\begin{array}{ccc}
X & \xrightarrow{\ \ \Phi\ \ } & X' \\
 & \text{birat} & \\
\downarrow \phi & & \downarrow \phi' V \\
Y & & Y'
\end{array}
$$

be a birational map between two Mori fiber spaces,

$$\phi : X \to Y,$$
$$\phi' : X' \to Y'.$$

*Then there is an algorithm, called the "**Sarkisov program**," to decompose Φ into a composite of the following four types of "**links**" (elementary transformations):*
   Type (I)

$$
\begin{array}{ccc}
 & Z & \dashrightarrow X_1 \\
 & \swarrow & \\
X & & \downarrow \\
\downarrow & & \\
Y & \leftarrow & Y_1
\end{array}
$$

Type (II)

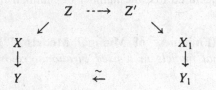

Type (III) *(Inverse of* Type (I))

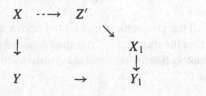

Type (IV)

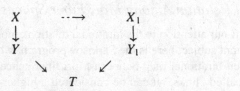

The birational morphisms $Z \to X$ and $Z' \to X'$ are divisorial contractions of extremal rays with respect to K_Z and $K_{Z'}$, respectively, in the category $\mathfrak{C}$, and the dotted arrows "$\dashrightarrow$" indicate sequences of codimension-two modifications called "log flips." We also have $\rho(Y_1/Y) = 1$ in Types (I), (II), (III) and $\rho(Y/T) = \rho(Y_1/T) = 1$ in Type (IV), where Y, Y_1, and T are normal projective varieties with only $\mathbb{Q}$-factorial singularities.

The main new feature as we go to higher dimension is that we have to allow certain codimension-2 operations called "log flips." The detail of the mechanism, which fits most naturally into the framework of the logarithmic category, will be given in Chapter 13 after we establish the logarithmic generalization of MMP in Chapter 11.

The Sarkisov program is an established theorem in dimension 3, thanks to the works of Sarkisov [3], Reid [6], and Corti [1]. It is expected to provide a powerful tool to study the birational structure of varieties with $\kappa = -\infty$, e.g., concerning the rationality question (cf. Section 3-2). We just mention a classical theorem of Iskovskikh–Manin [1] from this point of view:

Theorem 3-4-10. *Let $X_4 \subset \mathbb{P}^4$ be a smooth quartic 3-fold. Since $\rho(X_4) = 1$, the structure morphism $\phi; X_4 \to \operatorname{Spec} \mathbb{C}$ is a Mori fiber space. Then there is no nontrivial link starting from ϕ in the Sarkisov program, i.e., if there is a Mori fiber* *space $\phi' : X' \to Y'$ birational to ϕ,*

$$
\begin{array}{ccc}
X_4 & \xrightarrow[\text{birat}]{\Phi} & X' \\
\downarrow \phi & & \downarrow \phi' \\
\text{Spec}\,\mathbb{C} & & Y',
\end{array}
$$

then Φ is necessarily an isomorphism of Mori fiber spaces

$$
\begin{array}{ccc}
X_4 & \xrightarrow{\overset{\Phi}{\sim}} & X' \\
\downarrow \phi & & \downarrow \phi' \\
\text{Spec}\,\mathbb{C} & = & Y'.
\end{array}
$$

In particular, X_4 is not rational.

See also the argument in Section 3-2, where we showed that a linear pencil of cubic hypersurfaces in $\mathbb{P}^3$ as a Mori fiber space $\phi : X = \mathrm{Blp}_C \mathbb{P}^3 \to \mathbb{P}^1$ is not birationally trivial, as an application of the Sarkisov program in dimension 2 (over a nonalgebraically closed field).

3.5 Variations of the Mori Program

There are two major variations of the Mori program

the **logarithmic (log) Mori program**;
the **relative Mori program**.

The first one, the logarithmic Mori program, generalizes the Mori program according to Iitaka's philosophy (cf. Section 2-2). The key ingredients are the logarithmic generalizations of the main theorems as in Section 2-2, defining log terminal singularities in higher dimension, plus (existence of log flip) and (termination of log flips).

This will be discussed in some detail in Chapter 11.

The second one, the relative Mori program, deals with the category of varieties projective over some fixed variety S instead of the category of varieties projective over $\text{Spec}\,\mathbb{C}$. The argument for the relative Mori program goes almost verbatim as for the usual Mori progarm, putting the phrase "projective over S" in place of "projective over $\mathbb{C}$."

These two variations not only provide some generalizations of the Mori program but also play important and indispensable roles in the study of the original Mori program.

Though we will mostly restrict our presentation to the usual and original Mori program without making the logarithmic and/or relative generalizations explicit, the reader should not have much difficulty in working out the details of the generalizing processes, and we feel free to use some statements in the logarithmic and/or relative setting as needed in the study of the usual and original Mori program.

Example-Exercise 3-5-1 (cf. Kawamata–Matsuda–Matsuki [1]). *Let $\pi : X \to S$ be a variety projective over another variety S. There is the intersection pairing between the line bundles on X and the 1-cycles on X mapped to points by π,*

$$\mathrm{Pic}(X) \times Z_1(X/S) \to \mathbb{Z},$$

where the 1-cycles $\Sigma a_i C_i \in Z_1(X/S)$ satisfy the condition

$$\pi(C_i) = \mathrm{pt}. \quad \forall i.$$

We extend this pairing to the one over $\mathbb{R}$ between

$$N^1(X/S) := \{\mathrm{Pic}(X)/ \equiv_S\} \otimes_{\mathbb{Z}} \mathbb{R}$$

and

$$N_1(X/S) := \{Z_1(X/S)/ \equiv_S\} \otimes_{\mathbb{Z}} \mathbb{R},$$

where $\equiv_S$ refers to numerical equivalence with respect to the intersection pairing as above.

(i) *Show that $N^1(X/S)$ and $N_1(X/S)$ are finite-dimensional vector spaces dual to each other.*
   We set

$$\mathrm{NE}(X/S) := \text{the convex cone generated by effective 1-cycles on } X$$
$$\text{mapped to points by } \pi;$$
$$\overline{\mathrm{NE}}(X/S) := \text{ the closure of } \mathrm{NE}(X/S) \text{ in } N_1(X/S).$$

(ii) *Suppose π factors through another variety Y projective over S, i.e., we have morphisms $X \overset{\sigma}{\to} Y \overset{\tau}{\to} S$ with $\pi = \tau \circ \sigma$. Show that via the natural inclusion*

$$N_1(X/Y) \hookrightarrow N_1(X/S)$$

we can identify $\overline{\mathrm{NE}}(X/Y)$ with the following face F of $\overline{\mathrm{NE}}(X/S)$:

$$F := \{z \in \overline{\mathrm{NE}}(X/S); \tau^*H \cdot z = 0 \text{ for any } \tau\text{-ample line bundle } H \text{ on } Y\}$$
$$\subset \overline{\mathrm{NE}}(X/S).$$

CHAPTER 4

Singularities

This chapter is an attempt to give a unified and general account of **singularities** that are indispensable in the study of the (logarithmic) Mori program in higher dimensions, namely

> **terminal singularities;**
> **canonical singularities;**
> **log terminal singularities;**
> **log canonical singularities.**

In Section 4-1, we focus our attention on terminal singularities, recalling the definition (cf. Definition 3-1-5) and clarifying its subtleties. The importance of the notion of terminal singularities lies in the fact that together with $\mathbb{Q}$-**factoriality** it characterizes the singularities of the objects in the smallest category $\mathfrak{C}$ that contains the category $\mathfrak{S}$ of nonsingular projective varieties and in which MMP works.

In Section 4-2 we discuss canonical singularities. The importance of the notion of canonical singularities lies in the fact that it characterizes the singularities of the canonical models of varieties of general type.

In Section 4-3 we give the logarithmic counterparts of the notions of terminal and canonical singularities. Log terminal singularities together with $\mathbb{Q}$-factoriality should characterize the singularities of the objects in the smallest category $\mathfrak{D}$ that contains the category $\mathfrak{E}$ of log pairs consisting of nonsingular projective varieties with normal crossing boundary divisors and in which Log MMP works, while log canonical singularities should characterize those on the log canonical models of log pairs of log general type.

In Section 4-4 we try to give the definitions for the classes of singularities mentioned above in a "unified" way in terms of **discrepancies**." The charaterization using the notion of discrepancies is quite ideal from the viewpoint of purely dealing with the subject of singularities, as it is *local* in nature (even analytically local).

On the other hand, we have pursued the way to characterize those singularities that appear naturally in the course of carrying out the mininmal model program, which is *global* in nature. While the notions of terminal, canonical, and log canonical singularities find their natural positions in this scheme of characterizing them in terms of discrepancies and turn out to be analytically local, the notion of log terminal singularities does not and gives rise to several "spin-off" notions as below, caused by the gap between the *local* and *global* approaches:

kawamata log terminal singularities;
purely log terminal singularities;
divisorially log terminal singularities;
weakly kawamata log terminal singularities.

Though technical by nature of their origins, these spin-offs seem to play crucial roles in various parts of the logarithmic generalization of the Mori program.

In Section 4-5 we discuss the method of the **canonical cover**, which reduces the study of singularities of higher indices to that of index one. We also study the behavior of discrepancies under the canonical covers.

In Section 4-6 we give the (analytic) classification in dimension 2 of terminal canonical, log terminal, and log canonical singularities.

4.1   Terminal Singularities

First we recall the definition introduced in Section 3-1.

Definition 4-1-1 = Definition 3-1-5 (Terminal Singularities). *A normal variety X of dimension n has only* **terminal singularities** *if*

(1) *the canonical divisor K_X is $\mathbb{Q}$-Cartier, i.e., there exists $e \in \mathbb{N}$ such that eK_X is a Cartier divisor;*

(2) *there exists a projective birational morphism $f : V \to X$ from a nonsingular variety V such that in the ramification formula*

$$K_V = f^*K_X + \Sigma a_i E_i$$

all the coefficients for the exceptional divisors are strictly positive:

(∗) $a_i > 0 \; \forall E_i$ *exceptional.*

Remark 4-1-2. (i) Since X is singular in general, the definition of the canonical divisor K_X needs a little care. Over the nonsingular locus $X_{\text{reg}} = X - \text{Sing}(X)$ the sheaf of n-forms $\Omega^n_{X_{\text{reg}}}$ is a line bundle, and thus there exists a Weil divisor K_X on X_{reg} such that

$$\Omega^n_{X_{\text{reg}}} \cong \mathcal{O}_{X_{\text{reg}}}(K_X).$$

Since X is normal and hence $\operatorname{codim}_X \operatorname{Sing}(X) \geq 2$, the divisor K_X can be regarded as a Weil divisor on the whole of X. Note that

$$V - \operatorname{Exc}(f) \widetilde{\rightarrow} X - f(\operatorname{Exc}(f))$$

and that

$$\operatorname{Exc}(f) = f^{-1}(f(\operatorname{Exc}(f))), \ f(\operatorname{Exc}(f)) \supset \operatorname{Sing}(X) \text{ and } \operatorname{codim}_X f(\operatorname{Exc}(f)) \geq 2.$$

Therefore, there exists a unique Weil divisor $K_V|_{V-\operatorname{Exc}(f)}$ defined on $V - \operatorname{Exc}(f)$ such that

$$K_V|_{V-\operatorname{Exc}(f)} = K_X|_{X-f(\operatorname{Exc}(f))}.$$

Take a Weil divisor $\tilde{K}_V$ defined on the whole of V such that

$$\Omega_V^n \cong \mathcal{O}_V(K_V).$$

Then there exists a rational function $r \in \mathbb{C}(V) = \mathbb{C}(V - \operatorname{Exc}(f))$ such that

$$K_V|_{V-\operatorname{Exc}(f)} = \tilde{K}_V + \operatorname{div}(r)|_{V-\operatorname{Exc}(f)}.$$

Now, by setting

$$K_V = \tilde{K}_V + \operatorname{div}(r)$$

we see that $K_V|_{V-\operatorname{Exc}(f)}$ can be extended to a Weil divisor defined on the whole of V. Moreover, it is easy to check that this extension is unique, since there is no nonzero Weil divisor linearly equivalent to 0 supported only on $\operatorname{Exc}(f)$ (Exercise! Why?). This determines K_V uniquely.

We take the pullback $f^*(eK_S)$ of the Cartier divisor eK_X and obtain the formula

$$eK_V = f^*(eK_X) + \Sigma ea_i E_i \text{ with } ea_i \in \mathbb{Z},$$

where the support of the difference $\Sigma ea_i E_i$ necessarily lies in the exceptional locus $\operatorname{Exc}(f)$.

We just formally divide the above by e and obtain the ramification formula as the equality of $\mathbb{Q}$-divisors:

$$K_V = f^*K_X + \Sigma a_i E_i.$$

Note that though K_X is unique only up to linear equivalence, the coefficients a_i are independent of the choice of its representative as a Weil divisor and also independent of the multiplier e. (Exercise! Verify the last assertion!)

(ii) Once the condition $(*)$ is satisfied for one projective birational morphism $f : V \to X$ from a nonsingular variety V, it is satisfied by any proper birational morphism $g : W \to X$ from a nonsingular variety W. (Exercise! Why?) Therefore, we can replace condition (1) with the equivalent

(1') for *any* proper birational morphism $f : V \to X$ from a nonsingular variety V of dimension n in the ramification formula

$$K_V = f^*K_X + \Sigma a_i E_i$$

all the coefficients for the exceptional divisors are strictly positive:

(∗) $a_i > 0 \quad \forall E_i$ exceptional.

   Though it is not necessary to require in the definition for the exceptional locus of f to be of pure codimension one or consist of normal crossing divisors, in coherence with the logarithmic variations in Section 4-3 we could add these extra conditions on the exceptional locus and replace condition (1) with the equivalent

   (1″) there exists a projective birational morphism $f : V \to X$ from a nonsingular variety V such that the exceptional locus $\mathrm{Exc}(f)$ consists of normal crossing divisors and that in the ramification formula

$$K_V = f^* K_X + \Sigma a_i E_i$$

all the coefficients for the exceptional divisors are strictly positive:

(∗)   . $a_i > 0 \quad \forall E_i$ exceptional.

   (iii) The notion of terminal singularities is easily seen to be algebraically *local* in the following sense: Let $\{U_\alpha\}$ be a covering of X by algebraic Zariski open subsets. If X has only terminal singularities, then taking $f_\alpha = f|_{U_\alpha} : f^{-1}(U_\alpha) \to U_\alpha$ we see that each U_α has only terminal singularities. On the other hand, suppose each U_α has only terminal singularities with a projective birational morphism $f_\alpha : V_\alpha \to U_\alpha$ as required in condition (1) of the definition. Take any projective birational morphism $g : W \to X$ from a nonsingular variety W. Then by remark (ii) each $g_\alpha = g|_{U_\alpha} : g^{-1}(U_\alpha) \to U_\alpha$ satisfies condition (∗), since the resolution $f_\alpha : V_\alpha \to U_\alpha$ in the definition for U_α does, and hence so does $g : W \to X$. Thus X has also only terminal singularities.

   (iv) A more intrinsic definition, which does not depend on the existence of a nice and global resolution or require the condition on all the resolutions, will be given in Section 4-3 to show that the notion of terminal singularities is actually analytically local.

   The importance of the notion of terminal singularities lies in the fact that together with the notion of $\mathbb{Q}$-factoriality it characterizes the singularities that appear in the process of MMP starting from nonsingular projective varieties.

Theorem 4-1-3 (Characterization of $\mathbb{Q}$-factorial and Terminal Singularities).
$\mathbb{Q}$-factorial and terminal singularities are characterized as those that appear in the process of MMP starting from a nonsingular projective variety, assuming (existence of flip) and (termination of flips). More precisely, the category $\mathfrak{C}$ of normal projective varieites with $\mathbb{Q}$-factorial and terminal singularities is the smallest category that contains the category $\mathfrak{S}$ of nonsingular projective varieties and in which MMP works. That is to say, assuming (existence of flip) and (termination of flips), we have

   (i) *$\mathfrak{C}$ contains $\mathfrak{S}$,*
   (ii) *$\mathfrak{C}$ is closed under the operations of MMP, and*
   (iii) *any object $X \in \mathfrak{C}$ can be obtained through the process of MMP starting from some object in $\mathfrak{S}$.*

Moreover, any (algebraic) germ U of $\mathbb{Q}$-factorial and terminal singularities appears in a process of MMP starting from some nonsingular projective variety.

PROOF.   Condition (i) is obvious. Condition (ii) is a consequence of Proposition 8-2-1 and (existence of flip). In order to see condition (iii), take a birational morphism

$$f : V \rightarrow X$$

from a nonsingular projective variety by resolution of singularities. Carry out MMP over X to reach $g : X' \rightarrow X$. (It is easy to see that an extremal ray over X is also an extremal ray over Spec $\mathbb{C}$. (Exercise! Why? See Example-Exercise 3-5-1.) Thus these operations of MMP over X are also the operations of MMP over Spec $\mathbb{C}$.) Since X is $\mathbb{Q}$-factorial and so is X', the exceptional locus of g is of pure codimension one. (See Shafarevich [1], Theorem 2 of Section 4 in Chapter II. The proof for the purity of exceptional locus is given there assuming that all the varieties are nonsingular and thus factorial. But it is valid without any change assuming only $\mathbb{Q}$-factoriality.) Therefore, if g is not an isomorphism, then the ramification formula would read

$$K_{X'} = g^* K_X + \Sigma a_i E_i$$

with $\Sigma a_i E_i \neq 0$ and $a_i > 0$ $\forall i$, since X has only terminal singularities. But this would imply that $K_{X'}$ is not g-nef (Exercise! Why?), contradicting the fact that $g : X' \rightarrow X$ is an end result of MMP over X. Therefore, g is an isomorphism and hence $X' = X$ is obtained through a process of MMP starting from a nonsingular projective variety V.

The "moreover" part of the theorem is a *local* statement of the main assertion (iii). Therefore, we have only to show that any algebraic germ U of $\mathbb{Q}$-factorial and terminal singularities can be embedded into a normal projective variety with only $\mathbb{Q}$-factorial and terminal singularities. We may assume that U is affine and choose an embedding

$$U \hookrightarrow \overline{U}$$

into a projective variety $\overline{U}$ (with no restriction on singularities). Take a birational morphism

$$f : V \rightarrow \overline{U}$$

from a nonsingular projective variety V. Carry out MMP over $\overline{U}$ to obtain

$$g : X \rightarrow \overline{U}.$$

We look at the restriction

$$g|_{g^{-1}(U)} : g^{-1}(U) \rightarrow U.$$

As before, we conclude that $g|_{g^{-1}(U)}$ is an isomorphism and hence $g^{-1}(U) = U$ appears in a process X of MMP starting from a nonsingular projective variety V.

This completes the proof of Theorem 4-1-3. $\square$

Remark 4-1-4. We could use the relative MMP more directly to give the following simple characterization:

An algebraic variety X has $\mathbb{Q}$-factorial and terminal singularities if it is the end result of the process of the relative MMP over X itself starting from any nonsingular variety V projective and birational over X.

One may prefer the above characterization to that of Theorem 4-1-3, where the characterization of terminal singularities, which should be *local* in nature, was given in terms of the absolute MMP over Spec $\mathbb{C}$, which is *global* in nature.

4.2   Canonical Singularities

The purpose of this section is to discuss the class of singularities called "**canonical singularities**." Its definition is given by loosening the condition on terminal singularities just a little, but it is of no less fundamental importance, characterizing the singularities on the canonical models of varieties of general type.

Definition 4-2-1 (Canonical Singularities).   *A normal variety X of dimension n has only* **canonical singularities** *if*

(1) *the canonical divisor K_X is $\mathbb{Q}$-Cartier, i.e., there exists $e \in \mathbb{N}$ such that eK_X is a Cartier divisor,*
(2) *there exists a projective birational morphism $f : V \to X$ from a nonsingular variety V such that in the ramification formula*

$$K_V = f^* K_X + \Sigma a_i E_i$$

all the coefficients for the exceptional divisors are nonnegative:

(*) $a_i \geq 0$ $\forall E_i$ *exceptional.*

Remark 4-2-2.

(i) One of the more intrinsic ways of defining canonical singularities is that normal variety X of dimension n has only canonical singularities iff

   (1) the canonical divisor K_X is $\mathbb{Q}$-Cartier, i.e., there exists $e \in \mathbb{N}$ such that eK_X is a Cartier divisor,
   (2_P) for any proper birational morphism $f : V \to X$ from a nonsingular variety X, the natural inclusion

$$f_* \mathcal{O}_V(mK_V) \hookrightarrow \mathcal{O}_X(mK_X)$$

   is an isomorphism for any $m \in \mathbb{N}$.
   We leave it to the reader as an exercise to show that this intrinsic definition is equivalent to Definition 4-2-1.

(ii) The definition of canonical singularities is identical to that of terminal singularities except that we allow some of the coefficients a_i of the exceptional divisors to be 0, instead of requiring them all to be strictly positive.

(iii) Once condition $(*)$ is satisfied for one projective birational morphism $f :$ $V \to X$ from a nonsingular variety V, it is satisfied by any proper birational morphism $g : W \to X$ from a nonsingular variety W. Therefore, just as in the definition of terminal singularities, we can replace condition (2) with the equivalent

(2′) for *any* proper birational morphism $f : V \to X$ from a nonsingular variety V in the ramification formula

$$K_V = f^* K_X + \Sigma a_i E_i$$

all the coefficients for the exceptional divisors are nonnegative:

$(*)$ $\qquad\qquad\qquad a_i \geq 0 \quad \forall E_i$ exceptional.

Though it is not necessary to require in the definition for the exceptional locus of f to be of pure codimension one or consist of normal crossing divisors, in coherence with the logarithmic variations in Section 4-3 we could add these extra conditions on the exceptional locus and replace condition (2) with the equivalent

(2″) there exists a projective birational morphism $f : V \to X$ from a nonsingular variety V such that the exceptional locus $\mathrm{Exc}(f)$ consists of normal crossing divisors and that in the ramification formula

$$K_V = f^* K_X + \Sigma a_i E_i$$

all the coefficients for the exceptional divisors are nonnegative:

$(*)$ $\qquad\qquad\qquad a_i \geq 0 \quad \forall E_i$ exceptional.

(iv) The notion of canonical singularities is easily seen to be algebraically local, just as the notion of terminal singularities is.

(v) A more intrinsic definition, which does not depend on the existence of a nice and global resolution or require the condition on all the resolutions, will be given in Section 4-3 to show that the notion of canonical singularities is actually analytically local.

The importance of the notion of canonical singularities lies in the fact that it characterizes the singularities that appear on the canonical models of varieties of general type whose canonical rings are finitely generated as $\mathbb{C}$-algebras (which is conjecturally the case with any variety of general type (cf. Conjecture 3-1-17)).

Theorem 4-2-3 (Characterization of Canonical Singularities). *Canonical singularities are characterized as those that appear on the canonical models of varieties of general type whose canonical rings are finitely generated as $\mathbb{C}$-algebras. That is to say:*

(i) *A projective variety X has canonical singularities with ample canonical divisor K_X iff it is the canonical model*

$$X = \mathrm{Proj}\, R$$

of a nonsingular projective variety V of general type whose canonical ring

$$R := \oplus_{m \geq 0} H^0(V, \mathcal{O}_V(mK_V))$$

is finitely generated as a $\mathbb{C}$-algebra.

(ii) *Assume the finite generation conjecture of the relative canonical ring for a projective birational morphism: for a projective birational morphism $f : V \to X'$ from a nonsingular variety V to a variety X' the relative canonical ring*

$$\oplus_{m \geq 0} f_* \mathcal{O}_V(mK_V)$$

is finitely generated as an $\mathcal{O}_{X'}$-algebra (cf. Conjecture 3-1-17). Then for any (algebraic) germ $p \in U$ of canonical singularities there exists a nonsingular projective variety of general type whose canonical ring is finitely generated such that its canonical model has a germ of singularities analytically isomorphic to $p \in U$.

PROOF.   Here we present an argument by Reid [2] [7].

Suppose that a projective variety X has only canonical singularities with ample canonical divisor K_X. Let $f : V \to X$ be any projective birational morphism from a nonsingular variety V. Then the ramification formula reads

$$K_V = f^* K_X + \Sigma a_i E_i \text{ with } a_i \geq 0 \quad \forall E_i \text{ exceptional.}$$

Therefore, we have (cf. Remark 4-2-2 (i))

$$H^0(V, \mathcal{O}_V(mK_V)) \cong H^0(X, \mathcal{O}_X(mK_X)) \quad \forall m \in \mathbb{N}.$$

Thus we conclude by the assumption of K_X being ample that

$$X = \text{Proj} \oplus_{m \geq 0} H^0(X, \mathcal{O}_X(mK_X)) = \text{Proj} \oplus_{m \geq 0} H^0(V, \mathcal{O}_V(mK_V))$$

is the canonical model of V, which is of general type.

Conversely, suppose V is a nonsingular projective variety of general type whose canonical ring

$$R = \oplus_{m \geq 0} H^0(V, \mathcal{O}_V(mK_V))$$

is finitely generated as a $\mathbb{C}$-algebra. Then we can take $d \in \mathbb{N}$ such that

$$R_{[d]} := \oplus_{m \geq 0} H^0(V, \mathcal{O}_V(mdK_V))$$

is generated by the degree 1 part of the graded ring $R_{[d]}$. That is to say, for each $m \in \mathbb{N}$ the natural map

$$H^0(V, \mathcal{O}_V(dK_V))^{\otimes m} \to H^0(V, \mathcal{O}_V(mdK_V))$$

is surjective. In terms of linear systems, this is equivalent to saying

$$|mdK_V| = |mM| + mF \text{ for } m \in \mathbb{N},$$

where

$$|dK_V| = |M| + F$$

with $|M|$ the movable part and F the fixed part for the linear system $|dK_V|$. By replacing V with a further blowup, if necessary, we may assume that $|M|$ has no base point, that the associated morphism

$$\Phi_{|M|} : V \to X \subset \mathbb{P}(H^0(V, \mathcal{O}_V(dK_V)))$$

has the properties

$$\mathcal{O}_V(M) = \Phi_{|M|}^* \mathcal{O}_X(1),$$

and that

$$X = \mathrm{Proj}\ R_{[d]} = \mathrm{Proj}\ R \text{ and is normal by Proposition 1-2-16.}$$

In order to see that X has only canonical singularities, we first claim that

$$\mathrm{codim}_X(\Phi_{|M|}(G)) \geq 2 \text{ for any component } G \subset F.$$

In fact, suppose $\Phi_{|M|}(G)$ is a divisor of dim $\Phi_{|M|}(G) = n-1$ for some irreducible component $G \subset F$. Consider the exact sequence

$$0 \to \mathcal{O}_V(mM) \to \mathcal{O}_V(mM + G) \to \mathcal{O}_V(mM + G)|_G \to 0$$

and its associated cohomology sequence

$$H^0(V, \mathcal{O}_V(mM + G)) \to H^0(G, \mathcal{O}_V(mM + G)|_G) \to H^1(V, \mathcal{O}_V(mM)).$$

Since $G \subset F$, we see that the first map is identically zero for any $m \in \mathbb{N}$. We also see that $h^0(G, \mathcal{O}_V(mM + G)|_G)$ grows on the order of m^{n-1} under the assumption dim $\Phi_{|M|}(G) = n - 1$. Thus so would $h^1(V, \mathcal{O}_V(mM))$. On the other hand, since

$$H^i(X, \Phi_{|M|_*}\mathcal{O}_V(mM)) = H^i(X, \mathcal{O}_X(m)) = 0 \text{ for } i > 0 \text{ and } m \gg 0$$

by Serre's vanishing theorem, we conclude by the Leray spectral sequence

$$H^1(V, \mathcal{O}_V(mM)) = H^0(X, R^1\Phi_{|M|_*}\mathcal{O}_V(mM)) = H^0(X, R^1\Phi_{|M|_*}\mathcal{O}_V \otimes \mathcal{O}_X(m)),$$

where the last term grows at most on the order of m^{n-2}, since the support of the sheaf $R^1\Phi_{|M|_*}\mathcal{O}_V$ has at most dimension $n - 2$, since X is normal. This is a contradiction and completes the proof of the claim.

Now in order to finish the proof for the assertion that X has only canonical singularities and K_X is ample, we take an open subset $U \subset X - \Phi_{|M|}(F)$ over which $\Phi_{|M|}^{-1}$ is a well-defined morphism. Since X is normal and by the claim, we can take U such that

$$\mathrm{codim}_X(X - U) \geq 2.$$

Then

$$\mathcal{O}_X(1)|_U \cong \mathcal{O}_V(M)|_{\Phi_{|M|}^{-1}(U)} \cong \mathcal{O}_V(dK_V)|_{\Phi_{|M|}^{-1}(U)} \cong \mathcal{O}_X(dK_X)|_U.$$

This implies that dK_X coincides with the Cartier divisor $\mathcal{O}_X(1)$ and hence is ample and that

$$K_V = \Phi^*_{|M|} K_X + \frac{1}{d} F,$$

which implies that X has only canonical singularities.

This proves assertion (i).

In order to prove assertion (ii), we may assume first that U is affine and that it is embedded in a projective variety $\overline{U}$. Take a projective birational morphism

$$f : V \to \overline{U}$$

from a nonsingular variety V. Then by the assumption,

$$\oplus_{m \geq 0} f_* \mathcal{O}_V(mK_V)$$

is finitely generated as an $\mathcal{O}_{\overline{U}}$-algebra and we can take the morphism

$$g : X = Proj \oplus_{m \geq 0} f_* \mathcal{O}_V(mK_V) \to \overline{U},$$

where by a similar argument to the one for assertion (i) we see that X has only canonical singularities. Since

$$f_* \mathcal{O}_V(mK_V)|_U \cong \mathcal{O}_U(mK_U) \text{ for all } m \in \mathbb{N},$$

we conclude that

$$g|_{g^{-1}(U)} : g^{-1}(U) \tilde{\to} U$$

is an isomorphism. Now take a sufficiently ample divisor A on X and a general member $p \notin H \in |2A|$ so that the associated double cover

$$d : \tilde{X} \to X$$

has the property that $\tilde{X}$ has only canonical singularities and $K_{\tilde{X}}$ is ample. (Exercise! Check this assertion on the singularities of $\tilde{X}$ and the ampleness of $K_{\tilde{X}}$.) Now, since d is ramified only along $H \not\ni p$, the canonical model

$$\tilde{X} = \tilde{X}_{\text{can}}$$

has a germ of singularity analytically isomorphic to $p \in U$.

We note that we can claim only to find the canonical model (of a variety of general type whose canonical ring is finitely generated as a $\mathbb{C}$-algebra) that has a germ *analytically* isomorphic to $p \in U$ but *not algebraically* isomorphic to $p \in U$ in general.   □

Remark 4-2-4. As in Section 4-1, we could give the following simple characterization if we work with the relative category, assuming the finite generation conjecture of the relative canonical ring:

An algebraic variety has canonical singularities iff for any projective birational morphism $f : V \to X$ from a nonsingular variety, X itself is the (relative) canonical model of V over X.

This is an easy consequence of the characterization of canonical singularities as in Remark 4-2-2 (i), and its verification is left to the reader as an exercise.

4.3   Logarithmic Variations

In this section we give the logarithmic counterparts of the notions of terminal and canonical singularities.

Definition 4-3-1 (Logarithmic Pair). *A* **logarithmic pair** *(also called a* **log pair** *for short) (X, D) is a pair consisting of a normal variety X and a boundary $\mathbb{Q}$-divisor D. A boundary divisor $D = \Sigma d_k D_k$, by definition, has all the coefficients between 0 and 1, i.e., $0 \le d_k \le 1$, where the D_k are the distinct irreducible components. We call $K_X + D$ the* **log canonical divisor** *of the log pair (X, D).*

Definition 4-3-2 (Log Terminal and Log Canonical Singularities). *A logarithmic pair (X, D) has only* **log terminal** *(respectively* **log canonical***) singularities, or we say that the log pair (X, D) is log terminal (respectively log canonical) by abuse of language, if*

(1) *the log canonical divisor $K_X + D$ is $\mathbb{Q}$-Cartier, i.e., there exists $e \in \mathbb{N}$ such that $e(K_X + D)$ is a Cartier divisor, and*

(2) *there exists a projective birational morphism*

$$f : V \to X$$

from a nonsingular variety V such that

$$D_V = \Sigma E_i + f_*^{-1} D$$

is a divisor with only normal crossings, where the E_i are the exceptional divisors for f and where $f_*^{-1} D$ is the strict transform of D, and such that in the logarithmic ramification formula

$$K_V + D_V = f^*(K_X + D) + \Sigma b_i E_i$$

all the coefficients for the exceptional divisors are

(∗) $b_i > 0$ $\forall E_i$ *exceptional(respectively $b_i \ge 0$ $\forall E_i$ exceptional).*

Remark 4-3-3. (i) In the case of log canonical singularities, once condition (∗) is satisfied for one projective birational morphism $f : V \to X$ from a pair con- sisiting of nonsingular variety V and the boundary $\mathbb{Q}$-divisor D_V with only normal crossings, it is satisfied by any proper birational morphism $g : W \to X$ from a pair consisting of a nonsingular variety W and the boundary $\mathbb{Q}$-divisor D_W. Therefore, we can replace condition (2) with the equivalent

(2′) for *any* proper birational morphism

$$f : V \to X$$

from a nonsingular variety V such that

$$D_V = \Sigma E_i + f_*^{-1} D \text{ is a divisor with only normal crossings}$$

where the E_i are the exceptional divisors for f and where $f_*^{-1}D$ is the strict transform of D, we have in the logarithmic ramification formula

$$K_V + D_V = f^*(K_X + D) + \Sigma b_i E_i$$

all the coefficients for the exceptional divisors to be nonnegative:

$$(*)\qquad\qquad b_i \geq 0 \quad \forall E_i \text{ exceptional.}$$

On the other hand, in the case of log terminal singularities, condition $(*)$ *does* depend on the choice of a particular resolution $f : V \to X$: Let (X, D) be a log pair consisting of a nonsingular surface X with a boundary divisor $D = D_1 + D_2$, the irreducible components D_1 and D_2 being nonsingular and intersecting transversally at a point $p \in D_1 \cap D_2 \subset X$. Then, taking $f : V = X \to X$ to be the identity morphism condition (i) is satisfied, and hence (X, D) has only log terminal singularities. But if we take $f : V = \mathrm{Blp}_p X \to X$ to be the blowup of X at p, then $K_V + D_V = f^*(K_X + D)$, and we do not have the coefficient of the exceptional divisor strictly positive.

(ii) If we replace condition (2) with the following stronger condition (2″), then we get the notion of **divisorially log terminal singularities**

(2″) there exists a proper birational morphism

$$f : V \to X$$

from a nonsingular variety V such that the exceptional locus $\mathrm{Exc}(f)$ is of pure codimension one and

$$D_V = \Sigma E_i + f_*^{-1}D$$

is a divisor with only normal crossings, where the E_i are the exceptional divisors for f and where $f_*^{-1}D$ is the strict transform of D, and such that in the logarithmic ramification formula

$$K_V + D_V = f^*(K_X + D) + \Sigma b_i E_i$$

all the coefficients for the exceptional divisors are strictly positive:

$$(*)\qquad\qquad b_i > 0 \quad \forall E_i \text{ exceptional.}$$

If we further require f to possess an f-ample divisor $A = -\Sigma \delta_i E_i$ whose support coincides with the exceptional locus, i.e.,

$$A \quad \text{is } f\text{-ample and } \mathrm{Supp}A = \mathrm{SuppExc}(f) = \cup E_i,$$

then we get the notion of **weakly kawamata log terminal singularities**.

We discuss the relation among these notions at the end of the section .

(iii) We use the abbreviations **lc, lt, dlt, wklt** for log canonical, log terminal, divisorially log terminal, weakly kawamata log terminal singularities, respectively.

Now the characterizations of log terminal singularities (with $\mathbb{Q}$-factoriality) and log canonical singularities go parallel to those of terminal singularities (with $\mathbb{Q}$-factoriality) and canonical singularities.

Theorem 4-3-4 (Characterization of $\mathbb{Q}$-factorial and Log Terminal Singularities). *$\mathbb{Q}$-factorial and log terminal singularities are characterized as those that appear in the process of log MMP starting from logarithmic pairs consisting of nonsingular projective varieties and boundary $\mathbb{Q}$-divisors with only normal crossings, assuming (existence of log flip) and (termination of log flips). More precisely, the category $\mathfrak{D}$ of log pairs consisting of normal projective varieties and boundary $\mathbb{Q}$-divisors with only $\mathbb{Q}$-factorial and log terminal singularities is the smallest category that contains the category $\mathfrak{E}$ of log pairs consisting of nonsingular projective varieties and boundary $\mathbb{Q}$-divisors with only normal crossings and in which Log MMP works. That is to say, assuming (existence of log flip) and (termination of log flips), we have*

(i) *$\mathfrak{D}$ contains $\mathfrak{E}$,*
(ii) *$\mathfrak{D}$ is closed under the operations of log MMP, and*
(iii) *any object $(X, D) \in \mathfrak{D}$ can be obtained from some object in $\mathfrak{E}$ through the process of log MMP.*

Moreover, any (algebraic) germ (U, B) of $\mathbb{Q}$-factorial and log terminal singularities appears in a process of log MMP starting from some log pair consisting of a nonsingular projective variety and a boundary $\mathbb{Q}$-divisor with only normal crossings.

Theorem 4-3-5 (Characterization of Log Canonical Singularities). *Log canonical singularities are characterized as those that appear in the log canonical models of log pairs of log general type whose log canonical rings are finitely generated as $\mathbb{C}$-algebras. That is to say:*

(i) *A log pair (X, D) has log canonical singularities with ample log canonical divisor $K_X + D$ iff it is the log canonical model*

$$(X, D) = (\mathrm{Proj}\, R, \varphi_* B)$$

of a log pair (V, B) consisting of a nonsingular projective variety V and a boundary $\mathbb{Q}$-divisor B with only normal crossings of log general type whose log canonical ring

$$R = \oplus_{m \geq 0} H^0(V, \mathcal{O}_V(m(K_V + B)))$$

is finitely generated as a $\mathbb{C}$-algebra.
(We have the natural birational map $\varphi : V \dashrightarrow \mathrm{Proj}\, R$, which is defined on V over the complement of a closed subset whose codimension is at least 2. Since B is a divisor, φ is defined on the generic points of the irreducible components of B, and we take $\varphi_* B$ to be the image as a codimension-one cycle.)
(ii) *Assume the logarithmic finite generation conjecture for projective birational morphisms, which claims that the relative log canonical ring*

$$\oplus_{m \geq 0} f_* \mathcal{O}_V(m(K_V + D_V))$$

is finitely generated as an $\mathcal{O}_{X'}$-algebra for any projective birational morphism $f : (V, D_V) \to X'$ from a pair consisting of a nonsingular variety V and *a boundary $\mathbb{Q}$-divisor D_V with only normal crossings to a variety X'. Then for any (algebraic) germ $p \in (U, D_U)$ of log canonical singularities there exists a log pair of log general type consisting of a nonsingular projective variety and a boundary with only normal crossings whose log canonical ring is finitely generated as a $\mathbb{C}$-algebra such that its log canonical model has a germ of log canonical singularities analytically isomorphic to $p \in (U, D_U)$.*

(iii) *Again assuming the logarithmic finite generation conjecture for projective birational morphisms, a log pair (X, D) has only log canonical singularities iff for any projective birational morphism $f : (V, D_V) \to (X, D)$, from a log pair (V, D_V) consisting of a nonsingular variety V and a boundary $\mathbb{Q}$-divisor with only normal crossings*

$$D_V = \Sigma E_i + f_*^{-1} D,$$

where the E_i are the exceptional divisors for f and where $f_*^{-1} D$ is the strict transform of D, (X, D) itself is the (relative) log canonical model of (V, D_V) over X.

The proofs of Theorem 4-3-4 and Theorem 4-3-5 are completely parallel to those of Theorem 4-1-3 and Theorem 4-2-3, and hence details are left to the reader as an exercise.

Finally, we discuss the relation among the notions of log terminal, divisorially log terminal, and weakly Kawamata log terminal singularities.

Proposition 4-3-6.

 (i) *For a log pair (X, D), if we assume that X is $\mathbb{Q}$-factorial, then the three notions of log terminal, divisorially log terminal, and weakly kawamata log terminal coincide.*

(ii) *In general, we have the following implications:*

$$\text{dlt} \Longleftrightarrow \text{wklt} \Longrightarrow \text{lt.}$$

PROOF.   By definition the following implications are clear:

$$\text{dlt} \Longleftarrow \text{wklt} \Longrightarrow \text{lt.}$$

Therefore, in order to verify assertion (i) (assuming assertion (ii) for the moment), we have only to show under the assumption of $\mathbb{Q}$-factoriality the implication

$$\text{wklt} \Longleftarrow \text{lt.}$$

Let (X, D) be a log pair with only log terminal singularities with the assumption that X is $\mathbb{Q}$-factorial. By the definition of log terminal singularities, there exists a projective birational morphism

$$f : V \to X$$

from a nonsingular variety V such that

$$D_V = \Sigma E_i + f_*^{-1} D$$

is a divisor with only normal crossings, where the E_i are the exceptional divisors for f and such that in the logarithmic ramification formula

$$K_V + D_V = f^*(K_X + D) + \Sigma b_i E_i$$

all the coefficients for the exceptional divisors are strictly positive:

(∗) $b_i > 0.$

Now observe that since X is $\mathbb{Q}$-factorial and so is V, the exceptional locus $\mathrm{Exc}(f)$ for f is of pure codimension one (see the argument in the proof of Theorem 4-1-3) and hence that

$$\mathrm{SuppExc}(f) = \mathrm{Supp}\Sigma E_i.$$

Take an f-ample divisor A' on V. Then since X is $\mathbb{Q}$-factorial, $f_* A'$ is a $\mathbb{Q}$-Cartier divisor. Now,

$$A = A' - f^*(f_* A') = -\Sigma \delta_i E_i$$

is an f-ample divisor whose support lies in the exceptional locus $\mathrm{Exc}\,(f)$. Since A is f-ample, it is easy to see (Exercise! Why? See, for example, Lemma 13-1-4 known as the negativity lemma) that

$$\delta_i > 0 \quad \forall i.$$

Therefeore, we have

$$\mathrm{Supp}A = \mathrm{Supp}\Sigma E_i = \mathrm{SuppExc}(f),$$

proving assertion (i).

In order to verify assertion (ii), we have only to show the implication

$$\mathrm{dlt} \implies \mathrm{wklt}.$$

Let (X, D) be a log pair with only divisorially log terminal singularities. By definition, there exists a *proper* birational morphism

$$f : V \to X$$

from a nonsingular variety V such that the exceptional locus $\mathrm{Exc}\,(f)$ is of pure codimension one and

$$D_V = \Sigma E_i + f_*^{-1} D$$

is a divisor with only normal crossings, where the E_i are the exceptional divisors for f and such that in the logarithmic ramification formula

$$K_V + D_V = f^*(K_X + D) + \Sigma b_i E_i$$

all the coefficients for the exceptional divisors are strictly positive:

(∗) $b_i > 0.$

Note that

$$\mathrm{SuppExc}(f) = \mathrm{Supp}\Sigma E_i = \mathrm{Supp}\Sigma b_i E_i.$$

We take another resolution of singularities of (X, D) in the following way. First, by blowing up only over the locus where the birational map f^{-1} is not defined, we obtain a projective birational morphism

$$\sigma : W' \to X$$

from a nonsingular variety W' such that $f^{-1}\circ\sigma$ is a morphism. (See the elimination of points of indeterminacy by Hironaka [2][3].) Second, by blowing up only over the locus where $\sigma^{-1}(D) \cup \mathrm{Exc}(\sigma)$ is not a normal crossing divisor, we obtain a projective birational morphism

$$\tau : W \to W'$$

from a nonsingular variety W such that $\tau^{-1}(\sigma^{-1}(D)\cup\mathrm{Exc}(\sigma))\cup\mathrm{Exc}(\tau)$ is a divisor with only normal crossings. Now, $g = \sigma \circ \tau : W \to X$ is a projective birational morphism from a nonsingular variety W such that

$$D_W = \Sigma F_j + g_*^{-1}(D)$$

is a divisor with only normal crosings, where the F_j are the exceptional divisors for g and such that, since g is composed of blowups, we have a g-ample divisor A with

$$\mathrm{Supp}A = \mathrm{SuppExc}(g) = \mathrm{Supp}\Sigma F_j.$$

Moreover, setting $h = f^{-1}\circ\sigma\circ\tau : W \to V$, we have the logarithmic ramification formula

$$K_W + D_W = h^*(K_V + D_V) + R_{\log} = h^*(f^*(K_X + D) + \Sigma b_i E_i) + R_{\log}$$
$$= g^*(K_X + D) + \Sigma c_j F_j,$$

where $R_{\log}$ is the effective log ramification divisor (cf. Theorem 2-1-8). Noting that $\mathrm{Supp}\Sigma F_j \subset \mathrm{Supp}h^{-1}(\Sigma E_i)$ by construction, we have

$$c_j > 0 \quad \forall F_j \text{ exceptional for } g.$$

Thus g satisfies condition (ii) of the definition for wklt, and hence (X, D) has weakly kawamata log terminal singularities.

This completes the proof of Proposition 4-3-6. $\square$

Exercise 4-3-7 (cf. Hironaka [2][3], Bierstone–Milman [1], Encinas–Villamayor [1], Villamayor [1], Szábo [1], Kollár–Mori [2][3]). *For a log pair (X, D), the condition that (X, D) has only divisorially log terminal singularities (which is equivalent to the condition that (X, D) has only weakly kawamata log terminal singularities, as demonstrated in Proposition 4-3-6) is equivalent to the following conditions:*

(i) *The log canonical divisor $K_X + D$ is $\mathbb{Q}$-Cartier.*

(ii) *Let Z be the closed subset of X consisting of* $\text{Sing}(X)$ *and the locus inside of* $X - \text{Sing}(X)$, *where D is not a divisor with only normal crossings. (Exercise! Why is the locus Z closed?) Then any exceptional divisor E whose center lies over Z has discrepancy (with respect to the pair (X, D)) strictly bigger than* -1, *i.e.,*

$$a(E; X, D) > -1.$$

(See Proposition-Definition 4-4-1 for the definition of the discrepancy.)

4.4   Discrepancy and Singularities

Now we shift our attention to an attempt to give a more intrinsic definition using the notion of "**discrepancy**," unifying terminal, canonical, and log canonical singularities.

Proposition-Definition 4-4-1 (Discrepancy). *Let (X, D) be a log pair consisting of a normal variety X and a boundary $\mathbb{Q}$-divisor $D = \Sigma d_k D_k$. Assume that $K_X + D$ is $\mathbb{Q}$-Cartier. Let $f : V \to X$ be a birational morphism from a nonsingular variety V. Write the ramification formula*

$$K_V = f^*(K_X + D) - f_*^{-1}(D) + \Sigma a(E; X, D)E,$$

*where the E are the exceptional divisors for f. Then $a(E; X, D)$ is independent of the morphism f and depends only on the discrete valuation that corresponds to E. It is called the **discrepancy** of the pair (X, D) at E.*

   *The **log discrepancy** of the pair (X, D) at E is defined to be*

$$b(E; X, D) = a(E; X, D) + 1,$$

so that

$$K_V + D_V = f^*(K_X + D) + \Sigma b(E; X, D)E,$$

where

$$D_V = \Sigma E + f_*^{-1}(D).$$

   Moreover, we define

$$\text{discrep}(X, D) = \inf\{a(E; X, D); E \text{ corresponds to a discrete valuation}$$
$$\text{such that } \text{Center}_X(E) \neq \emptyset \text{ and } \text{codim}_X(\text{Center}_X(E)) \geq 2\}.$$

PROOF.   The proof is easy and left to the reader as an exercise.   □

Definition 4-4-2 (A Unifying Definition for Terminal, Canonical, Purely Log Terminal, and Log Canonical Singularities). *A log pair (X, D) consisting of a* *normal variety X and a boundary $\mathbb{Q}$-divisor $D = \Sigma d_k D_k$ has only*

$$
\begin{cases}
\text{terminal} \\
\text{canonical} \\
\text{purely log terminal} \\
\text{log canonical}
\end{cases}
\textit{singularities if discrep}(X, D) =
\begin{cases}
> 0, \\
\geq 0, \\
> -1, \\
\geq -1.
\end{cases}
$$

Remark 4-4-3. (i) It is straightforward to check that the above definition is compatible with the previous definitions, i.e.,

X has only terminal singularities in the sense of Definition 4-1-1

$\Longleftrightarrow (X, 0)$ has only terminal singularities in the sense of Definition 4-4-2;

X has only canonical singularities in the sense of Definition 4-2-1

$\Longleftrightarrow (X, 0)$ has only canonical singularities in the sense of Definition 4-4-2;

(X, D) has only log canonical singularities in the sense of Definition 4-3-1

$\Longleftrightarrow (X, D)$ has only log canonical singularities in the sense of Definition 4-4-2.

(ii) The notion of **kawamata log terminal singularities** is obtained by requiring the extra condition on the boundary $\mathbb{Q}$-divisor $D = \Sigma d_k D_k$ that all the coefficients be strictly less than 1:

$(\diamond)$ $\qquad\qquad\qquad\qquad\qquad 0 \leq d_k < 1.$

Thus we have the following implications (Exercise! Why?) among the singularities of the log pairs:

kawamata log terminal $\Longleftrightarrow$ purely log terminal and $(\diamond)$ $\Longleftrightarrow$ log terminal and $(\diamond)$.

Proposition 4-4-4. *The notion of*

> *terminal,*
>
> *canonical,*
>
> *purely log terminal,*
>
> *log canonical,*
>
> *kawamata log terminal,*

singularities is not only algebraically local but also analytically local in the following sense: Let $\{U_\alpha\}$ be a covering of (X, D) by open subsets in the usual topology. Then (X, D) has only

$$
\begin{cases}
\textit{terminal} \\
\textit{canonical} \\
\textit{purely log terminal} \qquad \textit{singularities in the algebraic category} \\
\textit{log canonical} \\
\textit{kawamata log terminal}
\end{cases}
$$

iff $(U_\alpha, D|_{U_\alpha})$ *has*

$$
\begin{cases}
\text{terminal} \\
\text{canonical} \\
\text{purely log terminal} \\
\text{log canonical} \\
\text{kawamata log terminal}
\end{cases}
\quad \text{singularities in the analytic category} \quad \forall \alpha.
$$

PROOF.   It is clear that the analytical *local* characterizations imply the corresponding algebraic ones. In fact, let E be an algebraic exceptional divisor whose center lies in U_α. Then since E is obviously an analytic exceptional divisor as well, we have

$$
a(E; X, D)
\begin{cases}
> 0 \\
\geq 0 \\
> -1 \\
\geq -1
\end{cases}
\Longleftarrow (U_\alpha, D|_{U_\alpha})
\begin{cases}
\text{terminal} \\
\text{canonical} \\
\text{purely log terminal (or klt)} \\
\text{log canonical}
\end{cases}
\begin{array}{l}
\text{in the analytic} \\
\text{category.}
\end{array}
$$

This verifies that the *local* analytic characterizations imply the algebraic ones. It is slightly more subtle to show that the algebraic characterizations imply the corresponding *local* analytic ones. The subtlety lies in the fact that an analytic exceptional divisor E_{an} (over U_α) may not be algebraic. We take a birational morphism $f : V \to X$ in the algebraic category from a nonsingular variety V such that $D_V = \mathrm{Exc}(f) \cup f^{-1}(D)$ is a divisor with only normal crossings. Then take a birational morphism $g : W \to V$ in the analytic category such that $D_W = \mathrm{Exc}(g) \cup f^{-1}(D_V)$ is a divisor with only normal crossings and such that E_{an} appears on W as a divisor (exceptional over X). Now it is easy to see (Exercise! Why?) via the logarithmic ramification formula (cf. Theorem 2-1-8) applied to the morphism of log pairs $g : (W, D_W) \to (V, D_V)$ in the analytic category that

$$
a(E_{\mathrm{an}}; U_\alpha, D|_{U_\alpha})
\begin{cases}
> 0 \\
\geq 0 \\
> -1 \\
\geq -1
\end{cases}
\Longleftarrow (X, D)
\begin{cases}
\text{terminal} \\
\text{canonical} \\
\text{purely log terminal (or klt)} \\
\text{log canonical}
\end{cases}
\begin{array}{l}
\text{in the algebraic} \\
\text{category.}
\end{array}
$$

This completes the proof.   □

Actually, with closer bookkeeping as in Kollár [8] or Masek [1] one can prove that

discrep(X, D) in the algebraic category $=$ discrep(X, D) in the analytic category,

which verifies both implications immediately.

Exercise 4-4-5 (cf. Exercise 4-3-6).   Show that the notion of divisorially log terminal or equivalently weakly kawamata log terminal singularities is also analytically *local* as well as algebraically local.

Question 4-4-6. *Is the notion of log terminal singularities algebraically local and/or analytically local?*

Summary 4-4-7. In Figure 4-4-8 we show the inclusion relation among the various notions of singularities in the algebraic category.

Figure 4-4-8.

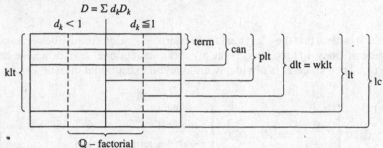

When all the coefficients d_k of the irreducible components of the boundary divisor $D = \Sigma d_k D_k$ are strictly less than 1, we have the implications

$$\text{terminal} \implies \text{canonical}$$
$$\implies \text{klt} \iff \text{plt} \iff \text{dlt} \iff \text{wklt} \iff \text{lt}$$
$$\implies \text{lc},$$

and $\mathbb{Q}$-factoriality does not affect the implication relation.

When some of the coefficients d_k of the irreducible components of the boundary divisor $D = \Sigma d_k D_k$ may be equal to 1, we have the implications

$$\text{terminal} \implies \text{canonical}$$
$$\implies \text{plt}$$
$$\implies \text{dlt} \iff \text{wklt}$$
$$\implies \text{lt}$$
$$\implies \text{lc}.$$

If we assume $\mathbb{Q}$-factoriality, the three notions of dlt, wklt, and lt coincide as in Proposition 4-3-6:

$$\text{dlt} \iff \text{wklt} \iff \text{lt}.$$

Finally, we end this section noting the difference between $\mathbb{Q}$-factoriality in the algebraic actegory and in the analytic category. The former insists that all the algebraic Weil divisors be $\mathbb{Q}$-Cartier and the latter all the analytic Weil divisors. As in general we have more analytic Weil divisors than algebraic ones, the latter implies the former but not conversely.

Example 4-4-9 (The Difference Between $\mathbb{Q}$-Factoriality in the Algebraic Category and $\mathbb{Q}$-Factoriality in the Analytic Category). First, we take $p \in U$ to be the vertex of the cone over a quadric:

$$p \in U \cong 0 \in \{xy - zw = 0\} \subset \mathbb{A}^4.$$

It is easy to see that the two Weil divisors D_1 and D_2 corresponding to the two different rulings of the quadric,

$$\{XY - ZW = 0\} \subset \mathbb{P}^3,$$

are never $\mathbb{Q}$-Cartier. In fact, if D_1 and D_2 were $\mathbb{Q}$-Cartier, then $D_1 \cap D_2$ would be of pure codimension 1, while $D_1 \cap D_2 = p$. Thus $p \in U$ is not $\mathbb{Q}$-factorial in the algebraic category, much less in the analytic category.

Second, we take $q \in X$ to be the singularity resulting from a Type (E_3) extremal contraction $c : V \to X$ (which will be discussed in Example 8-3-1) from a nonsingular projective variety V. Since it is obtained by contracting an extremal ray on a nonsingular projective variety, it is $\mathbb{Q}$-factorial in the algebraic category. On the other hand, $q \in X$ is analytically isomorphic to $p \in U$ as above and thus not $\mathbb{Q}$-factorial in the analytic category.

4.5   Canonical Cover

In this section we introduce the fundamental notion of the **canonical cover**, due to Reid and Wahl. The main virtue of the canonical cover is to reduce the study of singularities of higher indices to that of singularities of index one. We also discuss the behavior of discrepancies under the canonical cover.

Proposition-Definition 4-5-1 (Canonical Cover). *Let*

$$p \in X$$

be a germ of a normal variety such that K_X is $\mathbb{Q}$-Cartier. Set

$$r = \text{the index of } K_X := \min\{e \in \mathbb{N}; e K_X \text{ is Cartier}\}.$$

Then there exists a finite morphism

$$c : \tilde{X} \to X$$

such that

(i) *$\tilde{X}$ is a normal (irreducible) variety with $c^{-1}(p) = q$ a point,*

(ii) *the extension of the function fields $\mathbb{C}(\tilde{X})/\mathbb{C}(X)$ is Galois and the Galois group $G \cong \mathbb{Z}/(r)$ acts on $\tilde{X}$ over X,*

(iii) *c is étale in codimension one,*

(iv) *$\mathcal{O}_{\tilde{X}}(K_{\tilde{X}}) \cong \mathcal{O}_{\tilde{X}}$, i.e., $K_{\tilde{X}}$ is Cartier and the index of $\tilde{X}$ is 1, and*

(v) *there is a nonvanishing section $\tilde{s}$ of $\mathcal{O}_{\tilde{X}}(K_{\tilde{X}})$ around q that is semi-invariant for the action of the Galois group $G \cong \mathbb{Z}/(r)$ and on which G acts faithfully.*

It is uniquely determined by the conditions (i), (ii), (iii), and (iv) up to analytic isomorphism, and is called the **canonical cover** *of X.*

PROOF.   By taking a small affine open neighborhood of p, we may assume that

$$X = \operatorname{Spec} R \text{ and } \mathcal{O}_X(rK_X) \cong \mathcal{O}_X$$

and that there is a nowhere-vanishing section

$$s \in H^0(X, \mathcal{O}_X(rK_X)) \cong H^0(X, \mathcal{O}_X),$$

which corresponds to a rational function

$$f \in \mathbb{C}(X) \text{ with } \operatorname{div}(f) + K_X = 0.$$

We claim that the polynomial $T^r - f$ is irreducible in $\mathbb{C}(X)[T]$. (See Claim 5-1-2.) In fact, if not, then $(\sqrt[r]{f})^i \in \mathbb{C}(X)$ for some $1 \le i < r$ and we have (Exercise! Why?) $\operatorname{div}\{(\sqrt[r]{f})^i\} + iK_X = 0$, which implies that iK_X is a Cartier divisor, contradicting our choice of the index r.

Thus $\mathbb{C}(X)[\sqrt[r]{f}] = \mathbb{C}(X)[T]/(T^r - f)$ is a finite field extension, which is Galois, of $\mathbb{C}(X)$ with the Galois group $G \cong \mathbb{Z}/(r)$.

We take S to be the integral closure of R in $\mathbb{C}(X)[\sqrt[r]{f}]$ and set

$$c : \tilde{X} = \operatorname{Spec} S \to X = \operatorname{Spec} R$$

to be the finite morphism corresponding to the inclusion $c^* : R \hookrightarrow S$. Note that the Galois group G acts on S and induces the splitting into the eigenspaces

$$S = \oplus_{i=0}^{r-1} R_i \cdot (\sqrt[r]{f})^i \text{ with } R_0 = R \text{ and } R_i \subset \mathbb{C}(X).$$

We claim that

$$m_q := m_p \oplus \left\{ \oplus_{i=1}^{r-1} R_i \cdot (\sqrt[r]{f})^i \right\}$$

is the only maximal ideal in S lying over m_p, where $m_p \subset R = R_0$ is the maximal ideal corresponding to the point p.

In fact, take any element

$$r_j \cdot (\sqrt[r]{f})^j \in \left\{ \oplus_{i=1}^{r-1} R_i \cdot \left(\sqrt[r]{f}\right)^i \right\} \text{ with } r_j \in R_j, \quad 1 \le j < r.$$

Then $\{r_j \cdot (\sqrt[r]{f})^j\}^r = r_j^r \cdot f^j \in R$, and this vanishes at p. (If it did not vanish at p, then we would have in a neighborhood of p

$$0 = \operatorname{div}(r_j^r \cdot f^j) = r \cdot \operatorname{div}(r_j) + j \cdot \operatorname{div}(f) = r \cdot (\operatorname{div}(r_j) - jK_X),$$

which would imply

$$\operatorname{div}(r_j) = jK_X, \text{ where } 1 \le j < r,$$

again contradicting the choice of our index r.) Therefore, we have

$$m_q \subset \sqrt{c^{-1}m_p \cdot \mathcal{O}_{\tilde{X}}}.$$

On the other hand, we also have

$$S/m_q = (\oplus_{i=0}^{r-1} R_i \cdot (\sqrt[r]{f})^i)/(m_p \oplus \{\oplus_{i=1}^{r-1} R_i\}) = R_0/m_p = R/m_p \cong \mathbb{C}.$$

Thus we conclude that m_q is the unique maximal ideal lying over m_p.

This proves that $c : \tilde{X} \to X$ satisfies conditions (i) and (ii).

It is easy to see (Exercise! Why? See the proof of Claim 5-1-3) that

$$c|_{c^{-1}(X_{\text{reg}})} : c^{-1}(X_{\text{reg}}) \to X$$

is étale, proving condition (iii). This also implies that

$$K_{\tilde{X}}|_{c^{-1}(X_{\text{reg}})} = c^*(K_X|_{X_{\text{reg}}}).$$

Since $\text{codim}_{\tilde{X}} c^{-1}(X_{\text{reg}}) \geq 2$, we conclude that

$$r \cdot \{\text{div}(\sqrt[r]{f}) + K_{\tilde{X}}\} = \text{div}(f) + r K_{\tilde{X}}$$
$$= c^* \{\text{div}(f) + r K_X\} = 0.$$

This implies

$$\text{div}(\sqrt[r]{f}) + K_{\tilde{X}} = 0,$$

proving condition (iv) that $K_{\tilde{X}}$ is Cartier.

Moreover, the section $\tilde{s} \in H^0(\tilde{X}, \mathcal{O}_{\tilde{X}}(K_{\tilde{X}}))$ correspnding to the rational function $\sqrt[r]{f} \in \mathbb{C}(\tilde{X})$ is obviously semi-invariant for the action of the Galois group, and G acts faithfully on it. This proves condition (iv).

Next we verify that conditions (i), (ii), (iii), and (iv) characterize the canonical cover up to analytic isomorphism.

Let $\hat{c} : \hat{X} \to X$ be a finite morphism satisfying conditions (i), (ii), (iii), and (iv). By conditions (iii) and (iv) there exists a rational function $h \in \mathbb{C}(\hat{X})$ such that

$$0 = \text{div}(h) + K_{\hat{X}} = \text{div}(h) + \pi^* K_X.$$

Therefore, for $g \in G$ we have $g(h) = u_g \cdot h$, where u_g is a unit. By replacing h with $\sqrt[r]{\prod_{g \in G} u_g} \cdot h$, we may assume that

$$g(h) = \epsilon(g) \cdot h, \quad \text{where } \epsilon(g) \text{ is an } r\text{th root of unity.}$$

Now,

$$\hat{c}_*(0) = \hat{c}_*(\text{div}(h) + c^* K_X) = \text{div}(\text{Norm}(h)) + r K_X = \text{div}(h') + r K_X.$$

Thus we conclude that $\hat{c} : \hat{X} \to X$ factors as

$$\hat{X} \to \tilde{X} \xrightarrow{c} X,$$

where $\tilde{X}$ is the normalization of X in $\mathbb{C}(X)[h] = \mathbb{C}[\sqrt[r]{f}]$ with $h^r = f \in \mathbb{C}(X)$ and $\text{div}(f) + r K_X = 0$ as before. Now, since $[\mathbb{C}(\hat{X} : \mathbb{C}(X)] = r = [\mathbb{C}(\tilde{X}) : \mathbb{C}(X)]$, we conclude that $\hat{X} = \tilde{X}$.

As a rational function $f \in \mathbb{C}(X)$ with $\text{div}(f) + r K_X = 0$ is unique only unique up to units, and we replace h as in the proof with $\sqrt[r]{\prod_{g \in G} u_g} \cdot h$; the canonical cover is unique up to analytic isomorphism.

This completes the proof of Proposition-Definition 4-5-1.   □

Remark 4-5-2. There is a more intrinsic description of the canonical cover: Let

$$s \in H^0(X, \mathcal{O}_X(rK_X)) \cong H^0(X, \mathcal{O}_X)$$

be a nowhere-vanishing dection. Define an $\mathcal{O}_X$-algebra structure on

$$\oplus_{i=0}^{r-1} \mathcal{O}_X(-iK_X)$$

by the following multiplication rule:

$$\mathcal{O}_X(-iK_X) \times \mathcal{O}_X(-jK_X) \to \mathcal{O}_X((-(i+j)+r)K_X) \quad \text{if } i+j \geq r,$$
$$a \times b \mapsto a \cdot b \cdot s.$$

Then we take the natural morphism

$$c : \tilde{X} = Spec \oplus_{i=0}^{r-1} \mathcal{O}_X(-iK_X) \to X.$$

We leave it to the reader as an exercise to show that the canonical cover constructed in our proof does indeed coincide with this intrinsic one where the splitting $S = \oplus_{i=0}^{r-1} R_i \cdot (\sqrt[r]{f})^i$ corresponds to that of the $\mathcal{O}_X$-algebra $\oplus_{i=0}^{r-1} \mathcal{O}_X(-iK_X)$ with the description of R_i given by

$$R_i = \{r_i \in \mathbb{C}[X]; \operatorname{div}(r_i) + (-iK_X) \geq 0\}.$$

The next proposition compares the singularities of the canonical cover $\tilde{X}$ to those of X and vice versa by studying the behavior of the discrepancies under the canonical cover.

Proposition 4-5-3 (cf. Kawamata [9], Reid [2][3]). *Let*

$$p \in X$$

be a germ of a normal variety such that K_X *is* $\mathbb{Q}$-*Cartier, and*

$$c : \tilde{X} \to X$$

its canonical cover.
Then we have the following implications:

(i)

$$X \text{ has only } \begin{cases} \log \text{ terminal} \\ \log \text{ canonical} \end{cases} \text{ singularities}$$

$$\Longleftrightarrow$$

$$\tilde{X} \text{ has only } \begin{cases} \log \text{ terminal} \\ \log \text{ canonical} \end{cases} \text{ singularities.}$$

$$\Longleftrightarrow$$

$$\tilde{X} \text{ has only } \begin{cases} \text{canonical} \\ \log \text{ canonical} \end{cases} \text{ singularities.}$$

(Note that since there is no boundary the notions of log terminal, Kawamata log terminal, and purely log terminal coincide.)

(ii)

$$X \text{ has only} \begin{cases} \text{terminal} \\ \text{canonical} \end{cases} \text{singularities}$$

$$\Longrightarrow$$

$$\tilde{X} \text{ has only} \begin{cases} \text{terminal} \\ \text{canonical} \end{cases} \text{singularities.}$$

Though the converse in (ii) does not hold in general (counterexamples can easily be found among some surface quotient singularities, where the canonical cover $\tilde{X}$ has canonical singularities (respectively terminal singularities and hence is nonsingular) but the original space X does not have canonical (respectively terminal) singularities), we have the implication

X has only canonical singularities and $\tilde{X}$ has only terminal singularities

$$\Longrightarrow$$

X has only terminal singularities.

PROOF. (i) Take projective birational morphisms $f : V \to X$ and $\tilde{g} : \tilde{V} \to \tilde{X}$ from nonsingular varieties V and $\tilde{V}$ so that $\tilde{V}$ also dominates V by another projective generically finite morphism $\tau : \tilde{V} \to V$:

$$
\begin{array}{ccc}
\tilde{V} & \xrightarrow{\tilde{f}} & \tilde{X} \\
\tau \downarrow & & \downarrow c \\
V & \xrightarrow{f} & X.
\end{array}
$$

Write the ramification formulas

$$(1) \quad K_V = f^* K_X + \Sigma a_i E_i,$$
$$(2) \quad K_{\tilde{V}} = \tilde{f}^* K_{\tilde{X}} + \Sigma b_j F_j,$$

where the E_i (respectively the F_j) are f-exceptional (respectively $\tilde{f}$-exceptional) divisors. We may assume, by taking further blowups if necessary, that ΣE_i and ΣF_j are divisors with only normal crossings. Note that

$$\tau^{-1}(\cup E_i) \subset \cup F_j,$$

since any divisor $D \subset \tau^{-1}(\cup E_i)$ is $f \circ \tau$-exceptional, which implies that it is $\tilde{f}$-exceptional, since c is a finite morphism, and that

$$\tau_*(\cup F_j) \subset \cup E_i,$$

since any $\tilde{f}$-exceptional divisor F is either τ-exceptional or f-exceptional. (Note that τ_* denotes the cycle-theoretic image. Thus if F is τ-exceptional, then $\tau_*(F)$ is considered to be zero or empty.)

Therefore, by the logarithmic ramification formula (Theorem 2-1-9 and Exercise 2-1-10) we have

$$(3) \quad K_{\tilde{V}} + \Sigma F_j = \tau^*(K_V + \Sigma E_i) + R,$$

where R is some τ-exceptional effective divisor.

Substituting $K_{\tilde{X}} = c^* K_X$ into (1) and (2) and looking at (3), we finally have

$$\Sigma(b_j + 1)F_j = \tau^*(\Sigma(a_i + 1)E_i) + R.$$

Therefore, we conclude that

$$a_i \geq -1 \quad \forall i \iff b_j \geq -1 \quad \forall j$$

and that

$$a_i > -1 \quad \forall i \iff b_j > -1 \quad \forall j.$$

(Aside from $\tau^{-1}(\cup E_j) \subset \cup F_j$ and $\tau_*(\cup F_j) \subset \cup E_i$ we know that $f(F_j) \not\subset \cup E_i$ would imply $F_j \leq R$.) Moreover, since $K_{\tilde{X}}$ is a Cartier divisor and thus $b_j \in \mathbb{Z} \ \forall j$,

$$b_j > -1 \quad \forall j \iff b_j \geq 0 \quad \forall j.$$

This proves assertion (i).

(ii) First note that

$$(4) \quad K_{\tilde{V}} = \tau^* K_V + R_\tau,$$

where the support of the effective ramification divisor R_τ coincides with the locus where τ is not étale. From formula (1) and $K_{\tilde{X}} = c^* K_X$ we obtain

$$K_{\tilde{V}} = \tilde{f}^* K_{\tilde{X}} + \tau^*(\Sigma a_i E_i) + R_\tau.$$

It follows immediately from this that

$$X \text{ has canonical singularities} \implies \tilde{X} \text{ has canonical singularities.}$$

Furthermore, note that if a divisor F_j is $\tilde{f}$-exceptional, then it is either τ-exceptional or f-exceptional and thus F_j shows up either in R_τ or in $\Sigma a_i E_i$, assuming that X has terminal singularities. This proves

$$X \text{ has terminal singularities} \implies \tilde{X} \text{ has terminal singularities.}$$

The last implication is a more subtle one due to Reid [3].

Let E be an exceptional divisor for f, Γ a prime divisor on $\tilde{V}$ with $\tau(\Gamma) = E$. Let e be the ramification index of Γ over E, i.e.,

$$\tau^* t_E = u \cdot t_\Gamma^e \quad \text{with } u \in \mathcal{O}_{\tilde{V}, \Gamma} \text{ a unit,}$$

where t_E and t_Γ are the uniformizing parameters of the DVRs $\mathcal{O}_{V,E}$ and $\mathcal{O}_{\tilde{V}, \Gamma}$, respectively.

For a line bundle $\tilde{L}$ on $\tilde{V}$ and a (rational) section $\tilde{s}$ of $\tilde{L}$, we define

$$v_\Gamma(\tilde{s}) = n \text{ if } \tilde{s} = a \cdot b,$$

where

$$v_\Gamma(a) = n, \text{ i.e., } a = u_a \cdot t_\Gamma^n \text{ with } u_a \in \mathcal{O}_{\tilde{V},\Gamma} \text{ a unit for } a \in \mathbb{C}(\tilde{V})$$

and b is a generator of the line bundle at the generic point of Γ. The number $v_E(s)$ is similarly defined for a line bundle L on V and a (rational) section s.

Let s be the nowhere-vanishing section of rK_X taken for the construction of the canonical cover $c : \tilde{X} \to X$. Then by construction (cf. Proposition-Definition 4-5-1)

$$H^0(\tilde{X}, \mathcal{O}_{\tilde{X}}(rK_{\tilde{X}})) \ni c^*s = \tilde{s}^r$$

for some nowhere-vanishing section $\tilde{s} \in H^0(\tilde{X}, K_{\tilde{X}})$. By pulling back to $\tilde{V}$, regard $\tilde{f}^*(c^*s)$ (respectively f^*s) as a section of $H^0(\tilde{V}, \mathcal{O}_{\tilde{V}}(rK_{\tilde{V}}))$ (respectively $H^0(V, \mathcal{O}_V(K_V))$). Obviously, we have

$$(5) \quad v_\Gamma(\tilde{f}^*c^*s) = v_\Gamma(\tilde{f}^*\tilde{s}^r) = r \cdot v_\Gamma(\tilde{f}^*\tilde{s}).$$

On the other hand, if we choose a *local* coordinate sytem $(t_E, x_2, \ldots)$ at the generic point of E such that $(t_\Gamma, \tau^*x_2, \ldots)$ gives a *local* coordinate system at the generic point of Γ, then

$$f^*s = w \cdot t_E^{v_E(f^*s)}(dt_E \wedge dx_2 \wedge \cdots)^{\otimes r} \text{ with } w \in \mathcal{O}_{V.E} \text{ a unit,}$$

which leads to

$$\tau^*(f^*s) = \tau^*w \cdot (u \cdot t_\Gamma^e)^{v_E(f^*s)}(d(u \cdot t_\Gamma^e) \wedge d\tau^*x_2 \wedge \cdots)^{\otimes r}$$
$$= \tilde{w} \cdot t_\Gamma^{e \cdot v_E(f^*s) + r(e-1)}(dt_\Gamma \wedge d\tau^*x_2 \wedge \cdots)^{\otimes r} \text{ with } \tilde{w} \in \mathcal{O}_{\tilde{V},\Gamma} \text{ a unit.}$$

Therefore, we have

$$(6) \quad v_\Gamma(\tilde{f}^*c^*s) = v_\Gamma(\tau^*f^*s) = e \cdot v_E(f^*s) + r(e-1).$$

From formulas (5) and (6) we conclude that

$$(7) \quad v_\Gamma(\tilde{f}^*\tilde{s}) = \frac{e}{r}v_E(f^*s) + (e-1).$$

We claim that

$$(8) \quad \gcd(v_\Gamma(\tilde{f}^*\tilde{s}) + 1, e) = 1.$$

We finish the proof of assertion (ii), postponing the verification of claim (8) until the end.

In the ramification formulas (1) and (2) we would like to show that

$$a_i \geq 0 \quad \forall i \quad \text{and} \quad b_j > 0 \quad \forall j \Longrightarrow a_i > 0 \quad \forall i.$$

Suppose on the contrary that, say, $a_o = 0$ for $E_o = E$. We take a divisor $F_o = \Gamma$ lying over E_o. Then $a_o = 0$ is equivalent to $v_E(f^*s) = 0$, which implies $v_\Gamma(\tilde{f}^*\tilde{s}) = e - 1$ by formula (7). But then claim (8) would imply $e = v_\Gamma(\tilde{f}^*\tilde{s}) + 1 = 1$, hence $v_\Gamma(\tilde{f}^*\tilde{s}) = 0$. This is equivalent to $b_o = 0$, contradicting the assumption $b_j > 0 \quad \forall j$.

Now in order to complete the proof, we have only to prove claim (8).

Let G be the Galois group of the finite field extension $\mathbb{C}(\tilde{X})/\mathbb{C}(X)$, i.e., $G = \mathrm{Gal}(\mathbb{C}(\tilde{X})/\mathbb{C}(X)) = \mathrm{Gal}(\mathbb{C}(\tilde{V})/\mathbb{C}(V))$. Note that we may choose V and $\tilde{V}$ such that the Galois group G acts on $\tilde{V}$ over V (cf. the equivariant resolution of singularities, Hironaka [2], Bierstone–Milman [1], Encinas–Villamayor [1], Villamayor [1]). Let $G_\Gamma \subset G$ be the decomposition group of Γ.

Then over E we have the diagram

$$
\begin{array}{ccc}
\mathrm{Spec}\,(c_*\mathcal{O}_{\tilde{V}})_E & \ni & \Gamma \\
\downarrow{\scriptstyle\sigma} & & \downarrow \\
\mathrm{Spec}\,\{(c_*\mathcal{O}_{\tilde{V}})_E\}^{G_\Gamma} & \ni & \Delta \\
\downarrow & & \downarrow \\
\mathrm{Spec}\,\mathcal{O}_{V,E} & \ni & E
\end{array}
$$

Since Γ is the only prime lying over Δ, we have

$$\#G_\Gamma = e,$$

and (after taking some étale cover, which does not affect the computation of the valuations or working in the analytic category) we may choose a uniformizing parameter $t_\Gamma \in \mathcal{O}_{\tilde{V},\Gamma}$ such that

$$\mathcal{O}_{\tilde{V},\Gamma} = \oplus_{i=0}^{e-1} B \cdot t_\Gamma^i,$$

where

$$B = \{(c_*\mathcal{O}_{\tilde{V}})_E\}^{G_\Gamma},$$

and such that a generator $g_\Gamma \in G_\Gamma$ acts as multiplication by a primitive eth root of unity η on the uniformizing parameter t_Γ, i.e.,

$$g_\Gamma^*(t_\Gamma) = \eta \cdot t_\Gamma.$$

Choose a *local* coordinate system $(t_\Gamma, \sigma^*x_2, \sigma^*x_3, \ldots)$ at the generic point of Γ, where $x_2, x_3, \ldots \in B$. Then $g_\Gamma \in G_\Gamma$ acts as

$$g_\Gamma^*(dt_\Gamma \wedge d\sigma^*x_2 \wedge d\sigma^*x_3 \wedge \cdots) = \eta \cdot dt_\Gamma \wedge d\sigma^*x_2 \wedge d\sigma^*x_3 \wedge \cdots.$$

Since $\tilde{s}$ is an eigenvector for the action of G (and hence for the action of G_Γ), so is $\tilde{f}^*\tilde{s}$. Writing

$$\tilde{f}^*\tilde{s} = \tilde{u} \cdot t_\Gamma^{v_\Gamma(\tilde{f}^*\tilde{s})} \cdot dt_\Gamma \wedge d\sigma^*x_2 \wedge d\sigma^*x_3 \wedge \cdots,$$

we conclude that a unit $\tilde{u} \in \mathcal{O}_{\tilde{V},\Gamma}$ is also an eigenvector for the action of G_Γ and hence that $\tilde{u} \in B$. This implies

$$g_\Gamma^*(\tilde{f}^*\tilde{s}) = \eta^{v_\Gamma(\tilde{f}^*\tilde{s})+1} \cdot \tilde{f}^*\tilde{s}.$$

Since a generator $g \in G$ acts on $\tilde{s}$ as multiplication by a primitive rth root of unity $\epsilon(g)$, so does g on $\tilde{f}^*\tilde{s}$. Thus $\eta^{v_\Gamma(\tilde{f}^*\tilde{s})+1}$ has to be another primitive eth root of unity, which implies the claim

$$\gcd(v_\Gamma(\tilde{f}^*\tilde{s}) + 1, e) = 1.$$

This completes the proof of Proposition 4-5-3.   □

Remark 4-5-4. For a germ $p \in (X, D)$ of a log pair consisting of a normal variety X and an integral boundary divisor $D = \Sigma d_k D_k$, i.e., $d_k = 1$ $\forall k$, a similar argument shows that if $K_X + D$ is $\mathbb{Q}$-Cartier with the index

$$r = \min\{e \in \mathbb{N}; e(K_X + D) \text{ is Cartier}\},$$

then there exists a finite morphism

$$\text{lc} : \tilde{X} \to X$$

such that

(i) $\tilde{X}$ is a normal (irreducible) variety with $(\text{lc})^{-1}(p) = q$ is a point,
(ii) the extension of the function fields $\mathbb{C}(\tilde{X})/\mathbb{C}(X)$ is Galois and the Galois group $G \cong \mathbb{Z}/(r)$ acts on $\tilde{X}$ over X,
(iii) the morphism lc is étale in codimension one, and
(iv) $K_{\tilde{X}} + \tilde{D}$ is Cartier, where $\tilde{D} = \text{lc}^* D$, and such that it is uniquely determined by conditions (i), (ii), (iii), and (iv) up to analytic isomorphism and hence is called the log canonical cover of (X, D).

If the boundary $\mathbb{Q}$-divisor $D = \Sigma d_k D_k$ of the log pair (X, D) were not integral, i.e., $d_k \neq 1$ or 0 for some k, then a similar construction would lead to a finite Galois cover ramified along $\Sigma_{d_i \neq 1} D_i$. If D is of a special form $D = \Sigma \frac{m-1}{m} D_k$ for some $m \in \mathbb{N}$, then the constructed cover lc $: \tilde{X} \to X$ has the property $K_{\tilde{X}} = (\text{lc})^*(K_X + D)$. So the properties of the log canonical divisor $K_X + D$ downstairs should be reflected in the properties of the canonical divisor $K_{\tilde{X}}$ upstairs. This observation will be utilized in the proof of the Kawamata–Viehweg vanishing theorem in Chapter 5.

Remark 4-5-5. Although the argument throughout Section 4-5 is carried out in the language of the algebraic category, the notion of canonical cover, as well as the assertions in Proposition 4-5-3, is valid in the analytic category. We leave it to the reader as an exercise to make necessary modifications to go from the algebraic category to the analytic one.

4.6 Classification in Dimension 2

In this section we give the **analytic classification in dimension 2** of terminal, canonical, log terminal, and log canonical singularities. Accordingly, we work in the analytic category. We consider only the case without the boundary $\mathbb{Q}$-divisors.

Terminal singularities in dimension 2 are nothing but **smooth** (i.e., nonsingular) points. This is an immediate consequence of Castelnuovo's contractibility criterion (Theorem 1-1-6).

Our classification of **canonical singularities in dimension 2**, which turn out to coincide with the so-called **rational double points**, follows **Reid's method**:

In order to study a singularity $p \in S$, take a (general) hyperplane section $p \in H \subset S$ and analyze $p \in H$. This reduces the dimension by 1, and inductively we should obtain some crucial information about the original $p \in S$. This principle was originally applied to analyze 3-fold terminal singularities $p \in X$ using a 2-dimensional hyperplane section $p \in H \subset X$ (cf. Reid [1][2][7], Mori [4]). Our intention is to develop a feeling for this method in dimension 3 (or higher) by applying it to an easier case of dimension 2. All the results are classical and well known, as, e.g., our study of the hyperplane sections which is nothing but Artin's analysis of the fundamental cycles. On the other hand, we believe our approach to be somewhat different from the existing ones and simpler. For example, we derive the standard defining equations for rational double points from scratch.

The canonical covers of log terminal singularities are canonical singularities (cf. Proposition 4-5-2). Our first approach to classifying **log terminal singularities in dimension 2** follows that of Kawamata [9]. Since canonical singularities in dimension 2 are quotients of $\mathbb{C}^2$ by some finite subgroups of SL $(2, \mathbb{C})$, we characterize log terminal singularities in dimension 2 as **quotients of $\mathbb{C}^2$ by some finite subgroups of GL(2,$\mathbb{C}$)** (without quasi-reflection), which are already classified by Brieskorn [1]. Our second approach is to study directly the cyclic group actions on canonical singularities in dimension 2 by deriving equivariant forms of the standard equations for rational double points. Thus we classify log terminal singularities in dimension 2 as quotients of hypersurface singularities (**hyperquotients** for short) that are canonical. This allows us, e.g., to determine the dual graphs of log terminal singularities without referring to the results of Brieskorn [1]. (See also the article by Alexeev in Kollár et al. [1] for a purely combinatorial argument to determine the dual graphs of log terminal singularities.)

Finally, we give the classification of **log canonical singularities in dimension 2** at the end of the section , again following Kawamata [9].

Though the analysis of this section is crucial in the subsequent study of the detailed structures of terminal singularities in dimension 3, which culminates in Mori's proof of the existence of flip in dimension 3, we do not use any of the results in the rest of the book, *except* for the following two:

(i) The singular locus of terminal singularities has codimension at least 3.
(ii) Let $p \in (S, D)$ be a germ of a log terminal singularity in dimension 2, where the boundary divisor $D = D_1 + D_2$ consists of two irreducible components D_1 and D_2 with coefficient 1 intersecting at p. Then p is necessarily a smooth point of the surface S, while D_1 and D_2 intersect transversally at p.

First we mention the two fundamental theorems that form the basis of our analysis.

Theorem 4-6-1 (Negative Definiteness of Intersection Form of Contractible Curves). *Let $p \in S$ be a germ of normal surface,*

$$f : V \to S$$

a proper birational morphism from a nonsingular surface V. *Then* $f^{-1}(p)$ *is connected and has a negative definite intersection form, i.e.,*

$$D^2 < 0 \text{ for a divisor } 0 \neq D = \Sigma d_i E_i,$$

where the $E_i \subset f^{-1}(p)$ *are exceptional divisors.*

PROOF.   In the algebraic category this is an easy consequence of the Hodge index theorem (cf. Theorem 1-2-8). Here we present a proof by Mumford [3] (cf. Fulton [1]), which applies to the analytic category as well. First note that $f^{-1}(p)$ is connected by Zariski's Main Theorem. Take a regular function r on S that vanishes at p. Then

$$\mathrm{div}(f^*r) = \Sigma_{i=1}^n m_i E_i + G,$$

where $E_1, E_2, \ldots, E_n$ are the irreducible components of $f^{-1}(p)$, all the m_i are strictly positive, i.e., $m_i > 0$ $\forall i$, and G is a nonzero effective divisor not containing any of the E_i with $G \cdot E_i > 0$ for some i. Now,

$$D^2 = (\Sigma d_i E_i)^2 = \Sigma_i d_i E_i \cdot \left(\Sigma_j \frac{d_j}{m_j} m_j E_j - \frac{d_i}{m_i} \mathrm{div}(f^*r) \right)$$

$$= \Sigma_i \left\{ \Sigma_{j \neq i} \frac{d_i}{m_i} \left(\frac{d_j}{m_j} - \frac{d_i}{m_i} \right) m_i E_i \cdot m_j E_j - \left(\frac{d_i}{m_i} \right)^2 m_i E_i \cdot G \right\}$$

$$= -\Sigma_{i<j} \left(\frac{d_i}{m_i} - \frac{d_j}{m_j} \right)^2 m_i E_i \cdot m_j E_j - \Sigma_i \left(\frac{d_i}{m_i} \right)^2 m_i E_i \cdot G.$$

Since $f^{-1}(p)$ is connected and since $E_i \cdot G > 0$ for some i, the above formula implies

$$D^2 < 0 \text{ unless } D = \Sigma_i d_i E_i = 0. \qquad \square$$

Theorem 4-6-2 (Existence of the Minimal Resolution in Dimension 2).   *Let* S *be a normal surface. Then there exists a projective birational morphism*

$$\pi : \hat{S} \to S$$

from a nonsingular surface $\hat{S}$ *such that* $K_{\hat{S}}$ *is* π-*nef, i.e.,*

$$K_{\hat{S}} \cdot E \geq 0 \text{ for any curve } E \subset \hat{S} \text{ with } \pi(E) = \text{pt.}$$

It has the following properties:

 (i) *It is absolutely minimal in the sense that if*

$$g : W \to S$$

is any proper birational morphism from a nonsingular surface W, *then it factors through* $\hat{S}$, *i.e., there exists a proper birational morphism* $\tau : W \to \hat{S}$ *such that* $g = \pi \circ \tau$.

(ii) *In the ramification formula*

$$K_{\hat{S}} = \pi^* K_S + \Sigma a_i E_i$$

where the pull-back $\pi^* K_S$ is taken in the sense of Mumford (See Remark 4-6-3 (i).), all the coefficients are nonnegative, i.e.,

$$a_i \leq 0 \quad \forall i.$$

More precisely, either all the coefficients are zero,

$$a_i = 0 \quad \forall i,$$

or all the coefficients are strictly negative,

$$a_i < 0 \quad \forall i.$$

Property (i) *shows in particular that such* $\pi : \hat{S} \to S$ *is unique up to isomorphism. (It is thus called the* **minimal resolution** *of S.)*

PROOF.   (i) Take a projective birational morphism

$$f : V \to S$$

from a nonsingular surface V by resolution of singularities. If K_V is not f-nef, then there exists a curve $E \subset V$ such that

$$K_V \cdot E < 0 \text{ and } f(E) = \text{pt. and thus } E^2 < 0 \text{ by Theorem 4-6-1.}$$

Thus E is a (-1)-curve (cf. Lemma 1-1-4), which can be contracted to obtain $\mu : V = V_0 \to V_1$. (We note that though Castelnuovo's contractibility criterion was stated under the global projectivity condition, its proof applies to our *local* situation projective over S without any change.) By Lemma 1-8-1, the morphism f factors through μ to give a projective birational morphism

$$f_1 : V_1 \to S \text{ with } f = f_1 \circ \mu.$$

Arguing inductively with V_1, which has one fewer exceptional divisor over S, we finally obtain

$$\pi : \hat{S} \to S$$

with $K_{\hat{S}}$ being π-nef. (In short, we just carry out the relative MMP over S starting from V in order to obtain the relative minimal model $\hat{S}$ over S.)

The proof for the absolute minimality (i) is identical to the one for the absolute minimality of the minimal model in dimension 2 (cf. Theorem 1-8-6).

(ii) If in the ramification formula some coefficient a_i is bigger than 0, then by Theorem 4-6-1

$$K_{\hat{S}} \cdot \Sigma_{a_i > 0} a_i E_i = (\pi^* K_S + \Sigma a_i E_i) \cdot \Sigma_{a_i > 0} a_i E_i$$
$$= (\Sigma_{a_i > 0} a_i E_i)^2 + (\Sigma_{a_i > 0} a_i E_i) \cdot (\Sigma_{a_i \leq 0} a_i E_i) < 0.$$

Thus there exists some E_{i_o} (with $a_{i_o} > 0$) such that

$$K_{\hat{S}} \cdot E_{i_o} < 0,$$

which contradicts $K_{\hat{S}}$ being π-nef.

Therefore, we conclude that

$$a_i \leq 0 \quad \forall i.$$

If some coefficient a_i is equal to 0 and not all of them are zero, then since $f^{-1}(p)$ is connected, there exists an exceptional divisor E_{i_o} with $a_{i_o} = 0$ such that E_{i_o} is not disjoint from $\Sigma_{a_i < 0} E_i$. But this would imply

$$K_{\hat{S}} \cdot E_{i_o} = (\pi^* K_S + \Sigma a_i E_i) \cdot E_{i_o}, = (\pi^* K_S + \Sigma_{a_i < 0} a_i E_i) \cdot E_{i_o} < 0,$$

contradicting $K_{\hat{S}}$ being π-nef! This proves the "more precisely" part of assertion (ii).

This completes the proof of Theorem 4-6-2. □

Remark 4-6-3. (i) Let

$$f : V \to S$$

be a proper birational morphism from a nonsingular surface V to a normal surface S. Let D be a Weil divisor. Utilizing Theorem 4-6-1, we can define the pullback of D without assuming that D is $\mathbb{Q}$-Cartier, i.e.,

the pullback $f^* D$ in the sense of Mumford

as the $\mathbb{Q}$-divisor

$$f^* D = f_*^{-1} D + \Sigma d_i E_i, \quad d_i \in \mathbb{Q},$$

satisfying

$$(f_*^{-1} D + \Sigma d_i E_i) \cdot E_j = 0 \quad \forall E_j \quad \text{exceptional for } f,$$

where $f_*^{-1} D$ is by definition the strict transform of D. Negative definiteness of the intersection forms among the contractible (exceptional) curves (cf. Theorem 4-6-1) guarantees that there exists a unique set of rational solutions for d_i and hence uniquely determines the pullback $f^* D$.

(ii) Though in the definition of terminal or canonical (respectively log terminal or log canonical) singularities in higher dimension we require that the (respectively log) canonical divisor K_X (respectively $K_X + D$) be $\mathbb{Q}$-Cartier, we start our classification of terminal or canonical (respectively log terminal or log canonical) singularities in dimension 2 *without* a priori assuming the condition

(1) K_S (respectively $K_S + B$) is $\mathbb{Q}$-Cartier.

Thus in the ramification formula we take the pullback of the (respectively log) canonical divisor in the sense of Mumford as in (i).

It turns out as a consequence of the classification that the (log) canonical divisor for a terminal or canonical (respectively log terminal or log canonical) defined as above is always $\mathbb{Q}$-Cartier and that the conventional definition as above and the standard definition coincide in dimension 2.

Exercise 4-6-4. Show that any proper birational morphism $f : V \to S$ from a nonsingular surface V to a normal surface S, which is isomorphic Exc ept over a point $p \in S$, is a projective morphism in the sense of Hartshorne [3], i.e., there exists a commutative diagram

$$
\begin{array}{ccc}
V & \hookrightarrow & \mathbb{P}^N \times S \\
f \searrow & & \downarrow \\
& S.
\end{array}
$$

Terminal Singularities in Dimension 2

Theorem 4-6-5 (Terminal Singularities in dimension 2). *A normal surface S has only terminal singularities iff S is nonsingular.*

PROOF. If S is nonsingular, then it has by definition only terminal singularities. Conversely, suppose S is a normal surface with only terminal singularities and let $\pi : \hat{S} \to S$ be its minimal resolution (cf. Theorem 4-6-2). If π is not an isomorphism, then there would exist an exceptional divisor for π. But then in the ramification formula we would have

$$K_V = \pi^* K_S + \Sigma_{i \in I} a_i E_i,$$
$$a_i > 0 \quad \forall i \text{ and } I \neq \emptyset,$$

by the definition of terminal singularities. On the other hand, from property (ii) of the minimal resolution in Theorem 4-6-2 we would have

$$a_i \leq 0 \quad \forall i,$$

a contradiction! Therefore, π is an isomorphism, and hence S is nonsingular.   □

We remark that according to Theorem 4-6-5 the category $\mathfrak{C}$ of normal projective varieties with only $\mathbb{Q}$-factorial and terminal singularities coincides with that of nonsingular projective varieties in dimension 2. This is why in dimension 2 we could stick to the category of nonsingular projective varieties to establish the minimal model program (cf. Theorem 3-1-16). We observed this fact already in Chapter 1 via Castelnuovo's contractibility criterion, which is also the key to prove Theorem 4-6-2.

Corollary 4-6-6. *Let X be a normal variety of dimension ≥ 3 with only terminal singularities. Then the singular locus $\mathrm{Sing}(X)$ of X has codimension at least 3, i.e.,*

$$\mathrm{codim}_X \mathrm{Sing}(X) \geq 3.$$

In particular, terminal singularities in dimension 3 are isolated.

PROOF. For simplicity, we deal with the case $\dim X = 3$. Since the assertion is local, we may assume that X is affine. Take a projective birational morphism

$f : V \to X$ from a nonsingular variety V by resolution of singularities. By the definition of terminal singularities the ramification formula reads

$$(*) \qquad K_V = f^* K_X + \Sigma a_i E_i \text{ with } a_i > 0 \quad \forall E_i \text{ exceptional for } f.$$

Suppose $\text{codim}_X \text{Sing}(X) < 3$. Then there exists an irreducible component $C \subset \text{Sing}(X)$ of dimension at least $\dim X - \text{codim}_X \text{Sing}(X) \geq 1$. Thus we can take a general hyperplane section H such that $H \cap C \neq \emptyset$. By Bertini's theorem we may assume that $f^* H$ is irreducible and nonsingular and hence that the restriction of the formula $(*)$ to $f^* H$ gives the ramification formula for H,

$$(**)$$
$$K_{f \cdot H} = (K_V + f^* H)|_{f \cdot H} = f^*(K_X + H|_H) + \Sigma a_i E_i|_{f \cdot H} = f^* K_H + \Sigma a_i E_i|_{f \cdot H},$$

which implies that H has terminal singularities. By Theorem 4-6-5, H is nonsingular. On the other hand, since $H \cap \text{Sing}(X) \neq \emptyset$, H must be singular, a contradiction!

The verification of the assertion in the case $\dim X > 3$, where we have to take several hyperplane sections to apply Theorem 4-6-5 instead of just one hyperplane section as above, is almost identical, and we leave it to the reader as an exercise. $\qquad \square$

> ## Canonical Singularities in Dimension 2

Theorem 4-6-7 (Canonical Singularities in Dimension 2). *Let $p \in S$ be a germ of canonical singularities in dimension 2. (Assume that $p \in S$ is not nonsingular. If $p \in S$ is nonsingular, then it has clearly only canonical singularities.) Then the dual graph for the exceptional curves of the minimal resolution is one of the following 5 types. Moreover, $p \in S$ is analytically isomorphic to a hypersurface singularity $0 \in \{f = 0\} \subset \mathbb{A}^3$, where the equation f is given as below according to the type of its dual graph.*

$A_n : f = xy + z^{n+1}$

$$\circ - \circ - \cdots - \circ$$

$D_n : f = x^2 + y(z^2 + y^{n-2})$

$E_6 : f = x^2 + y^3 + z^4$

$E_7 : f = x^2 + y(y^2 + z^3)$

$$E_8 : f = x^2 + y^3 + z^5$$

o — o — o — o — o — o — o
|
o

Conversely, any of the hypersurface singularities defined by one of the equations above is a canonical singularity in dimension 2 with the corresponding dual graph.

PROOF.   The "conversely" part is a straightforward calculation, where we observe that the minimal resolution can be obtained by successive blowups of singular points (cf. Lemma 4-6-14). We leave it as a *fun* exercise for the reader.

Our argument for the characterization of canonical singularities in dimension 2 is a baby version of Reid's argument (cf. Reid [1][2][7]) to analyze terminal singularities in dimension 3:

> **(Reid's Method)**
> Take a general hyperplane section to reduce the dimension by one. Then utilize the information on the hyperplane section to analyze the original singularity.

The analysis of the hyperplane sections in our case turns out be nothing but the study of the classical notion of the fundamental cycle after Artin. The analysis almost immediately implies that a canonical surface singularity is a hypersurface singularity in $\mathbb{A}^3$ defined by a single equation. We use the Weierstrass preparation theorem to derive the standard form of the defining equation, studying the behavior of the singularity under blowups. We present the proof in three steps.

Step 1. Analysis of the dual graph

Let $\pi : \hat{S} \to S$ be the minimal resolution. We observe first that by property (ii) of the minimal resolution in Theorem 4-6-2 and by the definition of canonical singularities we have

$$K_{\hat{S}} = \pi^* K_S.$$

This together with Theorem 4-6-1 and the arithmetic genus formula implies that any exceptional divisor E_i for π is a (-2)-curve, i.e.,

$$K_{\hat{S}} \cdot E_i = 0, \ E_i^2 = -2 \text{ and } E_i \cong \mathbb{P}^1.$$

Then by the same argument as in Theorem 1-6-4 using the negative definiteness of the intersection form among the exceptional curves we conclude that the dual graph is one of the five types as above (cf. Section 1-6).

Step 2. Verification that canonical singularities in dimension 2 are hypersurface singularities

In order to see that canonical singularities in dimension 2 are hypersurface singularities, we prove the following lemma, whose assertion holds not only for canonical singularities but also for rational singularities in general.

Lemma 4-6-8.   *Let $p \in S$ be a germ of a normal surface having only canonical singularities (or more generally rational singularities) with its minimal resolution*

$\pi : \hat{S} \to S$. *Then for a line bundle* $\mathcal{L} \in \text{Pic}(\hat{S})$

$\mathcal{L}$ *is trivial, i.e.,* $\mathcal{L} = \pi^* \mathcal{M}$ *for some* $\mathcal{M} \in \text{Pic}(S) \iff \mathcal{L} \cdot E_i = 0 \quad \forall E_i,$

where the E_i *are the exceptional curves for* π.
Similarly, for a divisor D *on* $\hat{S}$

$$D = \pi^* G \text{ for some divisor } G \text{ on } S \iff D \cdot E_i = 0 \quad \forall E_i.$$

PROOF.   Consider the exponential sequence

$$0 \to \mathbb{Z} \to \mathcal{O}_{\hat{S}} \to \mathcal{O}_{\hat{S}}^* \to 0$$

and its associated long cohomology sequence

$$R^1 \pi_* \mathcal{O}_{\hat{S}} \to R^1 \pi_* \mathcal{O}_{\hat{S}}^* \to R^2 \pi_* \mathbb{Z} \cong H^2(\pi^{-1}(p), \mathbb{Z}) \cong \oplus_i \mathbb{Z},$$

where $R^1 \pi_* \mathcal{O}_{\hat{S}}^*$ is canonically identified with $\text{Pic}(\hat{S})$ and the last map is given by taking the intersections

$$R^1 \pi_* \mathcal{O}_{\hat{S}}^* \to \oplus_i \mathbb{Z},$$

$$\mathcal{L} \in \text{Pic}(\hat{S}) \mapsto \oplus_i \mathcal{L} \cdot E_i.$$

(Exercise! Check the claimed isomorphisms $R^1 \pi_* \mathcal{O}_{\hat{S}}^* \cong \text{Pic}(\hat{S})$ and $R^2 \pi_* \mathbb{Z} \cong H^2(\pi^{-1}(p), \mathbb{Z})$ together with the description of the map cf. Proposition 8.1 in Hartshorne [3]). Therefore, in order to verify the assertion for the line bundles it suffices to prove

$$R^1 \pi_* \mathcal{O}_{\hat{S}} = 0,$$

which follows by definition if $p \in S$ has rational singularities, and follows from the relative analytical version of the Kawamata–Viehweg vanishing theorem (cf. Theorem 5-2-8 and Remark 5-2-9). In fact, take a π-ample divisor of the form

$$-\Sigma \delta_i E_i \text{ where } \delta_i > 0 \quad \forall i.$$

(Exercise! Show that one can take such a π-ample divisor described as above.) Since for a sufficiently small ϵ the pair $(\hat{S}, \epsilon(\Sigma \delta_i E_i))$ has only $\mathbb{Q}$-factorial and log terminal singularities and since $-K_{\hat{S}} - \epsilon(\Sigma \delta_i E_i) = -\pi^* K_S - \epsilon(\Sigma \delta_i E_i)$ is π-ample, the analytic version of Theorem 5-2-8 implies

$$R^1 \pi_* \mathcal{O}_{\hat{S}}(K_{\hat{S}} + \epsilon(\Sigma \delta_i E_i) - K_{\hat{S}} - \epsilon(\Sigma \delta_i E_i)) = R^1 \pi_* \mathcal{O}_{\hat{S}} = 0.$$

The assertion for the divisors follows from the one for the line bundles. Details are left to the reader as an exercise.   □

The following notion of the fundamental cycle after Artin [1] [2] is an easy consequence of the above lemma. It analyzes the general hyperplane section of a canonical singularity in dimension 2, by describing its pullback to the minimal resolution.

Proposition 4-6-9 (Analysis of the Fundamental Cycle=Description of the General Hyperplane Section). *Let $p \in S$ be a germ of canonical singularities in dimension 2 (or more generally rational singularities in dimension 2) and $\pi : \hat{S} \to S$ its minimal resolution. Then there exists a unique nonzero effective divisor $Z = \Sigma b_i E_i$ (called the **fundamental cycle**) supported on the exceptional locus $\pi^{-1}(p)$ such that*

(i) *$-Z$ is π-nef, i.e.,*

$$Z \cdot E_i \le 0 \quad \forall E_i,$$

(ii) *Z is the minimum among all the nonzero effective divisors supported on $\pi^{-1}(p)$ satisfying condition (i).*
   Moreover, the fundamental cycle has the properties:
(iii) *for any regular function r on S that vanishes at p we have*

$$\mathrm{div}(\pi^* r) \ge Z,$$

and for a general hyperplane section $p \in H \subset S$ (i.e., $H = \mathrm{div}(r)$ where a regular function r corresponds to a general member of the vector space $m_{S,p}/m_{S,p}^2$)

$$\pi^* H = div(\pi^* r) = Z + G,$$

where G and $\pi^{-1}(p)$ have no common components,
(iv)

$$\pi^{-1}(m_{S,p}) \cdot \mathcal{O}_{\hat{S}} = \mathcal{O}_{\hat{S}}(-Z).$$

The fundamental cycle of a canonical singularity in dimension 2 with its dual graph of type A_n, D_n, E_6, E_7, or E_8 can be computed as in the following diagrams. The collection of the white circles $\circ$ with the associated coefficients on top indicates the fundamental cycle, while in order to obtain the description of the general hyperplane one has only to add the black vertices $\bullet$, with the coefficients equal to 1, which indicate the components intersecting $\pi^{-1}(p)$ transversally.

A_n:

D_n:

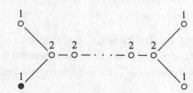

E_6:

$$
\begin{array}{ccccccccc}
\overset{1}{\circ} & - & \overset{2}{\circ} & - & \overset{3}{\circ} & - & \overset{2}{\circ} & - & \overset{1}{\circ} \\
& & & & | & & & & \\
& & & & \overset{}{\underset{2}{\circ}} & & & & \\
& & & & | & & & & \\
& & & & \underset{1}{\bullet} & & & &
\end{array}
$$

E_7:

$$
\begin{array}{ccccccccccccc}
\overset{1}{\circ} & - & \overset{2}{\circ} & - & \overset{3}{\circ} & - & \overset{4}{\circ} & - & \overset{3}{\circ} & - & \overset{2}{\circ} & - & \bullet \\
& & & & & & | & & & & & & \\
& & & & & & \underset{2}{\circ} & & & & & &
\end{array}
$$

E_8:

$$
\begin{array}{ccccccccccccc}
\overset{1}{\circ} & - & \overset{2}{\circ} & - & \overset{3}{\circ} & - & \overset{4}{\circ} & - & \overset{3}{\circ} & - & \overset{2}{\circ} & - & \underset{1}{\bullet} \\
& & & & & & | & & & & & & \\
& & & & & & \underset{2}{\circ} & & & & & &
\end{array}
$$

PROOF. Observe first that for a nonzero effective divisor $Z = \Sigma b_i E_i$ supported on $\pi^{-1}(p)$ an equivalent condition to (i) is given by

$$Z \cdot E_i \leq 0 \quad \forall i \iff \exists r \text{ a regular function on } S \quad \text{s.t.} \quad \text{div}(\pi^* r) = Z + G$$

$$\text{where } G \text{ is an effective divisor}$$

$$\text{having no common components with } \pi^{-1}(p).$$

In fact, if

$$Z \cdot E_i \leq 0 \quad \forall i,$$

then the divisor

$$D = Z + \Sigma_i \{-(Z \cdot E_i)\} C_i,$$

where C_i is an irreducible divisor (analytically defined in a neighborhood of $\pi^{-1}(p)$) that intersects E_i exactly once transversally at the general point, has the property

$$D \cdot E_i = 0 \quad \forall i.$$

Thus by Lemma 4-6-8 the line bundle $\mathcal{O}_{\hat{S}}(D)$ is trivial, and the divisor D corresponds to a section in

$$\Gamma(\hat{S}, \mathcal{O}_{\hat{S}}(D)) = \Gamma(\hat{S}, \mathcal{O}_{\hat{S}}) = \Gamma(S, \pi_* \mathcal{O}_{\hat{S}}) = \Gamma(S, \mathcal{O}_S),$$

that is

$$D = \text{div}(\pi^* r) \text{ for some } r \in \Gamma(S, \mathcal{O}_S).$$

Therefore, setting $G = \Sigma_i \{-(Z \cdot E_i)\} C_i$ we have $\text{div}(\pi^* r) = Z + G$ as desired.

On the other hand, if for a regular function r on S that vanishes at p we have

$$\operatorname{div}(\pi^*r) = Z + G,$$

where Z is supported on $\pi^{-1}(p)$ and G is an effective divisor having no common components with $\pi^{-1}(p)$, then

$$Z \cdot E_i = -G \cdot E_i \le 0 \quad \forall i.$$

This proves the observation.

Take a nonzero effective divisor $Z = \Sigma b_i E_i$ supported on $\pi^{-1}(p)$ such that

$$Z \cdot E_i \le 0 \quad \forall i.$$

Then it is easy to see (Exercise! Why?) that

$$Z \ge \Sigma E_i.$$

Let r be a regular function on S such that

$$\operatorname{div}(\pi^*r) = Z + G,$$

where G is an effective divisor having no common components with $\pi^{-1}(p)$.

Take another nonzero effective divisor $Z' = \Sigma b_i' E_i$ such that

$$Z' \cdot E_i \le 0 \quad \forall i$$

with a regular function r' on S such that

$$\operatorname{div}(\pi^*r') = Z' + G',$$

where G' is an effective divisor having no common components with $\pi^{-1}(p)$.

Then for general $\alpha, \beta \in \mathbb{C}$

$$\operatorname{div}(\pi^*(\alpha r + \beta r')) = \Sigma \min\{b_i, b_i'\} E_i + \hat{G},$$

where $\hat{G}$ is an effective divisor having no common components with $\pi^{-1}(p)$.

Thus we conclude that the divisor

$$\hat{Z} = \Sigma \min\{b_i, b_i'\} E_i$$

has the property

$$\hat{Z} \cdot E_i \le 0 \quad \forall i.$$

The argument above proves the existence of a nonzero effective divisor $Z \ge \Sigma E_i$ satisfying conditions (i) and (ii), and such Z is necessarily unique.

Property (iii) also follows from the argument above. Property (iv) follows from (iii) and the observation that G (in the expression $\pi^*H = Z + G$) has no fixed point as H varies among the general hyperplane sections, which generate the maximal ideal $m_{S,p}$.

It can be directly checked that the cycle given in the diagram for a canonical singularity is the minimum nonzero effective divisor Z such that

$$Z \cdot E_i \le 0 \quad \forall i.$$

For example, an algorithm to find the fundamental cycle Z, given a dual graph, can be described as follows: First recall that

$$Z \geq Z_0 = \Sigma E_i.$$

If $Z_0 \cdot E \leq 0$ $\forall i$, then $Z = Z_0$ is the fundamental cycle. If there is a component E such that $Z_0 \cdot E > 0$, then set $Z_1 = Z_0 + E$. Observe that $Z \geq Z_1$. (Exercise! Why?) At the kth stage of the process with $Z \geq Z_k$, if $Z_k \cdot E_i$ $\forall i$, then $Z = Z_k$ is the fundamental cycle. If there is a component E such that $Z_k \cdot E > 0$, then set $Z_{k+1} = Z_k + E$. Observe that $Z \geq Z_{k+1}$. Continue this process until we reach Z.

Once we know the fundamental cycle, the description of the pullback of a general hyperplane section $p \in H \subset S$ can be obtained easily by condition (iii),

$$\pi^* H = Z + G,$$

where G is an effective divisor having no common components with $\pi^{-1}(p)$, and by the condition

$$\pi^* H \cdot E_i = (Z + G) \cdot E_i = 0 \quad \forall i.$$

This completes the proof of Proposition 4-6-9.   □

The fundamental cycle, describing the general hyperplane section, makes it easy to see that canonical singularities in dimension 2 are hypersurface singularities of multiplicity 2.

Claim 4-6-10. *A canonical singularity $p \in S$ in dimension 2 is a hypersurface singularity of multiplicity 2, i.e.,*

$$p \in S \overset{analytically}{\cong} 0 \in \{f = 0\} \subset \mathbb{A}^3,$$

where f is a holomorphic function around the origin with

$$\mathrm{mult}_0 f = 2.$$

PROOF.   From the description of the pullback $\pi^* H = Z + G$ of a general hyperplane $p \in H \subset S$ in Proposition 4-6-9 it follows that

$$\pi : G = \hat{H} \to H$$

gives the normalization of the curve H.

First we deal with the case where the dual graph is of type D_n, E_6, E_7, or E_8.

In each of these cases, $\hat{H}$ is a nonsingular irreducible curve, intersecting the fundamental cycle Z at a point $q \in \hat{H}$ transversally. Observe that if we take another general hyperplane section $H' = \mathrm{div}(r')$, then

$$\pi^* H' = Z + \hat{H}' \text{ with } q \notin \hat{H}'.$$

Since the coefficient of the component of Z intersecting $\hat{H}$ is 2 (see the dual graph for the fundamental cycle) and since $q \notin \tilde{H}'$, we conclude that

$$\mathrm{mult}_q \pi^* H'|_{\hat{H}} = 2.$$

Therefore, for a suitable choice of an affine embedding $p \in S \subset \mathbb{A}^N$ with the first coordinate $x_1 = r'$ we see that π can be described as

$$\pi : \hat{H} \to H \subset \mathbb{A}^N,$$
$$t \mapsto (u_1 t^2, u_2 t^{n_2}, u_3 t^{n_3}, \dots, u_N t^{n_N}),$$

where

$$2 < n_2 < n_3 < \cdots < n_N,$$
$$\gcd(n_i, 2) = 1 \text{ for } i = 2, \dots, N,$$
$$u_1, \dots, u_N \text{ are units in } \mathcal{O}_{\hat{H}, q},$$
$$t \text{ is the generator for } m_{\hat{H}, q}.$$

Then

$$\dim_{\mathbb{C}} m_{H, p} / m_{H, p}^2 = \dim_{\mathbb{C}} \widehat{m_{H, p}} / \widehat{m_{H, p}}^2 \quad (\text{where } \widehat{m_{H, p}} \text{ is the completion of } m_{H, p})$$
$$= \dim_{\mathbb{C}} (u_1 t^2, u_2 t^{n_2}, \dots, u_N t^{n_N}) / (u_1 t^2, u_2 t^{n_2}, \dots, u_N t^{n_N})^2$$
$$= \dim_{\mathbb{C}} \cdot \mathbb{C} t^2 \oplus \mathbb{C} \cdot t^{n_2} = 2,$$

where $(u_1 t^2, u_2 t^{n_2}, \dots, u_N t^{n_N})$ denotes the maximal ideal for the formal power series ring

$$\mathbb{C}[[u_1 t^2, u_2 t^{n_2}, \dots, u_N t^{n_N}]] = \widehat{\mathcal{O}_{H, p}} \subset \widehat{\mathcal{O}_{\hat{H}, q}} = \mathbb{C}[[t]].$$

We conclude that

$$\text{embeddim}_p S = \dim_{\mathbb{C}} m_{S, p} / m_{S, p}^2 = \dim_{\mathbb{C}} m_{H, p} / m_{H, p}^2 + 1 = 3.$$

Thus $p \in S$ is a hypersurface singularity

$$0 \in \{f = 0\} \subset \mathbb{A}^3 \text{ with } \text{mult}_0 f = \text{mult}_0 f_H = 2.$$

Now we deal with the case where the dual graph is of type A_n.

In this case, the normalization $\hat{H}$ splits into two nonsingular irreducible components $\hat{H}_1$ and $\hat{H}_2$, each intersecting the fundamental cycle Z transversally at points q_1 and q_2, respectively. Observe that we can take two hyperplane sections $H'_1 = \text{div}(r'_1)$ and $H'_2 = \text{div}(r'_2)$ (cf. Lemma 4-6-8) such that

$$\pi^* H'_1 = Z + G'_1 \quad q_1 \notin G'_1 \text{ but } q_2 \in G'_1,$$
$$\pi^* H'_2 = Z + G'_2 \quad q_1 \in G'_2 \text{ but } q_2 \notin G'_2.$$

Therefore, for a suitable choice of an affine embedding $p \in S \subset \mathbb{A}^N$ with the first coordiante $x_1 = r'_1$ and the second $x_2 = r'_2$, we see that π can be described as

$$\pi : \hat{H} = \hat{H}_1 \coprod \hat{H}_2 \to H \subset \mathbb{A}^N,$$
$$t_1 \mapsto (u_1 t_1, v_1 t_1^2, \dots)$$
$$t_2 \mapsto (v_2 t_2^2, u_2 t_2, \dots)$$
$$u_1 \in \mathcal{O}_{\hat{H}_1, q_1} \text{ and } u_2 \in \mathcal{O}_{\hat{H}_2, q_2} \text{ are units,}$$
$$v_1 \in \mathcal{O}_{\hat{H}_1, q_1}, v_2 \in \mathcal{O}_{\hat{H}_2, q_2},$$

where t_1 and t_2 are the generators of the maximal ideals $m_{\hat{H}_1, q_1}$ and $m_{\hat{H}_2, q_2}$, respectively. Then

$$\dim_\mathbb{C} m_{H,p}/m_{H,p}^2$$

$$= \dim_\mathbb{C} \widehat{m_{H,p}}/\widehat{m_{H,p}}^2$$

$$= \dim_\mathbb{C}(u_1 t_1 \oplus v_2 t_2^2, v_1 t_1^2 \oplus u_2 t_1, \ldots)/(u_1 t_1 \oplus v_2 t_2^2, v_1 t_1^2 \oplus u_2 t_1, \ldots)^2$$

$$= \dim_\mathbb{C} \mathbb{C} \cdot t_1 \oplus \mathbb{C} \cdot t_2 = 2,$$

where $(u_1 t_1 \oplus v_2 t_2^2, v_1 t_1^2 \oplus u_2 t_1, \ldots)$ is the ideal of the formal power series ring

$$\mathbb{C}[[(u_1 t_1 \oplus v_2 t_2^2, v_1 t_1^2 \oplus u_2 t_1, \ldots)]] = \widehat{\mathcal{O}_{H,p}} \subset \widehat{\mathcal{O}_{\hat{H}_1, q_1}} \oplus \widehat{\mathcal{O}_{\hat{H}_2, q_2}} = \mathbb{C}[[t_1]] \oplus \mathbb{C}[[t_2]].$$

This leads to the same conclusion as before.

This completes the proof of Claim 4-6-10. $\qquad\qquad\square$

Step 3. Derivation of the standard form of the defining equation

The final task is to determine the defining equation of multiplicity 2 (up to multiplication by a unit around the origin)

$$0 \in \{f = 0\} \subset \mathbb{A}^3$$

of a canonical singularity in dimension 2 and derive its standard form.

We recall the following basic theorem in complex analysis in several variables.

Theorem 4-6-11 (Weierstrass Preparation Theorem). *Let f be a holomorphic function around the origin of $\mathbb{C}^n$ with coordinates $(x_1, \ldots, x_{n-1}, w)$ such that*

$$f(0, \cdots, 0, w) = w^m.$$

Then

$$f = u \cdot (w^m + \sigma_1 w^{m-1} + \cdots + \sigma_{m-1} w + \sigma_m),$$

where u, σ_i are holomorphic functions around the origin satisfying

$$u(0, \ldots, 0, 0) = 1$$

$$\sigma_i = \sigma_i(x_1, \ldots, x_{n-1}) \text{ and } \sigma_i(0, \cdots, 0) = 0.$$

Applying the Weierstrass preparation theorem we can take f in a suitable local coordinate system to be of the form

$$f = x^2 + a(y, z)x + b(y, z).$$

By completing the square, we may further assume that f is of the form

$$f = x^2 + g(y, z) \quad \text{mult}_0 g(y, z) \geq 2.$$

Claim 4-6-12. *Let*

$$f = x^2 + g(y, z)$$

be a defining equation of a canonical singularity in dimension 2 at $0 \in \mathbb{A}^3$ in a localcoordinate system (x, y, z). Then

$$\text{mult}_0 g(y, z) \leq 3.$$

PROOF.   Suppose

$$n := \text{mult}_0 g(y, z) > 3.$$

Then a general hyperplane section H of $\{f = 0\}$ can be written

$$H = \{f_H = 0\}, \quad \text{where } f_H = x^2 - y^n \text{ and } n \geq 4,$$

in a suitable *local*coordinate system, again first by using the Weierstrass preparation theorem and then by completing the square.

In the following we use the same notation as in the proof of Claim 4-6-10.

We first deal with the case where the dual graph is of type D_n, E_6, E_7, or E_8 and where $\hat{H}$ is irreducible and hence so is H. Therefore, n is odd in this case. Since then the normalization has the description

$$\pi : \hat{H} \to H \subset \mathbb{A}^2,$$
$$t \mapsto (t^n, t^2) = (x, y),$$

we conclude that there would be no elememt $e \in m_{H,p}$ such that $\text{mult}_q \pi^* e = 3$. On the other hand, we can easily construct an effective divisor $\tilde{Z} + \tilde{G}$ as below such that

$$(\tilde{Z} + \tilde{G}) \cdot E_i = 0 \quad \forall E_i \quad \text{and} \quad \text{mult}_q (\tilde{Z} + \tilde{G})|_{\hat{H}} = 3.$$

By Lemma 4-6-8 there exists $e \in m_{S,p}$ such that $\tilde{Z} + \tilde{G} = \text{div}(\pi^* e)$. But then for this $e \in m_{S,p}$ we have $\text{mult}_q \pi^* H = 3$, a contradiction!

Second, we deal with the case where the dual graph is of type A_n and where $\hat{H}$ is reducible and hence so is H. Therefore, $n = 2l$ $(l \geq 2)$ is even in this case. Since then the normalization has the description

$$\pi : \hat{H}_1 \to H \subset \mathbb{A}^2,$$
$$t_1 \mapsto (t_1^l, t_1) = (x, y),$$
$$\pi : \hat{H}_2 \to H \subset \mathbb{A}^2,$$
$$t_2 \mapsto (-t_2^l, t_2) = (x, y),$$

we conclude that there would be no elememt $e \in m_{H,p}$ such that $\text{mult}_{q_1} \pi^* e = 2$ and $\text{mult}_{q_2} \pi^* e = 1$. On the other hand, we can take the divisor $\tilde{G} + \tilde{Z} = \tilde{G}_1 + \tilde{G}_2 + \tilde{Z}$, where $\tilde{Z} = Z$ is the fundamental cycle and where $\tilde{G}_1$ intersects $\hat{H}_1$ at q_1 transversally, while $\tilde{G}_2$ intersects $\tilde{Z}$ transversally with the same component as $\hat{H}_2$ but *not* at q_2. By Lemma 4-6-8 there exists $e \in m_{S,p}$ such that $\tilde{G}_1 + \tilde{G}_2 + \tilde{Z} = \text{div}(\pi^* e)$. But then for this $e \in m_{H,p}$ we have $\text{mult}_{q_1} \pi^* e = 2$ and $\text{mult}_{q_2} \pi^* e = 1$, a contradiction! $\square$

We illustrate the divisors $\operatorname{div}(\pi^*e)$ below, used in the above argument, where the stars $\star$ indicate the components in $\tilde{G}$ and the white circles $\circ$ indicate the components in $\tilde{Z}$:.

A_n:

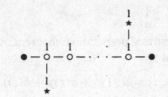

D_n:

E_6:

E_7:

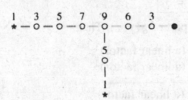

E_8:

By Claim 4-6-11 we have the following two cases to consider:

Case A:$\text{mult}_0 g(y, z) = 2$;
Case B:$\text{mult}_0 g(y, z) = 3$.

Case A: $\text{mult}_0 g(y, z) = 2$.

By applying the Weierstrass preparation theorem, we can choose a local coordinate system such that

$$g(y, z) = u \cdot (y^2 + a(z)y + b(z)).$$

By completing the square, we may further assume

$$g(y, z) = u \cdot (y^2 + vz^{n+1}),$$

where u, v are units and $n + 1 \geq 2$. Setting

$$x' = u^{-\frac{1}{2}}v^{-\frac{1}{2}} \cdot x, \quad y' = v^{-\frac{1}{2}} \cdot y, \quad z' = z,$$

we see that f takes the form

$$f = x'^2 + y'^2 + z'^{n+1}.$$

Finally, setting

$$X = x' + iy', \quad Y = x' - iy', \quad Z = z',$$

we see that f takes the form

$$\mathbf{A_n : f = xy + z^{n+1}}.$$

Case B: $\text{mult}_0 g(y, z) = 3$.

Let $C(g)$ be the cubic part of $g(y, z)$. Then

Subcase B.1: $C(g) \neq$ (a linear factor)^3;
Subcase B.2: $C(g) =$ (a linear factor)^3.

Subcase B.1: $C(g) \neq$ (a linear factor)^3.

In this case the cubic part decomposes into 3 distinct linear factors or into one linear factor multiplied by the square of another. Either way,

$$\{g(y, z) = 0\} \subset \mathbb{A}^2$$

has one nonsingular irreducible component defined by, say $Y' = 0$. Then taking a general coordinate Z' other than Y' and using the Weierstrass preparation theorem we see that g takes the form

$$g(y, z) = u \cdot Y'(Z'^2 + a(Y')Z' + b(Y'));$$

where u is a unit. By completing the square of the second factor, we may further assume

$$g(y, z) = u \cdot Y'(Z'^2 + vY'^{n-2}), \quad n \geq 4,$$

where u, v are units. Finally, setting

$$X = u^{-\frac{1}{2}} v^{-\frac{1}{2}} \cdot x, \quad Y = Y', \quad Z = v^{-\frac{1}{2}} \cdot Z',$$

we see that f takes the form

$$\mathbf{D_n} : \mathbf{f} = \mathbf{x}^2 + \mathbf{y}(\mathbf{z}^2 + \mathbf{y}^{n-2}).$$

Subcase B.2: $C(g) = $ (a linear factor)^3.

Replace y with the linear factor. Then again by applying the Weierstrass preparation theorem, we may assume that $g(y, z)$ is of the form

$$g(y, z) = u \cdot (y^3 + a(z)y^2 + b(z)y + c(z)),$$

where u is a unit. By killing the coefficient of the y^2 term, we may further assume that

$$g(y, z) = u \cdot (y^3 + b(z)y + c(z)) = uy^3 + vz^k y + wz^l.$$

Thus we may assume that f is of the form

$$f = x^2 + uy^3 + vz^k y + wz^l,$$

where u, v, and w are units and where $k \geq 3$ and $l \geq 4$.

Claim 4-6-13. *Let*

$$f = x^2 + uy^3 + vz^k y + wz^l$$

be the defining equation of a canonical singularity in dimension 2 at $0 \in \mathbb{A}^3$ in a localcoordinate system (x, y, z), where u, v, and w are units that are functions of y and z. Then

$$k \leq 3 \text{ or } l \leq 5.$$

PROOF.   Suppose $k \geq 4$ and $l \geq 6$. Then blowing up the origin we see in the affine open set defined by $\{z \neq 0\}$ a hypersurface singularity whose defining equation is of the form

$$f = x^2 + uzy^3 + vz^{k-1}y + wz^{l-2} = x^2 + g(y, z)$$

with $\text{mult}_0 g(y, z) \geq 4$. Claim 4-6-12 implies that this is not a canonical singularity. But this contradicts the following general property of canonical singularities in dimension 2 under blowups.   □

Lemma 4-6-14. *Let $p \in S$ be a germ of a canonical singularity in dimension 2. Let*

$$\mu : W \to S$$

be the blowup of S at p.

Then the minimal resolution $\pi : \hat{S} \to S$ factors through W:

$$
\begin{array}{ccc}
\hat{S} & \xrightarrow{\tau} & W \\
\pi \downarrow & \swarrow \mu & \\
S & &
\end{array}
$$

Moreover, W has again canonical singularities.

PROOF.   Since by Proposition 4-6-9 we have

$$\pi^{-1}(m_p) \cdot \mathcal{O}_{\hat{S}} = \mathcal{O}_{\hat{S}}(-Z) \text{ is invertible,}$$

the universal property of the blowup (cf. Hartshorne [3], Chapter II, Proposition 7.14 page 164) tells us that π factors through μ.

Let

$$0 \in \{f = 0\} \subset \mathbb{A}^3$$

be an expression of $p \in S$ as a hypersurface singularity. By embedding the blow up of S into the blowup of $\mathbb{A}^3$,

$$
\begin{array}{ccc}
W & \subset & \mathrm{Blp}_0 \mathbb{A}^3 \\
\mu \downarrow & & p \downarrow \\
S & \subset & \mathbb{A}^3,
\end{array}
$$

we observe that W has also hypersurface singularities and thus is Cohen–Macaulay and that by the adjunction formula (cf. Proposition 2.4 in Altman–Kleiman [1])

$$
\begin{aligned}
K_W &= K_{\mathrm{Blp}_0 \mathbb{A}^3} + W|_W \\
&= (p^* K_{\mathbb{A}^3} + 2E) + (p^* S - 2E)|_W \\
&= \mu^*(K_{\mathbb{A}^3} + S) = \mu^* K_S.
\end{aligned}
$$

Moreover, W has only isolated singularities. In fact, if W has nonisolated singularities along a curve, say D, then by taking a general hyperplane section C of W we observe that

$$\tau : \hat{C} = \tau^*(C) \to C$$

gives the normalization (around a neighborhood of a point q in $D \cap C$) and that

$$
\begin{aligned}
K_{\hat{C}} &= K_{\hat{S}} + \hat{C}|_{\hat{C}} = \tau^*(K_W + C)|_{\hat{C}} \\
&= \tau^*(K_W + C|_C) = \tau^* K_C.
\end{aligned}
$$

But C being singular at q, this is a contradiction (Exercise! Why?)!

Thus since W is regular in codimension one (R_1) and Cohen–Macaulay (thus S_2), W is normal by Serre's criterion. Since $\mathcal{O}_W(K_W)$ is invertible and hence K_W is a Cartier divisor and since

$$K_{\hat{S}} = \pi^* K_S = \tau^* K_W,$$

W has only canonical singularities.

This completes the proof of Lemma 4-6-14.   □

As a consequence of Claim 4-6-13, we have the following cases:

Subcase B.2.1: $k \geq 3$ and $l = 4$.
Subcase B.2.2: $k = 3$ and $l \geq 5$.
Subcase B.2.3: $k \geq 4$ and $l = 5$.

Subcase B.2.1: $k \geq 3$ and $l = 4$.

In this case, f is of the form

$$f = x^2 + uy^3 + vz^k y + wz^4 \text{ with } k \geq 3.$$

Replacing z with $z + \frac{1}{4w} vz^{k-3} y$, f takes the form

$$f = x^2 + \beta y^3 + py^2 z^2 + wz^4, \text{ where } \beta \text{ a unit}.$$

Replacing y with $y + \frac{1}{3\beta} pz^2$, f takes the form

$$f = x^2 + \beta y^3 + \gamma z^4, \text{ where } \beta, \gamma \text{ are units}.$$

Finally, by setting

$$X = x, \quad Y = \beta^{\frac{1}{3}} \cdot y, \quad Z = \gamma^{\frac{1}{4}} \cdot z,$$

we see that f takes the form

$$\mathbf{E_6} : \mathbf{x}^2 + \mathbf{y}^3 + \mathbf{z}^4.$$

Subcase B.2.2: $k = 3$ and $l \geq 5$.

Consider the curve in $\mathbb{A}^2$ defined by

$$g(y, z) = uy^3 + vz^3 y + wz^l.$$

By blowing up the origin we see in the affine open subset $\{z \neq 0\}$ the strict transform has an ordinary double point

$$uy^3 + vzy + wz^{l-3} = 0.$$

In particular, it has 2 irreducible components and hence so does $\{g(y, z) = 0\}$. One of them is necessarily nonsingular, since $\text{mult}_0 g(y, z) = 3$, and is defined by, say; $Y' = 0$, where

$$Y' = y + \text{higher terms}.$$

Then applying the Weierstrass preparation theorem again,

$$g(y, z) = u \cdot Y'(Y'^2 + pY'z^2 + sz^3),$$

where u, s are units. Now, by replacing Z' with $Z' + \frac{1}{3s} pY'$, f takes the form

$$f = x^2 + \beta y(y^2 + \gamma z^3), \text{ where } \beta, \gamma \text{ are units}.$$

Finally, by setting

$$X = \beta^{-\frac{1}{2}} \cdot x, \quad Y = y, \quad Z = \gamma^{\frac{1}{3}} \cdot z,$$

we see that f takes the form

$$\mathbf{E_7} : \mathbf{x^2 + y(y^2 + z^3)}.$$

Subcase B.2.3: $k \geq 4$ and $l = 5$.

In this case, f is of the form

$$f = x^2 + uy^3 + vz^k y + wz^5, \quad k \geq 4.$$

Replacing z with $z + \frac{1}{5w} vz^{k-4} y$, f takes the form

$$f = x^2 + \beta y^3 + py^2 z^3 + wz^5, \text{ where } \beta \text{ is a unit.}$$

Replacing y with $y + \frac{1}{3\beta} pz^3$, f takes the form

$$f = x^2 + \beta y^3 + \gamma z^5, \text{ where } \beta \text{ and } \gamma \text{ are units.}$$

Finally, by setting

$$X = x, \quad Y = \beta^{\frac{1}{3}} \cdot y, \quad Z = \gamma^{\frac{1}{5}} \cdot z,$$

we see that f takes the form

$$\mathbf{E_8} : \mathbf{x^2 + y^3 + z^5}.$$

This completes the proof of Theorem 4-6-7. □

The following corollary follows immediately from Theorem 4-6-7.

Corollary 4-6-15. *Canonical singularities in dimension 2 are determined analytically by the dual graphs of the exceptional curves appearing in their minimal resolutions. More precisely, if a germ of a normal surface singularity has only (-2)-curves as the exceptional curves in its minimal resolution and if the dual gragh is of Type A_n, D_n, E_6, E_7, or E_8, then it is analytically isomorphic to a hypersurface singularity in $\mathbb{A}^3$ defined by the corresponding equation described as in Theorem 4-6-7.*

Corollary 4-6-16. *For a germ of a normal surface singularity, the following are equivalent:*

(i) *it is a canonical singularity,*

(ii) *it is a rational (hypersurface) double point,*

(iii) *it is a rational Gorenstein point,*

(iv) *it is analytically isomorphic to a quotient singularity*

$$0 \in \mathbb{C}^2 / H,$$

where H is a finite subgroup of $\mathrm{SL}(2, \mathbb{C})$.

PROOF. Theorem 4-6-7 proves the implication (i) $\Longrightarrow$ (ii). The implication (ii) $\Longrightarrow$ (iii) is immediate. (Note that we do not need the assumption that it is a hypersurface singularity, since it is easy to see that a double point is a hypersurface singularity.) Suppose it is a rational Gorenstein singularity. Then the duality implies that $\pi : \hat{S} \to S$ being the minimal resolution, we have

$$0 = R^1\pi_*\mathcal{O}_{\hat{S}} = \mathcal{O}_S(K_S)/\pi_*\mathcal{O}_{\hat{S}}(K_{\hat{S}}) \text{ and hence } \pi^*K_S = K_{\hat{S}}.$$

Thus it has a canonical singularity, showing the implication (iii) $\Longrightarrow$ (i).

The classical result provides the following classification table of finite subgroups of SL$(2, \mathbb{C})$, whose quotients turn out to be hypersurface singularities defined by the equations below.

G	the description of the group	the equation
C_{n+1}	cyclic of order $n + 1$	$xy + z^{n+1}$
$\tilde{D}_{n-2}$	binary $(n - 2)$-dihedral	$x^2 + y(z^2 + y^{n-2})$
$\tilde{T}$	binary tetrahedral	$x^2 + y^3 + z^4$
$\tilde{O}$	binary octahedral	$x^2 + y(y^2 + z^3)$
$\tilde{I}$	binary icosahedral	$x^2 + y^3 + z^5$

Thus by Theorem 4-6-7 we see the implications (iv) $\Longleftrightarrow$ (i).   □

Remark 4-6-17. (i) The classification of finite subgroups of SL$(2, \mathbb{C})$ is straightforward. In fact, let G be a finite subgroup of SL$(2, \mathbb{C})$. We can take a G-invariant Hermitian metric on $\mathbb{C}^2$. Taking a unitary basis with respect to this Hermitian metric, G can be now considered a finite subgroup of SU$(2, \mathbb{C})$. Pay attention to the canonical homomorphism

$$h : \text{SU}(2, \mathbb{C}) \to \text{SO}(3, \mathbb{R}),$$

whose kernel is $\{\pm I_2\}$. Now, the image $h(G)$, being a finite subgroup of SO$(3, \mathbb{R})$, is the group of symmetries of a plane regular polygon or a solid regular polyhedron. That is to say, it is either a group of finite planar rotations, a dihedral group, the tetrahedral group, the octahedral group, or the icosahedral group. Now it is easy to determine G as above from this information on $h(G)$. The task of finding the generators of the invariant ring and determining the relation among them is more involved but is a classical subject, whose treatment can be found in some modern books such as Lamotke [1] and Mukai [1].

(ii) The subscripts for the types of singularities

$$A_n, D_n, E_6, E_7, E_8$$

refer to the number of the irreducible exceptional divisors for the minimal resolutions. On the other hand, as quotients by the finite subgroups of SL$(2, \mathbb{C})$, the group corresponding to the singularity of type D_n is the binary $(n - 2)$-dihedral group, i.e., the binary extension of the symmetries of the regular $(n - 2)$-polygon, the group often denoted by D_{n-2}. One should be aware of this discrepancy between the subscripts when referring to the types of singularities and when referring to the types of the corresponding dihedral groups.

(iii) For the standard forms of the equations of type A_n, one could use

$$x^2 + y^2 + z^{n+1} \text{ as well as } xy + z^{n+1}.$$

For the standard forms of the equations of type D_4, one could use

$$x^2 + y^3 + z^3 \text{ as well as } x^2 + y(z^2 + y^2).$$

We made our choice so that it will be consistent with the equivariant standard forms of the equations when we consider log terminal singularities later in this section and hence consider the cyclic group actions on canonical singularities in dimension 2. (See Theorem 4-6-22.)

(iii) The proof of Theorem 4-6-7 looks rather ad hoc toward the end (if not from the beginning), since we aimed at obtaining the standard form by changing analytical coordinates rather than aiming at a criterion for the equation f to define a canonical singularity, which should be invariant under the coordinate changes. M. Reid (cf. Reid [7]) gives a criterion for f to define a canonical singularity using the methods of toric geometry. His criterion is exactly the invariant kind of criterion we would like to have. In fact, if we used this criterion, then the key Claim 4-6-12 would follow immediately and we go directly to the business of deriving the standard form of the defining equation, which, though interesting, is not of central importance other than seeing that the canonical singularities are analytically determined by their dual graphs.

(iv) Our derivation of the standard forms of the defining equations for hypersurface singularities representing canonical singularities in dimension 2 is essentially characteristic-free, except for the specific places where we divide by 2, 3, or 5 to complete the square, killing the y^2 term in the cubic equation in y or killing the z^4 term in the quintic equation in z. By paying special attention to these places that *do* depend on the characteristics, we can actually derive the table by Artin [5] for the standard forms of the defining equations for hypersurface singularities representing canonical singularities in dimension 2 in all characteristics. (The table is the same for $\text{char}(k) = p > 5$ as the one for $\text{char}(k) = 0$, and the exotic ones occur only in $\text{char}(k) = 2, 3$, or 5. For details of the argument, we refer the reader to Matsuki [7].)

Log Terminal Singularities in Dimension 2

Here we present two different approaches, the first, **Approach 1**, after Kawamata [9] to characterize log terminal singularities in dimension 2 as the quotient singularities by finite subgroups of $GL(2, \mathbb{C})$ without quasi-reflection and resort to the table of such groups classified by Brieskorn [1], the second, **Approach 2**, to characterize them as the cyclic quotients of canonical singularieties in dimension 2 and classify such quotients directly by determining the equivariant standard forms of the defining equations of rational double points.

Approach 1

Theorem 4-6-18 (Characterization of Log Terminal Singularities in Dimension 2 as Quotient Singularities). *Let $p \in S$ be a germ of log terminal singularities in dimension 2.*

(i) *$p \in S$ is a rational singularity and the canonical divisor K_S is $\mathbb{Q}$-Cartier.*
(ii) *Let*

$$c : \tilde{S} \to S$$
$$\cup \qquad \cup$$
$$q \mapsto p$$

be the canonical cover with Galois group $\mathbb{Z}/(r)$, where r is the index of $p \in S$. Then $q \in \tilde{S}$ has only canonical singularities.
(iii) *As $q \in \tilde{S}$ is a germ of canonical singularities in dimension 2, it is a quotient of $\mathbb{C}^2$ by a finite subgroup $H \subset SL(2, \mathbb{C})$ without quasi-reflection (i.e., no nontrivial element of H has a fixed point except the origin). Therefore, we have a tower of étale coverings*

$$\mathbb{C}^2 - \{0\} \to \tilde{S} - \{q\} \to S - \{p\}$$

and the corresponding exact sequence of finite fundamental groups

$$\{e\} = \pi_1(\mathbb{C}^2 - \{0\}) \to \dot{H} = \pi_1(\tilde{S} - \{q\}) \to G = \pi_1(S - \{p\}) \to \mathbb{Z}/(r) \to 0.$$

The group G acts on $\mathbb{C}^2$ naturally and linearly, i.e., by choosing a suitable system of coordinates its action can be considered as that of a subgroup $G \subset GL(2, \mathbb{C})$ without quasi-reflection. The original germ $p \in S$ of log terminal singularities is analytically isomorphic to $(0 \in \mathbb{C}^2)/G$, i.e.,

$$p \in S \overset{\text{analytically}}{\cong} (0 \in \mathbb{C}^2)/G.$$

(iv) *Conversely, let G be a finite subgroup of $GL(2, \mathbb{C})$ without quasi-reflection. Then the quotient singularity $(0 \in \mathbb{C}^2)/G$ has only log terminal singularities.*

PROOF.   Let $\pi : \hat{S} \to S$ be the minimal resolution. In the ramification formula

$$K_{\hat{S}} = \pi^* K_S + \Sigma a_i E_i,$$

all the discrepancies a_i are strictly bigger than -1, as $p \in S$ is log terminal. Then taking a π-ample divisor of the form

$$-\Sigma \delta_i E_i, \text{ where } \delta_i > 0 \ \ \forall i,$$

we see that for sufficiently small $\epsilon > 0$, the log pair $(\hat{S}, D = \Sigma(-a_i)E_i + \epsilon \Sigma \delta_i E_i)$ has only $\mathbb{Q}$-factorial and log terminal singularities and that by the analytic version of Theorem 5-2-8

$$0 = R^1 \pi_* \mathcal{O}_{\hat{S}}(K_{\hat{S}} + D - K_{\hat{S}} - \Sigma(-a_i)E_i - \epsilon \Sigma \delta_i E_i) = R^1 \pi_* \mathcal{O}_{\hat{S}},$$

since

$$-K_{\hat{S}} - \Sigma(-a_i)E_i - \epsilon\Sigma\delta_i E_i \equiv -\pi^* K_S \to \epsilon\Sigma\delta_i E_i$$

is π-ample. Thus, $p \in S$ is rational.

Let $e \in \mathbb{N}$ be an integer such that

$$e(K_{\hat{S}} - \Sigma a_1 E_i) = e\pi^* K_S = \pi^* e K_S$$

is an integral divisor on $\hat{S}$. Since

$$e(K_{\hat{S}} - \Sigma a_i E_i) \cdot E_j = \pi^* e K_S \cdot E_j = 0 \quad \forall E_j,$$

we conclude by Lemma 4-6-8 that $e(K_{\hat{S}} - \Sigma a_i E_i) = \pi^* D$ for some Cartier divisor D on S and hence that $e K_S = \pi_* \pi^* e K_S = \pi_* \pi^* D = D$ is a Cartier divisor.

This proves assertion (i).

The assertion (ii) follows from Proposition 4-5-3 (i).

Now that $q \in \hat{S}$ is a germ of canonical singularites in dimension 2, Corollary 4-6-16 tells us that it is analytically isomorphic to a quotient singularity $0 \in \mathbb{C}^2/H$, where H is a finite subgroup of $SL(2, \mathbb{C})$ without quasi-reflection as described in the table of Corollary 4-6-16. Therefore, we have a tower of étale coverings

$$\mathbb{C}^2 - \{0\} \to \tilde{S} - \{q\} \to S - \{p\}$$

and the corresponding exact sequence of finite fundamental groups

$$\{e\} = \pi_1(\mathbb{C}^2 - \{0\}) \to H = \pi_1(\tilde{S} - \{q\}) \to G = \dot{\pi}_1(S - \{p\}) \to \mathbb{Z}/(r) \to 0.$$

From this it follows that $G = \pi_1(S - \{p\})$ acts naturally and holomorphically on $\mathbb{C}^2 - \{0\}$ and hence on $\mathbb{C}^2$ by Hartog's theorem and that its quotient is the original singularity $p \in S$. In order to see that the action of G can be linearized, we quote a standard lemma below.

Lemma 4-6-19 (Linearization of the Action) 9cf. Lemma 2.2 in Brieskorn [1] Lemma 9.9 in Kawamata [9]). *Let G be a finite group acting faithfully on the germ of $0 \in \mathbb{C}^n$ the origin of more generally acting faithfully on the germ of an affine variety $0 \in X \subset \mathbb{C}^n$ fixing the origin such that the Zariski tangent space is isomorphic to the ambient vector space $\mathbb{C}^n$, i.e.,*

$$\mathrm{Hom}_{\mathbb{C}}(m_{X,0}/m_{X,0}^2, \mathbb{C}) \cong \mathbb{C}^n.$$

Then by choosing a suitable system of coordinates $(z_1, \ldots, z_n)$, the action of G can be linearized, i.e., there exists an embedding

$$\alpha : G \hookrightarrow GL(n, \mathbb{C})$$

such that

$$g^*(^t(z_1, \ldots, z_n)) = \alpha(g)^t(z_1, \ldots, z_n) \quad \forall g \in G,$$

where

$$^t(z_1, \ldots, z_n) = \begin{pmatrix} z_1 \\ \cdots \\ z_n \end{pmatrix}.$$

PROOF.   Take an arbitrary system of coordinates $(z_1', \ldots, z_n')$. Since the action of G induces the (co)action on the vector space $m_{X,0}/m_{X,0}^2$, we have

$$g^*(\,^t(z_1', \ldots, z_n')) = \alpha(g)^t(z_n', \ldots, z_n') \bmod m_{X,0}^2,$$

where $\alpha(g) \in GL(n, \mathbb{C})$. We take the new system of coordinates to be

$$\,^t(z_1, \ldots, z_n) = \Sigma_{h \in G} h^*(\alpha(h)^{-1\,t}(z_1', \ldots, z_n')).$$

Then for $g \in G$ we have

$$
\begin{aligned}
g^*(\,^t(z_1, \ldots, z_n)) &= g^* \Sigma_{h \in G} h^*(\alpha(h)^{-1\,t}(z_1', \ldots, z_n')) \\
&= \Sigma_{hg \in G}(hg)^*(\alpha(g)\alpha(hg)^{-1\,t}(z_1', \ldots, z_n')) \\
&= \Sigma_{h \in G} h^*(\alpha(g)\alpha(h)^{-1\,t}(z_1', \ldots, z_n')) \\
&= \alpha(g)\{\Sigma_{h \in G} h^*(\alpha(h)^{-1\,t}(z_1', \ldots, z_n'))\} = \alpha(g)^t(z_1, \ldots, z_n).
\end{aligned}
$$

Since the (co)action on a system of coordinates determines the action analytically, $\alpha \colon G \to GL(n, \mathbb{C})$ is an embedding. This completes the proof of Lemma 4-6-19.   □

We go back to the proof of Theorem 4-6-18.

Note that since the action of G on $\mathbb{C}^2$ is induced from that of G as the fundamental group $\pi_1(S - \{0\})$ acting on the universal cover $\mathbb{C}^2$ [0], any element $g \in G$ different from the identity has no fixed points except for the origin. Thus $G \subset GL(2, \mathbb{C})$ is without quasi-reflection. This proves assertion (iii).

Observe that once the action of G is linearized and hence we can regard G as a subgroup of $GL(2, \mathbb{C})$ we have a homomorphism $\det \colon G \to \mathbb{C}^\times$ by taking the determinant of the matrices and hence that we have an exact sequence

$$\{e\} \to \mathrm{Ker} \to G \xrightarrow{\det} \mu_s \to 0,$$

where $\mu_s \subset \mathbb{C}^\times$ is the group of the sth roots of unity. We claim that this exact sequence coincides with our exact sequence of the fundamental groups

$$\{e\} = \pi_1(\mathbb{C}^2 - \{0\}) \to H = \pi_1(\tilde{S} - \{q\}) \to G = \pi_1(S - \{p\}) \to \mathbb{Z}/(r) \to 0$$

and hence $\mathrm{Ker} \cong H$ and $\mu_s = \mu_r \cong \mathbb{Z}/(r)$.

In order to see this claim, we have only to show that the natural map

$$
\begin{array}{ccc}
0 \in \mathbb{C}^2/\mathrm{Ker} & \xrightarrow{c'} & (0 \in \mathbb{C}^2)/G \\
\| & & \| \\
q' \in \tilde{S}' & \xrightarrow{c'} & p \in S
\end{array}
$$

coincides with the canonical cover $c \colon q \in \tilde{S} \to p \in S$. The germ $q' \in \tilde{S}'$ is normal by construction. Since the generator $dz_1 \wedge dz_2$ of the line bundle $\Omega_{\mathbb{C}^2}^2 \cong \mathcal{O}_{\mathbb{C}^2}(K_{\mathbb{C}^2})$ is invariant under the action of $\mathrm{Ker} \subset SL(2, \mathbb{C})$, it descends to a generator of the sheaf $\mathcal{O}_{\tilde{S}'}(K_{\tilde{S}'})$, which is hence a line bundle. The generator can be considered as a local section for $\mathcal{O}_{\tilde{S}'}$ and hence for the corresponding rational function $h' \in \mathbb{C}(\tilde{S}')$ we have

$$0 = \operatorname{div}(h') + K_{\tilde{S}'} = \operatorname{div}(h') + c'^* K_S.$$

Note that μ_s acts on $dz_1 \wedge dz_2$ as the multiplication and similarly on the generator for $\mathcal{O}_{\tilde{S}}$ and that μ_s acts on h' as the multiplication. Therefore,

$$0 = c'_*(\operatorname{div}(h') + c'^* K_S) = \operatorname{div}(\operatorname{Norm}(h')) + sK_S = \operatorname{div}(h'^s) + sK_S.$$

We conclude from this that the index r divides s and that $c' \colon \tilde{S}' \to S$ factors through the canonical cover, i.e., there exists a morphism $\sigma \colon \tilde{S}' \to \tilde{S}$ such that $c' = c \circ \sigma$. (See the argument for the characterization of the canonical cover in Proposition-Definition 4-5-1 again.) If σ is not an isomorphism, i.e., if the natural surjection $\mu_s \to \mu_r$ has a nontrivial kernel, then there would exist a nonvanishing section of $\mathcal{O}_{\tilde{S}'}(K_{\tilde{S}'})$ that is the pull-back of a nonvanishing section of $\mathcal{O}_{\tilde{S}}(K_{\tilde{S}})$ and hence is invariant under the action of this nontrivial kernel. But since the whole group μ_s acts faithfully on the generator descended from $dz_1 \wedge dz_2$, this is impossible. This proves the claim.

In order to see assertion (iv), let

$$f \colon (q \in \tilde{X}) = (0 \in \mathbb{C}^2) \to (p \in X) = (0 \in \mathbb{C}^2)/G$$

be the quotient map. Though assertion (i) of Proposition 4-5-3 was proved under the assumption that $c \colon \tilde{X} \to X$ is the canonical cover, we used only the fact $K_{\tilde{X}} = c^* K_X = f^* K_X$ in the proof. This clearly holds in our case, since f is étale outside of $0 = q = f^{-1}(p)$, since G is without quasi-reflection. Now that $(q \in \tilde{X}) = (0 \in \mathbb{C}^2)$ is nonsingular and hence has only canonical signularities, we conclude that $(p \in X) = (0 \in \mathbb{C}^2)/G$ has only log terminal singularities.

We note that by Chevalley's result (cf. Prill [1]) for an arbitrary finite subgroup G of $GL(2, \mathbb{C}^2)$ there exists a finite subgroup G' of $GL(2, \mathbb{C})$ without quasi-reflection such that

$$(0 \in \mathbb{C}^2)/G \cong (0 \in \mathbb{C}^2)/G'.$$

Thus one can get rid of the condition for G to be without quasi-reflection from (iv).

This completes the proof of Theorem 4-6-18.   □

Theorem 4-6-18 leads us to the complete classification of log terminal singularities in dimension 2, if we resort to the following table of finite subgroups of $GL(2, \mathbb{C})$ without quasi-reflection (up to conjugacy) by Brieskorn [1].

Theorem 4-6-20 (Classification of Finite Subgroups of $GL(2, \mathbb{C})$). *Let G be a finite subgroup of $GL(2, \mathbb{C})$ without quasi-reflection. Then G is given by the table below up to conjugacy.*

In the table the group H represents the kernel of the following exact sequence induced by the map taking the determinant:

$$\{e\} \to H = \operatorname{Ker} \to G \xrightarrow{\det} \mu_r \to 0.$$

In the column "Dual Graph" we describe the dual graph of the exceptional locus of the minimal resolution of the quotient singularity $(0 \in \mathbb{C}^2)/G$.

We use the following notation:

$$C_{n,q} := \left\{ \begin{pmatrix} \exp\left(\frac{2\pi\sqrt{-1}}{n}\right) & 0 \\ 0 & \exp\left(\frac{2\pi q\sqrt{-1}}{n}\right) \end{pmatrix} \right\}$$

with $0 < q < n$ and $\gcd(n, q) = 1$,

$Z_k :=$ *the cyclic group of order k in* $\mathbb{ZL}(2, \mathbb{C})$, *the center of* $GL(2, \mathbb{C})$,

$C_k :=$ *the cyclic group of order k in* $SL(2, \mathbb{C})$,

$D_k :=$ *the binary dihedral group of order $4k$,*

$T :=$ *the binary tetrahedral group,*

$O :=$ *the binary octahedral group,*

$I :=$ *the binary icosahedral group,*

$(H_1, N_1; H_2, N_2) = \{h \in GL(2, \mathbb{C}); h = h_1 h_2$ *for some* $h_1 \in H_1, h_2 \in H_2$

such that $h_1 \bmod N_1 = h_2 \bmod N_2\}$.

Note that $D_k = \tilde{D}_{2k+2}$, $T = \tilde{T}$, $O = \tilde{O}$, $I = \tilde{I}$ *in the notation of Corollary 4-6-16.*

Type	G	Dual Graph	H	r
A	$C_{n,q}$ $0 < q < n, \gcd(n,q) = 1$ $k := \gcd(n, q+1)$	$\langle n, q \rangle$	C_k	$\frac{n}{k}$
D-1	$(Z_{2m}, Z_{2m}; D_n, D_n)$ $\gcd(m, 2) = 1$ $\gcd(m, n) = 1$	$\langle b; 2, 1 : 2, 1; n, q \rangle$	D_n	m
D-2	$(Z_{4m}, Z_{2m}; D_n, C_{2n})$ $\gcd(m, 2) = 2$ $\gcd(m, n) = 1$	$\langle b; 2, 1; 2, 1; n, q \rangle$	C_{2n}	$2m$
E_6-1	$(Z_{2m}, Z_{2m}; T, T)$ $\gcd(m, 6) = 1$	$\langle b; 2, 1; 3, 1; 3, 1 \rangle$ $\langle b; 2, 1; 3, 2; 3, 2 \rangle$	T	m
E_6-2	$(Z_{6m}, Z_{2m}; T, D_2)$ $\gcd(m, 6) = 3$	$\langle b; 2, 1; 3, 1; 3, 2 \rangle$	D_2	$3m$
E_7	$(Z_{2m}, Z_{2m}; O, O)$ $\gcd(m, 6) = 1$	$\langle b; 2, 1; 3, 1; 4, 1 \rangle$ $\langle b; 2, 1; 3, 1; 4, 3 \rangle$ $\langle b; 2, 1; 3, 2; 4, 1 \rangle$ $\langle b; 2, 1; 3, 2; 4, 3 \rangle$	O	m
E_8	$(Z_{2m}, Z_{2m}; I, I)$ $\gcd(m, 30) = 1$	$\langle b; 2, 1; 3, 1; 5, 1 \rangle$ $\langle b; 2, 1; 3, 1; 5, 2 \rangle$ $\langle b; 2, 1; 3, 1; 5, 3 \rangle$ $\langle b; 2, 1; 3, 1; 5, 4 \rangle$ $\langle b; 2, 1; 3, 2; 5, 1 \rangle$ $\langle b; 2, 1; 3, 2; 5, 2 \rangle$ $\langle b; 2, 1; 3, 2; 5, 3 \rangle$ $\langle b; 2, 1; 3, 2; 5, 4 \rangle$	I	m

Remark 4-6-21 (Notation for the Dual Graph). (i) In the table of Theorem 4-6-20, the notation $\langle n, q \rangle$, where $0 < q < n$ and $\gcd(n, q) = 1$, represents the chain

$$\underset{(b_1)}{\bigcirc} - \underset{(b_2)}{\bigcirc} - \cdots - \underset{(b_m)}{\bigcirc}$$

such that $\frac{n}{q}$ has the continued fraction expansion

$$\frac{n}{q} = b_1 - \cfrac{1}{b_2 - \cdots} = [b_1, b_2, b_3, \ldots, b_m].$$

It is easy to see, as described in the table, that the exceptional locus of the minimal resolution of the quotient singularity by the cyclic group $C_{n,q}$ is the chain of rational curves E_i with self-intersection numbers $E_i^2 = -b_i$ and hence whose dual graph is given by $\langle n, q \rangle$. (See, e.g., Bart–Peters–Van de Ven [1], Fulton [2].)

(ii) The notation

$$\langle b; n_1, q_1; n_2, q_2; \ldots; n_s, q_s \rangle$$

refers to the tree where in the center we have a circle with the self-intersection number $-b$ and the branches of type $\langle n_i, q_i \rangle$ are attached to it:

The circle with the self-intersection $-b_{i,1}$ in the branch $\langle n_i, q_i \rangle$ is adjacent to the central circle, where

$$\frac{n_i}{q_i} = [b_{i,1}, b_{i,2}, \ldots].$$

For example,

Though **Approach 1** completes the classification of log terminal singularities in dimension 2 as in Kawamata [9] in an elegant way, we notice that we went against our principle of the book (a promise to derive the important results from scratch) referring to and depending on the results of Brieskorn [1].

Approach 2

Approach 2 starts with the same observation as in **Approach 1** that the canonical cover $c: q \in \tilde{S} \to p \in S$ of a germ $p \in S$ of log terminal singularities in dimension 2 has at most canonical singularities (cf. Theorem 4-6-18 (i) (ii)). This implies

$$p \in S \overset{\text{analytically}}{\cong} (q \in \tilde{S})/\mathbb{Z}/(r),$$

where $\mathbb{Z}/(r)$ is the cyclic group of order r, which is the Galois group of $q \in \tilde{S}$ over $p \in S$. Therefore, the classification of log terminal singularities in dimension 2 is reduced to the classification of the actions of cyclic groups on canonical singularities in dimension 2.

We remark that **Approach 2** is a baby version of the method by Reid–Mori to classify terminal singularities in dimension 3 as hyperquotients (cf. Reid [1][2][3], Mori [4]).

Theorem 4-6-22 (Classification of Log Terminal Singularities in Dimension 2 as Hyperquotients). *Let $p \in S$ be a germ of log terminal singularities in dimension 2. Then the canonical cover $c{:}q \in \tilde{S} \to p \in S$ is either nonsingular or a germ of canonical singularities in dimension 2. Accordingly, $p \in S$ is analytically isomorphic either to a cyclic quotient of $\mathbb{C}^2$,*

$$p \in S \overset{\text{analytically}}{\cong} (0 \in \mathbb{C}^2)/\mathbb{Z}/(r),$$

or to a cyclic quotient of a canonical singularity in dimension 2 expressed as a hypersurface singularity (called a **hyperquotient** *for short),*

$$p \in S \overset{\text{analytically}}{\cong} (0 \in \{f = 0\} \subset \mathbb{C}^3)/\mathbb{Z}/(r),$$

where for a suitable system of coordinates (x, y) for $\mathbb{C}^2$ in the nonsingular case and (x, y, z) of $\mathbb{C}^3$ in the hyperquotient case:

(i) *the cyclic group $\mathbb{Z}/(r)$ acts on $\mathbb{C}^2$ in the nonsingular case and on $\mathbb{C}^3$ in the hyperquotient case linearly and diagonally, i.e., for a generator $g \in \mathbb{Z}/(r)$ and a primitive rth root of unity ζ,*

$$g^* \begin{pmatrix} x \\ y \end{pmatrix} = \begin{pmatrix} \zeta^\alpha & 0 \\ 0 & \zeta^\beta \end{pmatrix} \begin{pmatrix} x \\ y \end{pmatrix}$$

and

$$g^* \begin{pmatrix} x \\ y \\ z \end{pmatrix} = \begin{pmatrix} \zeta^\alpha & 0 & 0 \\ 0 & \zeta^\beta & 0 \\ 0 & 0 & \zeta^\gamma \end{pmatrix} \begin{pmatrix} x \\ y \\ z \end{pmatrix},$$

(ii) *in the hyperquotient case the defining equation f of the canonical singularity is semi-invariant for the action of $\mathbb{Z}/(r)$ and it is one of the following types listed on the table below,*

Type	Equation	Conditions on $(\alpha, \beta, \gamma) \bmod (r)$
A_n ($n \geq 1$)	$xy + z^{n+1}$	$\alpha + \beta = (n+1)\gamma$
eA_n ($n \geq 2$)	$x^2 + y^2 + z^{n+1}$	$2\alpha = 2\beta = (n+1)\gamma$ $\alpha \neq \beta$
D_n ($n \geq 4$)	$x^2 + y(z^2 + y^{n-2})$	$2\alpha = \beta + 2\gamma = (n-1)\beta$
eD_4	$x^4 + y^3 + z^3$	$2\alpha = 3\beta = 3\gamma$ $\beta \neq \gamma$
E_6	$x^2 + y^3 + z^4$	$2\alpha = 3\beta = 4\gamma$
E_7	$x^2 + y(y^2 + z^3)$	$2\alpha = 3\beta = \beta + 3\gamma$
E_8	$x^2 + y^3 + z^5$	$2\alpha = 3\beta = 5\gamma$

(Note that we consider the nonsingular case to be of Type A_0. There is no Type eA_1.)

(iii) *any element of $\mathbb{Z}/(r)$ other than the identity has no fixed points except for the origin, a condition that translates into the combinatorial conditions in the nonsingular case,*

$$\gcd(\alpha, r) = \gcd(\beta, r) = 1,$$

and in the hyperquotient case

$$\gcd(\alpha, \beta, r) = \gcd(\beta, \gamma, r) = \gcd(\alpha, \gamma, r) = 1,$$

(iv) *the cyclic group $\mathbb{Z}/(r)$ acts faithfully on the semi-invariant generator of $\mathcal{O}_{\bar{S}}(K_{\bar{S}})$ (which is uniquely determined up to invariant units), i.e., in the nonsingular case*

$$dx \wedge dz$$

and in the hyperquotient case

$$\frac{dy \wedge dz}{\partial f/\partial x} = -\frac{dx \wedge dz}{\partial f/\partial y} = \frac{dx \wedge dy}{\partial f/\partial z}.$$

The condition translates into the combinatorial one in the nonsingular case,

$$\gcd(\alpha + \beta, r) = 1,$$

and in the hyperquotient case

$$\gcd(\gamma, r) = 1 \text{ for Type } A_n \ (n \geq 1),$$
$$\gcd(\beta + \gamma - \alpha, r) = 1 \text{ for the other types.}$$

This characterization of log terminal singularities in dimension 2 as hyperquotients determines their analytical structures and is canonical up to the choice of a generator of $\mathbb{Z}/(r)$ and a primitive rth root of unity ζ.

Moreover, the dual graphs of the exceptional loci of the minimal resultions of log terminal singularities in dimension 2 can be described as follows, according to the "types" of the singularities. All the dual graphs below with components nonsingular rational curves occur geometrically associated to some log terminal singularities, and they determine their analytic structures.

Equation Type	Dual Graph	Brieskorn's Type
$A_n \ (n \geq 0)$	$\langle m, q \rangle$	A
$eA_n \ (n \geq 2)$	$\langle b : 2, 1; 2, 1; m, q \rangle \ \ r$ even	D-2
$D_n \ (n \geq 4)$	$\langle b; 2, 1; 2, 1; m, q \rangle \ \ r$ odd	D-1
eD_4	$\langle b; 2, 1; 3, 1; 3, 2 \rangle$	E_6-2
E_6	$\langle b; 2, 1; 3, 1; 3, 1 \rangle$ $\langle b; 2, 1; 3, 2; 3, 2 \rangle$	E_6-1
E_7	$\langle b; 2, 1; 3, 1; 4, 1 \rangle$ $\langle b; 2, 1; 3, 1; 4, 3 \rangle$ $\langle b; 2, 1; 3, 2; 4, 1 \rangle$ $\langle b; 2, 1; 3, 2; 4, 3 \rangle$	E_7
E_8	$\langle b; 2, 1; 3, 1; 5, 1 \rangle$ $\langle b; 2, 1; 3, 1; 5, 2 \rangle$ $\langle b; 2, 1; 3, 1; 5, 3 \rangle$ $\langle b; 2, 1; 3, 1; 5, 4 \rangle$ $\langle b; 2, 1; 3, 2; 5, 1 \rangle$ $\langle b; 2, 1; 3, 2; 5, 2 \rangle$ $\langle b; 2, 1; 3, 2; 5, 3 \rangle$ $\langle b; 2, 1; 3, 2; 5, 4 \rangle$	E_8

PROOF.   Let $p \in S$ be a germ of log terminal singularities in dimension 2. Then by Proposition 4-5-3 (i), the canonical cover $q \in \tilde{S}(\xrightarrow{c} p \in S)$ has at most canonical singularities, i.e., it is either nonsingular or one of the hypersurface singularities as specified in Theorem 4-6-7. The Galois group $\mathbb{Z}/(r)$ acts naturally and faithfully on $q \in \tilde{S}$, and we have

$$p \in S \overset{\text{analytically}}{\cong} (q \in \tilde{S})/\mathbb{Z}/(r).$$

By Lemma 4-6-19 we may assume that the action of the Galois group $\mathbb{Z}/(r)$ is linear on $\mathbb{C}^2$ in the nonsingular case and on $\mathbb{C}^3$ in the hyperquotient case. Since it is commutative, we may further assume that the linear action is diagonal. Replacing an arbitrary generator f for the principal ideal $I_S = (f)$ with

$$\sqrt{\prod_{g \in \mathbb{Z}/(r)} u_g} \cdot f, \text{ where } g^* f = u_g \cdot f \text{ for } g \in \mathbb{Z}/(r) \text{ and a unit } u_g \in \mathcal{O}^\times_{S,p},$$

we find a semi-invariant generator, i.e., a semi-invariant defining equation for S in $\mathbb{C}^3$. (A smart reader may realize that by the construction of the canonical cover

$$\tilde{S} = \oplus_{i=0}^{r-1} \mathcal{O}_S(-iK_S),$$

where the direct summands are the eigenspaces for the Galois group, the action is automatically linear and diagonal and that a semi-invariant defining equation may be chosen from one of the eigenspaces.)

Now we show that we can choose the semi-invariant defining equations in the standard forms as specified in (iii) via an equivariant version os the Weierstrass preparation theorem, following exactly the line of argument for the classification of canonical singularities in Theorem 4-6-7 with the extra data on the cyclic group actions.

Theorem 4-6-23 (Equivariant Weierstrass Preparation Theorem). *Let $\mathbb{Z}/(r)$ be the cyclic group of order r acting on $\mathbb{C}^n$ with a system of semi-invariant coordinates $(x_1, \ldots, x_{n-1}, w)$. (Say, $g^*w = \zeta^a \cdot w$ for a generator $g \in \mathbb{Z}/(r)$, a primitive rth root of unity ζ, and $a \in \mathbb{N}$.) Let f be a holomorphic function around the origin that is semi-invariant under the action of $\mathbb{Z}/(r)$.*

If

$$f(0, \ldots, 0, w) = w^m \text{ for some } m \in \mathbb{N}$$

and

$$f = u \cdot f_W = u \cdot (w^m + \sigma_1 w^{m-1} + \cdots + \sigma_m)$$

is the decomposition of f into a unit u and the Weierstrass form f_W with

$$\sigma_i = \sigma_i(x_1, \ldots, x_{n-1}) \text{ with } \sigma_i(0, \ldots, 0) = 0,$$

then the functions u, f_W, $\sigma_1, \ldots, \sigma_m$ are all semi-invariants with

$$g^*(u) = u, \quad g^*(f_W) = \zeta^{am} \cdot f_W, \quad g^*(\sigma_i) = \zeta^{ai} \cdot \sigma_i \text{ for } 1 \le i \le m.$$

If in the coordinates $(x_1, \ldots, x_{n-2}, x_{n-2}, x_{n-1}, x_n) = (x_1, \ldots, x_{n-2}, v, w)$ f has the form

$$f = vw + Q(x_1, \ldots, x_{n-2}) + \text{ higher terms},$$

where $Q(x_1, \ldots, x_{n-2})$ is a quadratic form only in the variables $x_1, \ldots, x_{n-2}$, then one can choose semi-invariant coordinates of the form

$$v' = v + \text{ higher terms},$$
$$w' = w + \text{ higher terms},$$

such that in terms of the system of coordinates $(x_1, \ldots, x_{n-2}v', w')$ the function f can be written

$$f = v'w' + h(x_1, \ldots, x_{n-2}),$$

where h is a semi-invariant function of the variables $x_1, \ldots, x_{n-2}$ only and where the quadratic part of h is exactly $Q(x_1, \ldots, x_{n-2})$.

PROOF.   Note for the first part that the coefficients σ_i of the Weierstrass form f_W of f is nothing but the ith symmetric function of the roots $w_1, \ldots, w_m$ of $f = f(x_1, \ldots, x_{n-1}, w) = 0$, considered as an equation for w. Since w is semi-invariant, the roots are also semi-invariant with the same character. This proves the first part.

In order to prove the second part, first take the partial derivatives with respect to the system of coordinates $(x_1, \ldots, x_{n-2}, v, w)$:

$$\frac{\partial f}{\partial v} = w + \text{higher terms:} = w_1,$$

$$\frac{\partial f}{\partial w} = v + \text{higher terms:} = v_1.$$

Choose the new system of semi-invariant coordinates to be $(x_1, \ldots, x_{n-2}, v_1, w_1)$ and write f as

$$f + v_1 w_1 + a_{1,0} v_1 + a_{0,1} w_1 + \Sigma_{i+j \geq 2} a_{i,j} v_1^i + h(x_i, \ldots, x_{n-2}),$$

where the $a_{i,j}$ and h are semi-invariant holomorphic functions of the variables $x_1, \ldots, x_{n-2}$ only.

Observe that

$$\frac{\partial f}{\partial v_1} = \frac{\partial v}{\partial v_1} \cdot \frac{\partial f}{\partial v} + \frac{\partial w}{\partial v_1} \cdot \frac{\partial f}{\partial w} + \Sigma_{i=1}^{n-2} \frac{\partial x_i}{\partial v_1} \cdot \frac{\partial f}{\partial x_i}$$

$$= \frac{\partial v}{\partial v_1} \cdot w_1 + \frac{\partial w}{\partial v_1} \cdot v_1,$$

$$\frac{\partial f}{\partial w_1} = \frac{\partial v}{\partial w_1} \cdot \frac{\partial f}{\partial v} + \frac{\partial w}{\partial w_1} \cdot \frac{\partial f}{\partial w} + \Sigma_{i=1}^{n-2} \frac{\partial x_i}{\partial w_1} \cdot \frac{\partial f}{\partial x_i}$$

$$= \frac{\partial v}{\partial w_1} \cdot w_1 + \frac{\partial w}{\partial w_1} \cdot v_1.$$

Therefore, along the locus $\{v_1 = w_1 = 0\}$, we have $\frac{\partial f}{\partial v_1} = \frac{\partial f}{\partial w_1} = 0$. This implies

$$a_{1,0} = a_{0,1} = 0.$$

Set

$$f' = v_1 w_1 + \Sigma_{i+j \geq 2} a_{i,j} v_1^i w_1^j.$$

We claim that the locus defined by $\{f' = 0\}$ consists of two nonsingular components crossing normally at the origin. In fact, since $a_{1,0} = a_{0,1} = 0$, we see that after blowing up along $\{v_1 = w_1 = 0\}$ the strict transform of $\{f' = 0\}$ consists of two disjoint nonsingular components crossing the exceptional divisor transversally. The claim now follows immediately.

Observe that f' is semi-invariant and that each of the two nonsingular components $\{f' = 0\}$ is invariant under the action, since so is its tangent $\{v = 0\}$ or $\{w = 0\}$. Therefore, we can take some semi-invariant defining equations v' and w' for the two nonsingular components such that

$$v' = v + \text{higher terms},$$

$$w' = w + \text{higher terms, and such that}$$
$$f' = v'w'.$$

Therefore, we have

$$f = v'w' + h(x_1, \ldots, x_{n-2})$$

as desired. By construction, the quadratic part of h is exactly $Q(x_1, \ldots, x_{n-2})$.

This completes the proof of the equivariant weierstrass preparation theorem.

$\square$

We resume the process of deriving the standard form of the semi-invariant defining equation f for $q \in \tilde{S}$.

We take an arbitrary semi-inavariant defining equation f for the canonical singularity $q \in \tilde{S}$, i.e.,

$$q \in \tilde{S} \underset{\text{equivariant}}{\overset{\text{analytically}}{\cong}} 0 \in \{f = 0\} \subset \mathbb{C}^3$$

with their $\mathbb{Z}/(r)$ actions. Let $Q(f)$ be the quadratic part of f.

Case: The coefficients of x^2, y^2, and z^2 are all zero in $Q(f)$.

In this case, we can write

$$q(f) = xy + byz + czx, \text{ where } b, \ c \in \mathbb{C}.$$

If $b = 0$ or $c = 0$, then by replacing y with $y + cz$ or replacing x with $x + bz$, respectively, we see that f takes the form

$$f = xy + \text{higher terms.}$$

If $b \neq 0$ and $c \neq 0$, then by replacing y with $y + cz$, we see that f takes the form

$$f = xy + byz - bcz^2 + \text{higher terms,}$$

in which case we proceed to the next case, since the coefficient of z^2 is not zero.

By the equivariant Weierstrass preparation theorem in the former case, we see that f takes via some suitable coordinate change the form

$$f = xy + w \cdot z^{n+1}, \text{ where } w \text{ is a unit.}$$

By setting

$$X = \frac{1}{w} \cdot x, \quad Y = y, \quad Z = z,$$

we see that f takes the form

$$\mathbf{A_n} : \mathbf{xy} + \mathbf{z^{n+1}}.$$

Case: the coefficient of x^2 (or y^2, z^2) is not zero in $Q(f)$.

In this case, we proceed as in the proof of Theorem 4-6-7.

Case A: $\text{mult}_0 g(y, z) = 2$.

When in the quadratic part $Q(g)$ of g the coefficients of y^2 and z^2 are zero, f is via the equivariant Weierstrass perparation theorem of the form

$$f = x^2 + w \cdot yz, \text{ where } w \text{ is a unit.}$$

Thus by renaming the old variables $w \cdot y, z, x$ as the new x, y, z, respectively, we see that f takes the form

$$A_1 : xy + z^{l+1}.$$

When in $Q(g)$ the coefficient of y^2 (or z^2) is not zero, f is via the equivariant weierstrass preparation theorem of the form

$$f = x^2 + y^2 + w \cdot z^{n+1}, \text{ where } w \text{ is a unit.}$$

By setting

$$X = x^{-\frac{1}{2}} \cdot x, \quad Y = w^{-\frac{1}{2}} \cdot y, \quad Z = z,$$

we see that f takes the form

$$eA_n : x^2 + y^2 + z^{n+1}.$$

We note that in order to make Type A_n and Type eA_n mutually exclusive, we require that the action of $\mathbb{Z}/(r)$ on x and y be different in Type eA_n, i.e., $\alpha \neq \beta$. If $n = 1$, then it is easy to see that at least two of the variables x, y, and z have the same character with respect to the action of $\mathbb{Z}/(r)$ and that the equation can be brought into the form A_1. Thus, as a convention, we say that there is *no* Type eA_1.

Case B: $\text{mult}_0 g(y, z) = 3$.

Let

$$C(g) = c_3 y^3 + c_2 y^2 z + c_1 y z^2 + c_0 z^3$$

be the cubic part of g.

Subcase: $(c_3 \neq 0, c_0 \neq 0 \text{ and } c_2 = c_1 = 0)$.

In this subcase, via the equivariant Weierstrass preparation theorem f takes the form

$$f = x^2 + u \cdot (y^3 + a(z)y^2 + b(z)y + c(z)), \text{ where } u \text{ is a unit.}$$

By replacing y with $y + \frac{1}{3}a(z)$, we may further assume

$$g(y, z) = u \cdot y^3 + v \cdot z^k y + w \cdot z^3,$$

where u, v, and w are units and where $k \geq 3$. Replacing z with $z + \frac{1}{3w} v \cdot z^{k-2} y$, we see that f takes the form

$$f = x^2 + s \cdot y^3 + pz^2 y^2 + w \cdot z^3, \text{ where } s \text{ is a unit.}$$

Replacing y with $y + \frac{1}{3s} pz^2$, we see that f takes the form

$$f = x^2 + s \cdot y^3 + t \cdot z^3, \text{ where } t \text{ is a unit.}$$

Finally, by setting

$$X = x, \quad Y = s^{\frac{1}{2}} \cdot y, \quad Z = t^{\frac{1}{3}} \cdot z,$$

we see that f takes the form

$$\mathbf{eD_4} : \mathbf{x}^2 + \mathbf{y}^3 + \mathbf{z}^3.$$

We note that in order to make Type D_4 and Type eD_4 mutually exclusive, we require that the action of $\mathbb{Z}/(r)$ on y and z be different in Type eD_4, i.e., $\beta \neq \gamma$.

Subcase: $(c_1 \neq 0$ and $c_3 = c_2 = c_0 = 0)$ or $(c_2 \neq 0$ and $c_3 = c_1 = c_0 = 0)$.

In this subcase there is one nonsingular component of $\{g(y, z) = 0\}$ whose tangent is given by y (or z) and invariant under the action. Thus in the argument of Subcase B.1 we may assume that Y' is semi-invariant. The rest of the argument goes verbatim via the equivariant Weierstrass preparation theorem to lead f to Type D_n.

Subcase: $(c_3 \neq 0$ and $c_2 = c_1 = c_0 = 0)$ or $(c_0 \neq 0$ and $c_3 = c_2 = c_1 = 0)$.

In this subcase the argument in Subcase B.2 goes verbatim via the equivariant Weierstrass preparation theorem to lead f to Type D_n, E_6, E_7, or E_8.

That is to say, in the above three subcases, we see that f takes the form

$$\mathbf{D_n} : \mathbf{x}^2 + \mathbf{y}(\mathbf{z}^2 + \mathbf{y}^{n-2}),$$
$$\mathbf{E_6} : \mathbf{x}^2 + \mathbf{y}^3 + \mathbf{z}^4,$$
$$\mathbf{E_7} : \mathbf{x}^2 + \mathbf{y}(\mathbf{y}^2 + \mathbf{z}^3),$$
$$\mathbf{E_8} : \mathbf{x}^2 + \mathbf{y}^3 + \mathbf{z}^5,$$

This completes the proof of Theorem 4-6-22 (ii).

In order to see (iii) we just remark that by the construction of the canonical cover any element of the Galois group $\mathbb{Z}/(r)$ other than the identity has no fixed points other than the origin. The combinatorial translation of the condition follows immediately.

In order to see (iv), we note that obviously in the nonsingular case

$$dx \wedge dy$$

and in the hyperquotient case (Exercise! Why?)

$$\frac{dy \wedge dz}{\partial f / \partial x} = -\frac{dx \wedge dz}{\partial f / \partial y} = \frac{dx \wedge dy}{\partial f / \partial z}$$

are semi-invariant generators of $\mathcal{O}_{\tilde{5}}(K_{\tilde{5}})$. By Proposition-Definition 4-5-1 there is a generator of $\mathcal{O}_{\tilde{5}}(K_{\tilde{5}})$ that is semi-invariant and on which $\mathbb{Z}/(r)$ acts faithfully (up to invariant units). The combinatorial translation of the condition follows immediately.

Starting from a germ of log terminal singularity in dimension 2, the action of the Galois group $\mathbb{Z}/(r)$ of the canonical cover on its Zariski tangent space is canonical determined. The coordinates x, y, z form a semi-invariant basis of the Zariski tangent space (modulo $m_{X,0}^2$). From this it follows easily (Exercise! Why?)

that our derivation process leads to a unique standard form of the semi-invariant defining equation for the canonical cover

$$q \in \tilde{S} \overset{\text{analytically}}{\underset{\text{equivariant}}{\cong}} 0 \in \{f = 0\} \subset \mathbb{C}^3$$

up to the choice of a generator of $\mathbb{Z}/(r)$ and a primitive rth root of unity ζ. The analytic structure of $p \in S$ is clearly determined as

$$p \in S \overset{\text{analytically}}{\cong} (0 \in \{f = 0\} \subset \mathbb{C}^3)\mathbb{Z}/(r).$$

Conversely, if we have a hypersurface singularity $0 \in \{f = 0\} \subset \mathbb{C}^3$ with the $\mathbb{Z}/(r)$-action as specified above satisfying conditions (i), (ii), (iii), and (iv), then the quotient map

$$\sigma : 0 \in \{f = 0\} \subset \mathbb{C}^3 \to (0 \in \{f = 0\} \subset \mathbb{C}^3)\mathbb{Z}/(r)$$

is such that the original space $0 \in \{f = 0\} \subset \mathbb{C}^3$ is canonical of index 1, the map is étale outside of the origin, and there is a nonvanishing and semi-invariant section on which the Galois group $\mathbb{Z}/(r)$ acts faithfully. Thus we conclude by Proposition 4-5-3 (i) that $(0 \in \{f = 0\} \subset \mathbb{C}^3)/\mathbb{Z}/(r)$ is a germ of log terminal singularity and by Proposition-Definition 4-5-1 that the quotient map coincides with its canonical cover.

Next we show how to determine the dual graphs of the minimal resolutions of log terminal singularities $p \in S$ of Types A_n, eA_n, D_n, eD_n, E_6, E_7, and E_8.

In principle, this can be easily done by computing the action of $\mathbb{Z}/(r)$ on the minimal resolution $\tilde{\pi} : \hat{\tilde{S}} \to \tilde{S}$, which is obtained by a succession of blowups. In reality, the computation is rather involved and can be lengthy. In the following, we just demonstrate how to compute in the case of Type eD_4, leaving the remaining cases as an exercise to the reader.

Computation of the dual graph for Type eD_4

The standard form of the semi-invariant defining equation of the canonical cover $q \in \tilde{S}$ of Type eD_4 as a hypersurface is

$$f = x^2 + y^3 + z^3$$

with the action of $\mathbb{Z}/(r)$ specified as

$$g^*(x, y, z) = (\zeta^\alpha x, \zeta^\beta y, \zeta^\gamma z)$$

for a generator $g \in \mathbb{Z}/(r)$ and a primitive rth root of unity ζ. The integers α, β, and γ satisfy

$$2\alpha = 3\beta = 3\gamma, \quad \beta \neq \gamma \bmod (r)$$

and conditions (iii) and (iv).

We blow up the origin $0 \in \{f = 0\} \subset \mathbb{C}^3 \cong q \in \tilde{S}$ to obtain the morphism $\sigma : \tilde{S}' \to \tilde{S}$ with the exceptional locus $\tilde{E}' \cong \mathbb{P}^1$ over q. In the affine piece with the new coordinates (x', y', z), where

$$x = x'z, \quad y = y'z, \quad z = z,$$

the variety $\tilde{S}'$ is defined by

$$f' = z'^2 + y'^3 z + z = x'^2 + (y'^3 + 1)z = 0,$$

and $\tilde{E}'$ is defined by

$$z = 0 \text{ and } x' = 0.$$

On $\tilde{S}'$ we have three singular points, each of which is a canonical singularity of Type A_1, at

$$P_i = (x', y', z) = (0, -\omega^i, 0), \quad i = 0, 1, 2, \text{ where } \omega = \exp\left(\frac{2\pi\sqrt{-1}}{3}\right).$$

The minimal resolution $\tilde{\pi} : \hat{\tilde{S}} \to \tilde{S}$ of the canonical cover can be obtained if we further blow up these three singular points.

The action of $\mathbb{Z}/(r)$ on $\tilde{S}$ lifts to the action on $\tilde{S}'$ with

$$g^*(x', y', z) = (\zeta^{\alpha-\gamma} x', \zeta^{\beta-\gamma} y', \zeta^\gamma z).$$

From the numerical conditions

$$2\alpha = 3\beta = 3\gamma, \quad \beta \neq \gamma \bmod(r),$$

and

$$\gcd(\beta, \gamma, r) = 1,$$

it is easy to see that 3 divides both α and r and that r must be odd.

Consider the subgroup

$$G_3 = \langle g^3 \rangle \subset \mathbb{Z}/(r)$$

of index 3 in $\mathbb{Z}/(r)$. The quotient $\tilde{S}'/G_3$ is nonsingular everywhere except for the images of the P_i. In fact, outside of E the subgroup G_3 acts properly discontinuously and at $P \in E - \{P_i; i = 0, 1, 2\}$ the action can be described using the local coordinates $(x', y' - y'(P))$:

$$(g^3)^*(x', y' - y'(P)) = (\zeta^{3\alpha-3\beta} x', y' - y'(P)).$$

We claim that the images of the P_i are singular points of the quotient $\tilde{S}'/G_3$, each of which is a canonical singularity of type A_1.

Note that locally $\tilde{S}'$ is the quotient of $\mathbb{C}^2$ by the cyclic group $\langle \tau \rangle$ of order 2, where τ has no fixed point except for the origin and where $\tau^2 = $ identity, since $P_i \in S'$ is a canonical singularity of Type A_1. Note also that $\mathbb{C}^2 - \{0\}$ is the universal cover of $\tilde{S}' - \{P_i\}$ with $\pi_1(\tilde{S}' - \{P_i\}) = \langle \tau \rangle$ locally. Since P_i is a fixed point of the action of G_3, the subgroup G_3 acts on $\tilde{S}' - \{P_i\}$ and hence on the universal cover $\mathbb{C}^2 - \{0\}$. Observe that by construction the actions of the cyclic group $\langle \tau \rangle$ and G_3 on $\mathbb{C}^2 - \{0\}$ (and hence on $\mathbb{C}^2$) commute with each other. Therefore, by choosing suitable coordinates (u, v) on $\mathbb{C}^2$, we see that the actions of $\langle \tau \rangle$ and G_3 can be simultaneously linearized and diagonalized. More specifically, since $P_i \in \tilde{S}'$ is defined by $x'^2 - w_i \cdot (y' - y'(P_i))z = 0$, where w_i is a unit invariant under the action of G_3, and since the coordinates x', z, and $y' - y'(P_i)$ are semi-invariants under the action of G_3, we may assume (cf. Lemman 4-6-19)

$$y' - y'(P_i) = u^2, \quad z = v^2, \quad \frac{x'}{\sqrt{w_i}} = uv,$$

where in this system of coordinates we have

$$\tau = \begin{pmatrix} -1 & 0 \\ 0 & -1 \end{pmatrix} \text{ and } g^3 = \begin{pmatrix} a & 0 \\ 0 & b \end{pmatrix} = \begin{pmatrix} 1 & 0 \\ 0 & \xi \end{pmatrix}$$

with ξ a primitive $\frac{r}{3}$th root of unity. In fact, since $y' - y'(P_i) = u^2$ is invariant under the action of g^3, we have $a^2 = 1$. But if $a = -1$, then $(g^3)^{|G_3|} = (g^3)^{\frac{r}{3}} \neq$ identity, since r, and hence $\frac{r}{3}$, is an odd number. Thus $a = 1$. Since g^3 has order $|G_3| = \frac{r}{3}$, we have $b = \xi$ a primitive $\frac{r}{3}$ root of unity.

Therefore, we have

$$(P_i \in \tilde{S}')/G_3 \overset{\text{analytically}}{\cong} \{0 \in \mathrm{Spec}\,[u, v])/\langle \tau \rangle\}/G_3$$

$$\overset{\text{analytically}}{\cong} \{0 \in \mathrm{Spec}\,[u, v])/G_3\}/\langle \tau \rangle$$

$$\overset{\text{analytically}}{\cong} 0 \in \mathrm{Spec}\,[u, v^{\frac{r}{3}}])/\langle \tau \rangle,$$

where the last singularity is easily seen to be a canonical singularity of Type A_1, since

$$\tau^*(u, v^{\frac{r}{3}}) = (-u, -v^{\frac{r}{3}}),$$

since $\frac{r}{3}$ is an odd number.

Thus on $\tilde{S}'/G_3$ we have three singularities P_1, P_2, P_3 of Type A_1 on $\tilde{E}'$. (The quotient morphism $\tilde{S}' \to \tilde{S}'/G_3$ maps $\tilde{E}'$ onto its image isomorphically, since its coordinate y' is invariant under the action of G_3. By abuse of notation we use the same letters P_1, P_2, P_3, and $\tilde{E}'$ to denote their images.)

Finally, we study the quotient $\tilde{S}'/\mathbb{Z}/(r) = (\tilde{S}'/G_3)/\mathbb{Z}/(3)$, where

$$\mathbb{Z}/(3) = \{\mathbb{Z}/(r)\}/G_3,$$

The group $\mathbb{Z}/(3)$ acts properly discontinuously on $\tilde{S}'/G_3$ outside of the two fixed points $\{0, \infty\}$ on $\tilde{E}'$, permuting the three singularities P_1, P_2, P_3 of Type A_1. We claim that the two points $\{0, \infty\}$ give rise to two quotient singularities of distinct Types $\langle 3, 1 \rangle$ and $\langle 3, 2 \rangle$ collectively.

In order to see this claim, observe first that on $\tilde{S}'$ local coordinates are given by

$$\left(\frac{x}{z}, \frac{y}{z} \right) \text{ around } \{0\} \text{ with } (g^3)^* \left(\frac{x}{z}, \frac{y}{z} \right) = \left(\zeta^{3\alpha - 3\gamma} \frac{x}{z}, \frac{y}{z} \right) = \left(\zeta^{\alpha} \frac{x}{z}, \frac{y}{z} \right)$$

and

$$\left(\frac{x}{y},\frac{z}{y}\right) \text{ around } \{\infty\} \text{ with } (g^3)^* \left(\frac{x}{y},\frac{y}{y}\right) = \left(\zeta^{3\alpha-3\beta}\frac{x}{y},\frac{z}{y}\right) = \left(\zeta^\alpha\frac{x}{y},\frac{z}{y}\right)$$

and that hence on $\tilde{S}/G_3$ local coordinates are given by

$$\left(\left(\frac{x}{z}\right)^{o_\alpha},\frac{y}{z}\right) \text{ around } \{0\} \text{ with } g^* \left(\left(\frac{x}{z}\right)^{o_\alpha},\frac{y}{z}\right) = \left(\zeta^{(\alpha-\gamma)o_\alpha},\left(\frac{x}{z}\right)^{o_\alpha},\zeta^{\beta-\gamma}\frac{y}{z}\right)$$

and

$$\left(\left(\frac{x}{y}\right)^{o_\alpha},\frac{z}{y}\right) \text{ around } \{\infty\} \text{ with } g^* \left(\left(\frac{x}{y}\right)^{o_\alpha},\frac{z}{y}\right) = \left(\zeta^{(\alpha-\beta)o_\alpha},\left(\frac{x}{y}\right)^{o_\alpha},\zeta^{\gamma-\beta}\frac{z}{y}\right),$$

where o_α is the order of α in $\mathbb{Z}/(r)$ and where g is considered to be a generator for $\mathbb{Z}/(3)$.

Since $3\beta = 3\gamma \bmod (r)$ and since $\beta \neq \gamma \bmod (r)$, we conclude that $(*)\zeta^{\beta-\gamma}$ and $\zeta^{\gamma-\beta}$ are two distinct primitive cube roots of unity. Since $2\alpha = 3\beta = 3\gamma \bmod (r)$, we have

$$\left(\zeta^{(\alpha-\gamma)o_\alpha}\right)^3 = \zeta^{\alpha o_\alpha} = 1,$$
$$\left(\zeta^{(\alpha-\beta)o_\alpha}\right)^3 = \zeta^{\alpha o_\alpha} = 1$$

which implies that $\zeta^{(\alpha-\gamma)o_\alpha}$ and $\zeta^{(\alpha-\beta)o_\alpha}$ are cube roots of unity. Since $2\alpha = 3\beta = 3\gamma$ and since 3 divides r, the order o_α cannot be equal to r. Since $\gcd(\alpha,\beta,r) = \gcd(\alpha,\gamma,r) = 1$, neither the order of β nor that of γ divides o_α. Therefore, neither $\zeta^{(\alpha-\gamma)o_\alpha}$ nor $\zeta^{(\alpha-\beta)o_\alpha}$ is equal to 1. That is to say, they are both primitive cube roots of unity. Moreover, we claim that 3 divides o_α. In fact, assume that 3 does not divide o_α. If 3^2 divides r, then assumption implies that 3^2 divides α and hence that 3 divides both β and γ (since $2\alpha = 3\beta = 3\gamma \bmod (r)$), which is against the conditions in (iii). If 3^2 does not divide r, then $\beta - \gamma \neq 0 \bmod (3)$. Note that 3 does not divide β or γ, since 3 divides both α and r and since we have condition (iii). Therefore we should have $\beta = 1 \bmod (3)$, $\gamma = 2 \bmod (3)$, or $\beta = 2 \bmod (3)$, $\gamma = 1 \bmod (3)$. But this would imply $\beta + \gamma - \alpha = 0 \bmod (3)$, which contradicts condition (iv) $\gcd(\beta + \gamma - \alpha, r) = 1$, since 3 divides both α and r. Thus 3 divides o_α as claimed. Therefore, we have

$$\zeta^{(\alpha-\gamma)o_\alpha}/\zeta^{(\alpha-\beta)o_\alpha} = 1,$$

which implies that

$(**)$ $\zeta^{(\alpha-\gamma)o_\alpha}$ and $\zeta^{(\alpha-\beta)o_\alpha}$ are the same primitive cube roots of unity.

From $(*)$ and $(**)$ it follows that the two points $\{0,\infty\}$ give rise to two quotient singularities of distinct Types $\langle 3,1\rangle$ and $\langle 3,2\rangle$, collectively.

We have the natural birational morphism $\tilde{S}'/\mathbb{Z}/(r) \to S$, which is an isomorphism outside of p. The variety $\tilde{S}'/\mathbb{Z}/(r)$ is nonsingular except for the three singular points, one of Type $\langle 2,1\rangle$, which is the common image of P_1, P_2, P_3, and two of Types $\langle 3,1\rangle$ and $\langle 3,2\rangle$, collectively which are the images of $\{0,\infty\}$. They are on $\mathbb{P}^1$, which is the image of $\tilde{E}'$ and exceptional over p. If we take the minimal resolution $V \to \tilde{S}'/\mathbb{Z}/(r)$, then the exceptional locus of $V \to S$ is exactly as described in the table of Theorem 4-6-22 for eD_4. (The fact that the edge components of the resolution of Types $\langle 2, 1 \rangle, \langle 3, 1 \rangle, \langle 3, 2 \rangle$ cross transversally the central component follows from the above analysis using the coordinates.) The condition $-b \geq -2$ follows from the fact that this exceptional locus has the negative definite intersection form. Thus there is no (-1)-curve in the exceptional locus of $V \to S$, and hence $\pi : V = \tilde{S} \to S$ is the minimal resolution. Thus the dual graph is determined as described in the table.

Exercise 4-6-24. Compute and check that the dual graphs of log terminal singularities of Types A_n, eA_n, E_6, E_7, E_8 are given as described in the table of Theorem 4-6-19.

Finally, we verify that all the dual graphs listed above with the components being nonsingular rational curves do occur geometrically associated to some low log terminal singularities and they determine their analytic structures.

One way to show this is a cumbersome one: According to the table of the standard forms of the semi-invariant defining equation of the canonical singularities together with the description of the cyclic group actions we actually compute all the associated dual graphs with the self-intersection numbers and check that all the possibilities listed above do occur and that no two distinct standard forms give rise to the same dual graph with the same combination of the self-intersection numbers, unless the two defining equations and the descriptions of the actions of $\mathbb{Z}/(r)$ coincide up to the choice of a generator of $\mathbb{Z}/(r)$ and a primitive rth root of unity.

Another way to show this is a short one but rather heuristic using Grauert's contractibility criterion and Laufer's results:

First note that each of the possibilities for the dual graphs listed above can be realized as some configuration geometrically on a smooth surface. In fact, they can be obtained as a part of the configuration of the exceptional locus of the resolution of some canonical singularity, possibly blowing up further several times after the minimal resolution to reduce the self-intersection numbers. This also shows that the intersection form is negative definite by Theorem 4-6-1. Now Grauert's contractibility criterion (cf. Grauert [1]) tells us that the configuration can be contracted to an analytic point $p \in S$, so that this contraction is nothing but the minimal resolution $\pi : \hat{S} \to S$.

Theorem 4-6-25 (Grauert's Contractibility Criterion). *A configuration of a reduced, compact connected curve E with irreducible components E_i on a nonsingular surface $\hat{S}$ can be contracted analytically to a point $p \in S$ if and only if the intersection form is negative definite.*

Note that the discrepancies a_i in the ramification formula

$$K_{\hat{S}} = \pi^* K_S + \Sigma a_i E_i$$

can be computed only from the intersection form of the dual graph. Compute and check that for each of the possibilities listed above the discrepancies are strictly bigger than -1. (Though this may look like a difficult computation at first sight, we have a slightly slick way to confirm this as follows. Notice that each of the possible dual graphs is a subgraph of the dual graph for some log canonical but not log terminaly singularity. For example, the dual graph of Brieskorn's Type A for a log terminal singularity is a subgraph of the dual graph of the second case of $r = 2$ of some log canonical but not log terminal singularity, omitting the four circles on the two edges; See Theorem 4-6-28. In the latter, all the discrepancies for the circles in the middle row can be easily computed to be -1. Now it is easy to see that the process of omitting the four circles strictly increases the discrepancies to be strictly bigger than -1. Other cases can be computed similarly.) Thus $p \in S$ is log terminal. Therefore, all the possibilities do occur geometrically. Laufer [2] tells that these singularities are taut, i.e., analytically determined uniquely by their dual graphs with the specified intersection forms. Note that as in Remark 4-6-26 (ii) (cf. Artin [1][2][4]) the germ $p \in S$ can also be realized as an algebraic one.

This completes the proof of Theorem 4-6-22. $\qquad\qquad\qquad\qquad\qquad\qquad\square$

Remark 4-6-26. (i) **How to determine the standard form of the semi-invariant equation starting from a given dual graph**: In theory, as shown above, a dual graph determines the corresponding log terminal singularity and hence the type of the semi-invariant equation with the $\mathbb{Z}/(r)$-action. But how should one compute and determine these in practice, given a dual graph Γ?

We can first determine the discrepancies a_i of all the exceptional divisors E_i for the minimal resolution $\pi : \hat{S} \to S$ by the equations

$$(*) \qquad \begin{aligned} (K_{\hat{S}} + E_i) \cdot E_i &= -2, \\ (K_{\hat{S}} - \Sigma a_i E_i) \cdot E_j &= 0 \quad \forall j, \end{aligned}$$

since the intersection form is negative definite. The discrepancies a_i now determine the index r by the characterization (see the proof of Theorem 4-6-18 (i) and Lemma 4-6-8)

$$r = \min\{e \in \mathbb{N}; ea_i \in \mathbb{Z} \quad \forall E_i\}.$$

Let $f \in \mathbb{C}(S)$ be a rational function such that

$$\mathrm{div}\,(f) + rK_S = 0$$

and the canonical cover $\tilde{S}$ is the normalization of S in $\mathbb{C}(S)[t]/(t^r - f)$. Let $\tilde{\tilde{S}}$ be the normalization of $\hat{S}$ in $\mathbb{C}(\hat{S})[t]/(t^r - f) = \mathbb{C}(S)[t]/(t^r - f)$. Let V be the minimal resolution of $\tilde{\tilde{S}}$. We have the commutative diagram

$$\begin{array}{ccc} V \xrightarrow{\mu} \tilde{\tilde{S}} \xrightarrow{\tilde{\pi}} \tilde{S} \\ \downarrow \qquad\quad \downarrow \\ \hat{S} \xrightarrow{\pi} S. \end{array}$$

For an arbitrary point $\hat{q} \in \pi^{-1}(p)$, locally we can find a neighborhood U such that either U contains only one component E_i defined by $\{u = 0\}$ or two components E_i and E_{i+1} intersecting transversally at $\hat{q}$ and defined by $\{uv = 0\}$ and such that there is a rational function $g \in \mathbb{C}(\hat{S})$ with $\mathrm{div}(g)|_U = K_{\hat{S}} + E_i|_U$ in the former or with $\mathrm{div}(g)|_U = K_{\hat{S}} + E_i + E_{i+1}|_u$ in the latter. Then since

$$\mathrm{div}(f) + r\mathrm{div}(g)|_U = -r\pi^* K_S + r\mathrm{div}(g)|_U$$
$$= -r(K_{\hat{S}} - \Sigma a_i E_i) + r\mathrm{div}(g)|_U$$
$$= \begin{cases} r(1 + a_i)E_i|_U & \text{in the former case,} \\ r(1 + a_i)E_i + r(1 + a_{i+1})E_{i+1}|_U & \text{in the latter case,} \end{cases}$$

we conclude that locally $\tilde{\hat{S}}$ is described as the normalization of

$$\mathrm{Spec}\,\mathbb{C}[u, v][t']/(t'^r - u^{r(1+a_i)}) \text{ in the former case,}$$
$$\mathrm{Spec}\,\mathbb{C}[u, v][t']/(t'^r - u^{r(1+a_i)}v^{r(1+a_{i+1})}) \text{ in the latter case.}$$

Note that these cyclic covers of $\mathbb{A}^2 = \mathrm{Spec}\,\mathbb{C}[u, v]$ can be computed easily by, e.g., the method of toric geometry. By gluing these local pieces of information we can determine $\tilde{\hat{S}}$ globally, together with the $\mathbb{Z}/(r)$-action. Now we can compute the dual graph of $(\tilde{\pi} \circ \mu)^{-1}(q)$ and its intersection form (i.e., the self-intersection numbers of the components). Thus we can also determine $q \in \tilde{S}$ and hence the type of the semi-invariant defining equation A_n, eA_n, D_n, eD_4, E_6, E_7, or E_8. We can also determine the characters of the action of $\mathbb{Z}/(r)$ on the semi-invariant coordinates around the fixed points on $\tilde{\pi}^{-1}(q)$, which are the ratios of x, y, and z. This allows us finally to determine the characters of the semi-invariant coordinates x, y, z in the semi-invariant defining equation f.

Observe that the above algorithm for determining the standard of the semi-invariant defining equation starting from a given dual graph shows that the dual graph completely determines the analytic structure as

$$p \in S \overset{\text{analytically}}{\cong} (0 \in \{f = 0\} \subset \mathbb{C}^3)/\mathbb{Z}(r)$$

and hence that log terminal singularities are "taut," without referring to Laufer's results [2].

(ii) **How to show that all dual graphs occur geometrically without referring to Grauert's contractibility criterion:**

Let Γ be one of the dual graphs listed above. As in the proof, we remark that one can realize the configuration consisting of nonsingular rational curves as components $E_1, \ldots, E_n$ with the specified self-intersection numbers starting from the minimal resolution of a canonical singularity $\hat{T} \to T$ and taking further blowups $V \to \hat{T}$ if necessary. Now we apply (log) MMP on V over T with respect to the log canonical divisor $K_V + \Sigma E_i$.

By Theorem 4-6-1 applied to $V \to T$ we see immediately that there exists E_{i_1} such that

$$\{K_V + \Sigma E_{i_1}\} \cdot E_{i_1} = \{\Sigma(1 + a_i)E_i\} \cdot E_{i_1} < 0 \text{ and } E_{i_1}^2 < 0.$$

Therefore, E_{i_1} is an extremal rational curve with respect to $K_V + \Sigma E_i$ and can be contracted via $\mu_1 = \mathrm{cont}_{E_{i_1}} : V \to V_1$. By Thoerem 4-6-1 applied to $V_1 \to T$, we see that there exists (the image of) $E_{i_2}(i_2 \cong i_1)$ such that

$$\{K_{V_1} + \Sigma_{i\cong i_1} E_i\} \cdot E_{i_2} = \{\Sigma_{i\cong i_1}(1 + a_i)E_i\} \cdot E_{i_2} < 0 \text{ and } E_{i_2}^2 < 0.$$

Therefore, E_{i_2} is an extremal rational curve with respect to $K_V + \Sigma_{i\cong i_1} E_i$ and can be contracted via $\mu_2 = \mathrm{cont}_{E_{i_2}} : V_1 \to V_2$.

We repeat the process until all the components of Γ are contracted to a point $p \in S = V_n$, which is the desired log terminal singularity having Γ as the dual graph for the exceptional locus of the minimal resolution.

This completes the discussion of Remark 4-6-26.

Exercise 4-6-27. Determine the standard form of the semi-invariant defining equation for the canonical cover $q \in \tilde{S}(\overset{c}{\to} p \in S)$ when the dual graph of the exceptional locus $\pi^{-1}(p)$ of the minimal resolution $\pi : \hat{S} \to S$ is given by

Log Canonical Singularities in Dimension 2

Finally, we give the classification of log canonical singularities in dimension 2 (which are not log terminal).

A log canonical but not log terminal singularity of *index* 1 in dimension 2 is classically called a simple elliptic or cusp (or simple quasi-elliptic) singularity, having an elliptic curve or a cycle of rational curves as the exceptional locus for the minimal resolution. These singularities are a part of the larger class of singularities called elliptic Gorenstein, whose detailed study forms the core of the classification theory of terminal singularities in dimension 3 by Reid [2][3][7] and Mori [4].

The study of a log canonical but not log terminal singularity of index > 1 is reduced to that of index 1 via the canonical cover, thanks to Proposition 4-5-3.

The main portion of the argument here for the classification of log canonical singularities in dimension 2 is taken from Section 10 of Kawanata [9]. (See also Kawamata [1] [2].)

Theorem 4-6-28 (Classification of Log Canonical Singularities in Dimension 2). *Let $p \in S$ be a germ of log canonical but not log terminal singularities in dimension 2.*

Then the canonical divisor K_S is $\mathbb{Q}$-Cartier.

Let

$$r = \min\{e \in \mathbb{N}; eK_S \text{ is a Cartier divisor}\}$$

be the index. We denote by $\pi : \hat{S} \to S$ the minimal resolution and by E the exceptional locus for π. Let $c : q \in \tilde{S} \to p \in S$ be the canonical cover. We denote by $\tilde{\pi} : V \to \tilde{S}$ the minimal resolution of $\tilde{S}$ and by $\tilde{E}$ the exceptional locus for $\tilde{\pi}$.

The germ $p \in S$ is one of the following:

$$\boxed{r = 1}$$

The first case is that the exceptional locus $E = E_1$ is irreducible and is either a nonsingular elliptic curve or a nodal rational curve. The dual graph consists of a single point

$$\overset{\circ}{1(b_1)}$$
$$\cdots \quad -b_1 \leq -1.$$

The second case is that the exceptional locus E is a cycle of nonsingular rational curves $E_1 \cup \cdots \cup E_n$ where $n \geq 2$. The dual graph is thus a cycle

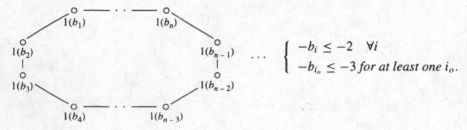

$$\cdots \quad \begin{cases} -b_i \leq -2 \quad \forall i \\ -b_{i_o} \leq -3 \text{ for at least one } i_o. \end{cases}$$

In the above two cases, the discrepancies a_i of the exceptional divisors E_i for the minimal resolution are all equal to -1, i.e., the ramification formula reads

$$K_{\hat{S}} = \pi^* K_S + \Sigma a_i E_i = \pi^* K_s - \Sigma 1 \cdot E_i.$$

*(When the exceptional locus is a nonsingular elliptic curve in the first case above, the singularity is called a **simple elliptic** singularity. When the exceptional locus is an irreducible nodal rational curve in the first case or a cycle of rational curves in the second case above, the singularity is called a **cusp** singularity.)*

In the dual graphs of the exceptional loci of the minimal resolutions above and below, the number $-a_i$ indicates the negative of the discrepancy, and the number $b_i = -E_i^2$ indicates the negative of the self-intersection number for the exceptional divisor E_i, and we write

$$\overset{\circ}{-a_i(b_i)}$$

where the circle $\circ$ represents the component E_i. In all the cases, the discrepancies are uniquely determined by the dual graphs, and hence we put specific numbers for the a_i. On the other hand, the b_i are not uniquely determined by the dual graphs only, and we give the conditions they have to satisfy next to the dual graph. All the combinations satisfying the conditions do occur geometrically assoiceated to some log canonical but not log terminal singularities.

$\boxed{r = 2}$

Since the index is 2, the canonical cover $c : q \in \tilde{S} \to p \in S$ has Galois group of order 2. Note that the Galois group also acts on the minimal resolution V of $\tilde{S}$ and hence also on its exceptional locus $\tilde{E}$.

The first case is that $\tilde{E}$ is an arbitrary irreducible nonsingular elliptic curve and the generator g of the Galois group acts on $\tilde{E}$ as the involution $g(x) = i(x) = -x$ for $x \in \tilde{E}$. We have four fixed points (the torsion points of order 2) whose stabilizer groups are the full group.

The dual graph for E in this case is

$$\langle b; 2, 1; 2, 1; 2, 1; 2, 1 \rangle \quad \cdots \quad -b \le -3.$$

The second case is that $\tilde{E}$ is a cycle of an even number $2m$ of rational curves and the generator g of the Galois group acts on $\tilde{E}$ as the reflection fixing two components.

The dual graph for E in this case is

$$\cdots \begin{cases} -b_1 \le -2 \text{ and } 3 \le i \le n - 2 \\ -b_{i_o} \le -3 \text{ for at least one } 3 \le i_o \le n - 2. \end{cases}$$

(Note that $n = m + 1 + 4$, and hence the number of circles in the middle row is equal to $n - 4 = m + 1 > 2$.)

$\boxed{r = 3}$

Since the index is 3, the canonical cover $c : q \in \tilde{S} \to p \in S$ has Galois group of order 3. The exceptional locus $\tilde{E}$ for the minimal resolution V of $\tilde{S}$ is a nonsingular elliptic curve

$$\tilde{E} \cong \mathbb{C}/(\mathbb{Z} + \omega \cdot \mathbb{Z}), \text{ where } \omega = \exp\left(\frac{2\pi\sqrt{-1}}{3}\right).$$

A generator g of the Galois group acts on $\tilde{E}$ as multiplication by ω, i.e., $g(x) = \omega \cdot x$ for $x \in \tilde{E}$. We have three fixed points $\{0, \frac{\sqrt{3}}{3}\exp(\frac{\pi\sqrt{-1}}{6}), \frac{\sqrt{3}}{3}\exp(\frac{\pi\sqrt{-1}}{4})\}$

The dual graph for E in this case is one of the following four:

$$\langle b; 3, 1; 3, 1; 3, 1 \rangle \quad \cdots \quad -b \le -2,$$
$$\langle b; 3, 1; 3, 1; 3, 2 \rangle \quad \cdots \quad -b \le -2,$$
$$\langle b; 3, 1; 3, 2; 3, 2 \rangle \quad \cdots \quad -b \le -2,$$
$$\langle b; 3, 2; 3, 2; 3, 2 \rangle \quad \cdots \quad -b \le -2.$$

(Note that the last condition $-b \le -3$ is different from the previous conditions of $-b \le -2$.)

$\boxed{r = 4}$

Since the index is 4, the canonical cover $c : q \in \tilde{S} \to p \in S$ has a cyclic Galois group of order 4. The exceptional locus $\tilde{E}$ for the minimal resolution V of $\tilde{S}$ is a nonsingular elliptic curve

$$\tilde{E} \cong \mathbb{C}/(\mathbb{Z} + \sqrt{-1} \cdot \mathbb{Z}).$$

A generator g of the Galois group acts on $\tilde{E}$ as multiplication by $\sqrt{-1}$, i.e., $g(x) = \sqrt{-1} \cdot x$ for $x \in \tilde{E}$. We have two fixed points $\{0, \frac{\sqrt{2}}{2}\exp(\frac{\pi\sqrt{-1}}{4})\}$ whose stabilizer groups are the full group and two others $\{\frac{1}{2}, \frac{1}{2}\sqrt{-1}\}$ whose stabilizer groups are the subgroup of order 2.

The dual graph for E in this case is one of the following three:

$$\langle b; 2, 1; 4, 1; 4, 1 \rangle \quad \cdots \quad -b \le -2,$$
$$\langle b; 2, 1; 4, 1; 4, 3 \rangle \quad \cdots \quad -b \le -2,$$
$$\langle b; 2, 1; 4, 3; 4, 3 \rangle \quad \cdots \quad -b \le -3,$$

$\boxed{r = 6}$

Since the index is 6, the canonical cover $c : q \in \tilde{S} \to p \in S$ has a cyclic Galois group of order 6. The exceptional locus $\tilde{E}$ for the minimal resolution V of $\tilde{S}$ is a nonsingular elliptic curve

$$\tilde{E} \cong \mathbb{C}/(\mathbb{Z} + \omega \cdot \mathbb{Z}), \text{ where } \omega = \exp\left(\frac{2\pi\sqrt{-1}}{3}\right).$$

A generator g of the Galois group acts on $\tilde{E}$ as multiplication by $\exp(\frac{\pi\sqrt{-1}}{3})$, i.e., $g(x) = \exp(\frac{\pi\sqrt{-1}}{3}) \cdot x$ for $x \in \tilde{E}$. We have one fixed point $\{0\}$ whose stabilizer group is the full group, two others $\left\{\frac{\sqrt{3}}{3}\exp(\frac{\pi\sqrt{-1}}{6}), \frac{\sqrt{3}}{3}\exp(\frac{\pi\sqrt{-1}}{4})\right\}$ whose stabilizer groups are the subgroup of order 3 and three more $\left\{\frac{1}{2}, \frac{1}{2}\exp(\frac{\pi\sqrt{-1}}{3}), \frac{1}{2}\exp(\frac{2\pi\sqrt{-1}}{3})\right\}$ whose stabilizer groups are the subgroups of order 2.

The dual graph for E in this case is one of the following four:

$$\langle b; 2, 1; 3, 1; 6, 1 \rangle \quad \cdots \quad -b \le -2,$$
$$\langle b; 2, 1; 3, 1; 6, 5 \rangle \quad \cdots \quad -b \le -2,$$
$$\langle b; 2, 1; 3, 2; 6, 1 \rangle \quad \cdots \quad -b \le -2,$$
$$\langle b; 2, 1; 3, 2; 6, 5 \rangle \quad \cdots \quad -b \le -3,$$

Moreover, the analytic structure of $p \in S$ is completely determined by its dual graph with the self-intersection numbers for the exceptional divisors of the minimal resolution, except for the following two cases where extra information on the analytic structure of the exceptional locus E is needed; the case where $r = 1$ and

*the exceptional locus is a nonsingular elliptic curve, and the first case of $r = 2$.
In the former, we need to know the isomorphic class of the elliptic curve E and in
the latter the cross ratio of the four intersection points of the central component
with the other components, which determines the analytic structure of E.*

PROOF.   In the ramification formula for the minimal resolution

$$K_{\tilde{S}} = \pi^* K_S + \Sigma a_i E_i,$$

we have by Theorem 4-6-2 (ii)

$$\begin{cases} 0 > a_i \geq -1 & \forall i, \\ a_{i_0} = -1 & \text{for at least one } i_o. \end{cases}$$

Set

$$E' = \Sigma_{a_i = -1} E_i, \text{ while } E = \Sigma F_i.$$

Case: $E' = E$.

Consider the exact sequence

$$0 \to \mathcal{O}_{\tilde{S}}(-E') \to \mathcal{O}_{\tilde{S}} \to \mathcal{O}_{E'} \to 0$$

and its associated cohomology sequence

$$R^1 \pi_* \mathcal{O}_{\tilde{S}}(-E') \to R^1 \pi_* \mathcal{O}_{\tilde{S}} \to R^1 \pi_* \mathcal{O}_{\tilde{E}} = H^1(E', \mathcal{O}_{E'})$$
$$\to R^2 \pi_* \mathcal{O}_{\tilde{S}}(-E') \to R^1 \pi_* \mathcal{O}_{\tilde{S}} \to 0.$$

If we take a π-ample divisor of the form $-\Sigma \delta_i E_i$ where $\delta_i > 0 \forall i$, then the pair
$(\hat{S}, \Sigma_{a_i \neq -1}(-a_i)E_i + \epsilon \Sigma_i E_i)$ has only log terminal singularities for sufficiently
small $\epsilon > 0$. By the analytic version of Theorem 5-2-8 we have

$$R^i \pi_* S_{\hat{S}}(-E') = R^i \pi_* \mathcal{O}_{\hat{S}}(K_{\hat{S}} + \Sigma_{a_i = -1}(-a_i)E_i + \epsilon \Sigma \delta_i E_i - \pi^* K_S - \epsilon \Sigma \delta_i E_i) = 0$$

for $i > 0$. Moreover, by the arithmetic genus formula (see Lemma 1-1-4; the
formula is valid as E' is a reduced connected curve; the connectedness of E follows
from $H^0(E', \mathcal{O}_{E'}) \cong \mathbb{C}$, which is a consequence of the surjection $\mathcal{O}_S = \pi_* \mathcal{O}_{\hat{S}} \to$
$\pi_* \mathcal{O}_{E'} = H^0(E', \mathcal{O}_{E'}))$ we have

$$h^1(E', \mathcal{O}_{E'}) = \frac{1}{2}(K_{\hat{S}} + E') \cdot E' + 1 = \frac{1}{2}(\Sigma_{a_i \neq -1} a_i E_i) \cdot E' + 1 < 1,$$

where the last strict inequality is the consequence of the case assumption $E' \neq E$.
That is to say,

$$H^1(E', \mathcal{O}_{E'}) = 0.$$

Therefore, we have

$$R^i \pi_* \mathcal{O}_{\hat{S}} = 0 \text{ for } i > 0,$$

and hence $p \in S$ is rational. Now the same argument as in the case of log terminal
singularities via Lemma 4-6-8 shows that K_S is $\mathbb{Q}$-Cartier.

Case: $E' = E$

Consider the exact sequence

$$0 \to \mathcal{O}_{\hat{S}}(K_{\hat{S}}) \to \mathcal{O}_{\hat{S}}(K_{\hat{S}} + E) \to \omega_E \to 0$$

and its associated cohomology sequence

$$\pi_* \mathcal{O}_{\hat{S}}(K_{\hat{S}} + E) \to \pi_* \omega_E = H^0(E, \omega_E)$$
$$\to R^1 \pi_* \mathcal{O}_{\hat{S}}(K_{\hat{S}}).$$

Take a π-ample divisor of the form $-\Sigma \delta_i E_i$ where $\delta_i > 0 \forall i$ as above. The pair $(\hat{S}, \in \Sigma \delta_i E_i)$ has only log terminal singularities for sufficiently small $\in > 0$. By the analytic version of Theorem 5-2-8 we have

$$R^1 \pi_* \mathcal{O}_{\hat{S}}(K_{\hat{S}}) = R^i \pi_* \mathcal{O}_{\hat{S}}(K_{\hat{S}} + \epsilon \Sigma \delta_i E_i - \epsilon \Sigma \delta_i E_i) = 0.$$

Moreover, by Serre duality and by the arthmetic genus formula,

$$h^0(E, \omega_E) = h^1(E, \mathcal{O}_E) = \frac{1}{2}(K_{\hat{S}} + E) \cdot E + 1 = \frac{1}{2}\pi^* K_S \cdot E + 1 = 1.$$

Since

$$\deg_E \omega_E = \deg_E(K_{\hat{S}} + E|_E) = \pi^* K_S \cdot E = 0,$$

this implies $\omega_E \cong \mathcal{O}_E$. Therefore, there exists a section of $\pi_* \mathcal{O}_{\hat{S}}(K_{\hat{S}} + E)$ that restricts to a nowhere-vanishing section of $H^0(E, \omega_E) = H^0(E, \mathcal{O}_E)$. Therefore, we have

$$K_{\hat{S}} + E \sim 0$$

and hence

$$K_S = \pi_*(K_{\hat{S}} + E) \sim 0.$$

That is to say, the canonical divisor K_S is trivial and of course Cartier.

Now we analyze the singularity $p \in S$ according to its index r.

$r = 1$

Since $r = 1$, we conclude that all the discrepancies a_i are equal to -1 in the ramification formula for the minimal resolution

$$K_{\hat{S}} = \pi^* K_S + \Sigma a_i E_i$$

thanks to Theorem 4-6-2 (ii) and that

$$\Sigma E_i \text{ is a divisor with only normal crossings.}$$

(Note that if ΣE_i had a singularity worse than normal crossings, then $p \in S$ would not be log canonical). Moreover,

$$\omega_{\Sigma E_i} = K_{\hat{S}} + \Sigma E_i|_{\Sigma E_i} = \pi^* K_s|_{\Sigma E_i} \cong \mathcal{O}_{\Sigma E_i}.$$

Therefore, there are two possibilities.

The first case is that $E = E_1$ is irreducible and that it is either a nonsinglar elliptic curve or a nodal rational curve with $b_1 = E_1^2 \leq -1$, which follows from Theorem 4-6-1.

The second case is that E is reducible and that it is a cycle of nonsingular rational curves $E_1 \cup \cdots \cup E_n$ where $n \geq 2$. Since π is the minimal resolution and hence there is no (-1)-cureve E, we have $b_i = E_i^2 \leq -2 \forall i$. Since the intersection form for E must be negative definite by Theorem 4-6-1, we see that there exists at least one i_o such that $b_{i_o} = E^2 +_{i_o} \leq -3$.

This finishes the analysis for the case $r = 1$.

$\boxed{r \geq 2}$

Now we analyze the case where the index r is greater than or equal to 2. By Proposition 4-5-3, the canonical cover $q \in \tilde{S}$ has only log canonical but not log terminal singularities, since so does $p \in S$. Since the index of $q \in \tilde{S}$ is 1, it is either a simple elliptic or a cusp singularity by the analysis of the case $r = 1$. Note that the Galois group $\mathbb{Z}/(r)$ acts on the minimal resolution $\tilde{r} : V \to \tilde{S}$ and on its exceptional locus $\tilde{E}$.

The essential point of the analysis in the case $r \geq 2$ is the following key observation:

By construction of the canonical cover the invertible sheaf $\mathcal{O}_{\tilde{S}}(K_{\tilde{S}})$ has a nonvanishing section $\tilde{s}$ at q that is semi-invariant for the action and on which the Galois group $\mathbb{Z}/(r)$ acts faithfully. (See Proposition-Definition 4-5-1.) That is to say, for a generator $g \in \mathbb{Z}/(r)$ and a primitive rth root of unity ζ,

$$g^* \tilde{s} = \zeta \cdot \tilde{s}.$$

The section $\tilde{s}$ can be considered a nowhere-vanishing section of $\mathcal{O}_V(K_V + \tilde{E}) = \pi^* \mathcal{O}_{\tilde{S}}(K_{\tilde{S}})$, and thus it restricts to that of $\omega_{\tilde{E}} = \mathcal{O}_V(K_V + \tilde{E})|_{\tilde{E}}$. The Galoris group $\mathbb{Z}/(r)$ acts faithfully on this section $\tilde{s}$ of $\omega_{\tilde{E}}$, and hence the map

$$\mathbb{Z}/(r) \to \mathrm{Aut}(\tilde{E})$$

is injective.

We utilize this key observation to the analysis of $p \in S$ when $r \geq 2$ in the following.

Suppose $q \in \tilde{S}$ is a simply elliptic singularity, i.e., $\tilde{E}$ is a nonsingular elliptic curve. It is straightforward to list all the possibilites for the cyclic finite subgroups $\mathbb{Z}/(r) \subset \mathrm{Aut}(\tilde{E})$.

(i) $\tilde{E}$ is an arbitrary elliptic curve and $\mathbb{Z}/(2) \subset \mathrm{Aut}(\tilde{E})$ is generated by the involution $g = i$ such that $g(x) = i(x) = -x$ for $x \in \tilde{E}$.

(ii)

$$\tilde{E} \cong \mathbb{C}/(\mathbb{Z} + \omega \cdot \mathbb{Z}), \text{ where } \omega = \exp\left(\frac{2\pi\sqrt{-1}}{3}\right)$$

and $\mathbb{Z}/(3) \subset \mathrm{Aut}(\tilde{E})$ is generated by g, which acts as multiplication by ω, i.e., $g(x) = \omega \cdot x$ for $x \in \tilde{E}$.

(iii)

$$\tilde{E} \cong \mathbb{C}/(\mathbb{Z} + \sqrt{-1} \cdot \mathbb{Z})$$

and $\mathbb{Z}/(4) \subset \mathrm{Aut}(\tilde{E})$ is generated by g, which acts as multiplication by $\sqrt{-1}$, i.e., $g(x) = \sqrt{-1} \cdot x$ for $x \in \tilde{E}$.

(iv)

$$\tilde{E} \cong \mathbb{C}/(\mathbb{Z} + \omega \cdot \mathbb{Z})$$

and $\mathbb{Z}/(6) \subset \mathrm{Aut}(\tilde{E})$ is generated by g, which acts as multiplication by $\exp\left(\frac{\pi\sqrt{-1}}{3}\right)$, i.e., $g(x) = \exp\left(\frac{\pi\sqrt{-1}}{3}\right) \cdot x$ for $x \in \tilde{E}$.

It is easy to see that in each of the above cases it has the specified fixed points with their stabilizer groups as described in the assertions of the theorem and that the cases (i), (ii), (iii), and (iv) lead to the first case of $r = 2$, $r = 3$, $r = 4$, and $r = 6$, respectively. Details are left to the reader as an exercise.

Suppose $q \in \tilde{S}$ is a cusp singularity, i.e., $\tilde{E}$ is either a nodal curve or a cycle of nonsingular rational curves.

The key observation implies that none of the double points of $\tilde{E}$ is a fixed point for a generator g. In fact, locally around a double point, if it is a fixed point for g, we can choose a system of semi-invariant corrdinates (x, y) (see the proof of Lemma 4-6-19) such that $\tilde{E}$ is defined by $\{xy = 0\}$. (Note that each of the components is invariant under the action of $\mathbb{Z}/(r)$, since otherwise some element other than the identity has fixed points outside of $\tilde{E}$, a contradiction.) Note that $\omega_{\tilde{E}}$ has a local generator

$$\frac{dx \wedge dy}{xy},$$

which is obviously invariant under the action of $\mathbb{Z}/(r)$. But then we should have

$$\tilde{s} = u \cdot \frac{dx \wedge dy}{xy},$$

where u is a unit and semi-inavariant and hence is also invariant for $\mathbb{Z}/(r)$. This contradicts the fact $\mathbb{Z}/(r)$ acts on $\tilde{s}$ faithfully.

Therefore, we conclude that $\tilde{E}$ cannot be a nodal rational curve and hence can only be a cycle of nonsingular rational curves $E_1 \cup \cdots \cup E_n$, where $n \geq 2$.

We choose an affine coordinate x_i on each $E_i \cong \mathbb{P}^1$ such that the two intersection points with the other components are $0_i, \infty_i$ and such that the cycle is obtained by gluing ∞_i with 0_{i+1}. (We understand $0_{n+1} = 0_1$.) If an element g^j with $0 < j < r$ of the cyclic group $\mathbb{Z}/(r)$, considered as an automorphism on $\tilde{E}$, is orientation-preserving, i.e.,

$$\exists k \in \mathbb{Z} \text{ such that } g^j(E_i) = E_{i+k} \quad \forall i,$$

then $g^j(0_i) = 0_{i+k}$ and $g^j(\infty_i) = (\infty_{i+k})$. Therefore, we have

$$(g^j)^*(x_{i+k}) = c_i \cdot x_i \text{ for some } c_i \in \mathbb{C}^\times.$$

But then the nowhere-vanishing section of $\omega_{\tilde{E}}$, which is represented by dx_i/x_i on each E_i, is invariant for g^j. This again leads to a contradiction to the fact that $\mathbb{Z}/(r)$ acts on $\tilde{s}$ faithfully.

Therefore, we conclude that g is orientation-reversing and g^2, which is orientation-preserving, must be the identity and hence that $r = 2$. This, together with the fact that no double point is a fixed point for g, implies that $\tilde{E}$ is a cycle of an even number $2m$ of nonsingular rational curves and that the Galois group acts on $\tilde{E}$ as the reflection fixing two components. Let $\tau : W \to V/\mathbb{Z}(2)$ be the minimal resolution of the quotient, which dominates S with the proper birational morphism $\sigma : V/\mathbb{Z}(2) \to S$. We claim that $\sigma \circ \tau : W \to S$ is nothing but the minimal resolution $\pi : \hat{S} \to S$. In fact, observe that the dual graph is exactly the one described in the second case of $r = 2$, since each of the two fixed components for g has two fixed points. Observe also that the self-intersections are

$$-b_1 = -b_2 = -2 = -b_{n-1} = -b_n$$

and

$$-b_i \leq -2 \qquad \text{for } 4 \leq i \leq n-3,$$

since the self-intersection of E_i for $4 \leq i \leq n-3$ coincides with that of its preimage on $\tilde{S}$. Now it is easy to see (Exercise! Why?) that in order for the entire exceptional locus for $\sigma \circ \tau$ to have the negative definite intersection form we must have

$$-b_3, -b_{n-2} \leq -2 \text{ and } -b_{i_o} \leq -3 \text{ for at least one } 3 \leq i_o \leq n-2.$$

Thus there is no (-1)-curve in the exceptional locus of $\sigma \circ \tau$, and hence it is the minimal resolution $\pi : \hat{S} \to S$. This case of $q \in \tilde{S}$ being a cusp singularity leads to the second case of $r = 2$.

Observe that in each of the cases above, at a fixed point $\tilde{p} \in \tilde{E}$ with the stabililzer group of order n, a generator acts locally around $\tilde{p}$ as left multiplication of the matrix on some suitable system of coordinates $'(x, y)$,

$$\begin{pmatrix} \exp\left(\frac{2\pi\sqrt{-1}}{n}\right) & 0 \\ 0 & \exp\left(\frac{2nq\sqrt{-1}}{n}\right) \end{pmatrix}, \qquad \gcd(n, q) = 1,$$

and we may further assume that $\tilde{E}$ is defined by $\{x = 0\}$. From this description it follows immediately that the central component crosses transversally one of the components at the edge of the dual graph of the minimal resolution of the quotient singularity of type (n, q) in E. (Exercise! Why? One of the ways to confirm this fact is to use the toric description of the minimal resolution. See, e.g., Fulton [2].)

The discrepancies a_i can be determined and computed directly from the dual graph with the intersection form of the components, which is negative definite, since they satisfy the equations

$(*)$
$$\begin{aligned} (K_V + E_i) \cdot E_i &= -2, \\ (K_V - \Sigma a_i E_i) \cdot E_j &= 0 \quad \forall j. \end{aligned}$$

In order to demonstrate how to compute them, let us pick up the second possiblilty for the dual graph in the case $r = 2$. The equations are

$$2a_1 - a_3 = 0,$$
$$2a_2 - a_3 = 0,$$
$$2a_{n-1} - a_{n-2} = 0,$$
(*)
$$2a_n - a_{n-2} = 0,$$
$$-2 - a_1 - a_2 + (1 + a_3)b_3 = 0,$$
$$-2 - a_{n-1} - a_n + (1 + a_{n-2})b_{n-2} = 0,$$
$$-2 - a_{i-1} - a_{i+1} + (1 + a_i)b_i = 0, \quad \text{for } 4 \le i \le n - 3.$$

If we use the condition that

(**) $0 > a_i \ge -1$ and $a_{i_o} = -1$ for at least one i_o,

then the a_i can be quickly determined as follows. First, none of the a_1, a_2, a_{n-1}, a_n can be equal to -1. Thus $a_{i_o} = -1$ for some $3 \le i_o \le n - 2$. But this leads us immediately to conclude that

$$a_1 = a_2 = a_{n-1} = a_n = -\frac{1}{2} \text{ and } a_i = -1 \text{ for } 3 \le i \le n - 2.$$

We should emphasize that though we use condition (**) for a quick computation, it is actually a consequence of equations (*) and that the a_i are uniquely determined, since the intersection form is negative definite, without reference to condition (**).

The discrepancies for the other cases can be computed similarly and are left to the reader as an exercise.

The above argument shows only that they are the only possibilities. In order to see that they actually occur geometrically associated to some log canonical but not log terminal singularities in dimension 2, we follow the argument as in the case of log terminal singularities. First note that every one of the configurations for E specified by the dual graphs and the self-intersection numbers (intersection forms) can be realized geometrically. (Start from the exceptional loci of the minimal resolutions of canonical singularities and blow up further if necessary.) Grauert's contractibility criterion (cf. Theorem 4-6-25) tells us that it can be contracted to an analytic point $p \in S$. Since the discrepancies a_i are completely determined by the intersection form of E, they are as described in the dual graph of the list above. Note that actually the dual graphs completely determine the a_i and that we do not have to know the b_i to compute the discrepancies. Especially, we have $a_i \ge -1$ for all i and $a_{i_o} = -1$ for at least one i_o. Therefore, $p \in S$ is a germ of log canonical but not log terminal singularities, realizing the possibility geometrically. Observe also that by Artin [1][2][4] the germ $p \in S$ can be realized as an algebraic one. Finally, according to Laufer [2], all the dual graphs with the specified intersection forms as above are those of "taut" singularities and hence determine the analytic structures of $p \in S$ completely, except for the two cases described at the end of the assertion where the singularities are "pseudotaut" and hence we have to specify the additional information of the analytic structures of the exceptional loci to determine the analytic structures of $p \in S$.

This completes the proof of Theorem 4-6-28 and the classification of log canonical but not log terminal singularities.   □

Remark 4-6-29. (i) The above proof shows that log canonical but not log terminal singularities are rational $r > 1$ and not rational if $r = 1$.

(ii) It is possible to provide a purely arithmetical method, as presented by Alexeev in Chapter 3 of Kollár et al. [1], to determine the possible dual graphs of log terminal and/or log canonical singularities in dimension 2. This arithmetical method has a powerful advantage of being simple and even valid in all characteristics. (As a consequence, it shows that the possibilities for the dual graphs remain the same through all the characteristics.) Its disadvantage may lie in the fact that it gives little information on the analytic structure of the singularity $p \in S$, without going into the analysis of the canonical cover. This is especially evident in positive characteristics where Laufer's results do not hold any longer even for log terminal singularities and some of these singularities may have the same dual graph but different analytic structures. (That is to say, they are not "taut." See Artin [5] for "exotic" canonical singularities in characteristics 2, 3, and 5.) It is then essential to go through the analysis of the canonical cover and look at the explicit defining equations.

Our method, following Kawamata [9], centers on the analysis of the canonical cover and determines the analytic structure. Though it may look lengthy at first sight, it works in principle in all characteristics with a more careful analysis of the canonical cover. We refer the reader to Kawamata [16] [19] and Matsuki [7] for the details of the analysis in positive characteristics.

(iii) We note that we could avoid the use of Laufer's results on taut and psedotaut singularities above to determine the analytic structures of log canonical (but not log terminal) singularities at the end, by directly determining the analytic structure of the canonical cover, which is simple elliptic or cusp, with its cyclic group action. The details are left to the reader as a fun and good exercise.

So far, we have classified log terminal and log canonical singularities in dimension 2 *without* boundary $\mathbb{Q}$-divisors. Though it is natural to consider the case *with* boundary $\mathbb{Q}$-divisors and though the method to classify them is in principle the same as the one *without* boundary $\mathbb{Q}$-divisors, the list of possibilities simply becomes huge. Here we are content only to list log terminal and log canonical singularities with the boundary $\mathbb{Q}$-divisors where all of the coefficients are equal to 1.

Exercises-Theorem 4-6-30. *Let $p \in (S, D)$ be a germ of log terminal or log canonical singularities in dimension 2 with a boundary $\mathbb{Q}$-divisor $D = \Sigma d_i D_i$, where all the coefficients d_i are equal to 1. Then it is analytically isomorphic to one of the following.*

> log terminal

(i) $p \in S$ is smooth and D consists of either a single component $D = D_1$ that is nonsingular at p or two components $D = D_1 + D_2$ intersection transversally at p.

(ii) $p \in S$ is a quotient singularity of Type A according to Brieskorn's notation (which is of Type A_n according to our notation) and $D = D_1$ consists of a single component that is nonsingular at $p \in S$. The dual graph of the exceptional locus of the minimal resolution is given as below.

$$\bullet - \circ - \circ - \cdots - \circ - \circ$$

> log canonical but not log terminal

(iii) $p \in S$ is a quotient singularity of Type $D - 1$ or Type $D - 2$ according to Brieskorn's notation (which is of Type eA_n or Type D_n according to our notation) and $D = D_1$ consists of a single component that is nonsingular at $p \in S$. The dual graph of the exceptional locus of the minimal resolution is given as below

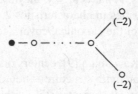

(iv) $p \in S$ is a quotient singularity of Type A according to Brieskorn's notation (which is of Type A_n according to our notation) and $D = D_1 + D_2$ consists of two components each of which is nonsingular at $p \in S$.
The dual graph of the exceptional locus of the minimal resolution is given as below.

$$\bullet - \circ - \circ - \cdots - \circ - \circ - \bullet$$

In the dual graphs above, $\bullet$ indicates the strict transform of D. The self-intersection number of an exceptional component indicated by $\circ$ can be any number less than or equal to -2, except for the two edge components in the dual graph of (iii), where their self-intersection numbers are both equal to -2.

Moreover, all the dual graphs in the above list do occur geometrically associated to some log terminal or log canonical singularities, and the analytic structure of $p \in (S, D)$ is completely determined by its dual graph together with the self-intersection numbers of the exceptional components for the minimal resolution.

PROOF.   As the proof is almost identical to that of the classification of log canonical but not log terminal singularities in dimension 2 in Theorem 4-6-28 without the boundary $\mathbb{Q}$-divisors, here we give only a proof of the following fact, which is an easy corollary of the statement above and which we use in Section 13-3 (cf. Fact 13-3-5):

Let $p \in (S, D)$ be a germ of a log terminal singularity in dimension 2, where the boundary divisor $D = D_1 + D_2$ consists of two irreducible components D_1 and D_2 with coefficient 1 intersecting at p. Then p is necessarily a smooth point of the surface S, with D_1 and D_2 are both nonsingular and intersect transversally at p.

Consider the point $p \in S$ without the boundary. Then since (cf. Proposition-Definition 4-4-1)

$$a(E; S.0) > a(E; S, D) \geq -1$$

for any exceptional divisor E over p, it is log terminal. By Theorem 4-6-18 the germ $p \in S$ is a quotient singularity, and hence there is a finite morphism

$$f : 0 \in \mathbb{C}^2 \to p \in S$$

that is étale outside of $f^{-1}(p) = 0$. Consider the log pair $(\mathbb{C}^2, f^{-1}(D))$. Then since $K_{\mathbb{C}^2} + f^{-1}(D) = f^*(K_S + D)$, the proof of Proposition 4-5-3 (i) applies to our finite morphism f with the boundary $\mathbb{Q}$-divisors, instead of the canonical cover, and leads us to conclude that $(\mathbb{C}^2, f^{-1}(D))$ has at most log canonical singularities. Therefore, $f^{-1}(D)$ consist of two nonsingular components intersecting transversally at $0 \in \mathbb{C}^2$. Again applying the proof of Proposition 4-5-3 (i) we see that if $p \in S$ is not smooth, then there are some exceptional divisors E for the minimal resolution of $p \in S$ and hence some exceptional divisors F over $0 \in \mathbb{C}^2$ such that

$$a(E; S, D) > -1 \quad \forall E \iff a(F; \mathbb{C}^2, f^{-1}(D)) > -1 \quad \forall F.$$

On the other hand, obviously we have

$$a(F; \mathbb{C}^2, f^{-1}(D)) = -1 \text{ for some } F,$$

and hence

$$a(E; S, D) \leq -1 \text{ for some } E.$$

This contradicts the assumption that $p \in (S, D)$ is a germ of log terminal singularities.

Therefore, we conlude that $p \in S$ is nonsingular and that D consists of two components intersecting transversally at p.

This completes the proof of the fact we use in Chapter 13. We refer the reader to Kawamata [9] for the verification of the details of all the possibilites (i) through (iv) in Exercise-Theorem 4-6-30.   □

This completes the presentation for Section 4-6, classifying terminal, canonical, log terminal, and log canonical singularities in dimension 2.

CHAPTER 5

Vanishing Theorems

The purpose of this chapter is to present several vanishing theorems of cohomology groups that form the technical backbone of our approach to the Mori program.

In Section 5-1 we discuss the celebrated **Kodaira vanishing theorem** with a couple of easy proofs, each of which indicates a slightly different view-point following Kollár [3][4] (cf. Clemens–Kollár–Mori [1]) and Esnault–Viehweg [1]. For the classical proof using harmonic analysis, we refer the reader to Kodaira [1] (cf. Wells [1]), and for the purely algebraic proof by Deligne–Illusie to Esnault–Viehweg [1].

In Section 5-2 we present the **Kawamata–Viehweg vanishing theorem**, which turns out to be one of the most powerful tools for proving the key theorems in establishing the Mori program. Its statements in various forms might look technical at first sight and even artificial at times. We would like, however, to convince the reader that all the technicalities converge to conceptual simplicity once we look at the theorem as the logarithmic generalization of the Kodaira vanishing theorem according to Iitaka's philosophy in Chapter 2. We emphasize this point in bold letters:

> **Kawamata–Viehweg Vanishing = Logarithmic Kodaira Vanishing**

Our proof of the Kawamata–Viehweg vanishing theorem reflects this point of view. We study carefully the logarithmic ramification of the covering constructed by Kawamata.

5.1   Kodaira Vanishing Theorem

Theorem 5-1-1 (Kodaira Vanishing Theorem).   *Let X be a nonsingular projective variety (over $\mathbb{C}$) and A an ample divisor on X. Then*

$$H^p(X, \mathcal{O}_X(K_X + A)) = 0 \text{ for } p > 0.$$

The theorem was originally proved by Kodaira [1] via harmonic analysis. Here we give two slightly different proofs, Proofs I and II, due to Kollár and Esnault–Viehweg.

Proof I: (Topological Vanishing + Hodge Theory + Serre Duality)

Step 1. Construction of a special Galois cover ramified along the complement of an affine space.

We take by Bertini's theorem $m \in \mathbb{N}$ $(m \gg 0)$ and $s \in H^0(X, \mathcal{O}_X(mA))$ such that the associated divisor

$$D = \operatorname{div}(s) = \operatorname{div}(f) + mA, \quad f \in \mathbb{C}(X),$$

is nonsingular and, in particular, reduced.

Claim 5-1-2. $T^m - f$ *is irreducible as a polynomial in $\mathbb{C}(X)[T]$.*

PROOF.   Take $\sqrt[m]{f}$ from the algebraic closure $\overline{\mathbb{C}(X)}$. If $T^m - f$ is not irreducible, then $(\sqrt[m]{f})^r \in \mathbb{C}(X)$ for some $0 < r < m$. By choosing r to be the minimum among such we may assume $r \mid m$. But then

$$D' = \operatorname{div}((\sqrt[m]{f})^r) + rA \geq 0,$$

which contradicts

$$D = \frac{m}{r} \cdot D'$$

being reduced.   □

By Claim 5-1-2 we see that $\mathbb{C}(X)[\sqrt[m]{f}] = \mathbb{C}(X)[T]/(T^m - f)$ is a finite field extension of $\mathbb{C}(X)$. Let

$$\pi : Z \to X$$

be the normalization of X in $\mathbb{C}(X)[\sqrt[m]{f}]$. Note that the Galois group $G \cong \mathbb{Z}/(m)$ acts on Z over X.

Claim 5-1-3. Z *is nonsingular and $\pi : Z \to X$ is étale outside of D.*

PROOF.   We study the local structure of

$$\pi : Z \to X.$$

Take a point $p \in X$ and its neighborhood $p \in U = \operatorname{Spec} R$ such that

$$A|_U = \operatorname{div}(g)|_U \text{ for some } g \in \mathbb{C}(X).$$

Then by construction $x = f \cdot g^m$ gives a local defining equation for $D|_U$. Note that $\mathbb{C}(X)[\sqrt[m]{f}] = \mathbb{C}(X)[\sqrt[m]{x}]$. Moreover, the integral closure S of R in $\mathbb{C}(X)[\sqrt[m]{x}]$ splits into the direct sums of eigenspaces

$$S = R_{m-1} \cdot (\sqrt[m]{x})^{m-1} \oplus R_{m-2} \cdot (\sqrt[m]{x})^{m-2} \oplus \cdots \oplus R_1 \cdot (\sqrt[m]{x}) \oplus R_0,$$

where

$$R_i \subset \mathbb{C}(X).$$

We check the criterion for $r_i \in \mathbb{C}(X)$ to be in R_i as follows:

$$
\begin{aligned}
r_i \in R_i &\iff r_i \cdot (\sqrt[m]{x})^i) \in S \\
&\iff \operatorname{div}(r_i \cdot (\sqrt[m]{x})^i) \geq 0 \\
&\iff m \cdot \operatorname{div}(r_i) + i D|_U \geq 0 \\
&\iff \operatorname{div}(r_i) \geq 0 \\
&\iff r_i \in R.
\end{aligned}
$$

We note that the second to last equivalence follows because D is reduced. Therefore,

$$S = R[\sqrt[m]{x}] = R[T]/(T^m - x).$$

Thus at the point p, if x is a unit, then there exist m nonsingular points over p and π is étale. If x is not a unit, then we can choose a local coordinate system $(x, x_2, \ldots, x_n)$ such that $x = x_2 = \cdots = x_n = 0$ defines p and such that over p there exists only one point q defined by $\sqrt[m]{x} = x_2 = \cdots = x_n = 0$, which hence is nonsingular.

This proves the claim.   □

Claim 5-1-4. *G acts on $\pi_* \mathcal{O}_Z$, whose splitting into eigenspaces coincides with*

$$\pi_* \mathcal{O}_Z = \oplus_{i=0}^{m-1} \pi_* \mathcal{O}_Z[\epsilon^i] = \oplus_{i=0}^{m-1} \mathcal{O}_X(-iA).$$

Also, the equality holds as an isomorphism of $\mathcal{O}_X$-algebras, where the $\mathcal{O}_X$-algebra structure on the right-hand side is given by (when $i + j \geq m$)

$$\mathcal{O}_X(-iA) \times \mathcal{O}_X(-jA) \to \mathcal{O}_X(-(i + j - m)A),$$
$$a \cdot b \mapsto a \cdot b \cdot s.$$

PROOF.   We use the notation of Claim 5-1-3.

Just as before, locally S splits into the direct summands of eigenspaces

$$S = R'_{m-1} \cdot (\sqrt[m]{f})^{m-1} \oplus R'_{m-2} \cdot (\sqrt[m]{f})^{m-2} \oplus \cdots \oplus R'_1 \cdot (\sqrt[m]{f}) \oplus R'_0,$$

where

$$R'_i \subset \mathbb{C}(X).$$

We check the criterion of $r'_i \in \mathbb{C}(X)$ to be in R'_i as follows:

$$r'_i \in R'_i \iff r'_i \cdot (\sqrt[m]{f})^i \in S$$
$$\iff \operatorname{div}(r'_i \cdot (\sqrt[m]{f})^i)|_U \geq 0$$
$$\iff m \cdot \operatorname{div}(r'_i) + i(D - mA)|_U \geq 0$$
$$\iff \operatorname{div}(r'_i) - iA|_U \geq 0, r'_i \in \mathcal{O}_X(-iA).$$

Note that the second-to-last implication follows from the fact that D is reduced by the choice, and thus all the coefficients of $\frac{i}{m}D$ ($0 \leq i \leq m - 1$) are less than 1. Therefore, the eigenspace is isomorphic to $\mathcal{O}_X(-iA)$, and this proves the first assertion. The second assertion follows immediately from the above description. □

Step 2. Study the action of G on $\pi_*\mathcal{O}_Z$ and $\pi_*\mathbb{C}_Z$.

First we choose the generator $g \in G$ such that g acts as multiplication by ϵ^i, where ϵ is a primitive mth root of unity, on the eigenspace $\pi_*\mathcal{O}_Z[\epsilon^i]$ in the decomposition of Claim 5-1-4. The Galois group G acts on $\pi_*\mathbb{C}_Z$, and let

$$\pi_*\mathbb{C}_Z = \oplus_{i=0}^{m-1} \pi_*\mathbb{C}_Z[\epsilon^i]$$

be its decomposition into eigenspaces with the same convention with respect to the action of the generator g. The eigenspaces $\pi_*\mathbb{C}_Z[\epsilon^i]$ can be visualized as in Figure 5-1-5:

Figure 5-1-5.

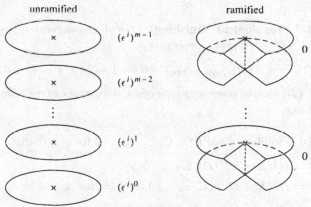

Namely, if π is not ramified over a point $p \in X$, then for a small connected neighborhood U,

$$\pi_*\mathbb{C}_Z[\epsilon^i](U) = \mathbb{C} \cdot (\oplus_{j=0}^{m-1}(\epsilon^i)^j),$$

where the direct sum $\oplus$ on the right-hand side of the above equation is taken over $U_Z, g(U_Z), g^2(U_Z), \cdots, g^{m-1}(U_Z)$, once we choose and fix one U_Z of the m sheets over U.

If π is ramified over $p \in X$, then

$$\pi_*\mathbb{C}_Z[\epsilon^0]_p = \mathbb{C} \text{ and } \pi_*\mathbb{C}_Z[\epsilon^i] = 0 \text{ for } i \neq 0.$$

Therefore, we conclude that

$$\pi_*\mathbb{C}_Z[\epsilon^0] = \mathbb{C}_X,$$
$$\pi_*\mathbb{C}_Z[\epsilon^i] = i_! \mathcal{C}_{X-D,i},$$

where $\mathcal{C}_{X-D,i}$ is a locally constant sheaf of rank 1 over $X - D$ and where $i_!$ denotes the extension of the constant sheaf $\mathcal{C}_{X-D,i}$ to the whole of X by setting every stalk at $p \in D$ to be 0. Note that the sheaf $\mathcal{C}_{X-D,i}$ is different from the constant sheaf $\mathbb{C}_{X-D}$, since the monodromy acts nontrivially on the former and trivially on the latter.

The action of G on $\pi_*\mathbb{C}_Z$ induces the action of G on the cohomology groups of the sheaf, and we conclude the following from the argument above.

Claim 5-1-6. *For $i \neq 0$, we have*

$$H^j(X, \pi_*\mathbb{C}_Z)[\epsilon^i] = H^j(X, \pi_*\mathbb{C}_Z[\epsilon^i])$$
$$= H^j(X, i_!\mathcal{C}_{X-D,i})$$
$$= H^j(X - D, \mathcal{C}_{X-D,i}).$$

Step 3. Topological Vanishing.

We recall the theorem below, which follows from Morse theorey (cf. Milnor [2]).

Theorem 5-1-7 (Topological Vanishing). *For a nonsingular affine algebraic variety $Z - \pi^{-1}(D)$ of complex dimension n,*

$$H^j(Z - \pi^{-1}(D), \mathbb{C}_{Z-\pi^{-1}(D)}) = 0 \text{ for } j = 2n - k, \quad k < n,$$

since $Z - \pi^{-1}(D)$ has the homotopy type of a CW-complex of real dimension less than or equal to n.

Theorem 5-1-7 implies, since $R^q\pi_*\mathbb{C}_{Z-\pi^{-1}(D)} = 0$ for $q > 0$, that

$$0 = H^j(Z - \pi^{-1}(D), \mathbb{C}_{Z-\pi^{-1}(D)})$$
$$= H^j(X - D, \pi_*\mathbb{C}_{Z-\pi^{-1}(D)}) \text{ for } j = 2n - k, \quad k < n.$$

Taking the eigenspace we conclude the following.

Claim 5-1-8.

$$0 = H^j(Z - \pi^{-1}(D), \mathbb{C}_{Z-\pi^{-1}(D)})[\epsilon^i]$$
$$= H^j(X - D, \pi_*\mathbb{C}_{Z-\pi^{-1}(D)})[\epsilon^i]$$
$$= H^j(X - D, \pi_*\mathbb{C}_{Z-\pi^{-1}(D)}[\epsilon^i])$$
$$= H^j(X - D, \pi_*\mathbb{C}_Z[\epsilon^i])$$
$$= H^j(X - D, \mathcal{C}_{X-D,i}) \text{ for } j = 2n - k, \quad k < n.$$

Exercise 5-1-9. Since $X - D$ is also affine and has the homotopy type of a CW-complex of real dimension less than or equal to n, it is easy to see that the cohomology groups vanish, i.e.,

$$H^j(X - D, \mathbb{C}_{X-D}) = 0 \text{ for } j = 2n - k, \quad k < n,$$

since they coincide with the usual singular cohomology groups.

For an arbitrary locally constant sheaf $\mathcal{C}_{X-D}$ of rank 1, the vanishing of the cohomology groups can be directly checked in the following way: Using the fact (cf. Milnor [2]) that a CW-complex of real dimension less than or equal to n can be chosen to be a deformation retract of $X - D$, take an open covering $\mathfrak{U} = \{U_\lambda\}$ of $X - D$ such that:

(a) the index set $\{\lambda\}$ is the same as the set of the 0-dimensional cells and U_λ is homotopic to the union of the open cells whose closures contain the 0-dimensional cell corresponding to λ,

(b) the intersection of any finite number of open subsets

$$U_{\lambda_0 \cdots \lambda_s} = U_{\lambda_0} \cap \cdots \cap U_{\lambda_s}$$

from the collection $\mathfrak{U}$ is a disjoint union of simply connected open subsets,

(c)

$$\mathcal{C}_{X-D}|_{U_{\lambda_0 \cdots \lambda_s}} \cong \mathbb{C}_{X-D}|_{U_{\lambda_0 \cdots \lambda_s}} .$$

and

$$H^i(U_{\lambda_0 \cdots \lambda_s}, \mathcal{C}_{X-D}|_{U_{\lambda_0 \cdots \lambda_s}}) = 0 \text{ for } i > 0,$$

(d) the Cech resolution with respect to the covering $\mathfrak{U}$ of the locally constant sheaf $\mathcal{C}_{X-D}$ of rank 1 with respect the open covering $\mathfrak{U}$

$$0 \to \mathcal{C}_{X-D} \to C^{\cdot}(\mathfrak{U}, \mathcal{C}_{X-D})$$

has length at most n, i.e.,

$$C^i(\mathfrak{U}, \mathcal{C}) = 0 \text{ for } i > n,$$

(e) thanks to (c) and (d) the Cech resolution computes the cohomology groups to be

$$H^j(X - D, \mathcal{C}_{X-D}) = H^j(C^{\cdot}(\mathfrak{U}, \mathcal{C}_{X-D})) = 0 \text{ for } j > n.$$

Finally, using Poincaré duality we have the following claim.

Claim 5-1-10. *For* $i \neq 0$

$$\begin{aligned}
0 &= H^j(X - D, \mathcal{C}_{X-D,i}) \\
&= H^j(X, \pi_* \mathbb{C}_Z)[\epsilon^i] \\
&= H^j(Z, \mathbb{C}_Z)[\epsilon^i] (\text{ since } R^q \pi_* \mathbb{C}_Z = 0 \text{ for } q > 0) \\
&= H^k(Z, \mathbb{C}_Z)[\epsilon^{m-i}] (\text{ by Poincáre duality } (j = 2n - k)) \\
&= H^k(X, \pi_* \mathbb{C}_Z)[\epsilon^{m-i}] \text{ for } k < n.
\end{aligned}$$

Note that the nondegenerate pairing that gives the Poincáre duality

$$H^j(Z, \mathbb{C}_Z) \times H^{2n-j}(Z, \mathbb{C}_Z) \to H^{2n}(Z, \mathbb{C}_Z)$$

is compatible with the group action and that

$$H^{2n}(Z, \mathbb{C}_Z) \cong \mathbb{C}$$

is invariant under the group action.

Step 4. Hodge Theory

By Hodge theory, we have a commutative diagram where the horizontal arrows are G-equivariant and surjective:

$$
\begin{array}{ccc}
H^k(Z, \mathbb{C}_Z) & \longrightarrow & H^k(Z, \mathcal{O}_Z) \\
\| & & \| \\
H^k(X, \pi_* \mathbb{C}_Z) & \longrightarrow & H^k(X, \pi_* \mathcal{O}_Z)
\end{array}
$$

(In order to see this, consider, e.g., the de Rham complex and observe that the group action is compatible with differentiation and projection onto the $(0, k)$-forms.) This leads to the surjection of the eigenspaces individually:

$$0 = H^k(X, \pi_* \mathbb{C}_Z)[\epsilon^{m-i}] \to H^k(X, \pi_* \mathcal{O}_Z)[\epsilon^{m-i}] = H^k(X, \mathcal{O}_X(-(m-i)A)).$$

Setting $i = m - 1$, we have

$$H^k(X, \mathcal{O}_X(-A)) = 0 \text{ for } k < n.$$

Step 5. Serre Duality

Finally, by Serre Duality, setting $k = n - p$, we have

$$H^p(X, \mathcal{O}_X(K_X + A)) \cong H^k(X, \mathcal{O}_X(-A)) = 0, \quad p > 0.$$

This completes Proof I.   □

Now we present Proof II.

Proof II: (Serre Vanishing + Hodge Theory + Serre Duality)

Step 1. Construction of a special Galois cover.

Take the Galois cover

$$\pi : Z \to X$$

as in Proof I.

Step 2. Study the action of G on $\pi_* \mathcal{O}_Z$ and $\pi_* \Omega_Z^1$.

We already gave the description of the decomposition of $\pi_* \mathcal{O}_Z$ into the eigenspaces under the action of G in Claim 5-1-4 of Proof I. The following claim gives the description of the decomposition of $\pi_* \Omega_Z^1$ into the eigenspaces under the action of G.

Claim 5-1-11. *The Galois group G acts on $\pi_* \Omega_Z^1$, whose splitting into the eigenspaces is given by*

$$\pi_* \Omega_Z^1 = \Omega_Z^1[\epsilon^0] \oplus \{\oplus_{i=1}^{m-1} \Omega_Z^1[\epsilon^i]\} = \Omega_X^1 \oplus \{\oplus_{i=1}^{m-1} \Omega_X^1(\log D) \otimes \mathcal{O}_X(-iA)\}.$$

The Galois group G also acts on $\pi_*\Omega^1_Z(\log G)$ (cf. Definition 2-1-3), where $G = (\pi^*D)_{red}$, and its splitting into the eigenspaces is given by

$$\pi_*\Omega^1_Z(\log G) = \oplus_{i=0}^{m-1}\Omega^1_Z(\log G)[\epsilon^i] = \oplus_{i=0}^{m-1}\Omega^1_X(\log D) \otimes \mathcal{O}_X(-iA).$$

PROOF.   We give the description of the decomposition locally around a point $p \in D$. (The description locally around a point $p \notin D$ is similar and is left to the reader.) We prove the statement about $\pi_*\Omega^1_Z(\log G)$ first. Let $x_1, x_2, \cdots, x_n$ be a local coordinate system around p such that D is defined by $x_1 = 0$. Then $\pi^{-1}(p)$ consists of only one point q, around which a local coordinate system can be taken to be $y_1 = \sqrt[m]{x_1}, x_2, \cdots, x_n$ (cf. Claim 5-1-3). Using this coordinate system, we can describe

$$\{\pi_*\Omega^1_Z(\log G)\}_p = \pi_* \left\{ \mathcal{O}_{Z,p} \cdot m\frac{dy_1}{y_1} \oplus \mathcal{O}_{Z,p} \cdot dx_2 \oplus \cdots \oplus \mathcal{O}_{Z,p} \cdot dx_n \right\}$$

$$= \pi_*\mathcal{O}_{Z,p} \cdot \frac{dx_1}{x_1} \oplus \pi_*\mathcal{O}_{Z,p} \cdot dx_2 \oplus \cdots \oplus \pi_*\mathcal{O}_{Z,p} \cdot dx_n$$

$$= \{\pi_*\mathcal{O}_Z \otimes \Omega^1_X(\log D)\}_p,$$

identifying $m \ldots \frac{dy_1}{y_1}, dx_2, \ldots, dx_n$ with $\frac{dx_1}{x_1}, dx_2, \cdots, dx_n$. Therefore,

$$\pi_*\Omega^1_Z(\log G) = \pi_*\mathcal{O}_Z \otimes \Omega^1_X(\log D)$$

$$= (\oplus_{i=0}^{m-1}\pi_*\mathcal{O}_Z[\epsilon^i]) \otimes \Omega^1_X(\log D)$$

$$= \oplus_{i=0}^{m-1}\Omega^1_X(\log D) \otimes \mathcal{O}_X(-iA).$$

As for the decomposition of $\pi_*\Omega^1_Z$, we observe that for $i > 0$ taking any element

$$\omega \in \{\pi_*\Omega^1_Z(\log G)[\epsilon^i]\}_p = b_1 \cdot y_1^i \cdot \frac{dy_1}{y_1} \oplus b_2 \cdot y_1^i \cdot dx_2 \oplus \cdots \oplus b_n \cdot y_1^i \cdot dx_n,$$

where

$$b_1, b_2, \cdots, b_n \in \mathcal{O}_{X,p},$$

we automatically have

$$\omega \in \pi_*\Omega^1_Z,$$

since

$$y_1^i \cdot \frac{1}{y_1} \in \mathcal{O}_{Z,p}.$$

Thus

$$\pi_*\Omega^1_Z[\epsilon^i] = \pi_*\Omega^1_Z(\log G)[\epsilon^i].$$

On the other hand, we observe that for $i = 0$ taking any element

$$\omega \in \{\pi_*\Omega^1_Z(\log G)[\epsilon^0]\}_p = b_1 \cdot \frac{dy_1}{y_1} \oplus b_2 \cdot dx_2 \oplus \cdots \oplus b_n \cdot dx_n$$

we have

$$\omega \in \pi_* \Omega_Z^1 \iff b_1 \cdot \frac{1}{y_1} \in \mathcal{O}_{Z,p}$$
$$\iff b_1 = b_1' \cdot y_1{}^m (\text{ where } b_1' \in \mathcal{O}_{X,p})$$
$$\iff \omega = \frac{b_1'}{m} dx_1 \oplus b_2 dx_2 \oplus \cdots \oplus b_n dx_n (\text{ where } b_1' \in \mathcal{O}_{X,p})$$
$$\iff \omega \in \Omega_X^1.$$

Thus

$$\pi_* \Omega_Z^1[\epsilon^0] = \Omega_X^1.$$

This completes the proof of the claim. $\qquad\qquad\square$

Now holomorphic differentiation gives the sheaf homomorphism

$$d : \mathcal{O}_Z \to \Omega_Z^1$$

and its direct image

$$\pi_* d : \pi_* \mathcal{O}_Z \to \pi_* \Omega_Z^1,$$

which is G-equivariant, since the group action and differentiation commute. By taking the eigenspaces separately, we obtain for $i > 0$

$$\begin{array}{ccc}
\nabla_i : \pi_* \mathcal{O}_Z[\epsilon^i] & \longrightarrow & \pi_* \Omega_Z^1[\epsilon^i] \\
\| & & \| \\
\mathcal{O}_X(-iA) & \longrightarrow & \Omega_X^1(\log D) \otimes \mathcal{O}_X(-iA)
\end{array}$$

Step 3. Hodge Theory.

Claim 5-1-12. *The map of cohomology*

$$d : H^i(Z, \mathcal{O}_Z) \to H^i(Z, \Omega_Z^1), \quad i \geq 0,$$

induced by holomorphic differentiation is the zero map, and so is

$$\nabla_1 : H^i(X, \mathcal{O}_X(-A)) \to H^i(X, \Omega_X^1(\log D) \otimes \mathcal{O}_X(-A)).$$

PROOF.   By considering the representatives of cohomology in (harmonic) forms via the Hodge theory we have the commutative diagram

$$\begin{array}{ccc}
H^i(Z, \mathbb{C}) & \xrightarrow{d} & H^{i+1}(Z, \mathbb{C}) \\
\downarrow & & \downarrow \\
H^i(Z, \mathcal{O}_Z) & \xrightarrow{d} & H^i(Z, \Omega_Z^1),
\end{array}$$

where the vertical arrows are surjections. Since the arrow on the first row is obviously *zero*, we obtain the claim. $\qquad\qquad\square$

Step 4. Serre's Vanishing and Serre Duality.
Composing ∇_1 with the residue map

$$\mathcal{O}_X(-A) \overset{\nabla_1}{\to} \Omega^1_X(\log D) \otimes \mathcal{O}_X(-A) \overset{\text{Residue} \otimes \text{id}}{\to} \mathcal{O}_D \otimes \mathcal{O}_X(-A)$$

$$r \cdot y_1^l \mapsto \left(dr + \frac{1}{m} \cdot r \cdot \frac{dx_1}{x_1} \right) \cdot y_1^l \mapsto \frac{1}{m} \cdot r|_D \cdot y_1^l$$

we obtain

$$(\text{Residue} \otimes \text{id}) \circ \nabla_1 = \frac{1}{m}(\text{Restriction on } D).$$

By Claim 5-1-12 we see that the map at the cohomology level

$$m \times (\text{Residue} \otimes \text{id}) \circ \nabla_1 : H^i(X, \mathcal{O}_X(-A)) \to H^i(D, \mathcal{O}_D \otimes \mathcal{O}_X(-A))$$

is *zero*. Therefore, from the exact sequence

$$0 \to \mathcal{O}_X(-D) \otimes \mathcal{O}_X(-A) \to \mathcal{O}_X(-A) \to \mathcal{O}_D \otimes \mathcal{O}_X(-A) \to 0$$

we conclude that

$$H^i(X, \mathcal{O}_X(-(m+1)A)) = H^i(X, \mathcal{O}_X(-D) \otimes \mathcal{O}_X(-A)) \to H^i(X, \mathcal{O}_X(-A))$$

is surjective. On the other hand, by Serre duality and Serre vanishing (we choose m sufficiently large from the beginning)

$$H^i(X, \mathcal{O}_X(-(m+1)A)) \cong H^{n-i}(X, \mathcal{O}_X(K_X + (m+1)A)) = 0 \quad \text{for } n - i > 0.$$

Thus

$$H^i(X, \mathcal{O}_X(-A)) = 0 \text{ for } n > i \geq 0.$$

Again by Serre duality

$$H^p(X, \mathcal{O}_X(K_X + A)) = 0 \text{ for } p = n - i > 0.$$

This completes the presentation of the two proofs of the Kodaira vanishing theorem.   □

5.2   Kawamata–Viehweg Vanishing Theorem

The goal of this section is to establish the **Kawamata–Viehweg vanishing theorem** as the logarithmic version of the Kodaira vanishing theorem according to Iitaka's philosophy in Chapter 2 (cf. Principle 2-1-4 and Observation 2-1-8).

Theorem 5-2-1 (Kawamata–Viehweg Vanishing Theorem = Logarithmic Kodaira Vanishing Theorem). *Let (X, D) be a log pair consisting of a nonsingular projective variety X and a boundary $\mathbb{Q}$-divisor $D = \Sigma d_i D_i$, $0 \leq d_i \leq 1$, with the D_i distinct irreducible components crossing normally so that (X, D) has only* *log terminal singularities. Then for an ample* $\mathbb{Q}$*-divisor A such that* $K_X + D + A$ *is an integral divisor, we have*

$$H^p(X, \mathcal{O}_X(K_X + D + A)) = 0 \, for \, p > 0.$$

Theorem 5-2-2 (Slight Modification of Theorem 5-2-1). *Let the conditions be as in Theorem 5-2-1, except that we require further that all the coefficients be strictly less than 1, i.e.,*

$$0 \le d_i < 1.$$

Then

$$H^p(X, K_X + D + A) = 0 \, for \, p > 0.$$

The implication Theorem 5-2-1 $\Longrightarrow$ Theorem 5-2-2 is obvious. In order to see the other implication Theorem 5-2-1 $\Longleftarrow$ Theorem 5-2-2, we replace D and A with $(1-\epsilon)D$ and $A+\epsilon D$ for sufficiently small $\epsilon > 0$, since $A+\epsilon D$ remains ample. Thus Theorem 5-2-1 and Theorem 5-2-2 are equivalent. Another equivalent formulation of the Kawamata–Viehweg vanishing theorem is given below as Theorem 5-2-3. We note that this is the form that is commonly used, and thus acquired the reputation of being technical, since the presentation makes it obscure that the theorem is the logarithmic version of the Kodaira vanishing theorem. The artificial-looking statement of taking the "round up" of a $\mathbb{Q}$-divisor and the assumption of the fractional part having only normal crossings start fitting in naturally after one looks at them in terms of the logarithmic category.

We use the following notation to denote the round up, or the integral part, of a $\mathbb{Q}$-divisor $A = \Sigma a_k A_k$ with the A_k distinct irreducible components

$$\lceil A \rceil := \Sigma \lceil a_k, \rceil A_k, \text{ where } a_k \le \lceil a_k \rceil < a_k + 1 \text{ and } \lceil a_k \rceil \in \mathbb{Z},$$
$$[A] := \Sigma [a_k]A_k, \text{ where } a_k - 1 < [a_k] \le a_k \text{ and } [a_k] \in \mathbb{Z}.$$

Theorem 5-2-3 (Another Formulation of Theorem 5-2-2). *Let X be a nonsingular projective variety, A an ample* $\mathbb{Q}$*-divisor such that the fractional part* $\lceil A \rceil - A$ *has a support with only normal crossings. Then*

$$H^p(X, K_X + \lceil A \rceil) = 0 \, for \, p > 0.$$

Equivalence between Theorem 5-2-2 and Theorem 5-2-3 is more than obvious once one puts

$$\lceil A \rceil - A = D \text{ and thus } K_X + \lceil A \rceil = K_X + D + A.$$

Since all the three theorems above are equivalent, we can prove any one of them. Here we prove the Kawamata–Viehweg vanishing theorem in the formulation of Theorem 5-2-3.

PROOF.   Proof of the Kawamata–Viehweg vanishing theorem in the formulation of Theorem 5-2-3 We explain the rough idea of our proof (in the formulation of

Theorem 5-2-2, however) as follows: We take an appropriate finite Galois cover

$$\pi : Y \to X$$

such that $K_Y = \pi^*(K_X + D)$ and $\pi^* A$ is an integral divisor. Then applying Kodaira vanishing on Y we get

$$H^p(Y, K_Y + \pi^* A) = 0 \text{ for } p > 0,$$

which in turn implies

$$H^p(X, K_X + D + A) = H^p(Y, K_Y + \pi^* A)^G = 0 \text{ for } p > 0$$

by taking the G-invariant part for the Galois group G.

We start by looking at the precise description of the covering we take.

Lemma 5-2-4 (Covering Lemma of Kawamata). *Let X and A be as in Theorem 5-2-3. Set $D = \Sigma d_i D_i = \lceil A \rceil - A$. Then for any sufficiently divisible and big $m \in \mathbb{N}$ there exists a finite Galois cover*

$$\pi : Y \to X$$

such that

(i) Y is nonsingular projective,
(ii) $\pi^* A$ is an integral divisor, and
(iii)

$$K_Y = \pi^* \left(K_X + \Sigma \frac{m-1}{m} D_i + \Sigma \frac{m-1}{m} G_j \right),$$

where $G = \Sigma G_j$ is some divisor, sharing no common component with D, such that

$$G \cup D \text{ has only normal crossings.}$$

(iv)

$$\frac{m-1}{m} + (1 - d_i) \geq 1 \quad \forall i.$$

Note that the description of the rough idea is imprecise in the sense that the actual ramification divisor is not D but $\Sigma \frac{m-1}{m} D_i + \Sigma \frac{m-1}{m} G_j$.

PROOF. Proof of the Covering Lemma

We fix a very ample divisor H on X.

We take $m \in \mathbb{N}$ to be sufficiently divisible and big so that

$$md_i \in \mathbb{N}(\text{and hence } m(1 - d_i) \in \mathbb{N}) \text{ and } \frac{m-1}{m} + (1 - d_i) \text{ for every } i$$

and such that

$$mH - D_i \text{ is very ample for every } i \in I. \qquad \square$$

For each $i \in I$ we take $n = \dim X$ of general elements

$$G_1^{(i)}, G_2^{(i)}, \ldots, G_n^{(i)} \in |mH - D_i|$$

such that

$$\cup_{i \in I} D_i \cup_{i \in I} \left\{ \cup_{k=1}^n G_k^{(i)} \right\}$$

is a divisor with only normal crossings. (Exercise! Show that we can actually take such $G_k^{(i)}$'s via the application of Bertini's theorem.) Let $X = \cup_{\alpha \in A} U_\alpha$ be an affine open cover of X with the transition functions for the line bundle $\mathcal{O}_X(H)$

$$\{a_{\alpha\beta}; a_{\alpha\beta} \in H^0(U_\alpha \cap U_\beta, \mathcal{O}_X^*)\}$$

and local sections for $\mathcal{O}_X(mH)$

$$\{\varphi_{k\alpha}^{(i)}; \varphi_{k\alpha}^{(i)} \in H^0(U_\alpha, \mathcal{O}_X)\}$$

such that

$$(G_k^{(i)} + D_i)|_{U_\alpha} = \operatorname{div}(\varphi_{k\alpha}^{(i)}) \text{ on } U_\alpha$$

and such that

$$\varphi_{k\alpha}^{(i)} = a_{\alpha\beta}^m \cdot \varphi_{k\beta}^{(i)}.$$

We claim that the normalization Y of X in $\mathbb{C}(X)[(\varphi_{k\alpha}^{(i)})^{\frac{1}{m}}]_{i,k}$ for some $\alpha \in A$ provides the desired cover. (Note that $\mathbb{C}(X)[(\varphi_{k\alpha}^{(i)})^{\frac{1}{m}}]_{i,k} = \mathbb{C}(X)[(\varphi_{k\beta}^{(i)})^{\frac{1}{m}}]_{i,k}$ for any $\alpha, \beta \in A$.)

Since $\mathbb{C}(Y)/\mathbb{C}(X)$ is a Kummer extension, $\pi : Y \to X$ is obviously a finite Galois cover.

We see that Y is nonsingular in the following way. Take any closed point $x \in U_\alpha$. Set

$$I_x := \{i \in I; x \in D_i\}.$$

Then for each $i \in I_x$ there exists k_i with $1 \le k_i \le n$ such that $x \notin G_{k_i}^{(i)}$. Now the set

$$R_x := \left\{ \varphi_{k_i\alpha}^{(i)}; i \in I_x \right\} \cup \left\{ \varphi_{k\alpha}^{(i)}; i \notin I_x, x \in G_k^{(i)} \right\} \cup \left\{ \varphi_{k\alpha}^{(i)}/\varphi_{k_i\alpha}^{(i)}; i \in I_x, x \in G_k^{(i)} \right\}$$

forms a part of a regular system of parameters of $\mathcal{O}_{X,x}$. Set

$$T_x := \left\{ \varphi_{k\alpha}^{(i)}/\varphi_{k_i\alpha}^{(i)}; i \in I_x, x \notin G_k^{(i)} \right\} \cup \left\{ \varphi_{k\alpha}^{(i)}; i \notin I_x, x \notin G_k^{(i)} \right\}.$$

Any element in T_x is a unit in $\mathcal{O}_{X,x}$. Now it is sufficient to prove the claim below to see that Y is nonsingular, i.e., $\mathcal{O}_{Y,y}$ is a regular local ring for any $y \in Y$ with $\pi(y) = x$.

Claim 5-2-5. *Let R be a regular local ring over $\mathbb{C}$ of dimension n with the maximal ideal m_R such that $R/m_R = \mathbb{C}$. Let*

$$z_1, z_2, \ldots, z_n$$

be a regular system of parameters and let

$$u_1, u_2, \ldots, u_s$$

be units in R. Let $m \in \mathbb{N}$ be a positive integer. Then for any maximal ideal $m_{R'}$ of

$$R' = R\left[z_1^{\frac{1}{m}}, z_2^{\frac{1}{m}}, \ldots, z_l^{\frac{1}{m}}, u_1^{\frac{1}{m}}, u_2^{\frac{1}{m}}, \ldots, u_s^{\frac{1}{m}}\right], \quad 1 \le l \le n,$$

the localization of R' by $m_{R'}$, denoted by R_1, is a regular local ring with a regular system of parameters

$$z_1^{\frac{1}{m}}, z_2^{\frac{1}{m}}, \ldots, z_l^{\frac{1}{m}}, z_{l+1} \ldots, z_n.$$

PROOF.   Proof of Claim 5-2-5 Since $m_{R'}$ is generated by

$$z_1^{\frac{1}{m}}, z_2^{\frac{1}{m}}, \ldots, z_l^{\frac{1}{m}}, z_{l+1}, \ldots, z_n, u_1^{\frac{1}{m}} - \alpha_1, \ldots, u_s^{\frac{1}{m}} - \alpha_s,$$

where

$$u_t^{\frac{1}{m}} = \alpha_t \bmod m_{R'}, \quad \alpha_t \in \mathbb{C},$$

it is sufficient to show that

$$u_t^{\frac{1}{m}} - \alpha_t \in \left(z_1^{\frac{1}{m}}, z_2^{\frac{1}{m}}, \ldots, z_l^{\frac{1}{m}}, z_{l+1}, \ldots, z_n\right) R_1 \quad \forall t.$$

This follows from the equation

$$u_t - \alpha_t^m = (u_t^{\frac{1}{m}} - \alpha_t^{\frac{1}{m}}) \cdot v_t, \text{ where } v_t \text{ is a unit in } R_1,$$

and from

$$u_t - \alpha_t^m \in m_R = (z_1, z_2, \ldots, z_n)R. \qquad \square$$

We apply the claim to our situation, setting

$$R = \mathcal{O}_{X,x},$$
$$R_x = \{z_1, \ldots, z_l\},$$
$$T_x = \{u_1, \ldots, u_s\}.$$

Since

$$\mathbb{C}(X)[\varphi_{k\alpha}^{(i)}]_{i,k} = \mathbb{C}(X)\left[z_1^{\frac{1}{m}}, z_2^{\frac{1}{m}}, \ldots, z_l^{\frac{1}{m}}, u_1^{\frac{1}{m}}, u_2^{\frac{1}{m}}, \ldots, u_s^{\frac{1}{m}}\right],$$

we conclude that $R_1 = \mathcal{O}_{Y,y}$ is a regular local ring.

Now that we know condition (i) that Y is nonsingular, condition (ii) follows immediately from the description of a regular system of parameters in Claim 5-2-5 and so does condition (iii) by setting $\cup G_j = \cup G_k^{(i)}$.

This completes the proof of the covering lemma. $\square$

Now we resume the proof of the Kawamata–Viehweg vanishing theorem.

First note that the sheaf $\mathcal{O}_Y(K_Y + \pi^*A)$ has a description of its sections over open subsets $V \subset Y$ as

$$\Gamma(V, \mathcal{O}_Y(K_Y + \pi^*A)) = \{f \in \mathbb{C}(Y); \operatorname{div}(f) + K_Y + \pi^*A|_V \geq 0\},$$

and accordingly, the sheaf $\pi_*\mathcal{O}_Y(K_Y + \pi^*A)$ has a description of its sections over open subsets $U \subset X$ as

$$\Gamma(U, \pi_*\mathcal{O}_Y(K_Y + \pi^*A)) = \Gamma(V = \pi^{-1}U, \mathcal{O}_Y(K_Y + \pi^*A))$$
$$= \{f \in \mathbb{C}(Y); \operatorname{div}(f) + K_Y + \pi^*A|_V \geq 0\}.$$

Therefore, there is a natural action of G on $\pi_*\mathcal{O}_Y(K_Y + \pi^*A)$, which is compatible with the action of G on $\mathbb{C}(Y)$. This leads to the following description of the G-invariant part $\{\pi_*\mathcal{O}_Y(K_Y + \pi^*A)\}^G$,

$$\Gamma(U, \{\pi_*\mathcal{O}_Y(K_Y + \pi^*A)\}^G)$$
$$= \{f \in \mathbb{C}(Y)^G; \operatorname{div}(f) + K_Y + \pi^*A|_{V=\pi^{-1}(U)} \geq 0\}$$
$$= \left\{f \in \mathbb{C}(X); \operatorname{div}(f) + K_X + \Sigma\frac{m-1}{m}D_i + \Sigma\frac{m-1}{m}G_j + A|_U \geq 0\right\}$$
$$= \{f \in \mathbb{C}(X); \operatorname{div}(f) + K_X + \lceil A\rceil|_U \geq 0\}$$

(Recall condition (iv) on m), i.e.,

$$\{\pi_*\mathcal{O}_Y(K_Y + \pi^*A)\}^G = \mathcal{O}_X(K_X + \lceil A\rceil).$$

By applying Kodaira vanishing on Y we get

$$H^p(Y, K_Y + \pi^*A) = 0 \text{ for } p > 0.$$

Note that the functor of taking global sections and that of taking the G-invariant part commute for the sheaves with G-action and hence so do the operation of taking cohomology and that of taking the G-invariant part. Therefore, we have

$$H^p(X, \mathcal{O}_X(K_X + \lceil A\rceil)) = H^p(X, \{\pi_*\mathcal{O}_Y(K_Y + \pi^*A)\}^G)$$
$$= H^p(X, \pi_*\mathcal{O}_Y(K_Y + \pi^*A))^G$$
$$= H^p(Y, \mathcal{O}_Y(K_Y + \pi^*A))^G = 0 \text{ for } p > 0.$$

This completes the proof of the Kawamata–Viehweg Vanishing theorem. $\square$

Theorem 5-2-6 (Relative Version of Kawamata–Viehweg Vanishing Theorem).
Let

$$f : X \to S$$

be a projective morphism from a nonsingular variety X onto a variety S, A an f-ample $\mathbb{Q}$-divisor such that the fractional part $\lceil A\rceil - A$ has a support with only normal crossings. Then

$$R^p f_*\mathcal{O}_X(K_X + \lceil A\rceil) = 0 \text{ for } p > 0.$$

PROOF.   Step 1. We may assume that both X and S are projective and A is an ample $\mathbb{Q}$-divisor.

We leave the verification of this reduction step to the reader as an exercise. (See the proof of Theorem 1-2-3 in Kawamata–Matsuda–Matsuki [1].)

Step 2. Reduction to the previous (nonrelative) case.

Take an ample divisor H on S.

Consider for $m \gg 0$ the spectral sequence

$$
\begin{aligned}
E_2^{q,p} &= H^q(S, R^p f_* \mathcal{O}_X(K_X + \lceil A \rceil + m f^* H)) \\
&= H^q(S, R^p f_* \mathcal{O}_X(K_X + \lceil A \rceil) \otimes \mathcal{O}_S(mH)) = 0 \\
&\qquad\qquad\qquad\qquad\qquad \text{for } q > 0 \text{ by Serre vanishing} \\
&\Longrightarrow H^{q+p}(X, \mathcal{O}_X(K_X + \lceil A \rceil + m f^* H)).
\end{aligned}
$$

Thus

$$
E_2^{q,p} = 0 \text{ for } q > 0 \text{ and hence } E_2^{0,p} = E_\infty^p,
$$

i.e.,

$$
H^0(S, R^p f_* \mathcal{O}_X(K_X + \lceil A \rceil) \otimes \mathcal{O}_S(mH)) = H^p(X, \mathcal{O}_X(K_X + \lceil A \rceil + m f^* H))
$$
$$
= 0 \text{ for } p > 0
$$

by the Kawamata–Viehweg vanishing theorem. On the other hand, the sheaf

$$
R^p f_* \mathcal{O}_X(K_X + \lceil A \rceil) \otimes \mathcal{O}_S(mH)
$$

is generated by global sections for $m \gg 0$ by Serre's theorem (cf. Theorem 1-1-7), and thus

$$
R^p f_* \mathcal{O}_X(K_X + \lceil A \rceil) \otimes \mathcal{O}_S(mH) = 0 \text{ as a sheaf for } p > 0.
$$

Therefore, we finally conclude that

$$
R^p f_* \mathcal{O}_X(K_X + \lceil A \rceil) = 0 \text{ for } p > 0. \qquad \square
$$

Using the relative version of the Kawamata–Viehweg vanishing theorem, we can now establish the Kawamata–Viehweg vanishing theorem for possibly singular logarithmic pairs (X, D) with only $\mathbb{Q}$-factorial and log terminal singularities.

Theorem 5-2-7 (Kawamata-Viehweg Vanishing Theorem for Singular Log Pairs). *Let (X, D) be a log pair consisting of a normal projective variety X with a boundary $\mathbb{Q}$-divisor $D = \Sigma d_i D_i$, $0 \le d_i \le 1$, such that it has only $\mathbb{Q}$-factorial and log terminal singularities. Let A be an ample $\mathbb{Q}$-divisor on X such that $K_X + D + A$ is an integral divisor. Then*

$$
H^p(X, \mathcal{O}_X(K_X + D + A)) = 0 \text{ for } p > 0.
$$

PROOF. By the definition of log terminal singularities, there exists a projective birational morphism

$$
f : V \to X
$$

from a nonsingular projective variety V such that

$$
D_V = f_*^{-1}(D) + \Sigma E_j
$$

is a divisor with only normal crossings, where the E_j are the exceptional divisors for f and in the ramification formula

$$K_Y + D_V = f^*(K_X + D) + \Sigma b_j E_j$$

we have

$$b_j > 0 \quad \forall j.$$

Moreover, since X is $\mathbb{Q}$-factorial, there exists an effective $\mathbb{Q}$-divisor $\Sigma \delta_j E_j, 0 < \delta_j \ll 1$, with its support the union of the exceptional divisors for f such that

$$f^*A - \Sigma \delta_j E_j \text{ is ample on } V.$$

(Exercise! Construct such a $\mathbb{Q}$-divisor on V. Why do we need the assumption of X being $\mathbb{Q}$-factorial?) Thus

$$K_V + D_V + f^*A - \Sigma \delta_j E_j = f^*(K_X + D + A) + \Sigma(b_j - \delta_j)E_j,$$

where

$$b_j - \delta_j > 0 \quad \forall j.$$

Since $K_X + D + A$ is an integral divisor, there exists a divisor $\Sigma e_j E_j, \quad 0 \le e_j < 1$, such that

$$K_V + D_V' + f^*A - \Sigma \delta_j E_j \text{ is an integral divisor,}$$

where

$$D_V' = D_V - \Sigma e_j E_j$$

is a boundary $\mathbb{Q}$-divisor with only normal crossings. Note that then

$$K_V + D_V' + f^*A - \Sigma \delta_j E_j = f^*(K_X + D + A) + \Sigma(b_j - \delta_j - e_j)E_j,$$

where

$$b_j - \delta_j - e_j > -1 \quad \forall j,$$

which implies

$$f_* \mathcal{O}_V(K_V + D_V' + f^*A - \Sigma \delta_j E_j) = \mathcal{O}_X(K_X + D + A).$$

Now Corollary 5-2-6 implies

$$R^k f_* \mathcal{O}_V(K_V + D_V' + f^*A - \Sigma \delta_j E_j) = 0 \text{ for } k > 0.$$

Therefore, we conclude via the Kawamata–Viehweg vanishing theorem on V and the Leray spectral sequence that

$$\begin{aligned} 0 &= H^p(V, \mathcal{O}_V(K_V + D_V' + f^*A - \Sigma \delta_j E_j)) \\ &= H^p(X, f_* \mathcal{O}_V(K_V + D_V' + f^*A - \Sigma \delta_j E_j)) \\ &= H^p(X, \mathcal{O}_X(K_X + D + A)) \qquad \text{for } p > 0. \end{aligned}$$

This completes the proof.   □

We leave the proof of the following relative version of Theorem 5-2-7 to the reader as an exercise.

Theorem 5-2-8 (Relative Version of Kawamata–Viehweg Vanishing Theorem for Singular Log Pairs). *Let (X, D) be a log pair consisting of a normal variety X with a boundary $\mathbb{Q}$-divisor $D = \Sigma d_i D_i$, $0 \leq d_i \leq 1$, such that it has only $\mathbb{Q}$-factorial and log terminal singularities. Let*

$$f : X \to S$$

be a projective morphism onto a variety S, A an f-ample $\mathbb{Q}$-divisor on X such that $K_X + D + A$ is an integral divisor. Then

$$R^p f_* \mathcal{O}_X(K_X + D + A) = 0 \text{ for } p > 0.$$

Remark 5-2-9. Theorem 5-2-8 holds in the analytic category as well as in the algebraic category. The analytic version is used, e.g., in the proof of Lemma 4-6-8. For the details of its proof (together with its precise statement), which is almost parallel to the one in the algebraic category, we refer the reader to Nakayama [2].

CHAPTER 6

Base Point Freeness of Adjoint Linear Systems

The purpose of this chapter is to discuss "**base point freeness**" of **linear systems of adjoint type**, i.e., linear systems of type

$$|K_X + B|,$$

where K_X denotes the canonical divisor of a variety X and B is a (boundary) divisor with some specific conditions depending on the situation. As is clear from the formulation, the most natural framework for linear systems of adjoing type is that of the logarithmic category discussed in Chapter 2. Our key tool is the logarithmic version of the Kodaira vanishing theorem, i.e., the Kawamata–Viehweg vanishing theorem. Our viewpoint centering on adjoint linear systems, is in the spirit of Ein–Lazarsfeld [1], which applied the **method of Kawamata–Reid–Shokurov** to solve Fujita's conjecture in dimension 3.

In Section 6-1 we demonstrate the typical arguments to show base point freeness of linear systems of adjoint type by discussing a theorem leading to a special case of Fujita's conjecture for ample divisors that are themselves base point free.

In Section 6-2 we present the **base point freeness theorem** in the form we need to establish the Mori program. Emphasis is made in our presentation so that the otherwise notoriously technical proofs are seen to be quite natural and based upon the same simple philosophical strategy, initiated by Kawamata and later developed by Benveniste, Reid, Shokurov, and others. We see that the original inspiration of Mori in the form toward the contraction theorem fuses in beautiful harmony with the arguments in the logarithmic category equipped with the powerful Kawamata–Viehweg vanishing theorem.

In Section 6-3 we complete our argument by providing a proof for the **nonvanishing theorem of Shokurov**. We emphasize that the arguments for its proof are identical to those for the proofs of the theorems of the previous sections, except at one crucial point where we observe an ingenious idea of Shokurov on how to concentrate the multiplicity of a divisor in order to create an appropriate log canonical adjoint linear system.

6.1   Relevance of Log Category to Base Point Freeness of Adjoint Linear Systems

In this section we demonstrate the relevance of the logarithmic category to base point freeness of the linear systems of adjoint type by providing a proof for the following theorem. All the key ingredients for the subsequent sections can be observed in a particularly simple form below.

Theorem 6-1-1. *Let X be a nonsingular projective variety with an ample divisor A. Let $x \in X$ be a (closed) point of X and $l \in \mathbb{N}$ a positive integer.*
Suppose there exists an effective $\mathbb{Q}$-divisor B with

$$B \sim_{\mathbb{Q}} lA$$

(i.e. $\exists n \in \mathbb{N}$ s.t. nB is integral and $nB \sim nlA$) such that the log pair (X, B) satisfies the conditions:

$$\begin{cases} (X, B) \text{ is log canonical over } U - \{x\}, \\ (X, B) \text{ is not log canonical at } x, \end{cases}$$

where U is a neighborhood of x, i.e., all the exceptional divisors whose centers lie over $U - \{x\}$ have discrepancies greater than or equal to -1 and there exists at least one exceptional divisor whose center on X is x with discrepancy less than -1.
    Then

$$|K_X + lA| \text{ is base point free at } x,$$

i.e.,

$$\exists s \in H^0(X, \mathcal{O}_X(K_X + lA)) \text{ s.t. } s(x) \neq 0.$$

Before giving a proof of this theorem, we state an interesting corollary, whose positive characteristic version is given a proof by Smith [1] using the method of tight closure.

Corollary 6-1-2 (Fujita's Conjecture for Base Point Free Ample Divisors).
Let X be a nonsingular projective variety of dimension n and A an ample divisor on X with the condition

$$(*) \qquad\qquad |A| \text{ is base point free.}$$

Then

$$|K_X + lA| \text{ is base point free for } l \geq n + 1.$$

PROOF.   Proof of Corollary 6-1-2

Fix an arbitrary (closed) point $x \in X$. Since $|A|$ is base point free and ample,

$$\Phi_{|A|} : X \to \mathbb{P}(H^0(X, \mathcal{O}_X(A)))$$

is a finite morphism onto its image. Thus the sublinear system

$$|A|_x = \{M \in |A|; x \in M\} \subset |A|$$

has no base points other than x, at least in a neighborhood U of x. Therefore, if we choose general members

$$B_1, B_2, \ldots, B_n, B_{n+1} \in |A|_x,$$

then by Bertini's Theorem,

$$B_1 + B_2 + \cdots + B_n + B_{n+1}|_{U-x}$$

has only normal crossings. Moreover, the exceptional divisor obtained by blowing up x has discrepancy at most

$$(n - 1) - (n + 1) = -2 < -1.$$

Therefore,

$$B = B_1 + B_2 + \cdots + B_n + B_{n+1} \in |(n + 1)A|$$

satisfies the conditions of Theorem 6-1-1 and thus

$$|K_X + (n + 1)A| \text{ is base point free at } x.$$

Since $x \in X$ is arbitrary, we conclude that

$$|K_X + (n + 1)A| \text{ is base point free.}$$

This completes the argument for $l = n + 1$, and the same argument works for $l > n+1$ (or argue that since $|K_X+(n+1)A|$ is base point free and since $|A|$ is base point free, $|K_X + lA|$ becomes automatically base point free for $l \geq n + 1$).   □

Now we go back to the proof of Theorem 6-1-1.

PROOF.   Proof of Theorem 6-1-1

First we explain the rough idea of the proof. We would like to construct a projective birational morphism

$$f : Y \to X$$

from another nonsingular projective variety with the exact sequence

$$0 \to \mathcal{O}_Y(f^*(K_X + lA) - F) \to \mathcal{O}_Y(f^*(K_X + lA)) \to \mathcal{O}_Y(f^*(K_X + lA))|_F \to 0,$$

where F is a reduced divisor mapping to the point x, i.e.,

$$f(F) = x.$$

If we could show that

$$H^1(Y, \mathcal{O}_Y(f^*(K_X + lA) - F)) = 0,$$

then the standard chase of the long cohomology sequence gives the assertion. In fact, the sequence

$$H^0(Y, \mathcal{O}_Y(f^*(K_X + lA))) \longrightarrow H^0(Y, \mathcal{O}_Y(f^*(K_X + lA))|_F) \longrightarrow 0$$
$$\| \qquad\qquad\qquad\qquad \| \qquad\qquad\qquad\qquad \|$$
$$H^0(X, \mathcal{O}_X(K_X + lA)) \longrightarrow H^0(F, \mathcal{O}_F) \cong \mathbb{C} \longrightarrow 0$$

shows that there must be a section

$$s \in H^0(X, \mathcal{O}_X(K_X + lA))$$

whose pullback on Y restricts to a nonzero element in $\mathbb{C}$ on F, and hence $s(x) \neq 0$.

For the construction of a resolution $f : Y \to X$, the behavior of singularities of the log pair (X, B) ($K_X + B$ being log canonical over $U - \{x\}$ and not log canonical at x) plays a major role, while the vanishing of H^1 is accomplished through the Kawamata–Viehweg vanishing theorem.

Step 1. Construction of a nice resolution $f : Y \to X$.

By Hironaka's theorem, we take a projective birational morphism

$$f : Y \to X$$

from a nonsingular projective variety Y such that

$$f_*^{-1}(B) \cup \mathrm{Exc}(f) \text{ is a divisor with normal crossings.}$$

We write the ramification formula

$$K_Y = f^*(K_X + B) + \Sigma a_i E_i - f_*^{-1}(B)$$
$$\text{where the } E_i \text{ are exceptional divisors for } f$$
$$= f^*(K_X + B) + \Sigma a_j F_j,$$
$$\text{where } \cup_j F_j = f_*^{-1}(B) \cup (\cup_l E_i).$$

The assumption of the log pair (X, B) being log canonical over $U - \{x\}$ and not being log canonical at x implies that

$$\Sigma_{a_j < -1} a_j F_j = \Sigma_{a_j < -1, f(F_j) = x} a_j F_j + \Sigma_{a_j < -1, f(F_j) \cap x = \emptyset} a_j F_j.$$

Step 2. Construction of a **"just" log canonical** divisor and a nice reduced divisor F.

From the log pair (X, B), we subtract an appropriate portion ϵB from the boundary divisor so that $(X, B - \epsilon B)$ is "just" log canonical at x, i.e.,

$$K_Y = f^*(K_X + B - \epsilon B) + \Sigma \alpha_j F_j$$
$$= f^*(K_X + B - \epsilon B) - F + \Sigma_{\alpha_j > -1} \alpha_j F_j + \Sigma_{\alpha_j \leq -1, f(F_j) \cap x = \emptyset} \alpha_j F_j,$$

where F is a reduced divisor with $f(F) = x$.

Step 3. Add an ample $\mathbb{Q}$-divisor to bring the "just" log canonical divisor of Step 2 back to the divisor of the concern.

We add ϵB to the divisor $K_X + B - \epsilon B$ in order to come back to the divisor $K_X + B$ to obtain the formula

$$K_Y + f^* \epsilon B = f^*(K_X + B) - F + \Sigma_{\alpha_j > -1} \alpha_j F_j + \Sigma_{\alpha_j \leq -1, f(F_j) \cap x = \emptyset} \alpha_j F_j,$$

i.e.,

$$f^*\epsilon B = f^*(K_X + B) - F + \Sigma_{\alpha_j > -1}\alpha_j F_j + \Sigma_{\alpha_j \le -1, f(F_j) \cap x = \emptyset}\alpha_j F_j - K_Y.$$

Note that since ϵB is ample, there exists $-\Sigma\delta_j F_j$, $\quad 0 < \delta_j \ll \epsilon$, consisting only of exceptional divisors such that $f^*\epsilon B - \Sigma\delta_j F_j$ is ample,

$$f^*\epsilon B - \Sigma\delta_j F_j \equiv f^*_t(K_X + lA) - F - \Sigma_{F_j \subset F}\delta_j F_j + \Sigma_{\alpha_j - \delta_j > -1}(\alpha_j - \delta_j)F_j$$
$$+ \Sigma_{\alpha_j - \delta_j \le -1, f(F_j) \cap x = \emptyset}(\alpha_j - \delta_j)F_j - K_Y := R.H.S.$$

Now we have

$$\lceil R.H.S. \rceil = f^*(K_X + lA) - F + E - G - K_Y,$$

where $f(G) \cap x = \emptyset$, $E = \lceil \Sigma_{\alpha_j - \delta_j > -1}(\alpha_j - \delta_j)F_j \rceil$ is exceptional for f and where E and F share no common components.

Step 4. Application of Kawamata–Viehweg vanishing theorem.

Consider the exact sequence

$$0 \to \mathcal{O}_Y(f^*(K_X + lA) + E - F - G)$$
$$\to \mathcal{O}_Y(f^*(K_X + lA) + E)$$
$$\to \mathcal{O}_Y(f^*(K_X + lA) + E)|_{F \coprod G}$$
$$\to 0.$$

The Kawamata–Viehweg vanishing theorem gives

$$0 = H^1(Y, \mathcal{O}_Y(K_Y + \lceil R.H.S. \rceil)) = H^1(Y, \mathcal{O}_Y(f^*(K_X + lA) + E - F - G)).$$

Thus the long cohomology sequence gives

$$H^0(Y, \mathcal{O}_Y(f^*(K_X + lA) + E)) \to H^0(F, \mathcal{O}_Y(E)|_F)$$
$$\oplus H^0(G, \mathcal{O}_Y(f^*(K_X + lA) + E)|_G)$$
$$\to H^1 = 0.$$

Now that

$$0 \ne \mathbb{C} = H^0(F, \mathcal{O}_F) \subset H^0(F, \mathcal{O}_Y(E)|_F),$$

we conclude that

$$\exists s \in H^0(Y, \mathcal{O}_Y(f^*(K_X + lA) + E)) \cong H^0(X, \mathcal{O}_X(K_X + lA))$$
$$\text{s.t.} \quad s(x) \ne 0.$$

This concludes the proof.   □

6.2   Base Point Freeness Theorem

In this section we prove the **base point freeness theorem** of Kawamata–Reid–Shokurov in the form needed to establish MMP and its logarithmic generalization according to Iitaka's philosophy.

Theorem 6-2-1 (Base Point Freeness Theorem). *Let X be a normal projective variety with only $\mathbb{Q}$-factorial and terminal singularities, L a nef Cartier divisor such that $A = aL - K_X$ is ample for some $a \in \mathbb{N}$. Then $|mL|$ is base point free for all sufficiently large $m \in \mathbb{N}$.*

Theorem 6-2-2 (Logarithmic Base Point Freeness Theorem). *Let (X, D) be a log pair consisting of a normal projective variety X with a boundary $\mathbb{Q}$-divisor $D = \Sigma d_i D_i$, $0 \le d_i \le 1$, such that it has only $\mathbb{Q}$-factorial and log terminal singularities, L a nef Cartier divisor such that $A = aL - (K_X + D)$ is ample for some $a \in \mathbb{N}$. Then $|mL|$ is base point free for all sufficiently large $m \in \mathbb{N}$.*

Remark 6-2-3. (i) (**Connection with the Mori program**) If we look at the cone of effective curves $\overline{NE}(X)$ as in Figure 6-2-4, we see that either the hyperplane $L^{\perp} = \{z \in N_1(X); L \cdot z = 0\}$ is away from $\overline{NE}(X)$ and hence L is ample (cf. Kleiman's criterion for ampleness), in which case the assertion is obvious, or L is the crucial nef divisor in the proof of the cone theorem, so that the hyperplane $L^{\perp}$ "touches" $\overline{NE}(X)$ and $aL - K_X$ is ample for some $a \in \mathbb{N}$. We expect the existence of the contraction morphism of this face $L^{\perp} \cap \overline{NE}(X)$ given by $|mL|$, which should be base point free. This is the motivation behind the theorem coming from the Mori Program.

Figure 6-2-4.

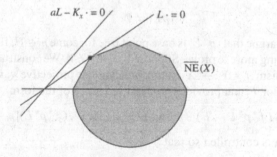

(ii) (**Connection with the linear systems of adjoint type**) We can express the divisor in the form of adjoint type

$$mL = K_X + A + (m - a)L = K_X + B,$$

where

$$B = A + (m - a)L \text{ is ample.}$$

From this point of view, the arguments for base point freeness of the linear system $|mL|$ follow almost the identical path of those of the previous section.

Before going into the discussion of base point freeness of the linear system $|mL|$ $(m \gg 0)$, we have to face the following more basic question:

Is the linear system $|mL|$ nonempty (nonvanishing), i.e., $|mL| \neq \emptyset$ for $m \gg 0$?

This fundamental question is affirmatively answered by Shokurov [1] in a slightly generalized form as the nonvanishing theorem below.

Theorem 6-2-5 (Nonvanishing Theorem of Shokurov). *Let (X, D) be a log pair consisting of a normal projective variety X with a boundary $\mathbb{Q}$-divisor $D = \Sigma d_i D_i$, $0 \leq d_i \leq 1$, such that it has only $\mathbb{Q}$-factorial and log terminal singularities, G an effective integral Cartier divisor, and L a nef Cartier divisor such that $A = aL + G - (K_X + D)$ is ample for some $a \in \mathbb{N}$. Then $|mL + G| \neq \emptyset$ for all sufficiently large $m \in \mathbb{N}$.*

We note not only that the nonvanishing theorem with $G = 0$ implies $|mL| \neq \emptyset$ under the conditions of the base point freeness theorem but also that the slight generalization of taking an effective divisor G into account will play a crucial role in the inductive argument of the proof.

In what follows we present the proof of the logarithmic base point freeness theorem (which of course implies theorem 6-2-1 as a special case) based upon the nonvanishing theorem, whose proof is postponed until the next section.

Before going into detail, we describe the outline of the proof.

First, by the nonvanishing theorem we know for $p \gg 0$ (which we choose to be a prime number) that

$$|pL| \neq \emptyset.$$

We would like to argue that $|p^n L|$ is base point free for some $n \in \mathbb{N}$. If $\mathrm{Bs}|pL| = \emptyset$, then there is nothing more to prove. Suppose $\mathrm{Bs}|pL| \neq \emptyset$. We construct a projective birational morphism $f : Y \to X$ from a nonsingular projective variety Y with a nonsingular divisor F that provides an exact sequence of the form

$$0 \to \mathcal{O}_Y(f^* p^n L - F) \to \mathcal{O}_Y(f^* p^n L) \to \mathcal{O}_Y(f^* p^n L)|_F \to 0.$$

Our construction is controlled so that

$$f(F) \subset \mathrm{Bs}|pL|,$$
$$H^1(Y, \mathcal{O}_Y(f^* p^n L - F)) = 0,$$
$$H^0(F, \mathcal{O}_Y(f^* p^n L)|_F) \neq 0.$$

These three conditions would imply $\mathrm{Bs}|pL| \supset f(F) \not\subset \mathrm{Bs}|p^n L|$ and hence

$$\mathrm{Bs}|p^n L| \overset{\neq}{\subset} \mathrm{Bs}|pL|.$$

If $\mathrm{Bs}|p^n L| \neq \emptyset$, then we repeat the argument. By noetherian induction, we finally conclude that

$$|p^n L| \text{ is base point free for some } n \in \mathbb{N}.$$

By the same argument starting with a different prime q we also conclude that

$$|q^{n'} L| \text{ is base point free for some } n' \in \mathbb{N}.$$

Finally, noting that for $m \gg 0$,

$$m = k p^n + l q^{n'} \text{ for some } k, l \in \mathbb{N} \cup \{0\},$$

we obtain the assertion of the theorem.

In the key part of the argument, which reduces the base locus by noetherian induction,

$H^1 = 0$ is achieved by the Kawamata–Viehweg vanishing theorem,

while

$H^0 \neq 0$ is achieved by the nonvanishing theorem.

Now we discuss the details.

Step 0. Reduction to the case with all the coefficients $d_i < 1$.

We may assume that all the coefficients d_i of the boundary $\mathbb{Q}$-divisor $D = \Sigma d_i D_i$ are strictly smaller than 1, i.e.,

$$d_i < 1 \quad \forall i.$$

In fact, since the ampleness is an open condition (cf. Kleiman's criterion for ampleness Theorem 1-2-5),

$$aL - (K_X + D) \text{ being ample implies } aL - (K_X + (1 - \epsilon)D) \text{ being ample}$$

for sufficiently small $\epsilon > 0$. Note that the $\mathbb{Q}$-factoriality of X guarantees that $K_X + (1 - \epsilon)D$ remains $\mathbb{Q}$-Cartier and that $(X, (1 - \epsilon)D)$ has only log terminal singularities.

By applying the nonvanishing theorem,

$$|pL| \neq \emptyset \text{ for some prime number } p \gg 0.$$

We claim that

$$\mathrm{Bs}|p^n L| = \emptyset \text{ for some } n \in \mathbb{N}.$$

Step 1. Construction of a nice resolution $f : Y \to X$.

By Hironaka's theorem, we take a projective birational morphism

$$f : Y \to X$$

from a nonsingular projective variety Y such that

$$\cup F_j = f_*^{-1}(D) \cup \mathrm{Exc}(f) \cup f^{-1}\mathrm{Bs}|pL|$$

is a divisor with only *simple* normal crossings. Here we require the crossings to be *simple*, i.e., all the irreducible components are required to be nonsingular as well as crossing normally with each other, in view of the future application of the nonvanishing theorem.

We write the ramification formula

$$K_Y = f^*(K_X + D) + \Sigma a_i E_i - f_*^{-1}(D)$$

where the E_i are exceptional divisors

$$= f^*(K_X + D) + \Sigma a_j F_j$$

where $a_j > -1 \quad \forall j$, thanks to Step 0

cf. logarithmic ramification formula theorem 2-1-9. We also take $f : Y \to X$ such that

$$|f^* pL| = f^*|pL| = |M| + \Sigma r_j F_j \text{ with } Bs|M| = \emptyset.$$

Step 2. Construction of a "just" log canonical divisor and a nice reduced divisor F.

We would like to make the log pair $(X, D + c'(pL))$ "just" log canonical. That is to say, by looking at the ramification formula

$$K_Y = f^*(K_X + D + c'(pL)) - c'M + \Sigma(-c' \cdot r_j + a_j)F_j$$

we would like to have

$$- c' \cdot r_j + a_j \geq -1 \quad \forall j, \text{ and}$$

$$- c' \cdot r_j + a_j = -1, \text{ for some } j.$$

Therefore, we choose

$$c' = \min_{r_j \neq 0} \left\{ \frac{a_j + 1}{r_j} \right\}.$$

Observe that the number c' is the **log canonical threshold** of the pair (X, D) with respect to the linear system $|pL|$ (cf. Chapters 13 and 14).

Note also that the part $-c'M$ can cause no trouble on the condition for the log pair $(X, D + c'(pL))$ to be log canonical and be ignored, since $|M|$ has no base points and thus can be split into effective $\mathbb{Q}$-divisors with only normal crossings (not only with each other but also with the F_j) and with very small coefficients. That is to say, we may choose members $M_k \in |lM|$ for $l \gg 0$ such that

$$c'M \sim_{\mathbb{Q}} \Sigma c_k M_k,$$

where $0 < c_k \ll 1$ and where $(\cup M_k) \cup (\cup F_j)$ is a divisor with only normal crossings.

Step 3. Add an ample $\mathbb{Q}$-divisor to bring the "just" log canonical divisor of Step 2 to the divisor of the concern.

Now add an ample divisor $aL - (K_X + D) + (p^n - a - c'p)L$ to $(K_X + D) + c'(pL)$, in order to come back to the divisor of the concern $p^n L$ to obtain the formula

$$K_Y + f^*(aL - (K_X + D) + (p^n - a - c'p)L)$$

$$= f^*(p^n L) - c'M + \Sigma(-c' \cdot r_j + a_j)F_j.$$

Therefore, we obtain

$$f^*(aL - (K_X + D) + (p^n - a - c'p)L) \equiv f^*(p^n L) - \Sigma c_k M_k + \Sigma(-c' \cdot r_j + a_j)F_j - K_Y.$$

As before, since $aL - (K_X + D)$ is ample, there exists $\Sigma \delta_j F_j$, $0 \leq \delta_j \ll 1$, consisting only of exceptional divisors such that

$$f^*(aL - (K_X + D)) - \Sigma \delta_j F_j$$

is ample and hence so is

$$f^*(aL - (K_X + D) + (p^n - a - c'p)L) - \Sigma \delta_j F_j$$
$$\equiv f^*(p^n L) - \Sigma c_k M_k + \Sigma(-c' \cdot r_j + a_j - \delta_j)F_j - K_Y.$$

Now we have

$$\lceil \text{R.H.S.} \rceil = f^*(p^n L) - F + E - K_Y,$$

where F is a reduced divisor with

$$f(F) \subset \text{Bs}|pL| \text{ and } E \text{ is exceptional for } f.$$

Step 4. Application of the Kawamata Viehweg vanishing theorem and the nonvanishing theorem of Shokurov.

Consider the exact sequence

$$0 \to \mathcal{O}_Y(f^*(p^n L) - F + E) \to \mathcal{O}_Y(f^*(p^n L) + E) \to \mathcal{O}_Y(f^*(p^n L) + E)|_F \to 0.$$

The Kawamata–Viehweg vanishing theorem gives

$$0 = H^1(Y, \mathcal{O}_Y(K_Y + \lceil \text{R.H.S.} \rceil)) = H^1(Y, \mathcal{O}_Y(f^*(p^n L) - F + E)).$$

Thus the long cohomology sequence gives a surjection

$$H^0(X, \mathcal{O}_X(p^n L)) \cong H^0(Y, \mathcal{O}_Y(f^*(p^n L) + E))$$
$$\to H^0(F, \mathcal{O}_Y(f^*(p^n L) + E)|_F) \to H^1 = 0.$$

We apply the nonvanishing theorem to conclude that

$$H^0(F, f^*(p^n L) + E|_F) \neq 0,$$

which would imply

$$f(F) \not\subset \text{Bs}|p^n L| \text{ and thus } \text{Bs}|p^n L| \stackrel{\neq}{\subset} \text{Bs}|pL|.$$

By noetherian induction (i.e., by repetition of the argument above we show that $\text{Bs}|p^{n'} L| \stackrel{\neq}{\subset} \text{Bs}|p^n L|$ for some $n' > n$ and so forth), we conclude that

$$\text{Bs}|p^n L| = \emptyset \text{ for some } n \in \mathbb{N}.$$

We remark that there is one small point in the above argument where we cheated. Let us check the conditions for applying the nonvanishing theorem.

Pretend that F is irreducible. Then F is a nonsingular projective variety with a nef divisor $f^*L|_F$, an effective divisor $E|_F$ such that

$$\text{R.H.S.}|_F = p^n f^*L|_F + E|_F - (K_F + B_F) \text{ is ample,}$$

where

$$B_F = (\lceil \text{R.H.S.} \rceil - \text{R.H.S.})|_F$$

is a divisor with only normal crossings and hence (F, B_F) is a log pair with only $\mathbb{Q}$-factorial and log terminal singularities. Thus we can apply the nonvanishing theorem to conclude (after taking a bigger n if necessary) that

$$H^0(F, f^*(p^n L) + E|_F) \neq 0.$$

We note that the above argument is valid only assuming that F is irreducible, which is not guaranteed if we stick to the choice of c' that we took. The next step takes care of this small technical problem.

Step 5. Tie Breaking.

When we take $\Sigma \delta_j F_j$ in Step 3 with the condition

$$f^*(aL - (K_X + D)) - \Sigma \delta_j F_j \text{ is ample,}$$

we perturb δ_j a little so that the minimum

$$c = \min_{r_j \neq 0} \left\{ \frac{a_j + 1 - \delta_j}{r_j} \right\}$$

is attained only at one unique irreducible component F_o. (This can be done, since the perturbation will not destroy the ampleness, which is an open condition.) Now we carry the whole argument above with c' replaced by c. This subtle change leaves the argument intact, except that now $F = F_o$ is irreducible, making the application of the nonvanishing theorem in Step 4 legitimate.

This completes the proof of the (logarithmic) base point freeness theorem, modulo the nonvanishing theorem of Shokurov, whose proof will be given in the next section.   □

Corollary 6-2-6 (Abundance Theorem for Minimal Models of General Type).
Let X be a minimal model of general type (of arbitrary dimension). Then the abundance conjecture holds for X, i.e., $|mK_X|$ is base point free for sufficiently divisible and large $m \in \mathbb{N}$.

PROOF.   Take an ample divisor A on X. Let e be the index of K_X such that eK_X is a Cartier divisor. We claim that there exists $l \in \mathbb{N}$ such that $|leK_X - A| \neq \emptyset$. This follows immediately from the following Kodaira's lemma. Now, by the claim, there exists a member $D \in |leK_X - A|$. Then for a sufficiently small rational number $\epsilon > 0$ the pair $(X, \epsilon D)$ has only log terminal singularities, eK_X is a nef Cartier divisor, and $2eK_X - (K_X + \epsilon D) \equiv (2e - 1 - le\epsilon)K_X + \epsilon A$ is ample. Thus by the Logarithmic base point freeness theorem, we conclude that $|m'eK_X|$ is base point free for all sufficiently large $m' \in \mathbb{N}$.   □

Lemma 6-2-7 (Kodaira's Lemma).   *Let M be a Cartier divisor on a projective variety X with $\kappa(X, M) = \dim X$. Then for any divisor A on X there exists $l \in \mathbb{N}$ such that*

$$|lM - A| \neq \emptyset.$$

PROOF.   Proof of Kodaira's Lemma

It suffices to prove the lemma when A is very ample. (Exercise! Why?) Take a general member $Y \in |A|$. Consider the exact sequence

$$0 \to H^0(X, \mathcal{O}_X(lM - A)) \to H^0(X, \mathcal{O}_X(lM)) \to H^0(Y, \mathcal{O}_Y(lM)).$$

Recall that there exist positive numbers $\alpha_X, \alpha_Y \in \mathbb{Q}$ and $m_o \in \mathbb{N}$ such that

$$h^0(X, mm_o M) \geq \alpha_X \cdot m^{\dim X} \text{ for } m \gg 0 \quad \text{(cf. Proposition 1-5-3),}$$
$$h^0(Y, mm_o M|_Y) \leq \alpha_Y \cdot m^{\dim X - 1} \text{ for } m \gg 0.$$

Thus if we set $l = mm_o$ for $m \gg 0$, then

$$H^0(X, \mathcal{O}_X(lM - A)) \neq 0.$$

This completes the proof of Kodaira's lemma and thus the proof of Corollary 6-2-6.

6.3   Nonvanishing Theorem of Shokurov

In this section we complete our line of argument by giving a proof of the **nonvanishing theorem of Shokurov**. Its argument is almost identical to that of the base point freeness theorem, except that at one crucial point we observe an ingenious idea of Shokurov [1] on how to concentrate the multiplicity of a divisor in order to create an appropriate log canonical adjoint linear system.

Theorem 6-3-1=Theorem 6-2-5 (Nonvanishing Theorem of Shokurov). *Let (X, D) be a log pair consisting of a normal projective variety X with a boundary $\mathbb{Q}$-divisor $D = \Sigma d_i D_i$, $0 \leq d_i \leq 1$, such that it has only $\mathbb{Q}$-factorial and log terminal singularities, G an effective integral Cartier divisor, and L a nef Cartier divisor such that $A = aL + G - (K_X + D)$ is ample for some $a \in \mathbb{N}$. Then $|mL + G| \neq \emptyset$ for all sufficiently large $m \in \mathbb{N}$.*

PROOF.   We prove the theorem separating the analysis into the two cases: $L \equiv 0$ and $L \not\equiv 0$.

> Case: $L \equiv 0$.

By the definition of log terminal singularities, there exists a projective birational morphism

$$f : Y \to X$$

from a nonsingular projective variety Y such that

$$f_*^{-1}(D) \cup \text{Exc}(f)$$

is a divisor with only normal crossings and the ramification formula reads

$$K_Y = f^*(K_X + D) + \Sigma a_i E_i \quad a_i > -1 \quad \forall E_i \text{ exceptional for } f.$$

Note that there exists $\Sigma \delta_i E_i$, $0 < \delta_i \ll 1$, consisting only of the exceptional divisors such that

$$f^* A - \Sigma \delta_i E_i \text{ is ample on } Y.$$

Set

$$D_Y = f_*^{-1}(D) + \Sigma_{0 > a_i > -1}(-a_i)E_i + \Sigma_{a_i \geq 0}(\lceil a_i \rceil - a_i)E_i + \Sigma \delta_i E_i,$$
$$G_Y = f^* G + \Sigma_{a_i \geq 0} \lceil a_i \rceil E_i.$$

Then the log pair (Y, D_Y) has only log terminal singularities while Y is nonsingular, G_Y is an effective integral Cartier divisor, and $L_Y = f^* L$ is a nef Cartier divisor such that $a L_Y + G_Y - (K_Y + D_Y) = f^*(A) - \Sigma \delta_i E_i$ is ample.

Moreover, we have

$$H^0(Y, m L_Y + G_Y) = H^0(Y, f^*(mL + G) + \Sigma_{a_i \geq 0} \lceil a_i \rceil E_i) \cong H^0(X, mL + G),$$

since $\Sigma_{a_i \geq 0} \lceil a_i \rceil E_i$ is effective and exceptional for f.

Thus we have to show the assertion only for (Y, D_Y), G_Y, L_Y.

Therefore, from the beginning, we may assume that X is nonsingular.

Under the case assumption of $L \equiv 0$,

$$mL + G - (K_X + D) \equiv aL + G - (K_X + D) \text{ is ample } \quad \forall m \in \mathbb{Z}.$$

Therefore, the Kawamata–Viehweg vanishing theorem gives

$$0 = H^p(X, \mathcal{O}_X(K_X + D + (mL + G - (K_X + D)))) = H^p(X, \mathcal{O}_X(mL + G)) \text{ for } p > 0.$$

Since X is nonsingular, we may use the Riemann–Roch theorem to compute

$$
\begin{aligned}
h^0(X, \mathcal{O}_X(mL + G)) &= \chi(\mathcal{O}_X(mL + G)) \\
&= \chi(\mathcal{O}_X(G)) \\
&\qquad \text{(The Euler charateristic depends only on the numerical class)} \\
&= h^0(X, \mathcal{O}_X(G)) \neq 0.
\end{aligned}
$$

Case: $L \not\equiv 0$.

Step 0. Reduction to the case with all the coefficients $d_i < 1$.

As in Step 0 of the proof for the base point freeness theorem, we may assume that all the coefficients d_i are less than 1.

Step 1. Construction of a nice resolution $f : Y \to X$.

Claim 6-3-2 (Concentration of Multiplicity). *Fix a point*

$$x_o \in X - (\text{Sing}(X) \cup D).$$

Then there exists $q \in \mathbb{N}$ $(q \geq a)$ and an effective $\mathbb{Q}$-divisor H such that

$$H \sim_{\mathbb{Q}} qL + G - (K_X + D)$$

and

$$\text{mult}_{x_o} H \geq n + 1,$$

where $n = \dim X$.

(Note that by taking q large, we may take H with $\text{mult}_{x_o} H$ *arbitrarily high. The number* $n + 1$ *in the inequality is chosen for the convention of the proof.)*

Postponing the proof of the claim until the end of this section, we will finish the proof of the nonvanishing theorem.

By Hironaka's theorem, we take a projective birational morphism

$$f : Y \to X$$

from a nonsingular projective variety Y such that

$$\cup F_j = f_*^{-1}(D) \cup \text{Exc}(f) \cup f^{-1}G \cup f^{-1}H$$

is a divisor with only *simple* normal crossings. We choose $f : Y \to X$ in such a way that it factors through $f_{x_o} : Y_{x_o} \to X$, which is the blowup of X at the point x_o.

We write the ramification formula

$$
\begin{aligned}
K_Y &= f^*(K_X + D) + \Sigma a_i E_i - f_*^{-1}(D) &&\text{where the } E_i \text{ are exceptional} \\
&= f^*(K_X + D) + \Sigma a_j F_j &&\text{where } a_j > -1 \ \ \forall j, \text{ thanks to Step 0} \\
&= f^*(K_X + D - G) + \Sigma \alpha_j F_j,
\end{aligned}
$$

where $\alpha_j = a_j + g_j > -1 \forall j$, since $a_j > -1 \forall j$ and since $f^*G = \Sigma g_j F_j$ is effective. Also, we write

$$f^*H = \Sigma r_j F_j.$$

Step 2. Construction of a "just" log canonical divisor and a nice reduced divisor F.

We would like to make $K_X + D - G + c'H$ "just" log canonical. That is to say, by looking at the ramification formula

$$K_Y = f^*(K_X + D - G + c'H) + \Sigma(-c' \cdot r_j + \alpha_j)F_j$$

we would like to have

$$
\begin{aligned}
-c' \cdot r_j + b_j &\geq -1 \ \ \forall j, \\
-c' \cdot r_j + b_j &= -1, \text{ for some } j.
\end{aligned}
$$

Therefore, we choose

$$c' = \min_{r_j \neq 0} \left\{ \frac{\alpha_j + 1}{r_j} \right\}.$$

Note that if we denote the strict transform of the exceptional divisor of the blowup f_{x_o} by F_{x_o}, then

$$\alpha_{x_o} = a_{x_o} = n - 1 \quad \text{and} \quad r_{x_o} \geq n + 1 \text{ by Claim 6-3-2.}$$

Thus

$$c' = \min_{r_j \neq 0} \left\{ \frac{\alpha_j + 1}{r_j} \right\} \leq \frac{\alpha_{x_o} + 1}{r_{x_o}} \leq \frac{n}{n + 1} < \frac{n + \frac{1}{2}}{n + 1} < 1.$$

Step 3. Add an ample divisor to bring the "just" log canonical divisor of Step 2 to the divisor of the concern.

Now add $(m - a)L + aL + G - (K_X + D) - c'H$ to $K_X + D - G + c'H$ to obtain

$$K_Y + f^*\{(m-a)L + aL + G - (K_X+D) - c'H\} \equiv f^*(mL+G) + \Sigma(-c' \cdot r_j + a_j)F_j.$$

Note that

$$(m - a)L + aL + G - (K_X + D) - c'H$$
$$\equiv (1 - c')\{(q - a)L + aL + G - (K_X + D)\} + (m - q)L$$
$$\equiv \left(1 - \frac{n + \frac{1}{2}}{n + 1}\right)\{(q - a)L + aL + G - (K_X + D)\}$$
$$+ \left(\frac{n + \frac{1}{2}}{n + 1} - c'\right)\{(q - a)L + aL + G - (K_X + D)\} + (m - q)L$$

is ample for $m \geq q$, since

$$1 - \frac{n + \frac{1}{2}}{n + 1} > 0, \quad \frac{n + \frac{1}{2}}{n + 1} - c' \geq 0,$$

since L is nef and since $aL + G - (K_X + D)$ is ample by assumption.

Thus we have

$$f^*\{(m - a)L + aL + G - (K_X + D) - c'H\} \equiv f^*(mL) + \Sigma(-c' \cdot r_j + \alpha_j)F_j - K_Y.$$

As before, there exists $\Sigma\delta_j F_j$, $0 < \delta_j \ll 1$, consisting of only the exceptional divisors such that

$$f^* \left(\left(1 - \frac{n + \frac{1}{2}}{n + 1}\right)\{(q - a)L + aL + G - (K_X + D)\}\right) - \Sigma\delta_j F_j$$

is ample and hence so is

$$f^*\{(m - a)L + aL + G - (K_X + D) + c'H\} - \Sigma\delta_j F_j$$
$$\equiv f^* \left(\left(1 - \frac{n + \frac{1}{2}}{n + 1}\right)\{(q - a)L + aL + G - (K_X + D)\}\right) - \Sigma\delta_j F_j$$
$$+ f^* \left(\left(\frac{n + \frac{1}{2}}{n + 1} - c'\right)\{(q - a)L + aL + G - (K_X + D)\} + (m - q)L\right)$$
$$\equiv f^*(mL) + \Sigma(-c' \cdot r_j + \alpha_j - \delta_j)F_j - K_Y = \text{R.H.S.}$$

Now we have

$$\lceil \text{R.H.S.} \rceil = f^*(mL) - F + E - K_Y,$$

where F is a reduced divisor with

$$F, E \subset \cup F_j \text{ and } E \text{ and } F \text{ share no common components.}$$

Step 4. Application of the Kawamata–Viehweg vanishing theorem and the nonvanishing in dimension one less by inductional hypothesis.

Consider the exact sequence

$$0 \to \mathcal{O}_Y(f^*(mL + G) - F + E) \to \mathcal{O}_Y(f^*(mL + G) + E)$$
$$\to \mathcal{O}_Y(f^*(mL + G) + E)|_F \to 0.$$

The Kawamata–Viehweg vanishing theorem gives

$$0 = H^1(Y, \mathcal{O}_Y(K_Y + \lceil \text{R.H.S.} \rceil)) = H^1(Y, \mathcal{O}_Y(f^*(mL) - F + E)).$$

Thus the long cohomology sequence gives a surjection

$$H^0(Y, \mathcal{O}_Y(f^*(mL) + E)) \to H^0(F, \mathcal{O}_Y(f^*(mL) + E)|_F) \to H^1 = 0.$$

We try to apply the statement of the nonvanishing theorem (in dimension one less) to F, which has dimension one less, to conclude

$$H^0(F, \mathcal{O}_Y(f^*(mL) + E)|_F) \neq 0,$$

which would imply

$$H^0(Y, \mathcal{O}_Y(f^*(mL) + E)) \neq 0.$$

Observe that

$$f^*(mL) + E = f^*(mL) + \Sigma_{-c' \cdot r_j + \alpha_j - \delta_j > -1} \lceil -c' \cdot r_j + \alpha_j - \delta_j \rceil F_j$$
$$\leq f^*(mL) + \Sigma_{\alpha_j = a_j + g_j > -1} \lceil \alpha_j \rceil F_j$$
$$\leq f^*(mL) + \Sigma g_j F_j + \Sigma_{a_j > 0} \lceil a_j \rceil F_j.$$

Since $\Sigma_{a_j > 0} \lceil a_j \rceil F_j$ is effective and exceptional for f, we finally conclude that

$$0 \neq H^0(Y, \mathcal{O}_Y(f^*(mL) + E))$$
$$\subset H^0(Y, \mathcal{O}_Y(f^*(mL) + \Sigma g_j F_j + \Sigma_{a_j > 0} \lceil a_j \rceil F_j)) \cong H^0(X, \mathcal{O}_X(mL + G)).$$

We check the conditions to apply the nonvanishing theorem to F, assuming that F is irreducible. Then F is a nonsingular projective variety with a nef divisor $f^*L|_F$ and an effective divisor $E|_F$ such that

$$\text{R.H.S.}|_F = mf^*L|_F + E|_F - (K_F + B_F) \text{ is ample,}$$

where

$$B_F = (\lceil \text{R.H.S.} \rceil - \text{R.H.S.})|_F.$$

If $\lceil \text{R.H.S.} \rceil - \text{R.H.S.}$ does not contain F, then B_F is a well-defined boundary $\mathbb{Q}$-divisor with only normal crossings, such that (F, B_F) is a log pair with only log terminal singularities.

Thus we can apply the nonvanishing theorem to F, which has dimension one less, by inductional hypothesis and conclude (after taking a bigger m if necessary) that

$$H^0(F, \mathcal{O}_F(f^*(mL) + E)|_F) \neq 0.$$

We note that the above argument is valid only if we assume that F is irreducible and that $\lceil$R.H.S.$\rceil -$ R.H.S. does not contain F. The next step takes care of these technical requirements.

Step 5. Tie Breaking.

When we take $\Sigma \delta_j F_j$ in Step 3 with the condition

$$f^*(aL + G - (K_X + D)) - \Sigma \delta_j F_j \text{ is ample,}$$

we perturb δ_j a little so that the minimum

$$c = \min_{r_j \neq 0} \left\{ \frac{b_j + 1 - \delta_j}{r_j} \right\}$$

is attained only at one unique irreducible component F_o. (This can be done, since the perturbation will not destroy the ampleness, which is an open condition.) Now we carry the whole argument above with c' replaced by c. This subtle change leaves the argument intact, except that now $F = F_o$ is irreducible and $-cr_o + \alpha_o - \delta_o = -1$ and hence $\lceil$R.H.S.$\rceil -$R.H.S. does not contain F_o, making the inductive application of the nonvanishing theorem in Step 4 legitimate.

PROOF.   Proof of Claim 6-3-2

Now the only remaining thing to be proved is Claim 6-3-2.

Since

$$L \not\equiv 0 \text{ and is nef}$$

and since

$$aL + G - (K_X + D) \text{ is ample,}$$

we have

$$L \cdot (aL + G - (K_X + D))^{n-1} > 0.$$

Therefore, for a sufficiently big $q \in \mathbb{N}$, $\quad q \geq a$, we have

$$(qL + G - (K_X + D))^n \geq (q - a)L \cdot (aL + G - (K_X + D))^{n-1} > (n + 1)^n.$$

Now take sufficiently divisible and big $l \in \mathbb{N}$ such that

$$l(qL + G - (K_X + D))$$

is an integral Cartier divisor, and

$$H^p(X, \mathcal{O}_X(l(qL + G - (K_X + D)))) = 0 \text{ for } p > 0$$

by Serre's vanishing. Thus we compute

$$h^0(X, \mathcal{O}_X(l(qL + G - (K_X + D)))) = \chi(\mathcal{O}_X(l(qL + G - (K_X + D))))$$
$$= \frac{\{(qL + G - (K_X + D))\}^n}{n!} l^n + (\text{ lower terms in } l)$$
$$> \frac{(n + 1)^n}{n!} l^n.$$

On the other hand, considering the power series expansion of the local defining equation for a member

$$M \in |l(qL + G - (K_X + D))|$$

at x_o, which is a nonsingular point of X by choice, we have the following number of conditions for M to have multiplicity greater than or equal to $l(n + 1)$ at x_o:

$$\#\{\text{monomials of degree } < l(n + 1) \text{ in } n \text{ variables}\}$$

$$= \begin{pmatrix} l(n + 1) - 1 + n \\ n \end{pmatrix}$$

$$= \frac{(n + 1)^n}{n!} l^n + (\text{ lower terms in } l).$$

Therefore, for sufficiently large $l \in \mathbb{N}$,

$$h^0(X, l(qL + G - (K_X + D))) > \#\text{of the conditions.}$$

Therefore,

$$\exists M \in |l(qL + G - (K_X + D))| \text{ such that } \text{mult}_{x_o} M \geq l(n + 1).$$

Thus

$$H = \frac{M}{l} \sim_{\mathbb{Q}} qL + G - (K_X + D) \text{ has mult}_{x_o} H \geq n + 1.$$

This completes the proof of Claim 6-3-2 and hence that of the nonvanishing theorem. $\square$

CHAPTER 7

Cone Theorem

The purpose of this chapter is to give a proof of the **cone theorem**, providing the necessary key ingredients: the **rationality theorem** and **boundedness of the denominator**. We prove these two theorems by applying the same cohomological arguments developed for the proofs of the base point freeness theorem and the non-vanishing theorem of the previous chapter. We note that our point of view for discussing the behavior of divisors following Kawamata–Reid–Shokurov–Kollár is "dual" to the original approach of Mori, who discusses the behavior of curves directly via deformation theory. We will study Mori's argument in Chapter 10.

In Section 7-1 we establish the rationality theorem together with boundedness of the denominator. Their proofs follow an almost identical path to those of the proofs of the base point freeness theorem and the nonvanishing theorem in Chapter 6.

In Section 7-2 we give the precise statement of the cone theorem and briefly review how the rationality theorem and boundedness of the denominator provide its proof. The argument is verbatim the same as given in Section 1-3 in dimension 2.

7.1 Rationality Theorem and Boundedness of the Denominator

Recall that there is an intersection pairing between the line bundles and the 1-cycles on a projective variety X

$$\mathrm{Pic}\,(X) \times Z_1(X) \to \mathbb{Z},$$

and hence between the Cartier divisors and the 1-cycles

$$\mathrm{Div}\,(X) \times Z_1(X) \to \mathbb{Z}.$$

We extend this to a perfect pairing over $\mathbb{R}$

$$N^1(X) \times N_1(X) \to \mathbb{R}$$

where

$$N^1(X):=\left\{ \text{Pic}\,(X)/\equiv \right\} \otimes_{\mathbb{Z}} \mathbb{R} = \left\{ \text{Div}\,(X)/\equiv \right\} \otimes_{\mathbb{Z}} \mathbb{R},$$
$$N_1(X):=\left\{ Z_1(X)/\equiv \right\} \otimes_{\mathbb{Z}} \mathbb{R}$$

with $\equiv$ denoting the numerical equivalence. The vector spaces $N^1(X)$ and its dual $N_1(X)$ are finite dimensional, since $\text{Pic}\,(X)/\ker \tau$ maps subjectively onto $\text{Pic}\,(X)/\equiv$, where $\tau : \text{Pic}\,(X) \to H^2(X, \mathbb{R})$ is the natural map into a finite dimensional vector space over $\mathbb{R}$.

We set

NE(X):=the convex cone generated by effective 1-cycles on X

$\overline{\text{NE}}(X)$:=the closure of NE(X) in $N_1(X)$.

Recall also that an element $L \in N^1(X)$ is said to be nef if

$$L \cdot C \geq 0 \text{ for any curve on } X,$$

or equivalently if

$$L \geq 0 \text{ on NE}(X)(\Longleftrightarrow L \geq 0 \text{ on } \overline{\text{NE}}(X)).$$

Theorem 7-1-1 (Rationality Theorem and Boundedness of the Denominator).
Let X be a normal projective variety with only $\mathbb{Q}$-factorial and terminal singularities, H an ample Cartier divisor on X. Suppose K_X is not nef. Then the positive number

$$r := \sup\{t \in \mathbb{R}; H + tK_X \text{ is nef}\}$$

is a rational number.

Furthermore, expressing

$$\frac{r}{e} = \frac{p}{q} \text{ with } p, q \in \mathbb{N} \quad \gcd(p, q) = 1,$$

where

$$e := \min\{i \in \mathbb{N}; iK_X \text{ is Cartier}\},$$

we have

$$q \leq e(\dim X + 1).$$

Theorem 7-1-2 (Logarithmic Rationality Theorem and Boundedness of the Denominator). *Let X be a normal projective variety with a boundary $\mathbb{Q}$-divisor $D = \Sigma d_i D_i$, $0 \leq d_i \leq 1$, such that the pair (X, D) has only $\mathbb{Q}$-factorial and log terminal singularities, H an ample Cartier divisor on X. Suppose $K_X + D$ is not nef. Then the positive number*

$$r := \sup\{t \in \mathbb{R}; H + t(K_X + D) \text{ is nef}\}$$

is a rational number.
Furthermore, expressing

$$\frac{r}{e} = \frac{p}{q} \text{ with } p, q \in \mathbb{N} \quad \gcd(p, q) = 1,$$

where

$$e := \min\{i \in \mathbb{N}; i(K_X + D) \text{ is Cartier}\},$$

we have

$$q \le e(\dim X + 1).$$

The rest of the section is devoted to the proof of the logarithmic rationality theorem and boundedness of the denominator (which of course imply the usual rationality theorem and boundedness of the denominator).

PROOF.   Proof of Theorem 7-1-2
We will derive a contradiction assuming either

(1) $r \notin \mathbb{Q}$, or
(2) $r \in \mathbb{Q}$ but $q > e(\dim X + 1)$.

Actually, we show that under the assumption (1) or (2)

$$\exists x, y \in \mathbb{N} \quad \text{s.t.}$$
$$0 < ye - xr(< 1) \quad \text{and} \quad Bs|xH + ye(K_X + D)| = \emptyset,$$

which would imply

$$xH + ye(K_X + D) \text{ is nef and } r < \frac{ye}{x},$$

a contradiction!
Step $-\infty$. We may assume that H is very ample.
In fact, take $c, d \in \mathbb{N}$ satisfying the conditions

$$e < cr(\text{and } \gcd(cd, q) = 1 \text{ if } r \in \mathbb{Q})$$

so large that

$$H' = d(cH + e(K_X + D)) \text{ is very ample.}$$

Then using H' instead of H, we have

$$r' := \sup\{t \in \mathbb{R}; H' + t(K_X + D) \text{ is nef}\}$$

and

$$\frac{r'}{e} = cd \cdot \frac{r}{e} - d.$$

Therefore,

$$r \in \mathbb{Q} \Longleftrightarrow r' \in \mathbb{Q},$$

and if $r \in \mathbb{Q}$ or equivalently if $r' \in \mathbb{Q}$, then

$$q = q'.$$

Thus we may argue replacing H with H'.

Claim 7-1-3. Under the assumption (1) or (2), there exist $y_o \in \mathbb{N}$ and sufficiently small $\epsilon' > 0$ such that

$$\text{for any } x, y \in \mathbb{N} \text{ with } y \geq y_o \text{ and } 0 < ye - xr < 1 - \epsilon'$$

we have

$$|xH + ye(K_X + D)| \neq \emptyset.$$

We proceed with the proof of Theorem 7-1-2, postponing the proof of Claim 7-1-3 until the end of this section. We note only (Exercise! Why?) that under assumption (1) or (2) there exist infinitely many $x, y \in \mathbb{N}$ that satisfy the condition

$$y \geq y_o \text{ and } 0 < ye - xr < 1 - \epsilon'.$$

We choose $x, y \in \mathbb{N}$ satisfying

$$y \geq y_o \text{ and } 0 < ye - xr < 1 - \epsilon'.$$

Then by Claim 7-1-3, we have

$$|xH + ye(K_X + D)| \neq \emptyset.$$

If

$$\text{Bs}|xH + ye(K_X + D)| = \emptyset,$$

then we are done (deriving a contradiction!).

In the following, assuming

$$\text{Bs}|xH + ye(K_X + D)| \neq \emptyset$$

we show that

$$\exists x', y' \in \mathbb{N} \quad \text{s.t.}$$
$$y' \geq y_o \text{ and } 0 < y'e - x'r < 1 - \epsilon',$$
$$\text{Bs}|x'H + y'e(K_X + D)| \overset{\neq}{\subset} \text{Bs}|xH + ye(K_X + D)|.$$

By noetherian induction, we would eventually have

$$x', y' \in \mathbb{N} \quad \text{s.t.}$$
$$y' \geq y_o \text{ and } 0 < y'e - x'r < 1 - \epsilon',$$
$$\text{Bs}|x'H + y'e(K_X + D)| = \emptyset,$$

which leads again to a contradiction.

Step 1. Construction of a nice resolution $f : Y \to X$.

By Hironaka's theorem, we take a projective birational morphism

$$f : Y \to X$$

from a nonsingular projective variety Y such that

$$\cup F_j = f_*^{-1}(D) \cup \mathrm{Exc}(f) \cup f^{-1}\mathrm{Bs}|xH + ye(K_X + D)|$$

is a divisor with simple normal crossings.

We write the ramification formula for sufficiently small $(1 \gg \epsilon' \gg)\epsilon > 0$:

$$K_Y = f^*(K_X + (1 - \epsilon)D) + \Sigma a_i E_i - f_*^{-1}((1 - \epsilon)D)$$
$$= f^*(K_X + (1 - \epsilon)D) + \Sigma a_j F_j,$$

where the E_i are exceptional and where $a_j > -1$ $\forall j$ thanks to taking $\epsilon > 0$.

(Note that the above ramification formula is the only place where we use the assumption of X being $\mathbb{Q}$-factorial, which implies that D is $\mathbb{Q}$-Cartier.)

We choose $(\epsilon' \gg)\epsilon > 0$ so small that

$$xH + (ye - 1)(K_X + D) + \epsilon D = xH + ye(K_X + D) - (K_X + (1 - \epsilon)D)$$

is ample. We also take the resolution $f : Y \to X$ such that

$$f^*|xH + ye(K_X + D)| = |M| + \Sigma r_j F_j \text{ with } \mathrm{Bs}|M| = \emptyset.$$

Step 2. Construction of a "just" log canonical divisor and a nice reduced divisor F.

We would like to make the log pair

$$(X, (1 - \epsilon)D + c'(xH + ye(K_X + D)))$$

"just" log canonical. That is to say, by looking at the ramification formula

$$K_Y = f^*(K_X + (1 - \epsilon)D + c'(xH + ye(K_X + D))) - c'M + \Sigma(-c' \cdot r_j + a_j)F_j$$

we would like to have

$$-c' \cdot r_j + a_j \geq -1 \quad \forall j,$$
$$-c' \cdot r_j + a_j = -1$$

for some j. Therefore, we choose

$$c' = \min_{r_j \neq 0} \left\{ \frac{a_j + 1}{r_j} \right\}.$$

Note again that the part $-c'M$ can cause no trouble on the condition for the log pair $(X, (1 - \epsilon)D + c'(xH + ye(K_X + D)))$ to be log canonical and be ignored, since $|M|$ has no base points and thus can be split into $\mathbb{Q}$-effective divisors with only normal crossings and very small coefficients (cf. the argument on page 285).

Step 4. Add an ample divisor to bring the "just" log canonical divisor of Step 2 to the divisor of the concern.

Now add

$$f^*((x' - (c' + 1)x)H + (y' - (c' + 1)y)e(K_X + D))$$
$$+ f^*(xH + (ye - 1)(K_X + D) + \epsilon D) + c'M$$

to obtain the formula

$$K_Y + f^*((x' - (c' + 1)x)H + (y' - (c' + 1)y)e(K_X + D))$$
$$+ f^*(xH + (ye - 1)(K_X + D) + \epsilon D) + c'M$$
$$= f^*(K_X + (1 - \epsilon)D + c'(xH + ye(K_X + D))) - c'M$$
$$+ \Sigma(-c' \cdot r_j + a_j)F_j$$
$$+ f^*((x' - (c' + 1)x)H + (y' - (c' + 1)y)e(K_X + D))$$
$$+ f^*(xH + (ye - 1)(K_X + D) + \epsilon D) + c'M,$$

i.e.,

$$f^*((x' - (c' + 1)x)H + (y' - (c' + 1)y)e(K_X + D))$$
$$+ f^*(xH + (ye - 1)(K_X + D) + \epsilon D) + c'M$$
$$= f^*(x'H + y'e(K_X + D)) + \Sigma(-c' \cdot r_j + a_j)F_j - K_Y.$$

We will choose $x', y' \in \mathbb{N}$ such that

$$0 < y'e - x'r \leq \min\{1 - \epsilon', (c' + 1)(ye - xr)\},$$

which implies that

$$(x' - (c' + 1)x)H + (y' - (c' + 1)y_o)e(K_X + D)$$

is nef. In fact, in case (1) we choose

$$x' = \left[\frac{lye}{r}\right] = lx + \left[l\left(\frac{ye}{r} - x\right)\right] \text{ and } y' = ly,$$

where l is a sufficiently large integer such that

$$r\left(\frac{lye}{r} - \left[\frac{lye}{r}\right]\right) < \min\{1 - \epsilon', (c' + 1)(ye - xr)\}.$$

In case (2), we choose

$$x' = x + lq \text{ and } y' = y + lp,$$

where l is a sufficiently large integer.

As before, since $xH + (ye - 1)(K_X + D) + \epsilon D$ is ample, there exists $-\Sigma\delta_j F_j$ with $0 < \delta_j \ll \epsilon \ll 1$ consisting only of exceptional divisors such that

$$f^*((x' - (c' + 1)x)H + (y' - (c' + 1)y)e(K_X + D))$$
$$+ f^*(xH + (ye - 1)(K_X + D) + \epsilon D) + c'M - \Sigma\delta_j F_j$$
$$= f^*(x'H + y'e(K_X + D)) + \Sigma(-c' \cdot r_j + a_j - \delta_j)F_j - K_Y$$

is ample. Now we have

$$\lceil \text{R.H.S.} \rceil = f^*(x'H + y'e(K_X + D)) - F + E - K_Y,$$

where F is a reduced divisor with

$$f(F) \subset \text{Bs}|xH + ye(K_X + D)| \text{ and } E \text{ is exceptional for } f.$$

Step 4. Application of the Kawamata–Viehweg Vanishing Theorem
Consider the exact sequence

$$0 \to \mathcal{O}_Y(f^*(x'H + y'e(K_X + D)) - F + E)$$
$$\to \mathcal{O}_Y(f^*(x'H + y'e(K_X + D)) + E)$$
$$\to \mathcal{O}_F(f^*(x'H + y'e(K_X + D)) + E|_F) \to 0.$$

The Kawamata–Viehweg vanishing theorem gives

$$0 = H^1(Y, \mathcal{O}_Y(K_Y + \lceil \text{R.H.S.} \rceil))$$
$$= H^1(Y, \mathcal{O}_Y(f^*(x'H + y'e(K_X + D)) - F + E)) = 0.$$

Thus the long cohomology sequence gives a surjection

$$H^0(X, \mathcal{O}_Y(x'H + y'e(K_X + D))) \cong H^0(Y, \mathcal{O}_Y(f^*(x'H + y'e(K_X + D)) + E))$$
$$\to H^0(F, \mathcal{O}_F(f^*(x'H + y'e(K_X + D)) + E|_F))$$
$$\to H^1 = 0.$$

Pretend for the moment that F is irreducible and that the divisors $\lceil \text{R.H.S.} \rceil - \text{R.H.S.}$ and F have no common components.

Claim 7-1-4.

$$H^0(F, f^*(x'H + y'e(K_X + D)) + E|_F) \neq 0.$$

We will prove Claim 7-1-4 together with Claim 7-1-3 at the end.
The claim implies

$$f(F) \not\subset \text{Bs}|x'H + y'e(K_X + D))|.$$

Observe that by the choice of x', y' we have

$$x'H + y'e(K_X + D) = l(xH + ye(K_X + D)) + \left[l\left(\frac{ye}{r} - x\right)\right]H \text{ in case (1),}$$

$$x'H + y'e(K_X + D) = l(qH + pe(K_X + D)) + (xH + ye(K_X + D)) \text{ in case (2).}$$

Since H is very ample by Step 0 in case (1) and since $\text{Bs}|l(qH + pe(K_X + D))| = \emptyset$ for $l \gg 0$ by the logarithmic base point freeness theorem (cf. Theorem 6-2-2) in case (2), we conclude that

$$\text{Bs}|x'H + y'e(K_X + D)| \overset{\neq}{\subset} \text{Bs}|xH + ye(K_X + D)|.$$

By noetherian induction, we conclude that

$$\exists x, y \in \mathbb{N} \quad s.t.$$
$$0 < ye - xr < 1 - \epsilon' \quad \text{and} \quad \text{Bs}|xH + ye(K_X + D)| = \emptyset,$$

a contradiction.

Step 5. Tie Breaking.

Here we discuss the technical problem of how to avoid the possibility that F may be reducible and that $\lceil R.H.S. \rceil - R.H.S.$ and F may have some common components.

In Step 4, we take a divisor $\Sigma \delta_j F_j$ with $0 < \delta_j \ll \epsilon \ll 1$ such that

$$f^*(xH + (ye - 1)(K_X + D) + \epsilon D) - \Sigma \delta_j F_j$$

is ample. Now instead of c' we choose

$$c = \min_{r_j \neq 0} \left\{ \frac{a_j + 1 - \delta_j}{r_j} \right\}$$

after perturbing the δ_j a little so that the minimum is attained only at one unique irreducible component F_o. (This can be done, since perturbation will not destroy the ampleness, which is an open condition.)

This subtle change leaves the rest of the argument intact, except that now $F = F_o$ is irreducible and $\lceil R.H.S. \rceil - R.H.S.$ and F have no common components and thus $(\lceil R.H.S. \rceil - R.H.S.)|_F$ is a well-defined effective $\mathbb{Q}$-divisor with only normal crossings, making the argument in Step 4 legitimate.

The only remaining task is to prove Claim 7-1-3 and Claim 7-1-4, both of which follow from the following lemma.

Lemma 7-1-5. *Let V be a nonsingular projective variety of dimension d, and let u and v be positive real numbers, D_1 and D_2 Cartier divisors, and A a $\mathbb{Q}$-divisor satisfying the following conditions:*

(i) $\lceil A \rceil \geq 0$, *and* $\lceil A \rceil - A$ *has support with only normal crossings,*
(ii) $aD_1 + bD_2 + A - K_V$ *is ample and*

$$H^0(V, \mathcal{O}_V(aD_1 + bD_2 + \lceil A \rceil)) \cong H^0(V, \mathcal{O}_V(aD_1 + bD_2))$$

for $a, b \in \mathbb{N}$ with $b - au < v$.

Assume further that either

(1) $u \notin \mathbb{Q}$, *or*
(2) $u \in \mathbb{Q}$ *and expressing*

$$u = \frac{p'}{q'} \text{ with } p', q' \in \mathbb{N} \text{ and } \gcd(p', q') = 1,$$

we have

$$\frac{v}{d+1} > \frac{1}{q'}.$$

Then

$$\exists b_o \in \mathbb{N} \quad s.t.$$
$$H^0(V, \mathcal{O}_V(aD_1 + bD_2 + \lceil A \rceil)) \neq 0,$$
$$\forall a, b \in \mathbb{N} \text{ with } b - au < v \text{ and } b \geq b_o.$$

How to prove Claim 7-1-3 via Lemma 7-1-5

In order to prove Claim 7-1-3 via Lemma 7-1-5, we set

$$V = Y,$$
$$D_1 = f^*H,$$
$$D_2 = f^*e(K_X + D),$$
$$A = K_Y - f^*(K_X + (1 - \epsilon)D) - \Sigma\delta'_j F_j,$$
$$u = \frac{r}{e}, \qquad v = \frac{1 - \epsilon'}{e},$$

where we choose $0 < \delta'_j \ll \epsilon \ll \epsilon' \ll 1$ such that

$$f^*\left(\frac{\epsilon'}{r}H + \epsilon D\right) - \Sigma\delta'_j F_j$$

is ample.

Condition (i) follows from the construction of $f : Y \to X$.

The first part of condition (ii) follows from the fact that for $a, b \in \mathbb{N}$ with $b - au < v$ the divisor,

$$\begin{aligned}
aD_1 + bD_2 + A - K_Y &= af^*H + bf^*e(K_X + D) - K_Y \\
&\quad + f^*(K_X + (1 - \epsilon)D) - \Sigma\delta'_j F_j \\
&= f^*\left\{\left(a - \frac{\epsilon'}{r}\right)H + (be - 1)(K_X + D)\right\} \\
&\quad + f^*\left(\frac{\epsilon'}{r}H + \epsilon D\right) - \Sigma\delta'_j F_j
\end{aligned}$$

is ample, since

$$\left(a - \frac{\epsilon'}{r}\right)H + (be - 1)(K_X + D) \text{ is nef as } be - 1 - \left(a - \frac{\epsilon'}{r}\right)r < 0,$$

and since

$$f^*\left(\frac{\epsilon'}{r}H + \epsilon D\right) - \Sigma\delta'_j F_j \text{ is ample.}$$

The second part of condition (ii) follows from

$$\begin{aligned}
H^0(V, \mathcal{O}_V(aD_1 + bD_2 + \lceil A\rceil)) &= H^0(Y, \mathcal{O}_Y(af^*H + bf^*e(K_X + D) + \lceil A\rceil)) \\
&\cong H^0(X, \mathcal{O}_X(aH + be(K_X + D))) \\
&\cong H^0(Y, \mathcal{O}_Y(af^*H + b^*e(K_X + D))) \\
&\cong H^0(V, \mathcal{O}_V(aD_1 + bD_2)),
\end{aligned}$$

since $\lceil A\rceil$ is an effective divisor that is exceptional for f.

Therefore, we have setting, $a = x, b = y$,

$$0 \neq H^0(Y, \mathcal{O}_Y(xf^*H + yf^*e(K_X + D) + \lceil A\rceil)) \cong H^0(X, \mathcal{O}_X(xH + ye(K_X + D)))$$

as desired.

How to prove Claim 7-1-4 via Lemma 7-1-5

In order to prove Claim 7-1-4 via Lemma 7-1-5, we set

$$V = F,$$
$$D_1 = f^*H|_F,$$
$$D_2 = f^*e(K_X + D)|_F,$$
$$A = \Sigma_{j \neq o}(-c \cdot r_j + a_j - \delta_j)F_j|_F, \text{ so that } \lceil A \rceil = E|_F,$$
$$u = \frac{r}{e}, \qquad v = \min\left\{\frac{1}{e}, (c+1)(y - \frac{r}{e}x)\right\}.$$

Condition (i) follows from the choice of c in Step 5.

The first part of condition (ii) follows from Step 4.

The second part of condition (ii) follows from the commutative diagram

$$
\begin{array}{ccc}
H^0(Y, \mathcal{O}_Y(f^*(aH + be(K_X + D)))) & \longrightarrow & H^0(F, \mathcal{O}_F(f^*(aH + be(K_X + D)|_F))) \\
\| & & \downarrow \\
H^0(Y, \mathcal{O}_Y(f^*(aH + be(K_X + D)) + E)) & \longrightarrow & H^0(F, \mathcal{O}_F(f^*(aH + be(K_X + D))|_F + \lceil A \rceil))
\end{array}
$$

In fact, the bottom homomorphism is surjective whenever $b - au < v$, since the Kawamata–Viehweg vanishing theorem implies

$$H^1(Y, \mathcal{O}_Y(f^*(aH + be(K_X + D)) - F + E)) = 0,$$

as shown in Step 4, and hence the natural inclusion on the right-hand side is also surjective.

Therefore, we have setting, $a = x', b = y'$,

$$0 \neq H^0(F, \mathcal{O}_F(xf^*H|_F + yf^*e(K_X + D)|_F + \lceil A \rceil))$$
$$= H^0(F, \mathcal{O}_F(x'H + y'e(K_X + D) + E|_F))$$

as desired.

Now we present the proof of Lemma 7-1-5.

PROOF.   Proof of Lemma 7-1-5

Set

$$P(a, b) = \chi(\mathcal{O}_V(aD_1 + bD_2 + \lceil A \rceil)),$$

which is a polynomial in a, b of degree at most d by the Riemann–Roch theorem. Since

$$aD_1 + bD_2 + A - K_V \text{ is ample for } a, b \in \mathbb{N} \text{ with } b - au < v,$$

we conclude by the Kawamata–Viehweg vanishing theorem that

$$P(a, b) = h^0(V, aD_1 + bD_2 + \lceil A \rceil) \text{ for } a, b \in \mathbb{N} \text{ with } b - au < v.$$

We claim that $P(a, b)$ is not identically zero as a polynomial in a, b, i.e.,

$$P(a, b) \not\equiv 0.$$

In fact, take

$$u_\alpha \in \mathbb{Q} \text{ with } 0 < u_\alpha < u.$$

Then on the line L defined by the equation $b = u_\alpha a$ we claim that $P(a, b)$ is not identically zero. Observe first that for $(a_\alpha, b_\alpha) \in \mathbb{N}^2$ on the line L

$$a_\alpha D_1 + b_\alpha D_2 = \lim_{m \to \infty} (m a_\alpha D_1 + m b_\alpha D_2 + A - K_V)$$

is nef, since

$$m a_\alpha D_1 + m b_\alpha D_2 + A - K_V \text{ is ample, since } b_\alpha - a_\alpha u < b_\alpha - a_\alpha u_\alpha = 0 < v.$$

Then the nonvanishing theorem implies that for $m \gg 0$ we have

$$P(m a_\alpha, m b_\alpha) = h^0(V, m(a_\alpha D_1 + b_\alpha D_2) + \lceil A \rceil) \neq 0.$$

Now set

$$U := \left\{ (a, b) \in \mathbb{N}^2; 0 < b - au < \frac{v}{d+1} \right\}.$$

Note that either in case (1) or (2)

$$\#U = \infty,$$

and that

$$\#\{L(a, b); L(a, b) \text{ is the line connecting } (0, 0) \text{ and } (a, b) \in U\} = \infty.$$

Since $P(a, b) \not\equiv 0$,

$$\exists L(a_\beta, b_\beta) \quad \text{with } (a_\beta, b_\beta) \in U \quad \text{s.t.} \quad P|_{L(a_\beta, b_\beta)} \not\equiv 0.$$

Since $P|_{L(a_\beta, b_\beta)} = P(j a_\beta, j b_\beta)$ is a polynomial of degree at most d in the variable j, we conclude that

$$1 \leq \exists j \leq d + 1 \quad \text{s.t.} \quad P(j a_\beta, j b_\beta) \neq 0.$$

By setting $a_\gamma = j a_\beta, b_\gamma = j b_\beta$, we obtain

$$a_\gamma, b_\gamma \text{ with } 0 < b_\gamma - a_\gamma u < v \quad \text{s.t.} P(a_\gamma, b_\gamma) \neq 0.$$

Now assume that such b_o as stated in the lemma does not exist. Then there exist infinitely many

$$(a_i, b_i) \in \mathbb{N}^2$$

such that

$$b_i - a_i u < v, a_i - d a_\gamma > 0, b_i - d b_\gamma > 0 \text{ and } P(a_i, b_i) = 0.$$

But then for $0 \leq k \leq d$ we have (Note that $(b_i - k b_\gamma) - (a_i - k a_\gamma) = (b_i - a_i u) - k(b_\gamma - a_\gamma u) < b_i - a_i u < v$.)

$$P(a_i - k a_\gamma, b_i - k b_\gamma) = h^0(V, (a_i - k a_\gamma)D_1 + (b_i - k a_\gamma)D_2 + \lceil A \rceil)$$
$$= h^0(V, (a_i - k a_\gamma)D_1 + (b_i - k a_\gamma)D_2) = 0,$$

since

$$P(a_\gamma, b_\gamma) = h^0(V, a_\gamma D_1 + b_\gamma D_2 + \lceil A \rceil) = h^0(V, a_\gamma D_1 + b_\gamma D_2) \neq 0.$$

This would imply

$$P(a, b)|_{\text{line passing } (a_i, b_i) \text{ with slope } \frac{b_\gamma}{a_\gamma}} \equiv 0.$$

Then since there are infinitely many such lines, it would imply

$$P(a, b) \equiv 0,$$

a contradiction!

This completes the proof of Lemma 7-1-5 and thus completes the proof of the rationality theorem and boundedness of the denominator.   □

7.2   Cone Theorem

Now that we have the rationality theorem and boundedness of the denominator at hand, we can derive the **cone theorem** in arbitrary dimension and its logarithmic version easily following the argument we used in dimension 2 (cf. Sections 1-2 and 1-3).

Theorem 7-2-1 (Cone Theorem). *Let X be a normal projective variety with only $\mathbb{Q}$-factorial and terminal singularities. Then*

$$\overline{NE}(X) = \overline{NE}(X)_{K_X \geq 0} + \Sigma R_l,$$

where

$$\overline{NE}(X)_{K_X \geq 0} := \{ z \in \overline{NE}(X); K_X \cdot z \geq 0 \}$$

and the R_l are the half-lines such that $R_l - \{0\}$ are in $\overline{NE}(X)_{K_X < 0}$ and they are of the form

$$R_l = \overline{NE}(X) \cap L^\perp$$

for some nef line bundles L.

Moreover, the R_l are discrete in the half-space $\overline{NE}(X)_{K_X < 0}$.

For each extremal ray R_l, there exists a curve $l \subset X$ such that $R_l = \mathbb{R}_+[l]$.

Theorem 7-2-2 (Logarithmic Cone Theorem). *Let (X, D) be a log pair consisting of a normal projective variety X with a boundary $\mathbb{Q}$-divisor $D = \Sigma d_i D_i$, $0 \leq d_i \leq 1$, such that it has only $\mathbb{Q}$-factorial and log terminal singularities. Then*

$$\overline{NE}(X) = \overline{NE}(X)_{K_X + D \geq 0} + \Sigma R_l,$$

where

$$\overline{NE}(X)_{K_X + D \geq 0} := \{ z \in \overline{NE}(X); (K_X + D) \cdot \geq 0 \},$$

and the R_l are the half-lines such that $R_l - \{0\}$ are in $\overline{\mathrm{NE}}(X)_{K_X+D<0}$ and they are of the form

$$R_l = \overline{\mathrm{NE}}(X) \cap L^\perp$$

for some nef line bundles L.

   Moreover, the R_l are discrete in the half-space $\overline{\mathrm{NE}}(X)_{K_X+D<0}$.

   For each extremal ray R_l, there exists a curve $l \subset X$ such that $R_l = \mathbb{R}_+[l]$.

We will go over the proof for the logarithmic cone theorem following exactly the argument in Section 1-3.

PROOF.   Proof of the Logarithmic Cone Theorem

Step 1. For any nef line bundle M with

$$\overline{\mathrm{NE}}(X)_{K_X+D<0} \cap M^\perp \neq \emptyset,$$

there exists an extremal ray R_l as above such that

$$R_l \subset \overline{\mathrm{NE}}(X) \cap M^\perp.$$

The basic idea is to perturb M by some (small portion of) an ample divisor B so that $M + \epsilon B + r(K_X + D)$ will cut out a face of smaller dimension from $\overline{\mathrm{NE}}(X)$.

Figure 7-2-3.

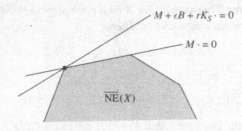

What makes this idea legitimate is boundedness of the denominator.

In the following instead of taking $M + \epsilon B$ with ϵ small, we consider $\nu M + B$ with ν big.

Define

$$r_M(\nu, B) := \sup\{t; \nu M + B + t(K_X + D) \text{ is nef}\}.$$

Since M is nef, $r_M(\nu, B)$ is a nondecreasing function of ν. Since the denominator of $r_M(\nu, B)$ is bounded and since $r_M(\nu, B)$ is bounded from above uniformly,

$$r_M(\nu, B) \leq \frac{B \cdot z}{-(K_X + D) \cdot z}$$

for some fixed

$$0 \neq z \in \overline{\mathrm{NE}}(X)_{K_X+D<0} \cap M^\perp$$

with the right-hand side independent of v, $r_M(v, B)$ stabilizes to a fixed $r_M(B)$ for $v \geq v_B$. Therefore,

$$0 \neq \overline{NE}(X) \cap ((v_B + 1)M + B + r_M(B)(K_X + D))^\perp \subset \overline{NE}(X) \cap M^\perp.$$

Take $B_1, B_2, \cdots, B_\rho$ to be ample divisors forming a basis of $N^1(X)$ with $\rho = \dim_{\mathbb{R}} N^1(X)$. Then

$$((v_{B_i} + 1)M + B_i + r_M(B_i)(K_X + D)) \cdot = 0, i = 1, 2, \cdots, \rho,$$

give at least $\rho - 1$ independent linear conditions (by increasing some v_{B_i} if necessary). Thus if

$$\dim \overline{NE}(X) \cap M^\perp > 1,$$

then there exists i such that

$$0 \neq \overline{NE}(X) \cap ((v_{B_i} + 1)M + B_i + r_M(B_i)(K_X + D))^\perp \overset{\neq}{\subset} \overline{NE}(X) \cap M^\perp.$$

(In fact, $(v_{B_i} + 1)M + B_i + r_M(B_i)(K_X + D)$ is nef but not ample by the choice of $r_M(B_i)$, and hence by Kleiman's criterion there exists $0 \neq z \in \overline{NE}(X)$ such that $\{(v_{B_i} + 1)M + B_i + r_M(B_i)(K_X + D)\} \cdot z = 0$.)

Note that

$$\overline{NE}(X)_{K_X + D < 0} \cap ((v_{B_i} + 1)M + B_i + r_M(B_i)(K_X + D))^\perp \neq \emptyset.$$

In order to see this, observe that since $(v_{B_i} + 1)M + B_i$ is ample and hence $(v_{B_i} + 1)M + B_i$ is > 0 on $\overline{NE}(X) - \{0\}$ by Kleiman's criterion, for $0 \neq z \in ((v_{B_i} + 1)M + B_i + r_M(B_i)(K_X + D))^\perp$ we have $(v_{B_i} + 1)M + B_i \cdot z > 0$ and hence $(K_X + D) \cdot z < 0$.

Now the claim follows by induction on the dimension of the face $\overline{NE}(X) \cap M^\perp$.

Step 2. $\overline{NE}(X) - \overline{NE}(X)_{K_X + D \geq 0} + \Sigma R_l$.

Suppose that

$$\overline{NE}(X) \overset{\neq}{\supset} \overline{NE}(X)_{K_X + D \geq 0} + \Sigma R_l.$$

Then there exists a line bundle G such that $G > 0$ on the R.H.S. but $G \cdot z < 0$ for some $z \in \overline{NE}(X)$. (See Figure 7-2-4.)

Figure 7-2-4.

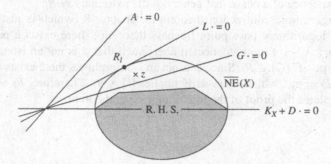

It follows immediately that $A = h(G + a(K_X + D))$ for some negative rational number a and sufficiently divisible h is an ample Cartier divisor. By the logarithmic rationality theorem, we obtain a nef line bundle $L = l(A + r(K_X + D))$, which has the property R.H.S. $\cap L^\perp = \{0\}$. But by Step 1, there exists $R_l \subset \overline{\text{NE}}(X) \cap L^\perp$ and $R_l \subset$ R.H.S., a contradiction!

Step 3. The R_l are discrete in the half-space $N_1(X)_{K_X+D<0}$.

If $K_X + D$ is numerically trivial, then there is no extremal ray (with respect to $K_X + D$) and nothing to prove. So we may assume that $K_X + D$ is not numerically trivial.

Choose ample Cartier divisors $A_1, A_2, \cdots, A_{\rho-1}$ such that

$$K_X + D, A_1, A_2, \cdots, A_{\rho-1}$$

form a basis of $N^1(X)$. For each R_l and a nef line bundle L such that

$$R_l = \overline{\text{NE}}(X) \cap L^\perp,$$

take $r_L(A_i)$ and ν_{A_i} as in Step 1. Then

$$0 \neq \overline{\text{NE}}(X) \cap ((\nu_{A_i} + 1)L + A_i + r_L(A_i)(K_X + D))^\perp = \overline{\text{NE}}(X) \cap L^\perp = R_l.$$

Therefore,

$$A_i + r_L(A_i)(K_X + D) \cdot \xi_l = 0 \text{ for } 0 \neq \xi_l \in R_l,$$

which implies

$$(*) \qquad\qquad \frac{A_i \cdot \xi_l}{-(K_X + D) \cdot \xi_l} = r_L(A_i).$$

Since we can take

$$\frac{A_i \cdot \{\}}{-(K_X + D) \cdot \{\}} \text{ for } i = 1, 2, \cdots, \rho - 1$$

as coordinates of the hyperplane $\{z \in N_1(X); -(K_X + D) \cdot z = 1\}$ and since the denominators of the $r_L(A_i)$ are bounded, we conclude from $(*)$ that the intersections of the R_l and the hyperplane are discrete, which is exactly the claim.

Step 4. $\overline{\text{NE}}(X) = \overline{\text{NE}}(X)_{K_X+D\geq 0} + \Sigma R_l$.

Since the R_l are discrete in the half-space $N_1(X)_{K_X+D<0}$, we can get rid of the closure sign "$^-$" from ΣR_l, and thus we have the claim.

Step 5. Existence of a curve that generates the extremal ray R_l.

By the logarithmic contraction theorem in Chapter 8 (which is just another form of the logarithmic base point freeness theorem), there exists a projective morphism $\phi : X \to Y$ with connected fibers such that ϕ is not an isomorphism and $\phi(C) = \text{pt.}$ iff $[C] \in R_l$. Since ϕ is not an isomorphism, there exists a curve l such that $\phi(l) = \text{pt.}$, which necessarily implies $[l] \in R_l$. Therefore, $R_l = \mathbb{R}_+[l]$.

This completes the proof of the logarithmic cone theorem.   □

CHAPTER 8

Contraction Theorem

This chapter starts with characterizing in Section 8-1 the **extremal contractions**, the key geometric operations in the process of the minimal model program, as what we call the **contractions of extremal rays** described in terms of the convex geometry of the cone of curves, thanks to the cone theorem of Chapter 7. Their existence is guaranteed by the base point freeness theorem of Chapter 6.

In Section 8-2, we discuss the basic properties of the contraction morphism of an extremal ray according to whether it is of **divisorial type**, **flipping type**, or **fibering type**.

In Section 8-3 we give some examples of extremal contractions. We discuss the contractions of extremal rays on smooth 3-folds, which have been completely classified by Mori [2]. As it turns out, there is no contraction of an extremal ray of flipping type on a smooth 3-fold. We will be content with some elementary examples of the contractions of extremal rays of flipping type on some singular 3-folds and on smooth varieties of dimensions greater than or equal to 4.

8.1 Contraction Theorem

In the process of the minimal model program we try to construct an extremal contraction as defined below whenever the canonical divisor K_X is not nef for a normal projective variety X with only $\mathbb{Q}$-factorial and terminal singularities.

Definition 8-1-1 (Extremal Contraction). *Let X be a normal projective variety X with only $\mathbb{Q}$-factorial and terminal singularities. Then a morphism $\phi : X \to Y$ is called an "extremal contraction" with respect to K_X if*

 (i) *ϕ is not an isomorphism,*

(ii) *for a curve $C \subset X$*

$$\phi(C) = \text{pt.} \implies K_X \cdot C < 0,$$

(iii) *all the curves contracted by ϕ are numerically proportional, i.e.,*

$$\phi(C) = \phi(D) = \text{pt.} \implies [D] = c[C] \text{ in } H_2(X, \mathbb{R}) \text{ for some } c \in \mathbb{Q}_{>0},$$

(iv) *ϕ has connected fibers with Y being normal and projective.*

Remark 8-1-2. We would like to emphasize that the conditions above required for an extremal contraction are natural from the viewpoint of running the minimal model program.

Condition (i) is to avoid the trivial morphism, i.e., the identity morphism from the extremal contractions, since it does not bring any progress in the process of running the minimal model program.

Condition (ii) is also natural, since an extremal contraction should be an exclusive feature when K_X is not nef.

Conditions (iii) and (iv) guarantee that the extremal contraction is the smallest possible building block for contraction morphisms and cannot be decomposed any more in the following sense: If $\psi : X \to V$ is a morphism (onto a projective variety V) that factors ϕ, i.e., $\phi = \tau \circ \psi$ for a morphism $\tau : V \to Y$, then either ψ is the identity morphism with $V = X$ or ψ coincides with ϕ. (Exercise!! Prove the last statement via Proposition 1-2-16. We also refer the reader to the remark right after Theorem 1-2-1.)

Theorem 8-1-3 (Contraction Theorem). *Let X be a normal projective variety with only $\mathbb{Q}$-factorial and terminal singularities. For each extremal ray R_l of $\overline{NE}(X)$ in the half space $N_1(X)_{K_X < 0}$, where*

$$R_l = \overline{NE}(X) \cap L^{\perp} \text{ for some nef line bundle } L,$$

there exists a morphism $\phi = \text{cont}_{R_l} : X \to Y$, called the **contraction of an extremal ray R_l** *with respect to K_X, such that*

(i) cont_{R_l} *is not an isomorphism,*
(ii) *for a curve $C \subset X$*

$$\text{cont}_{R_l}(C) = \text{pt.} \implies K_X \cdot C < 0,$$

(iii) *for a curve $C \subset X$*

$$\text{cont}_{R_l}(C) = \text{pt.} \iff [C] \in R_l,$$

(iv) cont_{R_l} *has connected fibers with Y normal and projective.*
Properties (i), (ii), (iii), and (iv) characterize the contraction of an extremal ray R_l with respect to K_X and show that cont_{R_l} is an extremal contraction with respect to K_X.
Moreover, the contraction of an extremal ray R_l with respect to K_X satisfies the following proerties:

(v) $\text{cont}_{R_l *} \mathcal{O}_X = \mathcal{O}_Y$,

(vi) $L = \text{cont}_{R_l}^* A$ *for some ample Cartier divisor A on Y,*

(vii) *for any Cartier divisor D on X*

$$D = \text{cont}_{R_l}^* G \text{ for some Cartier divisor G on Y}$$

$$\Longleftrightarrow$$

$$D \cdot C = 0 \quad \forall C \text{ with } [C] \in R_l.$$

In particular, we have the exact sequence

$$0 \to N^1(Y) \to N^1(X) \to N^1(X/Y) \to 0$$

and its dual exact sequence

$$0 \leftarrow N_1(Y) \leftarrow N_1(X) \leftarrow N_1(X/Y) \leftarrow 0.$$

Observe (cf. Example-Exercise!3-5-1) that the intersection pairing can be restricted to the one between the group of those 1-cycles whose supports are contracted to points on Y, which we denote by $Z_1(X/Y)$, and the group of the line bundles $\text{Pic}(X)$ *on X:*

$$Z_1(X/Y) \times \text{Pic}(X) \to \mathbb{Z}.$$

Denoting by $\equiv_Y$ the numerical equivalence induced by this intersection pairing, we define

$$N^1(X/Y) := \{\text{Pic}(X)/\equiv_Y\} \otimes \mathbb{R},$$

$$N_1(X/Y) := \{Z_1(X/Y)/\equiv_Y\} \otimes \mathbb{R}.$$

Conversely, any extremal contraction with respect to K_X is the contraction of some extremal ray with respect to K_X. That is to say, the extremal contractions with respect to K_X coincide with the contractions of the extremal rays with respect to K_X and hence enjoy all the properties (i) through (vii) above.

PROOF. Let L be a nef line bundle on X such that

$$R_l = \overline{\text{NE}}(X) \cap L^\perp.$$

Then by Kleiman's criterion for ampleness, it follows that $aL - K_X$ is ample for some sufficiently large $a \in \mathbb{N}$. Thus by the base point freeness theorem $|mL|$ is base point free for $m \gg 0$. Set

$$\text{cont}_{R_l} = \Phi_{|mL|} \text{ for } m \gg 0.$$

Now the assertions (i), (ii), (iii), (iv), (v), and (vi) follow directly from Proposition 1-2-16. We also see that properties (i), (ii), (iii), and (iv) characterize the contraction of an extremal ray R_l with respect to K_X and show that it is an extremal contraction with respect to K_X.

In order to establish the property (vii), let D be a Cartier divisor on X such that $D \cdot C = 0 \quad \forall C$ with $[C] \in R_l$. Thus $D = 0$ on R_l. Since L is nef and $R_l = \overline{\text{NE}}(X) \cap L^\perp$ and since the extremal rays are discrete in $\overline{\text{NE}}(X)_{K_X < 0}$ and make $\overline{NE}(X)$ "locally polyhedral" in shape around R_l, we conclude that for $l \in \mathbb{N}$ sufficiently large $lL + D$ is nef and that $\overline{NE}(X) \cap (lL+D)^\perp = R_l$. By property (vi) of the contraction morphism we see that $lL + D = (\text{cont}_{R_l})^* A'$ for some ample divisor A' on Y. We finally conclude that

$$D = (lL + D) - lL = (\text{cont}_{R_l})^*(A' - lA) = (\text{cont}_{R_l})^* G,$$

where we set $G = A' - lA$. The converse is obvious. This proves property (vii).

In order to demonstrate the exact sequence of the N^1's we have only to check the exactness in the middle (the other ones being obvious), which follows easily from the above argument. The exact sequence of the N_1's is now the dual consequence.

Conversely, suppose $\phi : X \to Y$ is an extremal contraction as defined in Definition 8-1-1. Take an ample Cartier divisor H on Y. Then $M = \phi^* H$ satisfies the conditions in step 1 of the proof of the cone theorem. Therefore, there exists an extremal ray R_l such that

$$R_l \subset \overline{NE}(X) \cap M^\perp.$$

Now it is easy to see (Exercise! Why?) that the contraction cont_{R_l} of the extremal ray R_l coincides with ϕ. $\square$

Exercise 8-1-4. (i) (**Contraction of an extremal face.**) The contraction theorem can be proved not only for the extremal rays but also for the extremal faces of dimension greater than 1: Let F_L be a face of $\overline{NE}(X)$ associated to some nef line bundle L, which sits entirely in the negative part of K_X, i.e.,

$$F_L = \overline{NE}(X) \cap L^\perp \subset \overline{NE}(X)_{K_X < 0}.$$

Then we have a morphism $\text{cont}_{F_L} : X \to Y$, called the contraction of an extremal face F_L, satisfying properties (i) through (vii) (replacing R_l with F_L in the statement).

(ii) (**Logarithmic contraction theorem.**) Make the precise statement of the logarithmic contraction theorem and provide its proof, which is almost identical to that of the contraction theorem above.

Remark 8-1-5. Property (vii) in Theorem 8-1-3 is a rather special feature of the contraction of an extremal ray (with respect to K_X), which does not hold for a general projective morphism $f : X \to Y$ even with connected fibers and even with the assumption that all the curves in the fibers are numerically proportional. The key fact is that the cone $\overline{NE}(X)$ is "locally polyhedral" around the extremal ray R_l (with respect to K_X), which may not hold for a general projective morphism, as we will see below.

Here is an example where the exact sequence fails to hold. We leave the verification of the several assertions in the example as an exercise to the reader.

Take $X = E \times E$, where E is a generic elliptic curve with no complex multiplication (cf. Hartshorne [1], Section IV. 4, Elliptic Curves, and Exercise 4.10) such that $\dim_\mathbb{R} N^1(X) = \dim_\mathbb{R} N_1(X) = 3$. Then the lattice in $N^1(X)$ generated by $\mathrm{Pic}(X)$ (we denote the lattice by $\mathrm{Pic}(X)$ by abuse of notation) is given by

$$\mathrm{Pic}(X) = \mathbb{Z} \cdot E_1 \oplus \mathbb{Z} \cdot E_2 \oplus \mathbb{Z} \cdot D \quad (\text{in } N_1(X) = N^1(X)),$$

where $E_1 = E \times \{p\}$, $E_2 = \{q\} \times E$, and D is the diagonal. Their intersection pairings are

$$E_1^2 = E_2^2 = D^2 = 0,$$
$$E_1 \cdot E_2 = E_1 \cdot D = E_2 \cdot D = 1,$$

and thus we have the intersection matrix

$$\begin{pmatrix} 0 & 1 & 1 \\ 1 & 0 & 1 \\ 1 & 1 & 0 \end{pmatrix}.$$

By diagonalizing the matrix we have the corresponding sublattice of rank 3 of $\mathrm{Pic}(X)$ generated by

$$U = E_1 + E_2 + D,$$
$$V = E_1 - D,$$
$$W = E_1 - 2E_2 - D.$$

Their intersection pairings are

$$U^2 = 6, \, V^2 = -2, \, W^2 = -6,$$
$$U \cdot V = V \cdot W = W \cdot U = 0.$$

If we introduce the coordinate system $(u, v, w) = uU + vV + wW$ in $N^1(X)$, then $\overline{\mathrm{NE}}(X)$, which happens to coincide with the closure of the ample cone $\overline{\mathrm{Amp}}(X)$ if we identify $N_1(X)$ with $N^1(X)$, is described as

$$\overline{\mathrm{NE}}(X) = \overline{\mathrm{Amp}}(X) = \{(u, v, w) \in N^1(X)_{\mathbb{Q}}; 6u^2 - 2v^2 - 6w^2 \geq 0, u \geq 0\}.$$

Consider the hyperplane $\mathcal{O}_X(E_1)^{\perp}$ in $N_1(X)$. Then $\overline{\mathrm{NE}}(X) \cap \mathcal{O}_X(E_1)^{\perp} = \mathbb{R}_{\geq 0}[E_1]$.

Choose a divisor D such that the hyperplane $D^{\perp}$ in $N_1(X)$ contains the point E_1 but is not tangent to $\overline{\mathrm{NE}}(X)$. Then observe that no matter how big we take $m \in \mathbb{N}$ the divisor $mE_1 + D$ is never nef, since $\overline{\mathrm{NE}}(X)$ is not "locally polyhedral" around the extremal ray $\mathbb{R}_{\geq 0}[E_1]$. (Note that $\mathbb{R}_{\geq 0}[E_1]$ has zero intersection with K_X.)

Figure 8-1-6.

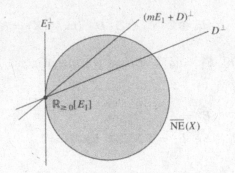

Take $f : X \to Y$ to be the second projection $p_2 : E \times E \to E$. If the exact sequence holds, then $D \cdot E_1 = 0$ would imply that $D \equiv p_2^* A$ for some divisor A on E. But this would lead to $mE_1 + D = p_2^*(mp + A)$ being nef for $m \gg 0$, a contradiction!

8.2   Contractions of Extremal Rays

In this section we discuss the basic properties of the contraction $\phi = \text{cont}_R : X \to Y$ of an extremal ray R (with respect to K_X) on a normal projective variety X with only $\mathbb{Q}$-factorial and terminal singularities. Note that the existence of such a contraction morphism and first few properties are given in the contraction theorem of the previous section.

According to the dimension of Y and the codimension of the exceptional locus of ϕ, the contraction morphism can be classified into the following 3 types:

○ **Divisorial Type**:	$\dim Y = \dim X$	and	$\text{codim}_X \text{Exc}(\phi) = 1$
○ **Flipping Type**:	$\dim Y = \dim X$	and	$\text{codim}_X \text{Exc}(\phi) \geq 2$
○ **Fibering Type**:	$\dim Y < \dim X$.		

In the following we discuss the basic properties of the contraction morphism according to its type.

Contraction of Divisorial Type

Proposition 8-2-1 (Basic Properties of Contraction of Divisorial Type). *Let* $\phi = \text{cont}_R : X \to Y$ *be the contraction of an extremal ray R (with respect to K_X) of divisorial type on a normal projective variety X with only $\mathbb{Q}$-factorial and terminal singularities. Then*

(i) *the exceptional locus $\text{Exc}(\phi)$ consists of a unique irreducible divisor E,*

(ii) *Y is a normal projective variety with only $\mathbb{Q}$-factorial and terminal singularities, and*

(iii) *the Picard number of Y drops by 1, i.e.,*

$$\rho(Y) = \rho(X) - 1.$$

PROOF.   (i) Let E be an irreducible component of $\text{Exc}(\phi)$ such that $\text{codim}_X E = 1$. Set

$$C = E \cap_{i=1}^{\dim \phi(E)} L_i \cap_{j=\dim \phi(E)+1}^{\dim X - 2} M_j,$$

where we choose $L_i \in |L|$ and $M_i \in |M|$ to be the general members with $L = f^* A$ the pullback of a very ample divisor A on Y and with M a very ample divisor on X. Then since C is a contractible curve on a normal (actually nonsingular) surface $S = \cap_{i=1}^{\dim \phi(E)} L_i \cap_{j=\dim \phi(E)+1}^{\dim X - 2} M_j$, we have

$$E \cdot C = (C^2)_S < 0 \quad \text{(by Theorem 4-6-1)} \quad \text{and} \quad [C] \in R.$$

If $E \overset{\neq}{\subset} \text{Exc}(\phi)$, then there would exist a curve $G \not\subset E$ and $[G] \in R$, which would imply

$$E \cdot G \geq 0 \quad \text{and} \quad [G] \in R,$$

a contradiction! Thus the exceptional locus $\text{Exc}(\phi) = E$ consists of a unique irreducible divisor.

(ii) Take a (Weil) divisor D on Y. For a curve C with $[C] \in R$, we have

$$E \cdot C < 0$$

by the proof of (i). Therefore, if we set $r = -\frac{\phi_*^{-1} D \cdot C}{E \cdot C}$, then

$$(\phi_*^{-1} D + rE) \cdot C = 0,$$

i.e.,

$$\phi_*^{-1} D + rE = 0 \text{ on } R.$$

Since X is $\mathbb{Q}$-factorial, there exists $d \in \mathbb{N}$ such that $d(\phi_*^{-1} D + rE)$ is a Cartier divisor. By Theorem 8-1-3 (vi) there exists a Cartier divisor G on Y such that

$$d(\phi_*^{-1} D + rE) = \phi^* G.$$

Now by taking the pushforward, we conclude that

$$dD = \phi_* d(\phi_*^{-1} D + rE) = \phi_* \phi^* G = G,$$

i.e., dD is a Cartier divisor. Since D is an arbitrary (Weil) divisor, Y is $\mathbb{Q}$-factorial.

Now that Y is $\mathbb{Q}$-factorial, we can write the ramification formula

$$K_X = \phi^* K_Y + aE \qquad \text{for some } a \in \mathbb{Q}.$$

By taking the intersection with a curve C with $[C] \in R$,

$$0 > K_X \cdot C = (\phi^* K_Y + aE) \cdot C = a(E \cdot C),$$

we conclude that $a > 0$, since $E \cdot C < 0$.

This implies, since X has only terminal singularities,

$$a(v, Y, \emptyset) \geq a(v, X, \emptyset) > 0$$

for the discrete valuations v exceptional over X, and

$$a(E, Y, \emptyset) = a > 0.$$

Therefore,

$$a(v, Y, \emptyset) > 0$$

for all the discrete valuations exceptional over Y. Thus Y has only terminal singularities.

(iii) This follows immediately from the exact sequence of Theorem 8-1-3 (vi).

$\square$

| Contraction of Flipping Type |

Here we note only the following very basic feature of the contractions of flipping type. For more details, we refer the reader to Chapter 9.

Proposition 8-2-2. Let $\phi = \mathrm{cont}_R : X \to Y$ be the contraction of an extremal ray R (with respect to K_X) of flipping type on a normal projective variety X with only $\mathbb{Q}$-factorial and terminal singularities. Then the canonical divisor K_Y is not $\mathbb{Q}$-Cartier, and hence Y has neither $\mathbb{Q}$-factorial nor terminal singularities.

PROOF.   Suppose K_Y is $\mathbb{Q}$-Cartier. Then since ϕ is small, i.e., $\mathrm{codim}_X \mathrm{Exc}(\phi) \geq 2$, we have $K_X = \phi^* K_Y$. But this would imply that for any curve $C \in R$, which is contracted to a point by ϕ we would have

$$K_X \cdot C = \phi^* K_Y \cdot C = K_Y \cdot \phi_* C = 0,$$

contradicting the negativity of the extremal ray R with respect to K_X. $\square$

| Contraction of Fibering Type |

The contraction of an extremal ray $\phi : X \to Y$ of fibering type is nothing but a Mori fiber space by definition, the study of which deserves a whole chapter if not an entire book. (See Section 1-4 for the discussion of Mori fiber spaces in dimension 2.) Here we prove only a small result about the target space Y.

Proposition 8-2-3. Let $\phi = \mathrm{cont}_R : X \to Y$ be the contraction of an extremal ray R of fibering type on a normal projective variety X with only $\mathbb{Q}$-factorial and terminal singularities. Then Y is $\mathbb{Q}$-factorial.

PROOF.   Take a prime divisor D on Y. Then there exists a prime divisor D' on X such that $\phi(D') = D$. (Exercise! Why?) Since X is $\mathbb{Q}$-factorial, there exists $d \in \mathbb{N}$ such that dD' is a Cartier divisor. Moreover, taking a curve C on X such that

$$\phi(C) = \mathrm{pt.} \quad \text{and} \quad \phi(C) \notin D,$$

we compute

$$dD' \cdot C = 0 \text{ and thus } dD' = 0 \text{ on } R.$$

Therefore, by Theorem 8-1-3 (vi) there exists a Cartier divisor G on Y such that

$$dD' = \phi^* G.$$

This implies

$$\text{Supp} D = \phi(D') = \phi(\phi^* G) = \text{Supp} G.$$

Since D is a prime divisor, G must be some multiple of D, and thus D is $\mathbb{Q}$-Cartier. This proves the proposition.   □

Exercise 8-2-4. State the basic properties of the contraction of an extremal ray on a log pair (X, D) with only $\mathbb{Q}$-factorial and log terminal singularities and prove them.

Remark 8-2-5. We will see in Chapter 10 that $\text{Exc}(\phi)$ is covered by rational curves, and thus every irreducible component of $\text{Exc}(\phi)$ is uniruled.

8.3   Examples

In this section we give some examples of the contractions of extremal rays (with respect to the canonical divisor K_X) discussed in the previous sections.

First, we present the five types of contractions of extremal rays of divisorial type on a nonsingular projective 3-fold, which, according to Mori [2], actually exhausts all the possiblilities if the contractions are birational.

Second, we give an example of the contraction of an extremal ray of flipping type on a nonsingular n-fold with $n \geq 4$. When $n = 4$, according to Kawamata [10], the example is analytically the only contraction of flipping type on a nonsingular n-fold. We note that there is no contraction of an extremal ray (with respect to the canonical divisor K_X) of flipping type on a nonsingular 3-fold.

Third, we give some examples of the contractions of extremal rays of fibering type on nonsingular projective 3-folds.

Example 8-3-1 (Contractions of Extremal Rays of Divisorial Type on Nonsingular Projective 3-Folds). There are five types, Types (E_1), (E_2), (E_3), (E_4), and (E_5), according to Mori [2], of contractions of extremal rays of divisorial type on nonsingular projective 3-folds.

> Type (E_1): Inverse of the blowup of a nonsingular projective 3-fold along a nonsingular curve.

Figure 8-3-2.

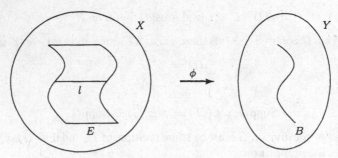

Take a nonsingular projective 3-fold Y and a nonsingular curve $B \subset Y$. Blow up Y along B to obtain $\phi : X \to Y$. The restriction of the morphism to the exceptional locus $\phi|_E : E = \phi^{-1}(B) \to B$ is a $\mathbb{P}^1$-bundle, and let l be one of the rulings. Now we claim that ϕ is the contraction of the extremal ray $R_l = \mathbb{R}_{\geq 0}[l]$ with respect to K_X. In fact, we check all the conditions in Definition 8-1-1 for $\phi : X \to Y$ to be an extremal contraction

(i) ϕ is not an isomorphism,

(ii) for a curve $C \subset X$ (cf. the proof of Proposition 8-2-1)

$$\phi(C) = \text{pt.} \implies K_X \cdot C = (\phi^* K_Y + E) \cdot C < 0$$

(iii) for curves $C, D \subset X$

$\phi(C) = \phi(D) = \text{pt.} \iff C$ and D are the rulings of the $\mathbb{P}^1$-bundle $\phi|_E : E \to B$

$\implies C$ is algebraically equialent to $D \implies [C] = [D]$ in $H_2(X, \mathbb{R})$,

(iv) ϕ has connected fibers with Y being nonsingular and projective.

Therefore, by the contraction theorem $\phi : X \to Y$ is the contraction of an extremal ray, which is nothing but $R_i = \mathbb{R}_{\geq 0}[l]$, with respect to K_X.

Type (E_2) : Inverse of the blowup of a nonsingular projective 3-fold at a point.

Figure 8-3-3.

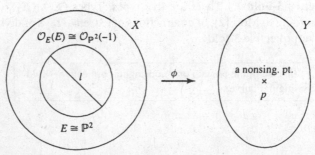

Take a nonsingular projective 3-fold Y and blow up a point $p \in Y$ to obtain $\phi : X \to Y$. The exceptional locus $E = \phi^{-1}(p)$ is isomorphic to $\mathbb{P}^2$, whose normal bundle in X is $\mathcal{N}_{E/X} \cong \mathcal{O}_{\mathbb{P}^2}(-1)$. Take a line l in E. Then just as in Type E_1 we see that ϕ is the contraction of the extremal ray $R_l = \mathbb{R}_{\geq 0}[l]$ generated by l.

Type (E_3) : Inverse of the blowup of a point analytically isomorphic to
$$0 \in \operatorname{Spec}\mathbb{C}[x, y, z, w]/(x^2 + y^2 + z^2 + w^2]),$$
where all the curves in the exceptional divisor $E \cong \mathbb{P}^1 \times \mathbb{P}^1$ are numerically proportional.

Take a nonsingular projective 3-fold Z that contains an irreducible curve C with one ordinary double point p, i.e., in a neighborhood of the singular point the curve is analytically given by

$$C = \{xy = z = 0\} \subset Z \overset{\text{analytically}}{\cong} \mathbb{A}^3 = \operatorname{Spec}\mathbb{C}[x, y, z].$$

Blow up Z along the curve C to obtain a birational morphism $\psi : Y \to Z$ from a normal projective 3-fold Y that has a uique singular point q over p. By direct computation we see that $q \in Y$ is analytically isomorphic to

$$0 \in \operatorname{Spec}\mathbb{C}[x, y, z, w]/(x^2 + y^2 + z^2 + w^2).$$

Blow up q to obtain

$$\phi : X \to Y.$$

Figure 8-3-4.

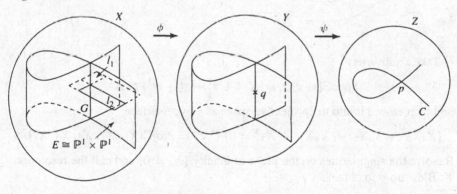

Observe that the exceptional locus E of ϕ is a divisor isomorphic to $\mathbb{P}^1 \times \mathbb{P}^1$; the two rulings l_1 and l_2 are numerically equivalent, since they (actually $G + l_1$ and $G + l_2$ as in Figure 8-3-4) are algebraically equivalent, parametrized by the same curve C, and hence any curve in E gives rise to the same 1-dimensional ray $R_l = \mathbb{R}_{\geq 0}[l_1] = \mathbb{R}_{\geq 0}[l_2]$. A direct computation shows that

$$K_X = \phi^* K_Y + 2E.$$

One can see as before ϕ is the contraction of the extremal ray R_l.

Exercise related to the example of Type (E_3).

(i) Compare the example of Type (E_3) with

$$f : X \to Y = \mathrm{Proj}\,\mathbb{C}[X_0, X_1, X_2, X_3, X_4]/(X_0^2 + X_1^2 + X_3^2) \subset \mathbb{P}^4,$$

the blowup of the vertex of the cone over the smooth quadric, where the two rulings of the exceptional locus $E \cong \mathbb{P}^1 \times \mathbb{P}^1$ are *not* numerically proportional.

(ii) The curve G (see Figure 8-3-4 in the example of Type (E_3)) generates a ray that is extremal but not negative with respect to K_X. Show that $\mathbb{R}_+[G]$ is an extremal ray of flopping type.

Type (E_4) : Inverse of the blowup of a point analytically isomorphic to
$$0 \in \mathrm{Spec}\,\mathbb{C}[x, y, z, w]/(x^2 + y^2 + z^2 + w^3]).$$

Figure 8-3-5.

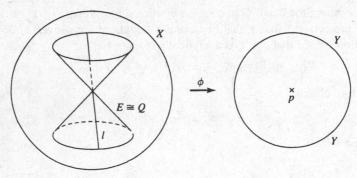

Take a subvariety

$$\mathrm{Spec}\,\mathbb{C}[x, y, z, w]/(x^2 + y^2 + z^2 + w^3) \subset \mathbb{A}^4$$

and then embed it into the projective space as a hypersurface

$$\{X_4 X_0^2 + X_4 X_1^2 + X_4 X_2^2 + X_3^3 = 0\} \subset \mathbb{P}^4 = \mathrm{Proj}\,\mathbb{C}[X_0, X_1, X_2, X_3, X_4].$$

Resolve the singularities on the plane at infinity $\{X_4 = 0\}$ and call the resolution Y. Blow up p to obtain

$$\phi : X \to Y.$$

By computing directly the blowup of the origin of $\mathrm{Spec}\,\mathbb{C}[x, y, z, w]/(x^2 + y^2 + z^2 + w^3)$, we see that X is nonsingular with the unique exceptional divisor E over P and that

$$E \cong Q \text{ the singular quadric} \subset \mathbb{P}^3 \text{ with } \mathcal{O}_E(E) \cong \mathcal{O}_Q(-1).$$

The ramification formula reads

$$K_X = \phi^* K_Y + E.$$

The same argument as before shows that ϕ is the contraction of the extremal ray $R_l = \mathbb{R}_+[l]$, where l is a line in the singular quadric $E \cong Q \subset \mathbb{P}^3$.

Type (E_5) : Inverse of the blowup of a point analytically isomorphic to $0 \in \mathrm{Spec}\mathbb{C}[x, y, z]^{(i)}$.

Figure 8-3-6.

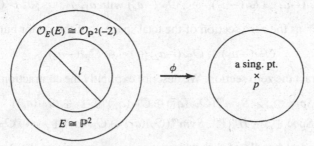

This contracts the exceptional divisor E isomorphic to $\mathbb{P}^2$ with normal bundle $\mathcal{O}_E(E) \cong \mathcal{O}_{\mathbb{P}^2}(-2)$ to the given singular point, which is the vertex of the cone over the Veronese surface. This is the type we already discussed in Example 3-1-3 in Chapter 3.

Remark 8-3-7. We remark that the contraction of Type (E_3) or Type (E_4) as well as Type (E_5) introduces a singularity on the resulting space Y, as discussed in Obstacle 3-1-2 (to carry out the minimal model program in the category of nonsingular projective varieties) in Chapter 3. This shows a clear contrast to the Castelnuovo's contractibility criterion in dimension 2, where the resulting space Y remains nonsingular after the contraction of an extremal ray of divisorial type.

Example 8-3-8 (Contractions of Extremal Rays of flipping type on Necessarily Singular Projective 3-Folds and on Nonsingular Projective n-Folds with $n \geq 4$). First we would like to bring the following basic fact to the attention of the reader:

There is *no* contraction of an extremal ray of flipping type on a nonsingular 3-fold.

This follows from deformation theory, which tells us that a curve whose class belongs to the extremal ray (with respect to K_X) has negative intersection with the canonical divisor and hence that it deforms and sweeps out the locus of at least dimension 2. We refer the reader to Chapter 10 for some more detail of the argument involving deformation theory.

(i) Francia–Hironaka example of contraction of an extremal ray of flipping type on a singular 3-fold.

We refer the reader to Example 3-1-9 in Chapter 3. Some more examples of contractions of extremal rays of flipping type on singular 3-folds will be given in

Chapter 14 in the framework of toric geometry. (Actually, the Francia–Hironaka example is just one of these toric examples.) For a more elaborate discussion of the description of the contractions of extremal rays of flipping type on singular 3-folds we refer the reader to the research papers Mori [5] and Kollár–Mori [1].

(ii) Examples of contractions of flipping type on nonsingular n-folds with $n \geq 4$. We start with the projective space $\mathbb{P}^d$ and negative line bundles on it

$$\mathcal{O}_{\mathbb{P}^d}(-a_1), \mathcal{O}_{\mathbb{P}^d}(-a_2), \cdots, \mathcal{O}_{\mathbb{P}^d}(-a_e) \text{ with } a_1, a_2, \cdots, a_e > 0.$$

We embed $\mathbb{P}^d$ as the zero section of the total space X of the vector bundle

$$\mathcal{O}_{\mathbb{P}^d}(-a_1) \oplus \mathcal{O}_{\mathbb{P}^d}(-a_2) \oplus \cdots \oplus \mathcal{O}_{\mathbb{P}^d}(-a_e).$$

We can contract the zero section. We describe explicitly the contraction as follows:

$$\phi : X = Spec \oplus_{m \geq 0} Sym^m(\mathcal{O}_{\mathbb{P}^d}(a_1) \oplus \mathcal{O}_{\mathbb{P}^d}(a_2) \oplus \cdots \oplus \mathcal{O}_{\mathbb{P}^d}(a_e))$$
$$\rightarrow Y = Spec \oplus_{m \geq 0} H^0(\mathbb{P}^d, Sym^m(\mathcal{O}_{\mathbb{P}^d}(a_1) \oplus \mathcal{O}_{\mathbb{P}^d}(a_2) \oplus \cdots \oplus \mathcal{O}_{\mathbb{P}^d}(a_e))).$$

Since the normal bundle of the zero section is isomorphic to the vector bundle itself, the adjunction formula tells us that

$$K_X \cdot l = (-d - 1 + \Sigma a_i),$$

where l is a line in $\mathbb{P}^d$.

Now we choose d and the a_i such that the above intersection number is negative. Note that since we require $\text{codim}_X \text{Exc}(\phi) \geq 2$ we must have $2 \leq e = n - d$ and that n must be at least 4 for the solution for d and the a_i to exist to make $K_X \cdot l < 0$.

When $n = 4$, the only solution is

$$d = 2, \qquad a_1 = a_2 = 1.$$

Actually, as in Kawamata [10], the contractions of extremal rays of flipping type on nonsingular 4-folds are all analytically isomorphic to this example.

(If we insist on working in the category of projective varieties, then we take some nonsingular projectivizations of X and Y.)

Example 8-3-9 (Contractions of fibering type on Nonsingular Projective 3-Folds).

We have the following five types: Types (C_1), (C_2), (D_1), (D_2), and (D_3), depending on the description of the fibers of the contraction as in Mori [2].

Case: dim $Y = 2$.

Type (C_1) : $\phi : X \rightarrow Y$ is a conic bundle with a singular fiber.

Take a divisor $X \subset \mathbb{P}^2 \times \mathbb{P}^2$ of bidegree $(2, 2)$ and $\phi : X \rightarrow Y = \mathbb{P}^2$ to be the first (or second) projection.

Type (C_2) : $\phi : X \rightarrow Y$ is a $\mathbb{P}^1$-bundle.

We leave it to the reader to show that such a projective morphism is indeed the contraction of an extremal ray and to give examples.

In general, Mori [2] shows that any contraction of an extremal ray of fibering type on a nonsingular projective 3-fold with dim $Y = 2$ is a conic bundle over a nonsingular surface Y.

Case: dim $Y = 1$.

Type (D_1): The general fiber of $\phi : X \to Y$ is a Del Pezzo surface of degree d with $1 \leq d \leq 6$.

Let $\tau : X \to \mathbb{P}^1 \times \mathbb{P}^2$ be a double cover whose branch locus is a smooth divisor of bidegree $(2, 4)$ in $\mathbb{P}^1 \times \mathbb{P}^2$. Then $\phi = p_1 \circ \tau : X \to Y = \mathbb{P}^1$ is a contraction of fibering type whose general fiber is a Del Pezzo surface of degree 1.

Let X be the blowup of V_2 with center an elliptic curve that is the intersection of two members of $|-\frac{1}{2}K_{V_2}|$, where V_2 is a double cover of $\mathbb{P}^3$ whose branch locus is a smooth quartic surface. Then $\phi : X \to Y = \mathbb{P}^1$, where $Y = \mathbb{P}^1$ parametrizes the pencil of $|-\frac{1}{2}K_{V_2}|$ containing the two members, gives a contraction of fibering type whose general fiber is a Del Pezzo surface of degree 2.

Let X be the blowup of $\mathbb{P}^3$ with center the intersection of two cubics. Then $\phi : X \to Y = \mathbb{P}^1$, where $Y = \mathbb{P}^1$ parametrizes the pencil of $|\mathcal{O}_{\mathbb{P}^3}(3)|$ containing the two cubics, gives a contraction of fibering type whose general fiber is a Del Pezzo surface of degree 3.

Let X be the blowup of the quadric $Q \subset \mathbb{P}^4$ with center the intersection of two members in $|\mathcal{O}_Q(2)|$. Then $\phi : X \to Y = \mathbb{P}^1$, where $Y = \mathbb{P}^1$ parametrizes the pencil of $|\mathcal{O}_Q(2)|$ containing the two members, gives a contraction of fibering type whose general fiber is a Del Pezzo surface of degree 4.

Let V_5 be a complete intersection of a linear subspace $\mathbb{P}^6$ in $\mathbb{P}^9$ and the Grassmann variety $\mathrm{Grass}(2,5)$ embedded in $\mathbb{P}^9$ by the Plücker embedding. Then $V_5 \subset \mathbb{P}^6$ is a Fano 3-fold whose general hyperplane section is a Del Pezzo surface of degree 5. Let X be the blowup of V_5 along an elliptic curve that is the intersection of two hyperplane sections. Then $\phi : X \to Y = \mathbb{P}^1$, where $Y = \mathbb{P}^1$ parametrizes the pencil of hyperplanes containing the elliptic curve, gives a contraction of fibering type whose general fiber is a Del Pezzo surface of degree 5.

The contractions of fibering type of Type (D_1) with the general fiber a Del Pezzo surface of degree $1 \leq d \leq 5$ can be found on some Fano 3-folds, but not the one of degree $d = 6$ (cf. Mori–Mukai [1], Matsuki [6]).

The following construction of an example of the contraction of an extremal ray of fibering type of Type (D_1) whose general fiber is a Del Pezzo surface of degree 6 was kindly communicated to the author by Professor Mori.

Let S be a Del Pezzo surface of degree 6. Since S is a 3-point blowup of $\mathbb{P}^2$, it has six (-1)-curves $E_0, \cdots, E_5$ aligned in a cyclic way (see Figure 8-3-10).

Figure 8-3-10.

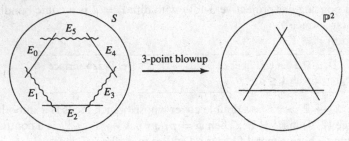

Observe that there is an automorphism $\sigma : S \to S$ (Exercise! Check!) such that it moves the (-1)-curves in such a way that

$$\sigma(E_i) = E_{i+1} \bmod 6$$

and that

$$\sigma^6 = \text{identity}.$$

Take an elliptic curve B with $\tau : B \to B$ a translation of order 6.

Now we take X to be the quotient of the product $S \times B$ by the automorphism (σ, τ) of order 6, i.e.,

$$X = S \times B / \langle (\sigma, \tau) \rangle,$$

and $\phi : X \to Y = B/\langle \tau \rangle$ to be the natural projection. Then $\phi : X \to Y$ is the contraction of fibering type of Type (D_1) whose fibers are all isomorphic to the Del Pezzo surface of degree 6. (Exercise! Check that all the curves in the fibers are numerically proportional.)

Mori [2] shows that there is no contraction of an extemal ray of fibering type from a nonsingular projective 3-fold to a curve whose general fiber is a Del Pezzo surface of degree 7 or a one-point blowup of $\mathbb{P}^2$.

Type (D_2) : ϕ is a quadric bundle.

Let $\tau : X \to \mathbb{P}^1 \times \mathbb{P}^2$ be a double cover of $\mathbb{P}^1 \times \mathbb{P}^2$ whose branch locus is a divisor of bidegree $(2, 2)$. Then $\phi = p_1 \circ \tau : X \to Y = \mathbb{P}^1$ is a contraction of fibering type, which is a quadric bundle.

Type (D_3) : ϕ is a $\mathbb{P}^2$-bundle.

We leave it to the reader to show that such a projective morphism is indeed the contraction of an extremal ray and to give examples.

CHAPTER 9

Flip

The purpose of this chapter is to study the main features of "**flip**," the key operation in the center of the minimal model program, surrounding the two major conjectures (**existence of flip**) and (**termination of flips**). This chapter, however, reveals very little of the most important operation "flip," discussing almost nothing deep in the direction toward (existence of flip). Our hope is that this introductory book will expose the reader to the subject without too much technical difficulty and motivate him to venture into the core of the theory afterwards.

In Section 9-1 we show that the conjecture (existence of flip) is equivalent to another basic conjecture that claims that the (local) **canonical ring** should be **finitely generated** (cf. Conjecture 3-1-17).

(Existence of flip) is a deep theorem of Mori [5] in dimension 3, via Kawamata's criterion [9] reducing the problem to (existence of flop) through Reid's "general elephant" conjecture (cf. Kollár–Mori [1]). This beautiful and intricate argument to show the existence of flip in dimension 3 would take up an entire book for itself. There are some new developments in the logarithmic category by Shokurov [2] and others. (See also Kollár et al. [1] and Takagi [1].) We will *not* discuss any of these exciting stories here in this chapter of our book. We also refer the reader to Kawamata [9] and Kollár–Mori [2][3] for the proofs of (existence of flip) in the relatively easier but still substantial case of semistable flips.

In Section 9-2 we verify (termination of flips) in dimension 3 introducing the notion of "**difficulty**" by Shokurov, which measures how much the operation of a flip improves the singularities by keeping a close eye on the behavior of discrepancies.

9.1   Existence of Flip

In this section we show that the *backbone* conjecture of the Mori program, which claims that the (local) canonical ring should be fintely generated, implies (existence of flip).

Definition 9-1-1 (Flip). *Let*

$$\phi : X \to Y$$

be a small contraction of an extremal ray with respect to K_X from a normal projective variety X with only $\mathbb{Q}$-factorial and terminal singularities, i.e., it satisfies the conditions:

(i) *ϕ is a birational morphism onto a normal projective variety Y,*
(ii) *ϕ is small, i.e., $\mathrm{codim}_X \mathrm{Exc}(\phi) \geq 2$,*
(iii) *$-K_X$ is ϕ-ample, and*
(iv) *all the curves in the fibers of ϕ are numerically proportional, i.e., $\rho(X/Y) = 1$.*

Then the morphism $\phi^+ : X^+ \to Y$ (if it exists) is called a **"flip"** *(of ϕ) if it satisfies the following conditions:*

(i$^+$) *ϕ^+ is a birational morphism from a normal projective variety X^+ with only $\mathbb{Q}$-factorial and terminal singularities onto Y,*
(ii$^+$) *ϕ^+ is small, i.e., $\mathrm{codim}_{X^+} \mathrm{Exc}(\phi^+) \geq 2$,*
(iii$^+$) *K_{X^+} is ϕ^+-ample, and*
(iv$^+$) *all the curves in the fibers of ϕ^+ are numerically proportional, i.e., $\rho(X^+/Y) = 1$.*

Once it exists, a flip of ϕ is unique, as the following proposition shows.

Proposition 9-1-2 ((Existence of Flip) $\iff$ Finite Generation of (Local) Canonical Ring). *Let*

$$\phi : X \to Y$$

be a small contraction of an extremal ray with respect to K_X from a normal projective variety X with only $\mathbb{Q}$-factorial and terminal singularities. Then a flip $\phi^+ : X^+ \to Y$ of ϕ exists iff the local canonical ring

$$R := \oplus_{m \geq 0} \phi_* \mathcal{O}_X(mK_X)$$

is finitely generated as an $\mathcal{O}_Y$-algebra, in which case the flip has the description

$$\phi^+ : X^+ = \mathrm{Proj}\, R \to Y$$

and hence is unique.

PROOF.   Suppose that the flip $\phi^+ : X^+ \to Y$ exists. Then

$$R := \oplus_{m \geq 0} \phi_* \mathcal{O}_X(mK_X)$$

$$= \oplus_{m \geq 0} \mathcal{O}_Y(m K_Y)$$
$$= \oplus_{m \geq 0} \phi^+_* \mathcal{O}_{X^+}(m K_{X^+})$$

is finitely generated as an $\mathcal{O}_Y$-algebra, since K_{X^+} is ϕ^+-ample.

Conversely, suppose that the local canonical ring

$$R := \oplus_{m \geq 0} \phi_* \mathcal{O}_X(m K_X)$$

is finitely generated as an $\mathcal{O}_Y$-algebra. Set

$$\phi^+ : X^+ := Proj \oplus_{m \geq 0} \phi_* \mathcal{O}_X(m K_X) \to Y.$$

By construction, ϕ^+ is a birational morphism from a normal variety X^+ onto Y. We claim it is also small. Assume that it is not small, i.e., there exists a divisor E on X^+ exceptional for ϕ^+. Then, on the one hand, we have the exact sequence

$$0 \to \mathcal{O}_{X^+} \to \mathcal{O}_{X^+}(E) \to \text{Coker} \to 0,$$

where Coker is clearly a nonzero sheaf. On the other hand, take a positive integer $r \in \mathbb{N}$ such that

$$R_{[r]} := \oplus_{m \geq 0} \phi_* \mathcal{O}_X(m r K_X)$$

is generated by the first-degree term $\phi_* \mathcal{O}_X(r K_X)$ as an $\mathcal{O}_Y$-algebra and let $\mathcal{O}_{X^+}(1)$ be the corresponding ϕ^+-very ample invertible sheaf on $X^1 = Proj\ R_{[r]}(= Proj\ R)$, such that

$$\phi^+_* \mathcal{O}_{X^+}(1) = \phi_* \mathcal{O}_X(r K_X).$$

Serre's theorem (cf. Theorem 1-1-7) implies that there exists a sufficiently divisible $l \in \mathbb{N}$ such that the natural homomorphism

$$\phi^{+*} \phi^+_* \{\text{Coker} \otimes \mathcal{O}_{X^+}(l)\} \to \text{Coker} \otimes \mathcal{O}_{X^+}(l) \text{ is surjective}$$

and

$$R^1 \phi^+_* \mathcal{O}_{X^+}(l) = 0.$$

But in the exact sequence

$$0 \to \phi^+_* \mathcal{O}_{X^+}(l) \xrightarrow{i} \phi^+_* \{\mathcal{O}_{X^+}(E) \otimes \mathcal{O}_{X^+}(l)\} \to \phi^+_* \{\text{Coker} \otimes \mathcal{O}_{X^+}(l)\}$$
$$\to R^1 \phi^+_* \mathcal{O}_{X^+}(l) = 0,$$

the map i is actually an isomorphism, since E is exceptional for ϕ^+. In fact, the commutative diagram for an open set $U \subset Y$ shows that i is an isomorphism

$$
\begin{array}{ccc}
\Gamma(U, \mathcal{O}_Y(lr K_Y)) & = & \Gamma(U - \phi^+(E), \mathcal{O}_Y(lr K_Y)) \\
\| & & \| \\
\Gamma(U, \phi^+_* \mathcal{O}_{X^+}(l)) & \hookrightarrow & \Gamma(U - \phi^+(E), \phi^+_* \mathcal{O}_{X^+}(l)) \\
| & & \| \\
\Gamma(U, \phi^+_* \mathcal{O}_{X^+}(E) \otimes \mathcal{O}_{X^+}(l)) & \hookrightarrow & \Gamma(U - \phi^+(E), \phi^+_* \mathcal{O}_{X^+}(E) \otimes \mathcal{O}_{X^+}(l))
\end{array}
$$

where the equality in the first row holds because $\mathcal{O}_Y(lrK_Y)$ is reflexive (or by definition). This would imply $\phi^+{}_*\text{Coker} \otimes \mathcal{O}_{X^+}(l) = 0$, which then would imply that $\text{Coker} \otimes \mathcal{O}_{X^+}(l)$ is a zero sheaf and hence so is Coker, a contradiction!

In the following we use the notation Z_{n-1} for the group of Weil divisors and Div for the group of Cartier divisors.

Now that ϕ and ϕ^+ have both been shown to be small by the above argument, taking the strict transform we have the isomorphism of $\mathbb{Q}$-Weil divisors

$$Z_{n-1}(X) \otimes \mathbb{Q} \xrightarrow{\sim} Z_{n-1}(X^+) \otimes \mathbb{Q}.$$

On the other hand, since X is $\mathbb{Q}$-factorial, Theorem 8-1-3 (vi) implies

$$Z_{n-1}(X) \otimes \mathbb{Q} = \phi^*(\text{Div}(Y) \otimes \mathbb{Q}) \oplus \mathbb{Q} \cdot K_X.$$

Therefore, we conclude that

$$Z_{n-1}(X^+) = \phi^{+*}(\text{Div}(Y) \otimes \mathbb{Q}) \oplus \mathbb{Q} \cdot K_{X^+}.$$

Since $\mathcal{O}_{X^+}(rK_{X^+}) \cong \mathcal{O}_{X^+}(1)$ is invertible, K_{X^+} is $\mathbb{Q}$-Cartier. This implies that X^+ is $\mathbb{Q}$-factorial and

$$\rho(X^+/Y) = 1.$$

Finally, we are left with the task of proving that X^+ has only terminal singularities. $\square$

Lemma 9-1-3 (Behavior of Discrepancies Under Flip). *Let*

$$
\begin{array}{ccc}
X & \dashrightarrow & X^+ \\
\phi \searrow & & \swarrow \phi^+ \\
& Y &
\end{array}
$$

be the commutative diagram of birational maps obtained by setting

$$X^+ = \text{Proj} \oplus_{m \geq 0} \phi_* \mathcal{O}_X(mK_X).$$

Then

$$a(\nu, X; \emptyset) \leq a(\nu, X^+; \emptyset)$$

for all discrete valuations ν. Moreover,

$$a(\nu, X; \emptyset) < a(\nu; X^+, \emptyset)$$

iff the center of ν on X is contained in $\text{Exc}(\phi)$ (or equivalently iff the center of ν on X^+ is contained in $\text{Exc}(\phi^+)$).

PROOF.   Proof of the lemma

Choose a sufficiently divisible $l \in \mathbb{N}$ so that lK_X is Cartier and that lK_{X^+} is ϕ^+-very ample. Take a common resolution

$$X \xleftarrow{\sigma} V \xrightarrow{\sigma^+} X^+$$

such that

$$\sigma^* lK_X = M + \Sigma r_i F_i,$$

where M is the $\sigma \circ \phi$-movable part of $\sigma^* l K_X$ and $\Sigma r_i F_i$ with $r_i \geq 0$ is the $\sigma \circ \phi$-fixed part. More precisely,

$$\sigma^* \mathcal{O}_X(l K_X) = \mathcal{O}_V(M) \otimes \mathcal{O}_V(\Sigma r_i F_i)$$
$$= \operatorname{im}\{(\phi \circ \sigma)^*(\phi \circ \sigma)_* \sigma^* \mathcal{O}_X(l K_X) \to \sigma^* \mathcal{O}_X(l K_X)\} \otimes \mathcal{O}_V(\Sigma r_i F_i),$$

where the common resolution is chosen such that

$$\operatorname{im}\{(\phi \circ \sigma)^*(\phi \circ \sigma)_* \sigma^* \mathcal{O}_X(l K_X) \to \sigma^* \mathcal{O}_X(l K_X)\} = \mathcal{O}_V(M)$$

is an invertible subsheaf of $\sigma^* \mathcal{O}_X(l K_X)$. Since $l K_{X^+}$ is ϕ^+-very ample, we have

$$\operatorname{im}\{(\phi \circ \sigma)^*(\phi \circ \sigma)_* \sigma^* \mathcal{O}_X(l K_X) \to \sigma^* \mathcal{O}_X(l K_X)\}$$
$$= \operatorname{im}\{(\phi \circ \sigma)^* \phi_* \mathcal{O}_X(l K_X) \to \sigma^* \mathcal{O}_X(l K_X)\}$$
$$= \operatorname{im}\{(\phi^+ \circ \sigma^+)^* \phi^+_* \mathcal{O}_{X^+}(l K_{X^+}) \to \sigma^* \mathcal{O}_X(l K_X)\}$$
$$= \sigma^{+*} \operatorname{im}\{\phi^{+*} \phi^+_* \mathcal{O}_{X^+}(l K_{X^+}) \to \mathcal{O}_{X^+}(l K_{X^+})\} \subset \sigma^* \mathcal{O}_X(l K_X)$$
$$= \sigma^{+*} \mathcal{O}_{X^+}(l K_{X^+}) \subset \sigma^* \mathcal{O}_X(l K_X).$$

That is to say,

$$M = \sigma^{+*} l K_{X^+}.$$

Moreover, the center of ν on V is contained in $\Sigma r_i F_i$ iff the center of ν on X is contained in $\operatorname{Exc}(\phi)$ (or equivalently iff the center of ν on X^+ is contained in $\operatorname{Exc}(\phi^+)$). In fact, since ϕ is an isomorphism outside of $\operatorname{Exc}(\phi)$, we have

$$\sigma(\Sigma r_i F_i) \subset \operatorname{Exc}(\phi).$$

If the above inclusion is not an equality, then there exists a curve C such that

$$C \not\subset \sigma(\Sigma r_i F_i), \ C \subset \operatorname{Exc}(\phi) \text{ and } \phi(C) = \text{pt}.$$

But then taking a curve G on V such that $\sigma(G) = C$, we compute

$$0 > l K_X \cdot C \geq l K_X \cdot \sigma_* D = \sigma^* l K_X \cdot D = (M + \Sigma r_i F_i) \cdot D \geq 0,$$

a contradiction. Therefore, we must have

$$\sigma(\Sigma r_i F_i) = \operatorname{Exc}(\phi).$$

Looking at the ramification formula

$$K_V = \sigma^* K_X + \Sigma a(\nu, X; \emptyset) E_\nu$$
$$= \sigma^{+*} K_{X^+} + \Sigma \frac{r_i}{l} F_i + \Sigma a(\nu, X; \emptyset) E_\nu$$
$$= \sigma^{+*} K_{X^+} + \Sigma a(\nu, X^+; \emptyset),$$

we see the assertion of the lemma:

$$\begin{cases} a(\nu, X; \emptyset) = a(\nu, X^+; \emptyset) & \text{if the center of } \nu \not\subset \operatorname{Exc}(\phi) = \sigma(\Sigma r_i F_i), \\ a(\nu, X; \emptyset) < a(\nu, X; \emptyset) + \dfrac{r_i}{l} = a(\nu, X^+; \emptyset) & \text{if the center of } \nu \subset \operatorname{Exc}(\phi) = \sigma(\Sigma r_i F_i). \end{cases}$$

$\square$

Now we go back to the proof that X has only terminal singularities.

If the discrete valuation ν is exceptional over X^+, then it is also exceptional over X, since both ϕ and ϕ^+ are small. Thus since X has only terminal singularities,

$$0 < a(\nu, X, \emptyset) \le a(\nu, X^+, \emptyset).$$

This completes the proof of Proposition 9-1-2.   □

9.2   Termination of Flips

In this section we prove **(termination of flips) in dimension 3** following the observation of Shokurov [1] on the behavior of discrepancies under flips:

Let

$$X = X_0\cdots\rightarrow X_0^+ = X_1\cdots\rightarrow X_1^+ = X_2\cdots\rightarrow \quad\cdots\quad X_{i-1}^+ = X_i\cdots\rightarrow X_i^+ = X_{i+1}\cdots$$
$$\searrow \quad\swarrow\searrow\quad\swarrow\searrow\quad\swarrow\searrow\cdots\swarrow\searrow\quad\swarrow\cdots$$
$$Y = Y_0 \qquad Y_1 \qquad\quad Y_2 \qquad\cdots\qquad Y_i \qquad\cdots$$

be a sequence of flips.

We observe by Lemma 9-1-3 that if we fix one discrete valuation ν, then the discrepancies form an increasing sequence under flips

$$a(\nu, X = X_0; \emptyset) \le a(\nu, X_1; \emptyset) \le \cdots \le -a(\nu, X_i; \emptyset) \le a(\nu, X_{i+1}; \emptyset)\cdots.$$

Shokurov realized that instead of focusing our attention on one discrete valuation if we consider the discrepancies for all the discrete valuations collectively, then we come up with the following notion of "**difficulty**," which measures effectively how much the singularities are improved by a flip in dimension 3.

Definition 9-2-1 (Difficulty).   *Let X be a normal projective variety of dimension 3 with only terminal singularities. Then the "**difficulty**" $d(X)$ is defined to be*

$$d(X) := \#\{\nu; a(\nu, X, \emptyset) < 1 \quad and \quad \nu \text{ exceptional over } X\}.$$

It is easy to see that $d(X)$ is a well-defined nonnegative integer, since any discrete valuation ν with $a(\nu, X, \emptyset) < 1$ and ν exceptional over X has to show up as an exceptional divisor for a resolution $f : Y \to X$ and there are only finitely many divisors on Y exceptional over X.

Now we can see that (termination of flips) holds in dimension 3 immediately by observing that the "difficulty" drops under each flip.

Theorem 9-2-2 (Termination of Flips in Dimension 3).   *(Termination of flips) in dimension 3 holds, since for each flip*

$$X \qquad\cdots\longrightarrow\qquad X^+$$
$$\phi\searrow \qquad\qquad \swarrow\phi^+$$
$$Y$$

the invariant "difficulty" strictly drops:

$$d(X) > d(X^+) \geq 0.$$

PROOF.   Obviously, we have to show only that the difficulty strictly drops under a flip, since a sequence of flips induces a sequence of strictly decreasing nonnegative integers

$$d(X = X_0) > d(X_1) > \cdots > d(X_i) > d(X_{i+1}) > \cdots \geq 0.$$

Let E^+ be an irreducible component of $\text{Exc}(\phi^+)$. Note that since ϕ^+ is a small morphism from a 3-fold X^+, the irreducible component E^+ is a curve. Since X^+ has only terminal singularities and since the singular locus of terminal singularities has codimension at least 3 (cf. Corollary 4-6-6), we conclude that the generic point of E^+ is sitting in the smooth locus of X^+. Let ν be the discrete valuation obtained by blowing up X^+ along the generic point of E^+. Then by Lemma 9-1-3,

$$a(\nu, X, \emptyset) < a(\nu, X^+, \emptyset) = (\dim X^+ - \dim E^+) - 1 = 1,$$

which implies

$$d(X) > d(X^+).$$

This completes the proof of (termination of flips) in dimension 3.   □

Remark 9-2-3.

(Termination of flips) in higher dimension becomes a more subtle problem. At present it is established only in dimension 4 by Kawamata–Matsuda–Matsuki [1] (and of course in dimension 3), where we not only follow the behavior of "difficulty" but also the behavior of the codimension-2 cycles. For the details, we refer the reader to the above-mentioned paper. There are also some developments concerning (termination of log flips) where we also seek some new invariants to measure what is improved by a flip, e.g., the notion of minimal discrepancy, etc. We refer the reader to Shokurov [2], Kawamata [15], Kollár et al. [1] for further details.

CHAPTER 10

Cone Theorem Revisited

The purpose of this chapter is to present Mori's original idea (cf. Mori [1][2]) to prove the cone theorem (in the smooth case) through his ingenious method of **"bend and break"** to produce rational curves of some bounded degree (with respect to an ample divisor or to the canonical divisor). This method leads to the result of Miyaoka–Mori [1] claiming the **uniruledness** of Mori fiber spaces, yielding the generalization by Kawamata [13] claiming that (every irreducible component of) the exceptional locus of an extremal contraction is also uniruled in general.

In Section 10-1 we discuss how to construct rational curves via **deformations** starting from a curve C having negative intersection with the canonical divisor. First we plunge into the world of positive characteristic to create lots of deformations of the curve via the Frobenius map. We observe that even after fixing some points $B \subset C$ there are still enough deformations (bend). Secondly, we observe that since we fix some points, the original curve must break into rational curves under deformation (break). Finally, we come back to the world of charactersitic zero by lifting the argument via modulo p reduction.

In Section 10-2 as an easy application of the results in Section 10-1 we give Mori's original proof of the cone theorem in the smooth case (without referring to the contraction theorem or to the rationality theorem).

Section 10-3 is devoted to Kawamata's elegant extension of the bend and break technique showing that each extremal ray is generated by a rational curve of low degree even in the singular and/or logarithmic case and that every irreducible component of the exceptional locus of an extremal contraction is also uniruled.

10.1 Mori's Bend and Break Technique

The purpose of this section is to construct rational curves through **Mori's bend and break technique**. We use the method of "**modulo p reduction**," and hence this is the only section in the book where we pay special attention to the characteristic $\mathrm{char}(k) = p$ of the base field k (which is assumed to be algebraically closed). We will actually switch back and forth between the world of positive characteristics and that of characteristic zero. We follow closely the original papers Miyaoka–Mori [1] and Mori [1][2].

Theorem 10-1-1 (Creating Rational Curves of Low Degree by Bend and Break). *Let X be a nonsingular projective variety over an algebraically closed field k of characteristic $\mathrm{char}(k) \geq 0$ and C a curve on X. Fix an ample divisor H on X. Suppose*

$$K_X \cdot C < 0.$$

Then for every closed point $x \in C$ there exists a rational curve l containing x such that

$$H \cdot l \leq 2 \dim X \frac{H \cdot C}{-K_X \cdot C}.$$

PROOF. Step 1 (**Bend**). Creating "lots" of deformations (and "enough" of them even after fixing some points) in positive characteristic.

In Step 1, we assume that

$$p = \mathrm{char}(k) > 0.$$

Let $\nu : \tilde{C} \to C$ be the normalization and $F^m : \tilde{C} \to \tilde{C}$ be the Frobenius morphism of degree $q = p^m$. Let $B \subset \tilde{C}$ be the closed subscheme of degree b. Then we estimate the dimension of the deformation space of the morphism $f = \nu \circ F^m : \tilde{C} \to X$ fixing the subscheme B. (The only specific information we use in the following argument is the estimate of the dimension of the deformation space below. The reader may take it as a black box if he wishes not to go into the details of deformation theory. For the basics and details of deformation theory we refer the reader to Mori's original papers [1][2] or Kollár [15].)

(Estimate of the dimension of the deformation space)

$$\dim_{[f]} \mathrm{Def}(f, B) \geq \alpha := \chi(\tilde{C}, \mathcal{I}_B f^* T_X)$$
$$= q(-K_X \cdot C) - \dim X(g(C) - 1) - \dim X \cdot b.$$

Note that if we take $m \gg 0$ (and hence $q \gg 0$) and

$$b = \left[\frac{1}{\dim X} \{q(-K_X \cdot C) - 1\} - (g(C) - 1) \right], \text{ where [ ] denotes the Gauss symbol,}$$

then

$$\dim_{[f]} \mathrm{Def}(f, B) \geq \alpha > 0,$$

and hence f has a nontrivial deformation fixing B.

Step 2 (**Break**). Observe the breaking of the original curve via the rigidity lemma.

Proposition 10-1-2. *Let $\tilde{C}$ be a nonsingular projective curve and $f : \tilde{C} \to C = f(\tilde{C}) \subset X$ a morphism onto a curve in a nonsingular projective variety X over an algebraically closed field k. (We put no condition on the characteristic of k in this proposition.) Fix an ample divisor H on X. Let b be a positive integer such that for any subscheme $B \subset C$ of degree b there exists a nontrivial deformation $\mathcal{F}$ of f parametrized by a variety D of positive dimension fixing B:*

$$\mathcal{F} : \tilde{C} \times D \to X \text{ with } \mathcal{F}|_{B \times D} = f|_B \circ p_1 : B \times D \to X.$$

(When $g(\tilde{C}) = 0$, we further assume $b \geq 3$.) Then for every closed point $x \in C$ there exists a rational curve l containing x such that

$$H \cdot l < 2 \cdot \frac{\deg f^* H}{b}.$$

PROOF. Let Δ be a smooth compactification (possibly after taking the normalization) of a general curve in D containing $[f]$. Then $\mathcal{F}$ induces a rational map

$$\tau : \tilde{C} \times \Delta \dashrightarrow X.$$

We eliminate the indeterminacy of the rational map τ by taking a sequence of blowups $\mu : S \to \tilde{C} \times \Delta$. The key point of our argument is to show that τ cannot be a morphism and thus there exists an exceptional curve (which is a rational curve) that maps onto the (almost) desired rational curve we are looking for. Let E_{ij} denote the total transform of the exceptional divisor obtained by blowing up an (infinitely near) point over $\{x_i\} \times \Delta$, $x_i \in B \subset \tilde{C}$. Note that since $b > 0$ in the case $g(\tilde{C}) = g(C) > 0$ and since $b \geq 3$ in the case $g(\tilde{C}) = 0$, we may assume that Δ does not arise from the automorphisms of $\tilde{C}$ and that hence the morphism $h = \tau \circ \mu$ maps S onto an irreducible surface (i.e., not onto C). On the other hand, the strict transform Δ'_i of $\{x_i\} \times \Delta$ is mapped onto a single point $f(x_i)$, since $x_i \in B$ is fixed by the deformation $\mathcal{F}$. Therefore,

$$(h^* H)^2 > 0, \tag{10.1}$$

$$h^* H \cdot \Delta'_i = 0. \tag{10.2}$$

Let $\tilde{C}_\delta$ and Δ_x denote the total transform of $\tilde{C} \times \{\delta\}$ and $\{x\} \times \Delta$ for general points $\delta \in \Delta$ and $x \in \tilde{C}$. Since

$$h^* H \cdot \tilde{C}_\delta = H \cdot h_* \tilde{C}_\delta = H \cdot f_* \tilde{C} = d = \deg f^* H, \tag{10.3}$$

we can write

$$h^* H \equiv a\tilde{C}_\delta + d\Delta_x - \Sigma_{i,j} e_{i,j} E_{i,j} + F, \tag{10.4}$$

where F is some divisor with

$$F \cdot \tilde{C}_\delta = F \cdot \Delta_x = F \cdot E_{i,j} = 0 \quad (\forall i, j) \qquad (10.5)$$

and where

$$a = h^* H \cdot \Delta_x \geq 0, \quad e_{i,j} = h^* H \cdot E_{i,j} \geq 0 \; \forall i, j. \qquad (10.6)$$

Note that

$$\Delta_i' \equiv \Delta_x - \Sigma \epsilon_{i,j} E_{i,j}, \; \text{where } \epsilon_{i,j} = 0 \text{ or } 1. \qquad (10.7)$$

Note also that

$$F^2 \leq 0 \qquad (10.8)$$

by the Hodge index theorem and by (5).

Plugging formulas (4), (5), (6), (7), and (8) into (1) and (2) we obtain

$$2ad - \Sigma_{i,j} e_{i,j}^2 > 0,$$
$$a - \Sigma_j \epsilon_{i,j} e_{i,j} = 0.$$

Therefore, we get

$$2d \Sigma_j \epsilon_{i,j} e_{i,j} - \Sigma_{i,j}(\epsilon_{i,j} e_{i,j})^2 \geq 2d \Sigma_j \epsilon_{i,j} e_{i,j} - \Sigma_{i,j} e_{i,j}^2 > 0.$$

Summing up the formulas above in the indices i, we get

$$2d \Sigma_{i,j} \epsilon_{i,j} e_{i,j} - b \Sigma_{i,j}(\epsilon_{i,j} e_{i,j})^2 > 0.$$

Therefore, for some indices i and j we have

$$0 < \epsilon_{i,j} e_{i,j} = e_{i,j} = h^* H \cdot E_{i,j} = H \cdot h_* E_{i,j} < \frac{2d}{b}.$$

(Note that this implies that g is never a morphism and the image of $E_{i,j}$ gives rise to the rational curves.) Since $E_{i,j} \cdot \Delta_i' = \epsilon_{i,j} > 0$, $h_* E_{i,j}$ contains $f(x_i)$. Pick an irreducible component $f(x_i) \in l \subset h_* E_{i,j}$. Then $f(x_i) \in l$ is a rational curve with

$$0 < H \cdot l \leq H \cdot h_* E_{i,j} < \frac{2d}{b}.$$

Now the above argument shows that for any choice of distinct b points $x_1, \ldots, x_b$ there exists at least one point x_i that is contained in a rational curve l of degree $H \cdot l < 2d/b$. Therefore, we conclude that there exists a set S of at most $b - 1$ distinct points on $\tilde{C}$ such that for each point $\tilde{x} \in U = \tilde{C} - S$ we can find a rational curve l with

$$x = f(\tilde{x}) \in l \text{ and } 0 < H \cdot l < \frac{2d}{b}.$$

On the other hand, the rational curves l with $0 < H \cdot l < 2d/b$ belong to a finite number of components in the Hilbert scheme over k. (In fact, since the boundedness of the degree also implies that the arithmetic genus $g_a(l) = h^1(l, \mathcal{O}_l)$ is bounded (Exercise! Why?), we have only finite possibilities for the Hilbert function. Therefore, such rational curves belong to a finite number of components in the Hilbert scheme.) Let u : Univ$_\mathcal{H} \to \mathcal{H}$ be the universal family over the union $\mathcal{H}$ of such components. Take the locus

$$\mathcal{H}_{rat} = \text{the closure of } \{p \in \mathcal{H}; u^{-1}(p) \text{ is an irreducible rational curve}\}.$$

It is easy to see that $u^{-1}(p)$ consists only of rational curves for any $p \in \mathcal{H}_{rat}$. There must be a component Z of $\mathcal{H}_{rat}$ such that im(Univ$_\mathcal{H} \times_{\mathcal{H}_{rat}} Z) \subset X$ contains $f(U)$ and thus contains C. This proves that for every point $x \in C$ there exists a rational curve $x \in l$ with

$$0 < H \cdot l < \frac{2d}{b}. \qquad \qquad \square$$

Remark 10-1-3. What is implicit in the above argument is the following rigidity lemma (cf. Mumford [5], page 43):

Lemma 10-1-4 (Rigidity Lemma). *Let Δ be a complete variety, $\tilde{C}$ and X any varieties, and $\tau : \tilde{C} \times \Delta \to X$ a morphism such that for some $x_o \in \tilde{C}$, $\tau(\{x_o\} \times \Delta)$ is a single point of X. Then there is a morphism $g : \tilde{C} \to X$ such that*

$$\tau = g \circ p_1,$$

where $p_1 : \tilde{C} \times \Delta$ is the first projection.

If $\tau : \tilde{C} \times \Delta \dashrightarrow X$ were a morphism fixing a point $x_o \in B$, then $\tau(\Delta_{x_o}) = \tau(\{x_o\} \times \Delta) = f(x_o)$ and hence by the rigidity lemma $\tau(\Delta_x) = f(x)$ $\forall x \in \tilde{C}$. But this contradicts the choice of a nontrivial deformation D. So τ cannot be a morphism, and somewhere on the surface $\tilde{C} \times \Delta$ we have to blow up to make τ a morphism. The curve $\tilde{C} = \tilde{C} \times [f]$ breaks up into a reducible curve where we blow up the surface and the image of the exceptional divisor is a rational curve. This is the reason why we want to fix a point $x_o \in B$ in order for bend and break to work.

Actually, if we use the rigidity lemma explicitly in the proof of the proposition, then without assuming $b \geq 3$ even when $g(\tilde{C}) = 0$ we can conclude that

$$\epsilon_{i,j} e_{i,j} > 0 \text{ for some indices } i \text{ and } j$$

and get the inequality

$$0 < \epsilon_{i,j} e_{i,j} \leq \frac{2d}{b} \text{ for some indices } i \text{ and } j,$$

where the second nonstrict inequality follows from $f^*H^2 \geq 0$. Therefore, without the extra assumption $b \geq 3$ even when $g(\tilde{C}) = 0$ we can conclude that for every point $x \in C$ there exists a rational curve $x \in l$ such that

$$H \cdot l \leq 2 \cdot \frac{\deg f^*H}{b}.$$

The argument also shows that $f_*(\tilde{C})$ breaks up as 1-cycles

$$f_*(\tilde{C}) \equiv l + Z,$$

where l is a rational curve and Z is some nonzero effective 1-cycle with

$$H \cdot f_*(\tilde{C}) > H \cdot l \text{ and } H \cdot f_*(\tilde{C}) > H \cdot Z$$

and

$$B \subset l \cup Z.$$

Now we carry the argument for Step 2 in positive characteristic based upon Proposition 10-1-2 and Step 1. By our choice of the Frobenius map f of deg $f = d = q = p^m$ and

$$b = \left[\frac{1}{\dim X} \{ q(-K_X \cdot C) - 1 \} - (g(C) - 1) \right],$$

we see for $m \gg 0$ that we have a rational curve $x \in l$ as described in Proposition 10-1-2 such that

$$0 < H \cdot l < \frac{2d}{b} = \frac{2q}{b} H \cdot C \approx 2 \dim X \frac{H \cdot C}{-K_X \cdot C}.$$

Since $H \cdot l$ is an integer, we have

$$0 < H \cdot l \leq 2 \dim X \frac{H \cdot C}{-K_X \cdot C}.$$

Step 3. Prove the assertion in characteristic 0 via modulo p reduction.

In Steps 1 and 2 we have already proved Theorem 10-1-1 in any positive characteristic. Here we lift the argument to characteristic zero by modulo p reduction.

Take a point $x \in C$. Then we may assume that X, H, x, and C are all defined over some integral domain R finitely generated over $\mathbb{Z}$. By eliminating finitely many primes, we may assume that X and C arc smooth over $\operatorname{Spec} R$. Then over an arbitrary prime of positive characteristic in $\operatorname{Spec} R$, we find a rational curve $x \in l$ with $H \cdot l \leq 2 \dim X \frac{H \cdot C}{-K_X \cdot C}$. On the other hand, such rational curves are contained in a finite number of components in the Hilbert scheme over R, since the degree is bounded. Thus the union of such components is projective over R. If we take the fiber of the universal family over the generic point, it is nonempty and provides the desired rational curve $x \in l$ in characteristic 0 over k.

Finally, we show how to break a rational curve having high intersection with $-K_X$ into rational curves having low intersection with $-K_X$ by a similar argument. This will be used in the proof of the cone theorem in the next section.

Theorem 10-1-4. *Let C be a rational curve on a nonsingular variety X over an algebraically closed field k (no condition on the characteristic of k). Then C breaks up into an effective sum of rational curves l_i as 1-cycles*

$$C \equiv \Sigma a_i l_i, \quad a_i \in \mathbb{N},$$

such that

$$-K_X \cdot l_i \leq \dim X + 1.$$

If we pick a closed point $x \in C$, then we may also require

$$x \in l_i \text{ for some } i.$$

PROOF. If $-K_X \cdot C \leq \dim X + 1$, there is nothing more to prove. Thus we may assume $-K_X \cdot C > \dim X + 1$. Let $\nu : \tilde{C} \to C$ be the normalization. Set $f = i \circ \nu : \tilde{C} \to X$. Then we estimate the dimension of the deformation space of the morphism f fixing one point $p = B \in \tilde{C}$:

$$\dim_{[f]} \text{Def}(f, B) \geq (-K_X \cdot C) - \dim X(g(C) - 1) - \dim X > \dim X + 1.$$

Under the assumption it cannot happen that $\dim X = 1$, since otherwise the rational curve $C = X$ is nonsingular and hence

$$-K_X \cdot C = -\deg K_C = 2 = \dim X + 1,$$

a contradiction. Therefore, we have $\dim X \geq 2$ and hence $\dim_{[f]} \text{Def}(f, B) > \dim X + 1 \geq 3$. Therefore, there is a curve $[f] \in D \in \text{Def}(f, B)$ not contained in the locus of deformations arising from the automorphism of $\tilde{C} \cong \mathbb{P}^1$ fixing p, which is of dimension 3. Therefore, by Remark 10-1-3, $f_*(\tilde{C}) = C$ breaks up as 1-cycles

$$C \equiv l + Z = \Sigma b_j l_j,$$

where the L_j are all rational curves with

$$H \cdot C > H \cdot l_j > 0.$$

Now by the induction on the degree of the curve with respect to H we see the assertion of the theorem. The assertion fixing a point is also immediate by carrying the argument choosing p with $\nu(p) = x$.   □

10.2   A Proof in the Smooth Case After Mori

In this section we present Mori's [2] original proof of the cone theorem (in the form modified by Kollár [15]) in the smooth case using the results of Section 10-1 obtained via the method of bend and break.

Theorem 10-2-1 (Cone Theorem for the Smooth Case). *Let X be a nonsingular projective variety over an algebraically closed field k (no condition on characteristic). Then*

$$\overline{NE}(X) = \overline{NE}(X)_{K_X \geq 0} + \Sigma \mathbb{R}_+[l],$$

where the l are rational curves with

$$0 < -K_X \cdot l \leq \dim X + 1$$

and the $\mathbb{R}_+[l]$ are discrete in the half-space $\overline{NE}(X)_{K_X < 0}$.

PROOF.   Set

$$\text{R.H.S.} = \overline{\text{NE}}(X)_{K_X \geq 0} + \Sigma \mathbb{R}_+[l].$$

Step 1. Show that

$$\overline{\text{NE}}(X) = \text{ the closure of R.H.S.}$$

Suppose

$$\overline{\text{NE}}(X) \overset{\neq}{\supset} \text{ the closure of R.H.S.}$$

Then there exists $0 \neq M \in N^1(X)$ and $0 \neq z \in \overline{\text{NE}}(X)$ such that

$$M \cdot z = 0,$$
$$M \geq 0 \text{ on } \overline{\text{NE}}(X),$$
$$M > 0 \text{ on the closure of R.H.S.} - \{0\}.$$

Note that by Kleiman's criterion for ampleness if we take an ample divisor H on X, then the set $C = \{\text{the closure of R.H.S.}\} \cap \{H \cdot = 1\}$ is compact and $H \cdot C \in \mathbb{N}$ for any curve C on X. Therefore (by taking some multiple of M if necessary), we may assume that $M \geq 1$ on C and hence that

$$M \cdot C \geq 1 \quad \forall \text{ a curve } C \subset X \text{ with } [C] \in \text{ the closure of R.H.S.}$$

Now take a sequence $\{M_i\}$ of ample $\mathbb{Q}$-divisors converging to M,

$$M_i = \frac{H_i}{m_i} \to M \text{ as } i \to \infty,$$

where the H_i are ample Cartier divisors with $m_i \in \mathbb{N}$, and a sequence $\{z_i\}$ of $\mathbb{Q}$-effective 1-cycles converging to z,

$$z_i = \Sigma a_{i,j} C_{i,j} \to z \text{ as } i \to \infty,$$

where the $C_{i,j}$ are curves on X and $a_{i,j} \in \mathbb{Q}_{>0}$. For each $i(\gg 0)$ there exists an index $j(i)$ such that

$$-K_X \cdot C_{i,j(i)} > 0 \text{ and } \frac{M_i \cdot C_{i,j(i)}}{-K_X \cdot C_{i,j(i)}} \leq \frac{M_i \cdot z_i}{-K_X \cdot z_i}.$$

Now applying Theorem 10-1-1 to H_i and $C_{i,j(i)}$ and then Theorem 10-1-4, we see that there exists a rational curve l_i such that

$$0 < M_i \cdot l_i \leq 2 \dim X \frac{M_i \cdot C_{i,j(i)}}{-K_X \cdot C_{i,j(i)}} \text{ and } -K_X \cdot l_i \leq \dim X + 1.$$

. The second inequality implies

$$[l_i] \in \text{ the closure of R.H.S.} - \{0\}.$$

Since $M \geq 1$ on C, we have for $i \gg 0$ that $M_i > \frac{1}{2}$ on C and hence $M_i \cdot l_i > \frac{1}{2}$.
 On the other hand, for $i \gg 0$ the first inequality implies

$$0 < M_i \cdot l_i \leq 2 \dim X \frac{M_i \cdot C_{i,j(i)}}{-K_X \cdot C_{i,j(i)}} \approx 2 \dim X \frac{M \cdot z}{-K_X \cdot z} = 0$$

and hence

$$0 < M_i \cdot l_i \leq 2 \dim X \frac{M_i \cdot C_{i,j(i)}}{-K_X \cdot C_{i,j(i)}} < \frac{1}{2},$$

a contradiction.

Step 2. Show that

$$\overline{NE}(X) = \text{R.H.S.}$$

Fix an ample divisor H on X. For $0 < \epsilon \ll 1$ we observe that

$$\overline{NE}(X) = \text{ the closure of R.H.S.}$$

$$= \text{ the closure of } (\overline{NE}(X)_{K_X+\epsilon H \geq 0} + \Sigma_l \text{ rational }_{,-K_X \cdot l \leq \dim X + 1, K_X + \epsilon H \cdot l < 0} \mathbb{R}_+[l])$$

$$= \overline{NE}(X)_{K_X+\epsilon H \geq 0} + \Sigma_l \text{ rational }_{,-K_X \cdot l \leq \dim X + 1, K_X + \epsilon H \cdot l < 0} \mathbb{R}_+[l],$$

where in the last equality we can eliminate the process of taking the closure, since the rational curves with

$$-K_X \cdot l \leq \dim X + 1 \text{ and } K_X + \epsilon H \cdot l < 0$$

have the bounded degree

$$0 < H \cdot l \leq \frac{\dim X + 1}{\epsilon}$$

and hence give rise to only a finite number of rays $\mathbb{R}_+[l]$. This also implies that the $\mathbb{R}_+[l]$ are discrete in the half-space $\overline{NE}(X)_{K_X<0}$, and now the equality $\overline{NE}(X) =$ R.H.S. holds by taking $\epsilon \to 0$.

This completes the proof.   □

10.3   Lengths of Extremal Rays

As an application of Mori's bend and break technique we present Kawamata's [13] result on the **lengths of extremal rays**, from which he elegantly derives two more important results: the **discreteness of extremal rays** (after Mori [2] and Kollár [1]) and the **uniruledness of the exceptional locus of a contraction morphism** (generalizing the result of Miyaoka–Mori [1]). We also remark that the rationality theorem holds not only for ample divisors but also for nef divisors, an observation made by Keel–McKernan that finds application in the proof of the log abundance theorem for 3-folds (cf. Keel–Matsuki-McKernan [1]).

Theorem 10-3-1 (Lengths of Extremal Rays). *Let X be a normal projective variety with a boundary $\mathbb{Q}$-divisor $D = \Sigma d_i D_i$, $0 \leq d_i \leq 1$, such that the log pair (X, D) has only $\mathbb{Q}$-factorial and log terminal singularities. Let $\phi : X \to Y$ be a morphism such that $-(K_X + D)$ is ϕ-ample. Then every irreducible component E of $\text{Exc}(\phi)$ is covered by rational curves l such that*

$$0 < -(K_X + D) \cdot l \leq 2(\dim E - \dim \phi(E)) \text{ with } \phi(l) = \text{ a point on } Y.$$

In particular, since an extremal ray R (with respect to $K_X + D$) gives rise to a contraction morphism $\phi = \text{cont}_R : X \to Y$ with $-(K_X + D)$ being ϕ-ample, it is generated by such a rational curve l whose length is bounded as above.

Postponing the proof of the above theorem, we present its immediate and important corollaries.

Corollary 10-3-2 (Discreteness of Extremal Rays Proved Again). *Let (X, D) be as in Theorem 10-3-1. Then the extremal rays are discrete in the half-space $\overline{NE}(X)_{K_X + D < 0}$.*

PROOF. Fix an ample divisor A on X. We have only to prove that for any $\epsilon > 0$ there are only finitely many extremal rays R in the half-space $\overline{NE}(X)_{K_X + D + \epsilon A < 0}$. If we take a rational curve l that generates R and satisfies the inequality as in Theorem 10-3-1,

$$0 < -(K_X + D) \cdot l \le 2(\dim E - \dim \phi(E)) \le 2 \dim X,$$

then

$$A \cdot l < \frac{-(K_X + D) \cdot l}{\epsilon} \le \frac{2 \dim X}{\epsilon},$$

meaning that the degree of l with respect to A is bounded. Thus such rational curves belong to a finite number of components in the Hilbert scheme, with only finite possibilities for the Hilbert function. Therefore, such rational curves can give rise to only finitely many numerical equivalence classes and hence only finitely many extremal rays.   □

Corollary 10-3-3 (Uniruledness of the Exceptional Locus of a Contraction Morphism). *Let (X, D) and $\phi : X \to Y$ be as in Theorem 10-3-1 (e.g., $\phi = \text{cont}_R : X \to Y$ is the contraction of an extremal ray R). Then every irreducible component E of $\text{Exc}(\phi)$ is uniruled. In particular, every (log) Mori fiber space is uniruled.*

PROOF. By Theorem 10-3-1 any irreducible component E of the exceptional locus $\text{Exc}(\phi)$ is covered by rational curves l such that

$$0 < -(K_X + D) \cdot l \le 2(\dim E - \dim \phi(E)) \le 2 \dim X \text{ with } \phi(l) = \text{ a point on } Y.$$

Fix an ample divisor A on X and B on Y. Since $-(K_X + D)$ is ϕ-ample, there exists $0 < \epsilon \ll 1$ such that

$$K_X + D + \epsilon A \cdot l < 0$$

for all such rational curves as above. Then as in the above corollary, we see that

$$A \cdot l < \frac{-(K_X + D) \cdot l}{\epsilon} \le \frac{2 \dim X}{\epsilon}$$

and that these rational curves belong to a finite number of components in the Hilbert scheme. Let $u : \text{Univ}_{\mathcal{H}} \to \mathcal{H}$ be the universal family over the union $\mathcal{H}$ of such components. Take the locus

$$\mathcal{H}_{\text{rat}} = \text{the closure of } \{p \in \mathcal{H}; u^{-1}(p) \text{ is an irreducible rational curve}\}.$$

It is easy to see (Exercise! Why?) that $u^{-1}(p)$ consists only of rational curves for any $p \in \mathcal{H}_{\text{rat}}$. Now there exists an irreducible component Z' (with the reduced structure) of $\mathcal{H}_{\text{rat}}$ such that

$$\text{Univ}_{\mathcal{H}} \times_{\mathcal{H}_{\text{rat}}} Z' \to X$$

is surjective. By taking some multiple hyperplane sections of Z', we may assume $\dim Z' = \dim X - 1$. (Exercise! Why?) Take the normalization U of $\{Univ_{\mathcal{H}} \times_{H_{\text{rat}}} Z'\}_{\text{red}}$. Then $U \to Z'$ is generically a conic bundle, and hence there exists a double cover $Z \to Z'$ such that

$$\mathbb{P}^1 \times Z \overset{\text{birational}}{\dashrightarrow} U \times_{Z'} Z \overset{\text{dominant}}{\to} X,$$

providing the necessary uniruledness structure to X.   □

Corollary 10-3-4 (Rationality Theorem and Boundedness of the Denominator for Nef Divisors). *The rationality theorem and boundedness of the denominator hold for* nef *divisors as well as* ample *divisors: Let* (X, D) *be as in Theorem 10-3-1 and let H be a* nef *Cartier divisor on X. (Note that in Theorem 7-1-1 H was assumed to be ample.) Suppose $K_X + D$ is not nef. Then the nonnegative number*

$$r := \sup\{t \in \mathbb{R}; H + t(K_X + D) \text{ is nef}\}$$

is a rational number. Moreover, there exists an extremal ray R with respect to $K_X + D$ such that

$$(H + r(K_X + D)) \cdot R = 0.$$

Furthermore, expressing

$$\frac{r}{e} = \frac{p}{q} \text{ with } p, q \in \mathbb{N} \quad g.c.d.(p, q) = 1,$$

where

$$e := \min\{i \in \mathbb{N}; i(K_X + D) \text{ is Cartier}\},$$

we have

$$q \leq 2e \dim X.$$

PROOF.   By Theorem 10-3-1 for each extremal ray R with respect to $K_X + D$ there exists a rational curve l that generates R and satisfies the inequality

$$0 < -(K_X + D) \cdot l \leq 2 \dim X.$$

On the other hand, the cone theorem implies

$$r = \inf \left\{ \frac{H \cdot l}{-(K_X + D) \cdot l} ; R = \mathbb{R}_+[l] \text{ is an extremal ray} \right\}.$$

Since

$$1 \leq -e(K_X + D) \cdot l \leq 2e \dim X \text{ and } -e(K_X + D) \cdot l \in \mathbb{N},$$

there exists an extremal ray R with a rational curve l as the generator such that

$$\frac{r}{e} = \frac{H \cdot l}{-e(K_X + D) \cdot l} = \frac{p}{q} \text{ with } p, q \in \mathbb{N}, \quad \gcd(p, q) = 1,$$

and

$$q \leq 2e \dim X.$$

The rest of the section is devoted to the proof of Theorem 10-3-1.

First note that by taking hyperplane cuts $\dim \phi(E)$ times via general members of a very ample divisor on Y and replacing X and D, we may assume that E maps to a point on Y. (Exercise! Justify this first reduction step. Why is it enough to prove the general hyperplane cuts to be covered by rational curves with the specified bounded degree with respect to $-(K_X + D)$?)

Fix a very ample divisor A on X. Let $\nu : \tilde{E} \to E$ be the normalization.

Lemma 10-3-5. *Let the situation be as above. Then*

$$A^{\dim E - 1} \cdot (K_X + D) \cdot E \geq (\nu^* A)^{\dim E - 1} \cdot K_{\tilde{E}}.$$

PROOF. Proof of Lemma 10-3-5 If we take a general member $X_1 \in |A|$, $E_1 = E \cap X_1$, $D_1 = D \cap X_1$ and $\nu_1 : \tilde{E}_1 \to E_1$, then since

$$K_{X_1} \sim K_X + A|_{X_1},$$
$$K_{\tilde{E}_1} \sim K_{\tilde{E}} + \nu^* A|_{\tilde{E}_1},$$

we have

$$A^{\dim E - 1} \cdot (K_X + D) \cdot E \geq (\nu^* A)^{\dim E - 1} \cdot K_{\tilde{E}}$$
$$\iff A^{\dim E_1 - 1} \cdot (K_{X_1} + D_1) \cdot E_1 \geq (\nu_1^* A)^{\dim E_1 - 1} \cdot K_{\tilde{E}_1}.$$

Therefore, by the repetition of the procedure above, we may assume $\dim E = 1$.

Now suppose the contrary, $(K_X + D) \cdot E < \deg K_{\tilde{E}}$. Then there exists a Cartier divisor H_E on E (Exercise! Why?) such that

$$(K_X + D) \cdot E < \deg H_E \leq \deg K_{\tilde{E}},$$
$$H^0(\tilde{E}, \mathcal{O}_{\tilde{E}}(K_{\tilde{E}} - \nu^* H_E)) \neq 0.$$

This implies via the trace map

$$H^0(E, \omega_E(-H_E)) \neq 0.$$

On the other hand, by restricting our consideration to a small analytic neighborhood of E in X, we can extend H_E to an analytic Cartier divisor H on X. Since $\deg\{H - (K_X + D)\}|_E > 0$, the divisor $H - (K_X + D)$ is ϕ-ample. Thus by the analytic version of Theorem 5-2-8 (cf. Remark 5-2-9) we have

$$R^1\phi_*\mathcal{O}_X(H) = 0,$$

which would imply

$$0 = H^1(E, \mathcal{O}_E(H_E)) \cong H^0(E, \omega_E(-H_E)) \text{ by Serre duality,}$$

leading to a contradiction. This proves the lemma.   □

Now we go back to the proof of Theorem 10-3-1.

Case: $\dim E = 1$.

By the lemma, we have

$$0 < -(K_X + D) \cdot E \le -\deg K_{\tilde{E}}.$$

Therefore, $\tilde{E}$ is a smooth rational curve. Setting $l = E$, we have

$$0 < -(K_X + D) \cdot E \le -\deg K_{\tilde{E}} = 2 = 2(\dim E - \dim \phi(E)).$$

Case: $\dim E > 1$.

In this case let C be the intersection curve of $\dim E - 1$ general members of $|\nu^* A|$ on $\tilde{E}$. We may assume that C is nonsingular irreducible and lies in the smooth locus of $\tilde{E}$. Then, by Theorem 10-1-1 of Miyaoka–Mori, for any point $x \in C$ we have a rational curve $\tilde{l} \ni x$ on $\tilde{E}$ such that

$$0 < -\nu^*(K_X + D) \cdot \tilde{l} \le 2 \dim \tilde{E} \cdot \frac{-(K_X + D) \cdot C}{-K_{\tilde{E}} \cdot C} \le 2 \dim \tilde{E},$$

where

$$0 < \frac{-(K_X + D) \cdot C}{-K_{\tilde{E}} \cdot C} \le 1$$

by the lemma. Therefore, E is covered by rational curves $l = \nu_*\tilde{l}$ with

$$0 < -(K_X + D) \cdot l \le 2(\dim E - \dim \phi(E)).$$

(Note that by the reduction step at the beginning we may assume $\dim \phi(E) = 0$.) This completes the proof of Theorem 10-3-1.   □

Remark-Question 10-3-6 (Optimal Estimate (!?) $2 \dim X$ vs. $\dim X + 1$).

While Kawamata shows as above that for each extremal ray R with respect to $K_X + D$ there exists a rational curve l that generates R with the estimate of the degree

$$-(K_X + D) \cdot l \le 2 \dim X$$

even with (log terminal) singularities, Mori's original estimate of the degree for a generating rational curve for the smooth case is

$$-K_X \cdot l \le \dim X + 1.$$

In contrast to the estimate as presented in Corollary 10-3-4,

$$\frac{H \cdot l}{e K_X \cdot l} = \frac{p}{q} \quad q \le 2e \dim X,$$

the estimate for boundedness of the denominator in the rationality theorem (with log terminal singularities) as presented in Theorem 7-1-1 is

$$\frac{H \cdot l}{e K_X \cdot l} = \frac{p}{q} \quad q \le e(\dim X + 1).$$

These may be taken as an indication that the optimal estimate for the degree may be $\dim X + 1$ and not $2 \dim X$. We ask whether we could improve the estimate of the degree from the present estimate $2 \dim X$ to the conjecturally optimal $\dim X + 1$.

CHAPTER 11

Logarithmic Mori Program

The purpose of this chapter is twofold. The first is to review what we have learned so far by going over the main ingredients of the **Mori program in the framework of the logarithmic category**. The reader who feels confident of and comfortable with the material so far should have no trouble understanding this part and should even skip this review chapter. Mastery and understanding of the **log minimal model program** will be rewarded when we utilize it as an essential part of the Sarkisov program in Chapter 13. The second is to present some of the subtleties that inevitably arise as we go from the usual category to the logarithmic category. There are some open conjectures even in dimension 3, though their statements are the natural generalizations in the logarithmic category, according to Iitaka's philosophy, of the corresponding ones in the usual category.

Note that we already discussed the log birational geometry of surfaces in Section 2-2, the logarithmic extension of the Mori program in dimension 2.

11.1 Log Minimal Model Program in Dimension 3 or Higher

According to Iitaka's philosophy, we would like to generalize the usual minimal model program (MMP) to the logarithmic minimal model program (called log MMP for short), which, given an input of a logarithmic pair (X, D) consisting of a nonsingular projective variety and a boundary $\mathbb{Q}$-divisor $D = \Sigma d_i D_i$ $(0 \le d_i \le 1)$ with only normal crossings, should produce a log minimal model or a log Mori fiber space:

$$(X, D)$$

$$\downarrow$$

$$\boxed{\text{Log MMP}}$$

log canonical divisor nef $\swarrow \quad \searrow$ otherwise

$(X, D)_{\text{min}}$: a log minimal model $(X, D)_{\text{mori}}$: a log Mori fiber
space.

The algorithm of log MMP works in the following way. If the log canonical divisor $K_X + D$ is nef, then the logarithmic pair (X, D) itself is a log minimal model and the program is over. If the log canonical divisor $K_X + D$ is not nef, then via the logarithmic cone theorem (Theorem 7-2-2) and the logarithmic contraction theorem (cf. Theorem 8-1-3, Exercise 8-1-4, and Theorem 6-2-2) we proceed as follows. If the contraction of an extremal ray $\phi : X \to Y$ with respect to $K_X + D$ has the property $\dim Y < \dim X$, then $\phi : (X, D) \to Y$ is a log Mori fiber space and we are done.

If the contraction of an extremal ray $\phi : X \to Y$ with respect to $K_X + D$ has the property $\dim Y = \dim X$, i.e., if ϕ is birational, then as in the case of the usual MMP we face the problem of not being able to stay in the category of logarithmic pairs (X, D) consisting only of nonsingular projective varieties X and boundary $\mathbb{Q}$-divisors D with only normal crossings (cf. Obstacle 3-1-2): There is a logarithmic pair consisting of a nonsingular 3-fold (or in general n-fold with $n \geq 3$) and a boundary $\mathbb{Q}$-divisor D such that the log canonical divisor $K_X + D$ is not nef, and any contraction of an extremal ray $\phi : X \to Y$ with respect to $K_X + D$ is birational and results in a logarithmic pair $(Y, D_Y = \phi_*(D))$ where either Y is not nonsingular or the boundary $\mathbb{Q}$-divisor D_Y does not have normal crossings only.

Therefore, in order to establish log MMP in dimension 3 or higher, it is impossible to stay in the category of logarithmic pairs consisting only of nonsingular projective varieties and boundary $\mathbb{Q}$-divisors with only normal crossings.

The notion of log terminal singularities (cf. Definition 4-3-2) together with the $\mathbb{Q}$-factoriality provides the right category for log MMP to work, as demonstrated in the following proposition.

Proposition 11-1-1 (cf. Observation 3-1-7). *Let (X, D) be a logarithmic pair consisting of a normal projective variety X of dimension n and a boundary $\mathbb{Q}$-divisor D with only $\mathbb{Q}$-factorial and log terminal singularities. Let*

$$\phi : X \to Y$$

be the contraction of an extremal ray with respect to $K_X + D$, which is birational.

If $\operatorname{codim}_X \operatorname{Exc}(\phi) = 1$, then the exceptional locus consists of a prime divisor and $(Y, D_Y = \phi_* D)$ has also only $\mathbb{Q}$-factorial and log terminal singularities. Moreover,

$$\rho(X) = \dim_{\mathbb{R}} N^1(X) = \rho(Y) + 1.$$

If $\mathrm{codim}_X \mathrm{Exc}(\phi) \geq 2$, *then the log canonical divisor* $K_Y + D_Y$ *is not* $\mathbb{Q}$*-Cartier, and hence* (Y, D_Y) *has neither* $\mathbb{Q}$*-factorial nor log terminal singularities.*

PROOF.   The proof is the same as those of Proposition 8-2-1 and Proposition 8-2-2, replacing the canonical divisor K_X with the log canonical divisor $K_X + D$.

So we enlarge the category to that of logarithmic pairs with only $\mathbb{Q}$-factorial and log terminal singularities. In this new category, as long as our contraction of an extremal ray is birational and not small (i.e., $\mathrm{codim}_X \mathrm{Exc}(\phi) = 1$), the resulting logarithmic pair will stay in the new category, and hence we can proceed with the algorithm of the program. Recall that as soon as the contraction is not birational, we obtain a log Mori fiber space by definition and come to an end of the program.

Now as you can see from the statement of Proposition 11-1-1, the existence of a small contraction (i.e., $\mathrm{codim}_X \mathrm{Exc}(\phi) \geq 2$) will cause a problem if we try to stay in the category of logarithmic pairs with only $\mathbb{Q}$-factorial and log terminal singularities. In fact, there is an example of a logarithmic pair (X, D) consisting of a normal projective 3-fold (or in general n-fold with $n \geq 3$) and a boundary $\mathbb{Q}$-divisor D with only $\mathbb{Q}$-factorial and log terminal singularities such that the log canonical divisor $K_X + D$ is not nef, and any contraction $\phi : X \to Y$ of an extremal ray with respect to $K_X + D$ is birational and small.

The problem caused by such examples is again resolved by the following conjecture $\square$

Conjecture 11-1-2 (Existence of Log Flip) (cf. Solution 3-1-11). *Let*

$$\phi : X \to Y$$

be a small *contraction of an extremal ray with respect to* $K_X + D$ *from a logarithmic pair* (X, D) *consisting of a normal projective variety* X *and boundary* $\mathbb{Q}$*-divisor with only* $\mathbb{Q}$*-factorial and log terminal singularities. Then there exists another morphism*

$$\phi^+ : X^+ \to Y$$

from a logarithmic pair (X^+, D^+) *consisting of a normal projective variety* X^+ *and a boundary* $\mathbb{Q}$*-divisor* D^+ *with only* $\mathbb{Q}$*-factorial and log terminal singularities such that* ϕ^+ *is also small and* $K_{X^+} + D^+$ *is* ϕ^+*-ample.* D^+ *is the strict transform of* D *under the birational map* $\phi^{+-1} \circ \phi$.

Once it exists, $\phi^+ : (X^+, D^+) \to Y$ has to be necessarily unique, since it has the presentation

$$X^+ = Proj \oplus_{m \geq 0} \phi_*^+(m(K_{X^+} + D^+)) = Proj \oplus_{m \geq 0} \phi_*(m(K_X + D))$$

and

$$D^+ = (\phi^{+-1} \circ \phi)_* D.$$

It is called the **log flip** of ϕ.

Whenever we have a small contraction $\phi : (X, D) \to Y$ of an extremal ray with respect to $K_X + D$, instead of trying to proceed with $(Y, D_Y = \phi_* D)$, which is not in the category, we replace (X, D) with (X^+, D^+) and proceed in the program with (X^+, D^+), which is in our category.

Now we have gathered all the necessary ingredients to present the flowchart for log MMP in dimension 3 or higher, except the last step, which guarantees that this flowchart does not contain an infinite loop. This can be achieved via the study of the behavior of the Picard number $\rho(X)$ (cf. Proposition 11-1-1) and by the following conjecture:

Conjecture 11-1-3 (Termination of Log Flips) (cf. Solution 3-1-14). *There is no infinite sequence of log flips*

$$(X, D) = (X_0, D_0) \dashrightarrow (X_0^+, D_0^+) = (X_1, D_1) \dashrightarrow (X_1^+, D_1^+) = (X_2, D_2^+) \dashrightarrow \cdots$$
$$\searrow \qquad \swarrow \searrow \qquad\qquad \swarrow \searrow \qquad \swarrow \searrow \cdots$$
$$Y = Y_0 \qquad\qquad Y_1 \qquad\qquad Y_2 \qquad \cdots$$

As a consequence we obtain the following.

Theorem 11-1-4 (Log MMP in Dimension 3 or Higher). *The log minimal model program in dimension 3 or higher works in the category $\mathfrak{D}$ of logarithmic pairs consisting of normal projective varieties and boundary $\mathbb{Q}$-divisors with only $\mathbb{Q}$-factorial and log terminal singularities, modulo the two conjectures (existence of log flip) and (termination of log flips). Moreover, it is the smallest category containing the category $\mathfrak{E}$ of logarithmic pairs consisting of nonsingular projective varieties and boundary $\mathbb{Q}$-divisors with only normal crossings in the following sense:*

(i) *$\mathfrak{D}$ satisfies requirements (a), (b), (c), (d), and (e) of the charaterization of log MMP given below and thus is closed under log MMP, i.e., any object in $\mathfrak{D}$ can be input into log MMP with an output in $\mathfrak{D}$ after finitely many steps, while all the intermediate objects in the process stay in the category $\mathfrak{D}$.*

(ii) *$\mathfrak{D}$ contains $\mathfrak{E}$, and the notion of the ($\mathbb{Q}$-Cartier) log canonical divisor is defined for $\mathfrak{D}$ (satisfying the required properties in (d) of the charaterization of log MMP), which coincides with that of the log canonical divisor in the category $\mathfrak{E}$ upon restriction.*

(iii) *Any object in $\mathfrak{D}$ can be reached through log MMP starting from some object in $\mathfrak{E}$.*

Thus any category satisfying (i), (ii), and (iii) contains the category $\mathfrak{D}$.

PROOF. We just recall the key ingredients (a), (b), (c), (d), and (e) of the mechanism of the (log) MMP:

(a) Any object $(X, D) \in \mathfrak{D}$ in the category has the well-defined $\mathbb{Q}$-Cartier (log) canonical divisor $K_X + D$, so that the intersection numbers with curves are also well-defined.

(b) The (log) cone theorem holds for objects $X \in \mathfrak{D}$, which states that

$$\overline{NE}(X) = \overline{NE}(X)_{K_X+D \geq 0} + \Sigma R_l,$$

where the R_l are discrete one-dimensional rays with negative intersection with the (log) canonical divisor $K_X + D$.

(c) The (log) contraction theorem holds for objects $(X, D) \in \mathfrak{D}$, which states that for each extremal ray R_l there exists a morphism with connected fibers

$$\phi = \mathrm{cont}_{R_l} : X \to Y$$

from X to a normal projective variety Y such that

$$\phi(C) = \mathrm{pt.} \iff [C] \in R_l \quad \forall C \subset X.$$

(d) We require that when the contraction of an extremal ray $\phi = \mathrm{cont}_{R_l} :$ $(X, D) \to (Y, D_Y = \phi_* D)$ is divisorial (i.e., $\mathrm{codim}_X \mathrm{Exc}(\phi) = 1$), the resulting log pair is again in the category $(Y, D_Y) \in \mathfrak{D}$ and that the (log) canonical divisor $K_Y + D_Y$ of (Y, D_Y) is the pushforward of the (log) canonical divisor $K_X + D_X$ of X. We remark, thanks to (existence of log flip), that when $\phi = \mathrm{cont}_{R_l} : (X, D) \to Y$ is small (i.e., $\mathrm{codim}_X \mathrm{Exc}(\phi) \geq 2$) the local log canonical ring

$$\oplus_{m \geq 0} \phi_*(m(K_X + D))$$

is a finitely generated $\mathcal{O}_Y$-algebra and the induced morphism

$$\phi^+ : X^+ = Proj \oplus_{m \geq 0} \phi_*(m(K_X + D)) \to Y$$

is again small with $(X^+, D^+) \in \mathfrak{D}$ and that the (log) canonical divisor of (X^+, D^+) is the strict transform of the (log) canonical divisor of (X, D) (cf. Proposition 9-1-2 and its logarithmic generalization).

(e) Finally, we note that the program terminates after finitely many steps, thanks to (termination of log flips) and bookkeeping of the Picard number.

Once all the ingredients have been established, the general mechanism for the log MMP works just like that for MMP.

We remark that (existence of log flip) and (termination of log flips) in dimension 3 are both theorems established in Shokurov [2], Kawamata [15], Kollár et al. [1], and Takagi [1].

11.2   Basic Properties of Log Minimal Models and Log Mori Fiber Spaces in Dimension 3 or Higher

Once the log minimal model program is established, we have **log minimal models** or **log Mori fiber spaces** as its end results. The purpose of this section is to browse through their (conjectural) basic properties, which are very similar to those of minimal models and Mori fiber spaces.

Definition 11-2-1 (Log Minimal Model and Log Mori Fiber Space). *A logarithmic pair* (X, D) *consisting of a normal projective variety* X *and a boundary* $\mathbb{Q}$-*divisor* D *with only* $\mathbb{Q}$-*factorial and log terminal singularities is a* **log minimal model** *if the log canonical divisor* $K_X + D$ *is nef.*

It is a **log Mori fiber space** *if it has a morphism* $\phi : (X, D) \to Y$ *such that*

(i) *ϕ is a morphism with connected fibers onto a normal projective variety Y of* $\dim Y < \dim X$, *and*

(ii) *all the curves F in fibers of ϕ are numerically proportional and*

$$-(K_X + D) \cdot F > 0.$$

Remark 11-2-2.

(i) It is easy to see that $\phi : (X, D) \to Y$ is a log Mori fiber space if and only if $\phi = \mathrm{cont}_{R_l} : (X, D) \to Y$ is the contraction of an extremal ray R_l with respect to $K_X + D$ on a logarithmic pair (X, D) consisting of a normal projective variety X and a boundary $\mathbb{Q}$-divisor D with only $\mathbb{Q}$-factorial and log terminal singularities, and $\dim Y < \dim X$.

(ii) Assuming that log MMP in dimension n holds via the conjectures (existence of log flip) and (termination of log flips), we see that log minimal models or log Mori fiber spaces are characterized as the end results of log MMP in dimension n. That is to say, starting from any log pair with only $\mathbb{Q}$-factorial and log terminal singularities (in particular, starting from a logarithmic pair consisting of a nonsingular projective variety and a boundary $\mathbb{Q}$-divisor with normal crossings) as its input, log MMP in dimension n produces as an end result either a log minimal model or a log Mori fiber space defined as above. Conversely, any log minimal model or a log Mori fiber space defined as above is an end result of log MMP in dimension n, starting from some appropriate logarithmic pair consisting of a nonsingular projective variety and a boundary $\mathbb{Q}$-divisor with normal crossings.

First note the following easy but fundamental property of log Mori fiber spaces.

Theorem 11-2-3. *Let $\phi : (X, D) \to Y$ be a log Mori fiber space. Then for any $p \in X$ there exists an irreducible curve G passing through p with $(K_X + D) \cdot G < 0$, and thus*

$$H^0(X, m(K_X + D)) = 0 \quad \forall m \in \mathbb{N} \quad (i.e., \kappa(X, D) = -\infty).$$

The easy dichotomy theorem also holds just as in the usual case.

Theorem 11-2-4 (Easy Dichotomy Theorem of log MMP). *Assume that log MMP in dimension n holds via the conjectures (existence of log flip) and (termination of log flips). Then an end result of log MMP starting from a logarithmic pair consisting of a normal projective variety of dimension n and a boundary $\mathbb{Q}$-divisor with only $\mathbb{Q}$-factorial and log terminal singularities is a log Mori fiber space iff* *there exists a nonempty Zariski open set $U \subset X$ such that for any point $p \in U$ there is an irreducible curve G passing through p and $(K_X + D) \cdot G < 0$.*

In contrast to the log Mori fiber spaces, whose log Kodaira dimension $\kappa(X, D)$ is $-\infty$, one of the most basic (conjectural) properties of log minimal models is that they have nonnegative Kodaira dimension:

Conjecture 11-2-5 (Existence of an Effective Pluri-Log Canonical Divisor). *Let (X, D) be a log minimal model. Then*

$$\kappa(X, D) \geq 0.$$

Theorems 11-2-3 together with Conjecture 11-2-5 leads to the following:

Conjecture 11-2-6 (Hard Dichotomy Conjecture of Log MMP). *Let (X, D) be a logarithmic pair consisting of a nonsingular projective variety of dimension n and a boundary $\mathbb{Q}$-divisor with only normal crossings (or more generally, a logarithmic pair consisting of a normal projective variety of dimension n and a boundary $\mathbb{Q}$-divisor with only $\mathbb{Q}$-factorial and log terminal singularities). Assume that log MMP in dimension n holds via the conjectures (existence of log flip) and (termination of log flips). Then an end result of log MMP in dimension n is a log minimal model (respectively a log Mori fiber space) iff $\kappa(X, D) \geq 0$ (respectively $\kappa(X, D) = -\infty$).*

It is easy to see via the theorem of Miyaoka–Mori [1] and Kawamata [13] that a log Mori fiber space is covered by rational curves just as well a Mori fiber space is and that hence it is uniruled. But this does not reflect the log structure, and it is desirable to come up with the logarithmic version of the notion of "ruled" or "uniruled."

Definition 11-2-7 (Log Ruled and Log Uniruled). *A logarithmic pair (X, D) with only $\mathbb{Q}$-factorial and log terminal singularities is "**log ruled**" (respectively **log uniruled**) if there exists a commutative triangle of birational maps (respectively generically finite maps with the vertical arrow being a birational morphism)*

$$
\begin{array}{ccc}
(V, D_V) & & f \\
\downarrow & & \searrow \\
Y \times \mathbb{P}^1 & \dashrightarrow & (X, D)
\end{array}
$$

where (V, D_V) is a logarithmic pair consisting of a nonsingular projective variety V and a boundary $\mathbb{Q}$-divisor D_V with only normal crossings such that $f^{-1}(D) + \mathrm{Exc}(f) \subset D_V$ and we have the logarithmic ramification formula $K_V + D_V = f^*(K_X + D) + R$ for some effective divisor R, and $(K_V + D_V) \cdot G < 0$ for the strict transform G of the general fiber $\{y\} \times \mathbb{P}^1$.

Question 11-2-8. *Is a log Mori fiber space log uniruled? When is a log Mori fiber space log ruled?*

Exercise 11-2-9. In Definition 11-2-7 for the notion of "log uniruled" if we require all the coefficients of D_V to be equal to 1, then we obtain the notion of "properly log uniruled." Show that

(X, D) properly log uniruled

$\Longleftrightarrow \exists \emptyset \neq U \subset X - D$ s.t.

$\forall x \in U \quad \exists$ a nontrivial map $f : \mathbb{A}^1 \to X - D$ with $x \in f(\mathbb{A}^1)$.

Keel–McKernan [1] has recently shown that all the log Mori fiber spaces (X, D) with coefficients of D equal to 0 or 1 are properly log uniruled in dimension 2.

Now we shift our attention to minimal models.

Toward more detailed descriptions of the structure of a log minimal model, we have the following fundamental conjecture:

Conjecture 11-2-10 (Log Abundance Conjecture). *Let (X, D) be a log minimal model. Then the pluri-log canonical system $|m(K_X + D)|$ is base point free for sufficiently divisible and large $m \in \mathbb{N}$.*

Note that the log abundance conjecture gives the morphism from a log minimal model onto its log canonical model,

$$\Phi = \Phi_{|m(K_X+D)|} : (X, D) \to (X, D)_{\mathrm{can}},$$

where

$$(X, D)_{\mathrm{can}} = \mathrm{Proj} \oplus_{m \geq 0} H^0(X, m(K_X + D)).$$

This morphism in turn should give crucial information about the global structure of (X, D).

We note that the hard dichotomy conjecture and the log abundance conjecture in dimension 3 are both theorems established in Keel–Matsuki–McKernan [1]. (See also Kawamata [12], Kollár et al. [1].)

11.3 Log Birational Relations Among Log Minimal Models and Log Mori Fiber Spaces in Dimension 3 or Higher

For the usual category of nonsingular projective varieties, or more generally that of normal projective varieties with only $\mathbb{Q}$-factorial and terminal singularities, the notion of "birational equivalence" works very well to describe their relation.

On the other hand, in order to discuss "**log birational relation among log pairs**," the mere notion of birational equivalence does not refeflect the features of log pairs well, and we have to introduce the notion of "**log MMP relation**" as in Section 2-2.

Definition 11-3-1 (Log MMP Relation). *A finite number of projective logarithmic pairs* (X_i, D_i), $i = 1, \ldots, k$ *with only $\mathbb{Q}$-factorial and log terminal singularities are said to be log MMP-related if there exists a logarithmic pair* (V, D_V) *consisting of a nonsingular projective variety V and a boundary $\mathbb{Q}$-divisor D_V with only normal crossings such that all logarithmic pairs (X_i, D_i) are obtained from (V, B_V) via processes of log MMP.*

> Log Birational Relation Among Log Minimal Models

Proposition 11-3-2. *Let* (X_1, D_1) *and* (X_2, D_2) *be two log minimal models that are log MMP related. Then they are isomorphic in codimension 2.*

The proof goes exactly like that of Proposition 12-1-2, claiming that any two minimal models that are birational are isomorphic in codimension one.

The difference in codimension 2 among log minimal models that are log MMP related can be described using a special codimension-2 operation called "log flop."

Conjecture 11-3-3 (Existence of Log Flop). *Let*

$$\phi : X \to Y$$

be a small *contraction of an extremal ray R_l (with respect to $K_X + D + \epsilon B$ where $1 \gg \epsilon > 0$, B is an effective divisor, and $(X, D + \epsilon B)$ has only log terminal singularities) from a projective logarithmic pair (X, D) with only $\mathbb{Q}$-factorial and log terminal singularities such that*

$$(K_X + D) \cdot l = 0 \text{ and } - B \text{ is } \phi\text{-ample}.$$

Then there exists another morphism $\phi^+ : X^+ \to Y$, called a **"log flop"** *of ϕ, from a projective logarithmic pair (X^+, D^+) with only $\mathbb{Q}$-factorial and log terminal singularities such that ϕ^+ is also small (D^+ is the strict transform of D) and for any curve l^+ contracted by ϕ^+,*

$$(K_{X^+} + D^+) \cdot l^+ = 0 \text{ and } B^+ \text{ is } \phi^+\text{-ample},$$

where B^+ is the strict transform of B.

Once it exists, the log flop $\phi^+ : (X^+, D^+) \to Y$ has to be necessarily unique, since B^+ is ϕ^+-ample and hence

$$X^+ = Proj \oplus_{m \geq 0} \phi^+_*(mB^+) = Proj \oplus_{m \geq 0} \phi_*(mB).$$

It is called the "B-log flop" (of ϕ). The word "B-log flop" also refers to the following commutative triangle of birational maps:

$$
\begin{array}{ccc}
B \subset (X, D) & \dashrightarrow & (X^+, D^+) \supset B^+ \\
\phi \searrow & & \swarrow \phi^+ \\
& Y &
\end{array}
$$

We remark that a B-log flop is actually a log flip with respect to a logarithmic pair $(X, D + \epsilon B)$ for some $0 \ll \epsilon \ll 1$ that has only $\mathbb{Q}$-factorial and log terminal singularities. Thus a log flop is a special case of a log flip.

We also need the statement of termination.

Conjecture 11-3-4 (Termination of Log Flops). *There is no infinite sequence of B-log flops for a fixed divisor B:*

$$(X, D) = (X_0, D_0) \dashrightarrow (X_0^+, D_0^+) = (X_1, D_1) \dashrightarrow (X_1^+, D_1^+) = (X_2, D_2^+) \dashrightarrow \cdots$$

$$Y = Y_0 \qquad\qquad Y_1 \qquad\qquad Y_2 \qquad \cdots$$

Again by the remark above we note that the conjecture (termination of Log flops) is a special case of the conjecture (termination of log flips).

Theorem 11-3-5. *Log minimal models that are log MMP related are connected by sequences of log flops, assuming (existence of log flop) and (termination of log flops).*

For the proof, which is parallel to that of the relation among minimal models, we refer the reader to Theorem 12-1-8.

Log Birational Relation Among Log Mori Fiber Spaces

We hope to give an algorithm that factors birational maps between two log Mori fiber spaces (inducing log MMP relations) in the form of the log Sarkisov program.

Question 11-3-6 (Log Sarkisov Program in Dimension 3 or Higher). *Let*

$$(X, D) \overset{\Phi}{\dashrightarrow} (X', D')$$
$$\downarrow\phi \qquad\qquad \downarrow\phi'$$
$$Y \qquad\qquad\quad Y'$$

be a birational map between two log Mori fiber spaces that induces a log MMP relation between them. Then Φ is a composite of the following four types of "links":
Type (I)

$$(Z, D_Z) \dashrightarrow (X_1, D_{X_1})$$
$$\swarrow \qquad\qquad\qquad \downarrow$$
$$(X, D)$$
$$\downarrow$$
$$Y \qquad\qquad \leftarrow \qquad\qquad Y_1$$

Type (II)

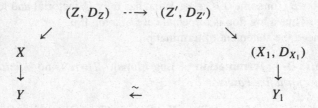

Type (III) (*Inverse of Type* (I))

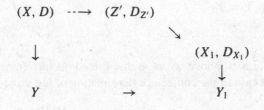

Type (IV)

$$(X, D) \quad \dashrightarrow \quad (X_1, D_{X_1})$$
$$\downarrow \qquad\qquad\qquad \downarrow$$
$$Y \qquad\qquad\qquad Y_1$$
$$\searrow \qquad \swarrow$$
$$T$$

The birational morphisms $(Z, D_Z) \to (X, D)$ *and* $(Z', D_{Z'}) \to (X_1, D_{X_1})$ *are divisorial contractions of extremal rays with respect to certain log canonical divisors, and the dashed arrows "*$\dashrightarrow$*" indicate sequences of log flips. We also have* $\rho(Y_1/Y) = 1$ *in Types* (I), (II), (III) *and* $\rho(Y/T) = \rho(Y_1/T) = 1$ *in Type* (IV), *where* $Y, Y_1,$ *and* T *are all normal projective varieties with only* $\mathbb{Q}$*-factorial singularities.*

Most importantly, we require that all intermediate logarithmic pairs (i.e., $(X, D), (Z, D_Z), (Z', D_{Z'}), (X_1, D_{X_1})$ *etc.) be log MMP-related.*

We remark that all the conjectures in this section are established theorems in dimension 3 and that Question 11-3-6 is established affirmatively in dimension 2 and remains open even in dimension 3 (cf. Bruno–Matsuki [1]).

CHAPTER 12

Birational Relation among Minimal Models

The purpose of this chapter is to discuss the **birational relation among minimal models** (cf. the strategic scheme MP 3 in the Introduction). In Section 1-8 we observed that a minimal model in dimension 2 in a fixed birational equivalence class is unique. This is no longer true in dimension 3 or higher, i.e., there may exist many minimal models in general even in a fixed birational equivalence class, and here arises a need to study the birational relation among them.

The most basic feature of the birational relation among minimal models (in a fixed birational equivalence class) is that they are isomorphic in codimension one. In Section 12-1 we describe the difference in codimension 2 (or more) more specifically. Namely, we show that any two minimal models in a birational equivalence class are connected by a sequence of "**flops**." A "flop" is a special but fundamental codimension-2 (or more) operation, which can be called a "brother" of a "flip." The main issues are the two conjectures (**existence of flop**) and (**termination of flops**). We will show only (termination of flops) in dimension 3, leaving the verification of the (existence of flop) in dimension 3 to the original research papers Kawamata [9], Kollár [8].

The description of the birational relation among minimal models in Section 12-1 is *regional*, i.e., it tells us how any *two* minimal models are related to each otheer. In Section 12-2 we give a more *global* description telling how *all* the minimal models (in a fixed birational equivalence class) are related to each other. We describe the relation in terms of the **chamber structure**. We observe that the ample cone of each minimal model occupies a chamber in a certain convex cone in a finite-dimensional vector space, each flop corresponds to a reflection between two chambers, and the ample cones of all the minimal models fill up this convex cone.

In Section 12-3 we discuss **how to count the number of minimal models** (in a fixed birational equivalence class). For a variety of general type we prove, via (existence of flop) and (termination of flops), that the number of chambers discussed in Section 12-2 is finite and thus the number of minimal models is finite. Though it is not difficult to give an example of a variety of nongeneral type for which the number of chambers is infinite, all known examples produce only a finite number of minimal models after they are identified through some automorphisms (which may differ from the prescribed birational maps). It remains as a question (a conjecture !?) whether *modulo automorphisms* there may still exist only a finite number of minimal models in general.

12.1   Flops Among Minimal Models

One of the main features of the minimal model in dimension 2 is that it is unique in a fixed birational class. This is no longer true in dimension 3 (or higher), as the following example indicates.

Example 12-1-1.   Take a cone over a smooth quadric

$$C_Q = \{XY - ZW = 0\} \subset \{(X : Y : Z : W : T)\} = \mathbb{P}^4.$$

If we blow up the unique singular point $(0 : 0 : 0 : 0 : 1)$ to obtain $f : V \to C_Q$, then the exceptional locus is a smooth quadric $E \cong Q \cong \mathbb{P}^1 \times \mathbb{P}^1$. The two rulings $\{p\} \times \mathbb{P}^1$ and $\mathbb{P}^1 \times \{q\}$ give rise to two different extremal rays on V whose contractions $\sigma_1 : V \to X_1$ and $\sigma_2 : V \to X_2$ give two nonsingular projective 3-folds $\tau_1 : X \to C_Q$ and $\tau_2 : X_2 \to C_Q$, respectively (cf. Atiyah's Flop Example 3-4-3).

Since both τ_1 and τ_2 are small and since K_{C_Q} is a Cartier divisor, we conclude that

$$K_{X_1} = \tau_1^* K_{C_Q} \quad \text{and} \quad K_{X_2} = \tau_2^* K_{C_Q}.$$

Take a general member $A \in |\mathcal{O}_{C_Q}(a)|$ for $a > 6$ and take a double cover $d : C_Q' \to C_Q$ ramified over A.

Now take the Cartesian product for $i = 1, 2$:

$$
\begin{array}{ccc}
X_i' & \longrightarrow & C_Q' \\
\downarrow & & \downarrow \\
X_i & \longrightarrow & C_Q.
\end{array}
$$

Then X_1' and X_2' are nonsingular minimal models, since $K_{X_1'}$ and $K_{X_2'}$ are both nef (and actually big), but the given birational map $X_1' \dashrightarrow C_Q' \dashleftarrow X_2'$ is not an isomorphism.

The above gives a counterexample to the uniqueness of minimal models in dimension 3 (or higher) in the sense that a birational map between two minimal models $X_1 \dashrightarrow X_2$ may fail to be an isomorphism in general.

But the extent to which it fails to be an isomorphism is *small*, as we observe in the next proposition.

Proposition 12-1-2 (Minimal Models Are Isomorphic in Codimension One.).
Let X_1 and X_2 be two minimal models (of dimension n) with a birational map

$$\phi : X_1 \cdots \rightarrow X_2.$$

Then X_1 and X_2 are isomorphic in codimension one, i.e., there exist closed subsets $B_1 \subset X_1$ and $B_2 \subset X_2$ of codimension at least two such that ϕ induces an isomorphism

$$X_1 - B_1 \overset{\phi}{\underset{\rightarrow}{\sim}} X_2 - B_2.$$

PROOF.   Take a nonsingular projective variety V that dominates both X_1 and X_2 by birational morphisms

$$X_1 \overset{f_1}{\leftarrow} V \overset{f_2}{\rightarrow} X_2,$$

where $\phi = f_2 \circ f_1^{-1}$. By taking a further blowup if necessary, we may assume that $\mathrm{Exc}(f_1)$ and $\mathrm{Exc}(f_2)$ are of pure codimension one. Write down the ramification formulas

$$K_V = f_1^* K_{X_1} + \Sigma a_i E_i = f_2^* K_{X_2} + \Sigma a_j E_j,$$

where the E_i (respectively the E_j) are exceptional divisors with respect to f_1 (respectively f_2). We note that since both X_1 and X_2 have only terminal singularities, we have

$$a_i > 0 \forall E_i \quad \text{and} \quad a_j > 0 \forall E_j,$$

and that since the exceptional loci are of pure codimension one, we have

$$V - \cup_i E_i \overset{\sim}{\rightarrow} X_1 - B_1,$$
$$V - \cup_j E_j \overset{\sim}{\rightarrow} X_2 - B_2,$$

where $B_1 = f_1(\cup_i E_i)$ and $B_2 = f_2(\cup_j E_j)$ are closed subsets of codimension at least two. Therefore, in order to verify the assertion we have only to prove

$$\cup_i E_i = \cup_j E_j.$$

Suppose this is not the case. Then there exists an f_1-exceptional but not f_2-exceptional divisor E_{i_o} (or f_2-exceptional but not f_1-exceptional divisor E_{j_o}, in which case the proof is identical and left to the reader). From the ramification formulas it follows then that

$$f_1^* K_{X_1} = f_2^* K_{X_2} - a_{i_o} E_{i_o} - \Sigma_{i \neq i_o} a_i E_i + \Sigma a_j E_j$$
$$= f_2^* K_{X_2} - F_1 + F_2 + G,$$

where F_1, F_2, and G share no common components and

$$F_1 \text{ is } f_1\text{-exceptional and } > 0, \text{ since } a_{i_o} > 0,$$
$$F_2 \text{ is } f_1\text{-exceptional and } \geq 0,$$
$$G \text{ is } f_2\text{-exceptional and } \geq 0.$$

Take a very ample divisor A on X_1 and a very ample diviosr B on V. Set

$$C = F_1 \cap \{\cap_{k=1}^{\dim f_1(F_1)} D_k\} \cap \{\cap_{h=\dim f_1(F_1)+1}^{n-2} L_h\},$$

where $D_k \in |f_1^* A|$ and $L_h \in |B|$ are the general members. Then

$$F_1 \cdot C < 0 \text{ (Exercise! Why? cf. Theorem 4-6-1)}$$

and

$$f_2^* K_{X_2} \cdot C \geq 0, F_2 \cdot C \geq 0 \text{ and } G \cdot C \geq 0,$$

which would imply

$$0 = f_1^* K_{X_1} \cdot C = (f_2^* K_{X_2} - F_1 + F_2 + G) \cdot C > 0,$$

a contradiction.

This completes the proof of Proposition 12-1-2.   □

Now the question is how these small differences in codimension 2 (or more) among minimal models arise and how they can be described. A special codimension-2 (or more) operation called "flop," a brother of "flip," answers the question.

Conjecture 12-1-3 (Existence of Flop). *Let*

$$\phi : X \to Y$$

be a contraction of an extremal ray $R_l = \mathbb{R}_+[l]$ *(with respect to* $K_X + \epsilon D$, *where* $1 \gg \epsilon > 0$ *and* D *is an effective* $\mathbb{Q}$-*divisor) of flopping type from a normal projective variety* X *with only* $\mathbb{Q}$-*factorial and terminal singularities, i.e.,*

  (i) ϕ *is a birational morphism onto a normal projective variety* Y,
  (ii) ϕ *is small, i.e.,* $\text{codim}_X \text{Exc}(\phi) \geq 2$,
  (iii) K_X *is* ϕ-*trivial and* $-D$ *is* ϕ-*ample*,
  (iv) *all the curves in the fibers of* ϕ *are numerically proportional, i.e.,* $\rho(X/Y) = 1$.
  Then there exists the morphism $\phi^+ : X^+ \to Y$, *called a* **"flop"** *of* ϕ, *i.e.:*

   (i$^+$) ϕ^+ *is a birational morphism from a normal projective variety* X^+ *with only* $\mathbb{Q}$-*factorial and terminal singularities onto* Y,
   (ii$^+$) ϕ^+ *is small, i.e.,* $\text{codim}_{X^+} \text{Exc}(\phi^+) \geq 2$,
   (iii$^+$) K_{X^+} *is* ϕ^+-*trivial and* D^+ *is* ϕ^+-*ample, where* D^+ *is the strict transform of* D,
   (iv$^+$) *all the curves in the fibers of* ϕ^+ *are numerically proportional, i.e.,* $\rho(X^+/Y) = 1$.

Once it exists, $\phi^+ : X^+ \to Y$ has to be necessarily unique, since D^+ is ϕ^+-ample and hence

$$X^+ = Proj \oplus_{m \geq 0} \phi^+{}_*(m D^+) = Proj \oplus_{m \geq 0} \phi_*(m D).$$

It is also called a "D-flop" (of ϕ). The word "flop" or "D-flop" also refers to the following commutative triangle of birational maps:

$$D \subset X \dashrightarrow X^+ \supset D^+$$
$$\phi \searrow \swarrow \phi^+$$
$$Y$$

We note that the triangle $X_1 \to C_Q \leftarrow X_2$ in Example 12-1-1 gives an example of a flop in dimension 3 called "Atiyah's flop."

(Existence of flop) is a theorem in dimension 3, thanks to Kawamata [9] and Kollár [8], whose proofs rely on the analysis of the structure of terminal singularities in dimension 3 by Reid [2][3][7], Mori [4]. Though it is not as difficult as showing (existence of flip) in dimension 3, we refer the reader to the original research papers cited above for details.

We also need the statement of termination.

Conjecture 12-1-4 (termination of flops). *There is no infinite sequence of D-flops for a fixed effective divisor D:*

$$X = X_0 \dashrightarrow X^+ = X_1 \dashrightarrow X_1^+ = X_2 \dashrightarrow \cdots \qquad X_{i-1}^+ = X_i \dashrightarrow X_i^+ = X_{i+1} \cdots$$
$$\searrow \quad \swarrow \searrow \quad \swarrow \searrow \quad \swarrow \searrow \cdots \quad \swarrow \searrow \quad \swarrow \cdots$$
$$Y = Y_0 \qquad Y_1 \qquad Y_2 \qquad \cdots \qquad Y_i \qquad \cdots$$

We prove (termination of flops) in dimension 3 following Kollár [8]. The proof is very similar to that of (termination of flips) in dimension 3 by Shokurov. The difference is that here we have not only to keep track of the discrepancies as we did with the "difficulty" argument of Shokurov but also to pay attention to the behavior of the 1-cycles on the reference divisor D.

Theorem 12-1-5 ((Termination of Flops) in Dimension 3). *(Termination of flops) holds in dimension 3.*

PROOF. We derive a contradiction assuming that there is an infinite sequence of D-flops starting from a normal projective 3-fold X with only $\mathbb{Q}$-factorial and terminal singularities, where D is an effective $\mathbb{Q}$-divisor on X.

One of the key points is the observation about the behavior of the discrepancies under a flop. □

Lemma 12-1-6 (cf. Lemma 9-1-3). *Let*

$$D \subset X \dashrightarrow X^+ \supset D^+$$
$$\phi \searrow \swarrow \phi^+$$
$$Y$$

be a D-flop. Then

$$a(v, X; D) \leq a(v, X^+; D^+)$$

for all discrete valuations v. Moreover,

$$a(v, X; D) < a(v, X^+; D)$$

iff the center of v on X is contained in $\mathrm{Exc}(\phi)$ (or equivalently iff the center of v on X^+ is contained in $\mathrm{Exc}(\phi^+)$).

The proof is identical to the observation of Shokurov about the behavior of the discrepancies under a flip (cf. Lemma 9-1-3) and is left to the reader as an exercise.

Lemma 12-1-7. *Let X and D be as above. Then for sufficiently small $\epsilon > 0$, there exist a finite number of discrete valuations $v_1, \ldots, v_m$ such that if*

$$a(v, X; \epsilon D) < 1,$$

then either v coincides with one of the finitely many discrete valuations $v_1, \ldots, v_m$ or v arises from blowing up a curve C_v whose generic point sits in the smooth locus of the divisor D and the smooth locus of X.

In the latter case, if the generic point of C_v sits in the smooth locus of the irreducible component D_k of D with the coefficient d_k, then

$$a(v, X; \epsilon D) = 1 - \epsilon d_k.$$

PROOF. Take a resolution of singularities $f : V \to X$ from a nonsingular projective variety such that $f^{-1}(D) \cup \mathrm{Exc}(f)$ is a divisor with only normal crossings. Since the ramification formula reads

$$K_V = f^* K_X + \Sigma a_i E_i \quad a_i > 0 \text{ for all the } E_i \text{ exceptional for } f,$$

and since any discrete valuation other than the ones corresponding to the E_i can be realized as an exceptional divisor after a sequence of blowups with smooth centers starting from V, it is easy to see that for sufficiently small $0 < \epsilon \ll \min\{a_i\}$ if $a(v, X; \epsilon D) < 1$, then either the discrete valuation v coincides with one of the finitely many discrete valuations corresponding to the exceptional divisors E_i or v arises from blowing up a curve C_v whose generic point sits in the smooth locus of the divisor D and the smooth locus of X. □

A sequence of D-flops is a sequence of ϵD-flops for any $\epsilon > 0$. Therefore, by replacing D with ϵD, we may assume by the above lemma that there exist a finite number of discrete valuations $v_1, \ldots, v_m$ such that if

$$a(v, X; D) < 1,$$

then either v coincides with one of the finitely many discrete valuations $v_1, \ldots, v_m$ or v arises from blowing up a curve C_v whose generic point sits in the smooth locus of the divisor D and the smooth locus of X. In the latter case, if the generic point of C_v sits in the smooth locus of the irreducible component D_k of D with the coefficient d_k, then

$$a(v, X; D) = 1 - d_k.$$

We take a positive integer N such that

$$Nd_k \in \mathbb{N}$$

for all the coefficients d_k of irreducible components D_k in D. We use the same letter D for denoting the strict transforms of D by abuse of notation.

We set

$$b(D, \phi_i) := \min\{a(\nu, X_i; D); \nu \text{ is a discrete valuation obtained by}$$

$$\text{blowing up the generic point of a curve } C_{\nu,i} \subset \text{Exc}(\phi_i)\}.$$

Note that if $C_{\nu,i} \subset \text{Exc}(\phi_i)$, then the generic point of $C_{\nu,i}$ is in the smooth locus of X_i, since X_i has only terminal singularities and hence the singular locus of X_i has codimension at least 3 (cf. Corollary 4-6-6). Note also that $C_{\nu,i} \subset D$, since $D \cdot C_{\nu,i} < 0$ as $-D$ is ϕ-ample. Therefore,

$$b(D, \phi_i) = \min\{a(\nu, X_i; D) = 1 - \text{mult}_{C_{\nu,i}} D\} < 1 \quad \text{and} \quad b(D, \phi_i) \in \frac{1}{N}\mathbb{Z}.$$

We consider the number

$$b_{\min} = \min_i\{b(D, \phi_i)\}.$$

Note that by Lemma 12-1-6 and Lemma 12-1-7 we have

$$-\infty < \min_{p=1,\ldots,m, D=\Sigma d_k D_k}\{a(\nu_p, X; D), 1 - d_k\} \le b(D, \phi_i)$$

and hence

$$-\infty < b_{\min}.$$

By truncating a finite number of D-flops at the beginning of the sequence, we may assume that the minimum $b_{\min}$ is attained by infinitely many $b(D, \phi_i)$'s, i.e.,

$$b_{\min} = b(D, \phi_i)$$

for infinitely many i.

Step 1. Show that there exists i_1 such that any discrete valuation ν obtained by blowing up the generic point of a curve $C_{\nu,i} \subset \text{Exc}(\phi_i)$ with $i \ge i_1$ and $a(\nu, X_i; D) = b_{\min}$ arises from blowing up a curve $C_{\nu,0}$ whose generic point sits in the smooth locus of the divisor D and the smooth locus of $X = X_0$. Moreover, X_0 and X_i are isomorphic around the generic point of $C_{\nu,0}$ (which is hence the generic point of $C_{\nu,i}$).

If $a(\nu, X_i; D) = b_{\min} < 1$, then by Lemma 12-1-6, Lemma 12-1-7, and the remark immediately after them we conclude that ν must coincide either with one of the finitely many discrete valuations $\nu_1, \ldots, \nu_m$ or with the one arising from blowing up a curve $C_{\nu,0}$ whose generic point sits in the smooth locus of the divisor D and the smooth locus of $X = X_0$.

By Lemma 12-1-6 there exists i_1 such that the discrete valuation ν with $a(\nu, X_i; D) = b_{\min}$ and $i \ge i_1$ cannot coincide with any one of the finitely many prescribed discrete valuations $\nu_1, \ldots, \nu_m$.

If X_0 and X_i are not isomorphic around the generic point of $C_{\nu,0}$, then the curve must be contracted by some flopping contraction ϕ_h with $h < i$. But then by Lemma 12-1-6 again we compute

$$b_{\min} \leq b(D, \phi_h) \leq a(\nu, X_h; D) < a(\nu, X_{h+1}; D) \leq a(\nu, X_i; D) = b_{\min},$$

a contradiction. This accomplishes Step 1.

Step 2. Show there exists $i_2(\geq i_1)$ such that no irreducible component of $\text{Exc}(\phi_i{}^+)$ with $i \geq i_2$ is contained in $D^{b_{\min}} = \Sigma_{b_{\min}=1-d_k} D_k$.

Let ν be the discrete valuation obtained by blowing up the generic point of an irreducible component of $\text{Exc}(\phi_i{}^+)$ contained in $D^{b_{\min}}$. Then since by Lemma 12-1-6 we have

$$a(\nu, X_0; D) \leq a(\nu, X_i^+; D) \leq 1 - d_k = b_{\min},$$

we conclude by Lemma 12-1-7 and by the following remark that ν must coincide either with one of the finitely many discrete valuations $\nu_1, \ldots, \nu_m$ or with the one arising from blowing up a curve $C_{\nu,0}$ whose generic point sits in the smooth locus of the divisor D and the smooth locus of $X = X_0$.

By Lemma 12-1-6 and by the fact

$$a(\nu, X_i^+; D) < 1 \quad \text{and} \quad a(\nu, X_i^+; D) \in \frac{1}{N}\mathbb{N}$$

there exists $i_2(\geq i_1)$ such that the discrete valuation ν as above cannot coincide with any one of the finitely many prescribed discrete valuations $\nu_1, \ldots, \nu_m$.

Therefore, ν arises from blowing up a curve $C_{\nu,0}$ whose generic point sits in the smooth locus of the divisor D and the smooth locus of $X = X_0$. Again the curve must be contracted by some flopping contraction ϕ_h with $h < i$, since $X = X_0$ and X_i^+ cannot be isomorphic around the generic point of $C_{\nu,0}$. But then by Lemma 12-1-6 we have

$$b_{\min} \leq a(\nu, X_h; D) < a(\nu, X_{h+1}; D) \leq a(\nu, X_i; D) \leq b_{\min},$$

a contradiction.

Step 3. Conclusion of the proof.

We consider the flopping contraction $\phi_i : X_i \to Y_i$ with $i \geq i_2$.

If $b(D, \phi_i) > b_{\min}$, then no irreducible component of $\text{Exc}(\phi_i)$ can be contained in $D^{b_{\min}}$ in X_i. No irreducible component of $\text{Exc}(\phi_i^+)$ can be contained in $D_{b_{\min}}$ in $X_i^+ = X_{i+1}$ by Step 2.

If $b(D, \phi_i) = b_{\min}$, then by Step 1 all the irreducible components of $\text{Exc}(\phi_i)$ are contained in $D^{b_{\min}}$. Again no irreducible component of $\text{Exc}(\phi_i^+)$ can be contained in $D^{b_{\min}}$ in $X_i^+ = X_{i+1}$ by Step 2.

Therefore, while there is no change in 1-cycles of $D^{b_{\min}}$ whenever $b(D, \phi_i) > b_{\min}$, some 1-cycles disappear from $D^{b_{\min}}$ whenever $b(D, \phi_i) = b_{\min}$. On the other hand, there are infinitely many indices i such that $b(D, \phi_i) = b_{\min}$. This is a contradiction! (Exercise! Check this part of the argument, claiming that from a fixed divisor $D^{b_{\min}}$ 2-cycles cannot disappear infinitely many times.)

This completes the proof of (termination of flops) in dimension 3.   □

Now (existence of flop) and (termination of flops) give the description of the modification in codimension 2 (or more) from one minimal model to another as a sequence of flops.

Theorem 12-1-8 (Minimal Models Are Connected by Flops.). *Minimal models (in a fixed birational class) are connected by sequences of flops, assuming (existence of flop) and (termination of flops).*

In particular, minimal models (in a fixed birational equivalence class) are connected by sequences of flops in dimension 3, where (existence of flop) and (termination of flops) are theorems.

PROOF.   Let X_1 and X_2 be two minimal models in a fixed birational class. Take a very ample divisor A on X_2 and its general member $D_2 \in |A_2|$. Set D_1 to be the strict transform of D_2 on X_1. We will show that X_1 is connected to X_2 by a sequence of D_1-flops.

Take a nonsingular projective variety V that dominates both X_1 and X_2 as in the proof of Theorem 12-1-2 (we also use the same notation). Then the same argument as in the proof of Theorem 12-1-2 (Exercise! See also the proof of Lemma 13-1-4) proves that

$$f_1^* K_{X_1} = f_2^* K_{X_2},$$
$$f_1^* D_1 = f_2^* D_2 + \Sigma b_j E_j \text{ with } b_j \geq 0 \forall E_j.$$

If $\Sigma b_j E_j = 0$, then the linear system $\phi_*^{-1}|A_2|$ would have no base points, and it induces a birational morphism

$$\phi = \Phi_{\phi_*^{-1}|A_2|} : X_1 \to X_2.$$

Then since both X_1 and X_2 are isomorphic in codimension one by Proposition 12-1-2 and since they are both $\mathbb{Q}$-factorial and projective, we conclude that $\phi : X_1 \xrightarrow{\sim} X_2$.

If $\Sigma b_j E_j \neq 0$, then we claim that there is a D_1-flop. In fact, take a very ample divisor B on V. Set

$$C = (\Sigma b_j E_j) \cap \{\cap_{k=1}^{\dim f_2(\Sigma b_j E_j)} M_k\} \cap \{\cap_{h=\dim f_2(\Sigma b_j E_j)+1}^{n-2} L_h\},$$

where $M_k \in |f_2^* D_2|$ and $L_h \in |B|$ are the general members. Then (Exercise! Why?)

$$f_{1*}(C) \cdot K_{X_1} = C \cdot f_1^* K_{X_1} = C \cdot f_2^* K_{X_2} = f_{2*}(C) \cdot K_{X_2} = 0,$$
$$f_{1*}(C) \cdot D_1 = C \cdot f_1^* D_1 = C \cdot (f_2^* D_2 + \Sigma b_j E_j) < 0.$$

Therefore, we conclude for sufficiently small $\epsilon > 0$ that $(X_1, \epsilon D_1)$ has only log terminal singularities (cf. Section 4-3), $K_{X_1} + \epsilon D_1$ is not nef, and there exists an extremal ray R_l such that $K_{X_1} \cdot l = 0$ and $D_1 \cdot l < 0$.

We remark that the contraction of R_l must be small, since any curve l whose class belongs to R_l must be contained in $B_1 = f_1(\cup E_i)$, i.e., b_i, which follows from $D_1 \cdot l < 0$. Thus, thanks to (existence of flop), there exists a D_1-flop of this contraction.

We continue this process with $X_1 = Z_1, Z_1{}^+ = Z_2, \ldots, Z_i{}^+ = Z_{i+1}, \ldots$ until, thanks to (termination of flops), we end up with Z_t, where for the strict transform D_t of D_1 on Z_t we have

$$f_t^* D_t = f_2^* D_2$$

and thus $Z_t \overset{\sim}{\to} X_2$. (See the argument at the beginning of the proof.) This proves our assertion that $X_1 = Z_1$ and $X_2 = Z_t$ are connected by a sequence of flops.   □

We remark that (existence of flops) and (termination of flops) are both theorems in dimension 3, and thus we do not need to "assume" these two for Theorem 12-1-5 in dimension 3. We also remark that we can give a fairly explicit description of flops in dimension 3 as in Kollár [11].

12.2   Chamber Structure of Ample Cones of Minimal Models

Section 12-1 describes how *two* different (but still in the same birational equivalence class) minimal models are related: They are connected by a sequence of flops. In this section we would like to seek a more global description of how *all* the minimal models (in a fixed birational equivalence class) are related to each other. The purpose of this section is to observe that the ample cone of each minimal model occupies a chamber in a convex cone called $\overline{\mathrm{Mov}}$ in some finite-dimensional vector space called N^1, each flop corresponds to a reflection between two chambers, and the ample cones of all the minimal models fill up the covex cone $\overline{\mathrm{Mov}}$ and hence form the **chamber structure**.

We start with the following basic lemma.

Lemma 12-2-1 (Strict Transform Preserves Numerical Equivalence.).

Let X_1 and X_2 be two normal projective varieties with only $\mathbb{Q}$-factorial singularities, which are isomorphic in codimension one via a birational map $\phi : X_1 \dashrightarrow X_2$. Then the map of taking the strict transforms

$$\phi_* : \mathrm{Div}(X_1) \otimes \mathbb{Q} \to \mathrm{Div}(X_2) \otimes \mathbb{Q}$$

preserves numerical equivalence and hence induces an isomorphism

$$N^1(X_1) \overset{\sim}{\to} N^1(X_2).$$

PROOF.   This follows easily from the negativity lemma (cf. Lemma 13-1-4) and is left as an exercise for the reader. The reader may refer to Lemma 13-1-5, which includes Lemma 12-2-1 as a special case.   □

Via Lemma 12-2-1 and Proposition 12-1-2 we see that all the minimal models X in a fixed birational equivalence class have $N^1(X)$ isomorphic to each other, which we call accordingly N^1 without referring to the specific minimal model X we choose.

The following proposition describes how the ample cones of two minimal models are located in N^1, if they are obtained from each other by one flop.

Proposition 12-2-2 (Flop Corresponds to Reflection.). *Let X and X^+ be two minimal models, which result from one flop with respect to an extremal ray $R_l = \mathbb{R}_+[l]$, i.e., we have a triangle of the form as described in Conjecture 12-1-3 (existence of flop):*

$$D \subset X \quad \dashrightarrow \quad X^+ \supset D^+$$
$$\phi \searrow \qquad \swarrow \phi^+$$
$$Y$$

Then:

(i) *The descriptions of the spaces $N^1(X)$ and $N^1(X^+)$ are given by*

$$N^1(X) = \phi^* N^1(Y) \oplus \mathbb{R}[D],$$
$$N^1(X^+) = \phi^{+*} N^1(Y) \oplus \mathbb{R}[D^+],$$

where $\phi^* N^1(Y) = \phi^{+*} N^1(Y) = \{l = 0\}$ is a hyperplane and $\mathbb{R}[D] = \mathbb{R}[D^+]$ through the identification $N^1(X) = N^1(X^+) = N^1$ taking the strict transforms. Thus we write

$$N^1 = N^1(Y) \oplus \mathbb{R}[D].$$

(ii) *The ample cones (or to be more precise, the closures of the ample cones, which are called the nef cones) $\overline{\mathrm{Amp}}(X)$ and $\overline{\mathrm{Amp}}(X^+)$ share the codimension-one face in N^1*

$$\overline{\mathrm{Amp}}(X) \cap N^1(Y) = \overline{\mathrm{Amp}}(X^+) \cap N^1(Y) = \overline{\mathrm{Amp}}(Y),$$

and they are located in N^1 opposite each other with respect to this hyperplane $N^1(Y)$ as

$$\overline{\mathrm{Amp}}(X) \ni (-D + nA) \cdot l > 0,$$
$$\overline{\mathrm{Amp}}(X^+) \ni (D^+ + nA) \cdot l = (D + nA) \cdot l < 0,$$

for an ample divisor A on Y and for sufficiently large $n \in \mathbb{N}$.

We illustrate assertions (i) and (ii) in Figure 12-2-3.

Figure 12-2-3.

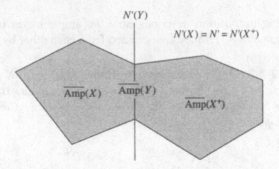

Observe that though we say that the flop corresponds to a "reflection," since $\overline{\mathrm{Amp}}(X)$ and $\overline{\mathrm{Amp}}(X^+)$ share a codimension-one face, the shapes of the two cones are *not* symmetric with respect to the hyperplane $N^1(Y)$ in any sense and may even have different numbers of edges.

PROOF.   Assertion (i) is a consequence of the (logarithmic) contraction theorem in Chapter 8 (cf. Theorem 8-1-3 and Exercise 8-1-4). Assertion (ii) follows immediately from (i).   □

Proposition 12-2-2 indicates that (the closures of) the ample cones of two minimal models are "reflections" of each other if they are obtained via a flop from each other. We try to extend this picture to involve all the minimal models.

Remark 12-2-4.

From here till the end of this section, for the simplicity of the presentation, **we work only in dimension 3**, where

(existence of flop)
(termination of flops)
the abundance conjecture

are all established theorems.

Since the abundance conjecture holds in dimension 3, each minimal model X has a semiample canonical divisor K_X, and thus it has a morphism to its canonical model

$$\Phi_X = \Phi_{|mK_X|} : X \to X_{\mathrm{can}}$$

for sufficiently divisible $m \in \mathbb{N}$, where the canonical model X_{can} is independent of the choice of a minimal model in a fixed birational equivalence class.

We work with the category over the canonical model X_{can}.

For example,

$$N^1 = N^1(X/X_{\mathrm{can}})$$

denotes the space of divisors modulo numerical equivalence over X_{can}. Lemma 12-2-1 and Proposition 12-2-2 hold without any change in this category of varieties projective over X_{can}, (cf. Section 3-5).

Definition 12-2-5. *Let N^1 be the space of divisors modulo numerical equivalence (over the canonical model) of all the minimal models in a fixed birational equivalence class. We define*

$$\text{Mov} := \text{ the convex cone generated by the classes}$$
$$\text{represented by the movable divisors over } X_{\text{can}},$$
$$\overline{\text{Mov}} := \text{ the closure of Mov,}$$

where a divisor D on X is said to be movable over X_{can} if

$$\text{codim}_X \text{Coker}(\Phi_X{}^* \Phi_{X*} \mathcal{O}_X(D) \to \mathcal{O}_X(D)) \geq 2.$$

Note that if D is movable over X_{can} on a minimal model X, then its strict transform D' on another minimal model X' is also movable over $X'_{\text{can}} = X_{\text{can}}$, since they are isomorphic in codimension one (cf. Proposition 12-1-2).

Remark 12-2-6.

We remark that in the notation of Kawamata [9] "Mov" refers to the interior of the closed cone $\overline{\text{Mov}}$, whereas in our notation "Mov" simply refers to the cone generated by the movable divisors (over X_{can}). Since the condition of being movable is not open, the cone Mov in our notation may contain a point that is not in the interior of $\overline{\text{Mov}}$. This is in contrast to the situation where Amp, the cone generated by the ample divisors, is the interior of the closed cone $\overline{\text{Amp}}$, which is the closure of Amp, since the condition of being ample is open.

Now we present the main theorem of this section.

Theorem 12-2-7 (Chamber Structure). *We have the decomposition*

$$\overline{\text{Mov}} = \text{ the closure of } \left\{ \bigsqcup \overline{\text{Amp}}(X_i / X_{\text{can}}) \right\},$$

where the X_i vary among all the minimal models in the birational equivalence class and where the symbol $\bigsqcup$ indicates that no two cones $\overline{\text{Amp}}(X_i / X_{\text{can}})$ and $\overline{\text{Amp}}(X_j / X_{\text{can}})$ intersect in their interiors when $i \neq j$.

Furthermore, each cone $\overline{\text{Amp}}(X_i / X_{\text{can}})$ is locally polyhedral in Mov in the following sense: For any point $p \in \text{Mov}_{\mathbb{Q}}$ with $p \notin \overline{\text{Amp}}(X_i / X_{\text{can}})$, the boundary of $\overline{\text{Amp}}(X_i / X_{\text{can}})$ is defined by a finite number of hyperplanes inside of the convex cone $\mathbb{R}_{\geq 0} \cdot p + \overline{\text{Amp}}(X_i / X_{\text{can}})$.

PROOF.   Take an arbitrary rational point $p \in \text{Mov}_{\mathbb{Q}}$. Fix a minimal model X_o. Then on X_o (some multiple of) p is represented by a divisor D movable over X_{can}. If D is nef over X_{can}, then $p \in \overline{\text{Amp}}(X_o / X_{\text{can}})$. If D is not nef over X_{can}, then D necessarily has some base points over X_{can}, i.e., $\text{Coker}(\Phi_{X_o}{}^* \Phi_{X_o*} \mathcal{O}_{X_o}(D) \to \mathcal{O}_{X_o}(D)) \neq \emptyset$. By the same argument as at the beginning of Theorem 12-1-8, we conclude that there exists an extremal ray $R_l \in \overline{NE}(X_o/X_{\mathrm{can}})$ generated by a curve l with $D \cdot l < 0$. (Note that $K_{X_o} \cdot l = 0$ is automatic.) It also follows that the contraction of R_l must be small and hence of flopping type, since any curve l contracted by the morphism must be contained in the support of $\mathrm{Coker}(\Phi_{X_o}{}^* \Phi_{X_o *} \mathcal{O}_{X_o}(D) \to \mathcal{O}_{X_o}(D))$, which has codimension at least two. Thus, thanks to (existence of flop), we reach another minimal model $(X_o)^+$ via the flop. We continue this process with $X_o = Z_1, Z_1^+ = Z_2, \ldots, Z_s^+ = Z_{s+1}, \ldots$ until, thanks to (termination of flops), we end up with another minimal model $Z_t = X_i$ such that the strict transform D_t (of D) is nef over X_{can}, i.e., $p \in \overline{\mathrm{Amp}}(X_i/X_{\mathrm{can}})$. Moreover, if two cones $\overline{\mathrm{Amp}}(X_i/X_{\mathrm{can}})$ and $\overline{\mathrm{Amp}}(X_j/X_{\mathrm{can}})$ intersect in their interiors, then there is a divisor A_i, ample on X_i, whose strict transform A_j on X_j is also ample on X_j. But this would imply

$$X_i = \mathrm{Proj} \oplus_{m \geq 0} H^0(X_i, \mathcal{O}_{X_i}(mA_i)) \cong \mathrm{Proj} \oplus_{m \geq 0} H^0(X_j, \mathcal{O}_{X_j}(mA_j)) = X_j,$$

where the isomorphism coincides with the given birational map, i.e., $X_i = X_j$. Therefore, we have the desired decomposition.

Now we verify the "furthermore" part of the assertion.

Let $\overline{\mathrm{Amp}}(X_i/X_{\mathrm{can}})$ be the nef cone corresponding to one minimal model X_i. Take a point $p \in \mathrm{Mov}_{\mathbb{Q}}$ with $p \notin \overline{\mathrm{Amp}}(X_i/X_{\mathrm{can}})$, which is represented by an effective $\mathbb{Q}$-divisor D on a minimal model X_i. Then by the logarithmic cone theorem over X_{can} for the logarithmic pair $(X_i, \epsilon D)$ with sufficiently small $\epsilon > 0$ we have

$$\overline{\mathrm{NE}}(X_i/X_{\mathrm{can}}) = \overline{\mathrm{NE}}(X_i/X_{\mathrm{can}})_{K_X + \epsilon D \geq 0} + \Sigma R_l,$$
$$= \overline{\mathrm{NE}}(X/X_{\mathrm{can}})_{D \geq 0} + \Sigma R_l,$$

where we note that K_{X_i} is numerically trivial over X_{can}. Note also that the number of the extremal rays R_l (having negative intersection with D) is finite, since D is taken from $\mathrm{Mov}_{\mathbb{Q}}$ and since we are in dimension 3.

Since

$$\overline{\mathrm{Amp}}(X_i/X_{\mathrm{can}}) = \{x \in N^1; x \geq 0 \text{ on } \overline{\mathrm{NE}}(X_i/X_{\mathrm{can}})\},$$

we claim that in the convex cone $\mathbb{R}_{\geq 0} \cdot p + \mathrm{Amp}(X_i/X_{\mathrm{can}})$ the boundary of $\overline{\mathrm{Amp}}(X_i/X_{\mathrm{can}})$ is defined by the hyperplanes $\{\cdot R_l = 0\}$. In fact, $\overline{\mathrm{Amp}}(X_i/X_{\mathrm{can}})$ is separated from $p = [D]$ by the hyperplanes $\{\cdot R_l = 0\}$. If these hyperplanes do not form the boundary in the convex cone $\mathbb{R}_{\geq 0} \cdot p + \overline{\mathrm{Amp}}(X_i/X_{\mathrm{can}})$, then there exists a point q in this cone (represented by a movable $\mathbb{Q}$-divisor G) such that q is on the same side as $\overline{\mathrm{Amp}}(X_i/X_{\mathrm{can}})$ with respect to each hyperplane $R_l^{\perp}$ but $q \notin \overline{\mathrm{Amp}}(X_i/X_{\mathrm{can}})$. (See Figure 12-2-8, which illustrates this part of the argument.)

Figure 12-2-8.

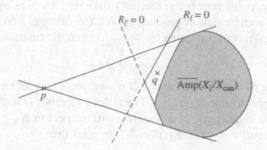

But then there exists an extremal ray R_G of G-flopping type on X_i. Since the hyperplane $R_G^\perp$ separates G and $\overline{\mathrm{Amp}}(X_i/X_{\mathrm{can}})$, D is negative on $R_G - \{0\}$. Therefore, R_G is an extremal ray with respect to $K_{X_i} + \epsilon D$, which is different from any of the R_l's, a contradiction. Therefore, $\overline{\mathrm{Amp}}(X_i/X_{\mathrm{can}})$ is locally polyhedral in Mov.

This completes the proof of Theorem 12-2-7.   □

Question 12-2-9. *Is the above decomposition locally finite in* Mov *(or in the interior of* $\overline{\mathrm{Mov}}$*), i.e., for any* $p \in$ Mov *(or* $p \in$ *the interior of* $\overline{\mathrm{Mov}}$*) does there exist a neighborhood* U_p *of* p *such that* $\#\{i\,;\,\overline{\mathrm{Amp}}(X_i/X_{\mathrm{can}}) \cap U_p \neq \emptyset\} < \infty$?

The author does not know any example that gives a negative answer to the above question or a rigorous proof to the affirmative. (Note that though Matsuki [6] claims locally finiteness in the interior of Mov, its proof in Remark II-5 is not rigorous.)

12.3   The Number of Minimal Models Is Finite (?!)

In this section we discuss **how to count the number of minimal models** in a fixed birational equivalence class. For a variety of general type we prove, via (existence of flop) and (termination of flops), that the number of minimal models is finite, a result that is implied by the fact that the number of the chambers discussed in the previous section is finite for a variety of general type. We also give an example of a variety of nongeneral type for which the number of chambers is infinite. So it might seem that we have an infinite number of minimal models in general. But here we have to be slightly more careful about the way(s) we count the number of minimal models. After clarifying how we count the number of minimal models, we present a conjecture that modulo automorphisms there may still exist only a finite number of minimal models.

Theorem 12-3-1 (Finiteness of Minimal Models of General Type) (cf. Kawamata–Matsuki [1], Kollár [8]). *There are only finitely many minimal models for a variety* X *of general type, assuming (existence of flop) and (termination of flops).*

PROOF. If a minimal model $X_{\min}$ exists, then $K_{X_{\min}}$ is nef and big, and thus $|mK_{X_{\min}}|$ is base point free for sufficiently divisible $m \in \mathbb{N}$ by the abundance theorem for minimal models of general type (cf. Corollary 6-2-6). Therefore, there exists a birational morphism given by $\Phi_{|mK_{X_{\min}}|}$ onto the canonical model

$$c : X_{\min} \to X_{\mathrm{can}}.$$

Take an effective divisor D such that $-D$ is c-ample. (Take an ample divisor A on $X_{\min}$ and an ample divisor B on X_{can}. Then take a member $D \in |-A + bc^*B|$ for $b \gg 0$.) Now the logarithmic cone theorem with respect to $K_{X_{\min}} + \epsilon D$ over X_{can} for sufficienly small $\epsilon > 0$ (cf. Section 7-2) implies that

$$\overline{\mathrm{NE}}(X_{\min}/X_{\mathrm{can}}) = \overline{\mathrm{NE}}(X_{\min}/X_{\mathrm{can}})_{K_{X_{\min}}+\epsilon D<0} = \Sigma R_l,$$

is polyhedral in shape and thus there are only finitely many extremal rays R_l (over X_{can}).

Fix a minimal model X_o.

Then the proof of Theorem 12-1-8 shows that starting from any other minimal model we can come back to X_o by a sequence of G-flops (over X_{can}) where G is the strict transfrom of a very ample divisor G_o on X_o. Take an effective divisor $D \equiv -G_o$ over X_{can}. Then by the symmetric feature of "flops," the sequence of G-flops can be reversed as a sequence of D-flops (over X_{can}).

Therefore,

(i) any minimal model can be reached by a sequnce of D-flops starting from X_o, and

(ii) for each minimal model there are only finitely many (D-flopping) extremal rays.

Then if there existed infinitely many minimal models, conditions (i) and (ii) would imply that there exists an infinite sequence of D-flops, contradicting (termination of D-flops).

In fact, if there exist infinitely minimal models, then first it follows from (i) and (ii) that the lengths of D-flops starting from $X_o = Z_1$ must be unbounded. Now using (ii) there exists Z_2 such that the length of D-flops starting with $Z_1 \dashrightarrow Z_2$ are unbounded. Now inductively we find a sequence of D-flops $Z_1 \dashrightarrow Z_2 \dashrightarrow \cdots \dashrightarrow Z_n$ and Z_{n+1} such that the lengths of D-flops starting with $Z_1 \dashrightarrow Z_2 \dashrightarrow \cdots \dashrightarrow Z_n \dashrightarrow Z_{n+1}$ are unbounded. This way we construct an infinite sequence of D-flops.   □

Corollary 12-3-2. *Assume the conjectures (existence of flop) and (termination of flops). Then for a variety of general type, we have the decomposition*

$$\overline{\mathrm{Mov}} = \coprod \overline{\mathrm{Amp}}(X_i/X_{\mathrm{can}}),$$

where we have only a finite number of chambers $\overline{\mathrm{Amp}}(X_i/X_{\mathrm{can}})$.

PROOF. The proof is identical to that of Theorem 12-2-7, where we had the extra assumption of working in dimension 3. Note that the assumption of the dimension was used only to guarantee (existence of flop), (termination of flops), the abundance conjecture and the finiteness of extremal rays of D-flopping type for a movable divisor D. (existence of flop) and (termination of flops) are assumed in the above theorem: the abundance conjecture is established for minimal models of general type; and the finiteness of extremal rays is automoatic, since the cone of effective curves $\overline{\text{NE}}(X_i / X_{\text{can}})$ is polyhedral in shape for each minimal model X_i.   □

Remark 12-3-3 (How to Count the Number of Minimal Models) How many minimal models do we have in a fixed birational equivalence class in general?.

Before trying to answer this question, we should clarify the way(s) we count the "number" of minimal models.

Consider the following situation: Starting from a nonsingular projective variety V we reach a minimal model X_1 through one process of MMP and another minimal model X_2 through another process of MMP:

$$X_1 \xleftarrow{\;f_1\;} V \xdashrightarrow{\;f_2\;} X_2.$$

The birational map $f_2 \circ f_1^{-1} : X_1 \dashrightarrow X_2$ is not an isomorphism, but there is an isomorphism $\sigma(\neq f_2 \circ f_1^{-1}) : X_1 \xrightarrow{\sim} X_2$ as an abstract variety.

Method 1

We count X_1 and X_2 as two different minimal models. More generally, under this Method 1, if we have a collection of minimal models $\{X_i\}_{i \in I}$ with a prescribed set of birational maps

$$\phi_{ij} : X_i \dashrightarrow X_j$$

such that

$$\phi_{ii} = Id_{X_i},$$
$$\phi_{ij} \circ \phi_{jk} = \phi_{ik},$$

and

$$\phi_{ij} \text{ is not an isomorphism if } i \neq j,$$

then we count the number of different minimal models in the collection $\{X_i\}_{i \in I}$ to be the cardinality #I. Note that the number of chambers in the decomposition of Theorem 12-2-7 is the number of all the birational minimal models counted under Method 1.

Method 2

If we have a collection of minimal models birational to each other $\{X_i\}_{i \in I}$, we count the number of different minimal models in the collection to be the number of isomorphic classes in $\{X_i\}_{i \in I}$ as abstract varieties.

Obviously,

# of minimal models counted under Method 1

$\geq$ # of minimal models counted under Method 2.

The theorems (uniqueness of minimal models in dimension 2) and (finiteness of minimal models of general type) hold in counting the number under Method 1, and hence also under Method 2.

Reid [3] gives an example where there exists an infinite number of minimal models counted under Method 1 (for a 3-fold of nongeneral type).

Example 12-3-4 (cf. Reid [3]). Let $f : X \to \mathbb{A}^2$ be the family of elliptic curves defined by

$$z_1^2 = ((z_2 - a)^2 - x)((z_2 - b)^2 - y), \quad a \neq b,$$

where x, y are the coordinates for $\mathbb{A}^2$, and z_1, z_2 the coordinates in the "vertical" direction. Then the fiber $f^{-1}(p)$ is a nonsingular elliptic curve when $p \in \{xy \neq 0\}$, a nodal rational curve when $p \in \{xy = 0\}$, $p \neq (0, 0)$, and two rational curves meeting transversally at two points when $p = (0, 0)$. We regard it as a family over the analytic germ $p \in \mathbb{A}^2$. Each of the two rational curves in the central fiber can be flopped consecutively (Exercise! Why?) to give rise to infinitely many (counted under Method 1) minimal models (over the germ $p \in \mathbb{A}^2$).

The chamber structure of the ample cones of minimal models is illustrated by the reflections of the group $\tilde{A}_1$, as in Figure 12-2-5.

However, these minimal models are isomorphic to each other over the germ $p \in \mathbb{A}^2$ (Exercise! Why?), and hence there is only one minimal model counted under Method 2.

Figure 12-3-5.

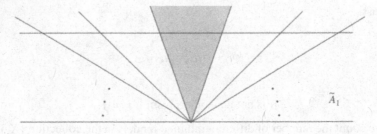

$\tilde{A}_1$

(Observe that we "cheated" a little in the description of the example above. We described the fibers only in an affine neighborhood in terms of z_1 and z_2. We leave it as an exercise to the reader to make this part rigorous and give the description of the entire family projective over $p \in \mathbb{A}^2$. Note also that the minimal models discussed above are only relative over the analytic germ $p \in \mathbb{A}^2$. We leave it as an exercise to the reader to construct an example where there exist an infinite number of minimal models over Spec$k$ (k is the base field) in the algebraic category counted under Method 1, modifying the above example.)

We learned the following conjecture from Professor Kawamata in conjunction with Morrison's conjecture on Calabi–Yau 3-folds.

Conjecture 12-3-6.   *The number of minimal models (in a fixed birational equivalence class) is always finite when counted under Method* 2.

Kawamata [17] shows that the above conjecture holds for 3-folds X with strictly positive Kodaira dimension $\kappa(X) > 0$.

CHAPTER 13

Birational Relation Among Mori Fiber Spaces

The purpose of this chapter is to discuss the **birational relation among Mori fiber spaces**. Here we focus our attention on the most important subject, the **Sarkisov program**, due to Sarkisov [3], Reid [6], Corti [1], which gives an algorithm for factoring a given birational map between Mori fiber spaces into a sequence of certain elementary transformations called "**links**." While it is a higher-dimensional analogue of the Castelnuovo–Noether theorem (cf. Theorem 1-8-8), its true meaning becomes clearer in the framework of the logarithmic category, with the main machinery of the program working under the log MMP discussed in Chapter 11. Our presentation is mostly in dimension 3, where all the necessary ingredients are established (with the most subtle part of showing "termination of Sarkisov program" ingeniously settled by Corti [1], as discussed in Section 13-2), leaving the details of the higher-dimensional case to the reader, where the general mechanism goes almost verbatim but some key ingredients still remain conjectural. (See Section 14-5 for the toric Sarkisov program, where we have all the necessary ingredients established in all dimensions.)

As a possible application we expect to have a better understanding of the group of self-birational maps of a Mori fiber space, e.g., the Cremona transformations of the projective space $\phi : \mathbb{P}^n \to \operatorname{Spec} \mathbb{C}$, by analyzing the structures of the links. This turns out to be, however, a formidable task even in dimension 3. In Section 13-3 we present Takahashi's work deriving the classical theorem of Jung on the structure of the automorphism group of the affine space $\mathbb{A}^2$ in dimension 2 as an application of the logarithmic generalization of the Sarkisov program. We hope that this will at least give the reader a taste of what a future application of the Sarkisov program might look like.

13.1   Sarkisov Program

In Section 1-8 we already discussed the Castelnuovo–Noether theorem, factoring birational maps among Mori fiber spaces in dimension 2, in the form of the Sarkisov program in dimension 2. The main idea of Sarkisov is to untwist a birational map between Mori fiber spaces into certain elementary transformations, called "links." We measure how far a given Mori fiber space is from the reference Mori fiber space via the Sarkisov degree. This main idea and the basic strategy using the Sarkisov degree remain the same in higher dimension. In dimension 2 our description of each link is quite explicit, though its construction in Section 1-8 is ad hoc based upon the classification of extremal rays in dimension 2. In this section we give a systematic treatment of the construction of each link in the framework of the log MMP. Our description of each link, however, becomes less explicit in dimension 3 or higher. Giving a better and more satisfactory description is one of the tasks of future research.

In this section we denote by $\mathfrak{C}$ the category of normal projective varieties with only $\mathbb{Q}$-factorial and terminal singularities.

Theorem 13-1-1 (Sarkisov Program in Dimension 3). *Let*

$$
\begin{array}{ccc}
X & \xrightarrow[\text{birat}]{\Phi} & X' \\[4pt]
\downarrow \phi & & \downarrow \phi' \\[4pt]
Y & & Y'
\end{array}
$$

be a birational map between two Mori fiber spaces in dimension 3:

$$
\phi : X \to Y,
$$
$$
\phi' : X' \to Y'.
$$

Then there exists an algorithm, called the **Sarkisov program,** *for decomposing* Φ *as a composite of the following four types of* **"links"***:*

Type (I): *A link of* Type (I) *consists of a diagram of the form*

$$
\begin{array}{ccc}
 & Z & \dashrightarrow \quad X_1 \\
 & \swarrow & \\
X & & \downarrow \\
\downarrow & & \\
Y & \longleftarrow & Y_1
\end{array}
$$

where $Z \to X$ *is a divisorial contraction of an extremal ray with respect to* K_Z *with* $Z \in \mathfrak{C}$*, the birational map* $Z \dashrightarrow X_1$ *is a sequence of log flips,* $\phi_1 : X_1 \to Y_1$ *is a new Mori fiber space with* $X_1 \in \mathfrak{C}$*, and* $Y_1 \to Y$ *is a morphism with connected fibers between normal projective varieties with only* $\mathbb{Q}$-factorial singularities and $\rho(Y_1/Y) = 1$.

Type (II): *A link of* Type (II) *consists of a diagram of the form*

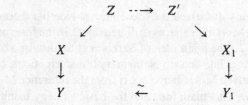

where $Z \to X$ *is a divisorial contraction of an extremal ray with respect to* K_Z *with* $Z \in \mathfrak{C}$, *the birational map* $Z \dashrightarrow Z'$ *is a sequence of log flips,* $Z' \to X_1$ *is a divisorial contraction of an extremal ray with respect to* $K_{Z'}$ *with* $Z' \in \mathfrak{C}$, *and* $\phi_1 : X_1 \to Y_1$ *is a new Mori fiber space with* $X_1 \in \mathfrak{C}$.

Type (III): *A link of* Type (III) *is the inverse of a link of* Type (I) *and hence consists of a diagram of the form*

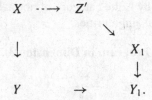

Type (IV): *A link of* Type (IV) *consists of a diagram of the form*

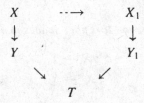

where the birational map $X \dashrightarrow X_1$ *is a sequence of log flips,* $\phi_1 : X_1 \to Y_1$ *is a new Mori fiber space with* $X_1 \in \mathfrak{C}$, *while* $Y \to T$ *and* $Y_1 \to T$ *are morphisms with connected fibers from normal projective varieties* Y *and* Y_1 *with only* $\mathbb{Q}$-*factorial singularities to a normal projective variety* T *with* $\rho(Y/T) = \rho(Y_1/T) = 1$.

We remark that the importance of the theorem lies in the mechanism of how the algorithm works, rather than in the mere fact that a birational map can be decomposed into a composite of links. At present, the description of the types of links is not satisfactory enough to characterize them as those that necessarily appear in some process of untwisting a birational map between Mori fiber spaces via the Sarkisov program. A more detailed study of links is indispensable for and crucial to the future application of the program.

The rest of the section will be spent in showing the general mechanism of the Sarkisov program, leaving the verification of its termination to the next section.

Strategy for Untwisting

The strategy for decomposing Φ into a composite of links, which we call "untwisting" of Φ, is to set up a good invariant of Φ with reference to the fixed Mori fiber space $\phi' : X' \to Y'$ (called the **Sarkisov degree**, the triplets of numbers (μ, λ, e) in lexicographical order, as will be defined below), which should tell us how far ϕ is from being an isomorphism of Mori fiber spaces.

It is the **Noether–Fano–Iskovskikh criterion** that allows us to judge precisely when Φ is an isomorphism of Mori fiber spaces in terms of the Sarkisov degree and the canonical divisor of X.

Starting with a given birational map Φ between two Mori fiber spaces, we ask whether Φ is actually an isomorphism of Mori fiber spaces via the NFI criterion. If the answer is *yes*, then there is nothing more to do and we proceed to *end* of the program. If the answer is *no*, then we "untwist" Φ by an appropriate link of Type (I), (II), (III), or (IV) to obtain a new birational map Φ_1. If Φ_1 is an isomorphism of Mori fiber spaces via the NFI criterion, we proceed to *end*. If not, we repeat the process with Φ_1. Each time we "untwist," the Sarkisov degree should strictly drop, i.e.,

$$(\mu, \lambda, e) = (\mu_0, \lambda_0, e_0) > (\mu_1, \lambda_1, e_1) > (\mu_2, \lambda_2, e_2) > \cdots,$$

where these are the Sarkisov degrees of $\Phi = \Phi_0, \Phi_1, \Phi_2, \ldots$ with respect to the fixed reference Mori fiber space $\phi' : X' \to Y'$:

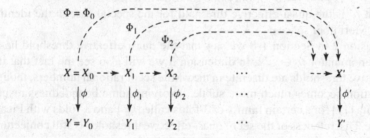

Finally by observing that this process has to come to an end after finitely many repetitions, the property that we verify in the next section via the analysis of the Sarkisov degree, we reach the Mori fiber space $\phi' : X' \to Y'$ expressing Φ as a composite of links.

Sarkisov degree.

First we choose and fix $\mu' \in \mathbb{N}$ and an ample divisor A' on Y' such that

$$H_{X'} = -\mu' K_{X'} + \phi'^* A'$$

is a very ample divisor on X'. (Note that $-K_{X'}$ is relatively ample. We refer the reader to Hartshorne [3], Chapter II, Proposition 7.10, or Iitaka [5], Theorem 7.11.)

We take a nonsingular projective variety V that dominates both X and X' by birational morphisms (which are compatible with Φ) via Hironaka's elimination of points of indeterminacy,

$$X \xleftarrow{\sigma} V \xrightarrow{\sigma'} X' \text{ with } \sigma' \circ \sigma^{-1} = \Phi.$$

For a member

$$\mathcal{H}_{X'} \in |H_{X'}|$$

we define the **"homaloidal" transform** $\mathcal{H}_X$ on X of $\mathcal{H}_{X'}$ to be

$$\mathcal{H}_X = \sigma_* \sigma'^* \mathcal{H}_{X'}.$$

We note that the homaloidal transform does not depend on the choice of V. We are ready to define the Sarkisov degree.

(i) μ : **the quasi-effective threshold**

The first of the triplet, the quasi-effective threshold μ, is defined to be a rational number (necessarily positive) such that

$$\mu K_X + \mathcal{H}_X \equiv_Y 0,$$

that is to say,

$$(\mu K_Y + \mathcal{H}_X) \cdot F = 0 \text{ for any curve } F \text{ in a fiber of } \phi : X \to Y.$$

Note that μ is independent of the choice of a member $\mathcal{H}_X$ (cf. Lemma 13-1-5) and that since all the curves contracted by ϕ are numerically proportional, we have to check $(\mu K_X + \mathcal{H}_X) \cdot F = 0$ for only one curve F in a fiber of ϕ.

Note that μ' is the quasi-effective threshold for the special case Φ the identity map of the Mori fiber space $\phi' : X' \to S'$.

In dimension 2 in Section 1-8 we saw that the quasi-effective threshold has a bounded denominator $\mu \in \frac{1}{3!}\mathbb{N}$. In dimension 3 we will also see the fact that the quasi-effective thresholds are discrete in the whole set of rational numbers, though the verification becomes much more subtle, involving some boundedness results of Kawamata [11] for a certain family of 3-folds called $\mathbb{Q}$-Fano 3-folds with Picard number one. Discreteness of the set of quasi-effective thresholds is still conjectural in higher dimensions.

(ii) λ : **the maximal multiplicity**

In order to define the second member of the triplet, the maximal multiplicity λ, we consider the linear system consisting of the homaloidal transforms $\mathcal{H}_X$ for $\mathcal{H}_{X'} \in |H_{X'}|$, which we denote by $\Phi^{-1}_{\text{homaloidal}}|H_{X'}|$. (We note that this linear system may be smaller than the complete linear system $|\mathcal{H}_X|$.) In dimension 2, λ was simply the maximum multiplicity of a general member $\mathcal{H}_X$ at the base points of $\Phi^{-1}_{\text{homaloidal}}|H_{X'}|$. In dimension 3 or higher, it is the reciprocal $1/\lambda$ that has the more natural and intrinsic description, which we use as the definition.

Definition 13-1-2 (Log Pair $(X, c\mathcal{H}_X)$ Being Canonical) (cf. Definition 4-4-2).
Let the situation be as above. We write the ramification formula

$$K_V = \sigma^* K_X + \Sigma a_k E_k,$$
$$\mathcal{H}_V = \sigma'^* \mathcal{H}_{X'} = \sigma^* \mathcal{H}_X - \Sigma b_k E_k,$$

for the homaloidal transform $\mathcal{H}_X$ of a general member $\mathcal{H}_{X'} \in |H_{X'}|$. For $c \in \mathbb{Q}_{\geq 0}$ we say that the log pair $(X, c\mathcal{H}_X)$ is **canonical** *if writing down the log ramification formula*

$$K_V + c\mathcal{H}_V = \sigma'^*(K_X + c\mathcal{H}_X) + \Sigma(a_k - cb_k)E_k,$$

we have

$$a_k - cb_k \geq 0 \quad \forall E_k.$$

Noting that the number $a_k - cb_k$ coincides with the usual log discrepancy

$$a_k - cb_k = a(E_k; X, c\mathcal{H}_X),$$

which does not depend on the common resolution V but only on E_k, the log pair $(X, c\mathcal{H}_X)$ is canonical if

$$a(E; X, c\mathcal{H}_X) \geq 0 \quad \forall E \text{ exceptional over } X.$$

Now we are ready to define the invariant λ.

If $\Phi_{\text{homaloidal}}^{-1}|H_{X'}|$ does not have any base point, then $\lambda = 0$ by definition. If $\Phi_{\text{homaloidal}}^{-1}|H_{X'}|$ has some base points, by taking the common resolution as before and writing the ramification formulae we define

$$\frac{1}{\lambda} = \max\{c \in \mathbb{Q}_{\geq 0}; a_k - cb_k \geq 0 \quad \forall k\} = \min\left\{\frac{a_k}{b_k}\right\}$$

Note that since $\Phi_{\text{homaloidal}}^{-1}|H_{X'}|$ has some base points, we have $b_k > 0$ for some E_k. Note also that the number $1/\lambda$ depends only on the linear system $\Phi_{\text{homaloidal}}^{-1}|H_{X'}|$ and is independent of the choice of a general member $\mathcal{H}_{X'}$ and hence of $\mathcal{H}_X$, since the numbers b_k stay the same for general members. It has a description free of the particular choice of the common resolution using the notion of log pairs being canonical:

$$\frac{1}{\lambda} = \max\{c \in \mathbb{Q}_{\geq 0}; (X, c\mathcal{H}_X) \text{ is canonical}\}$$
$$= \max\{c \in \mathbb{Q}_{\geq 0}; a(E; X, c\mathcal{H}_X) \geq 0 \quad \forall E \text{ exceptional over } X\}.$$

The reciprocal $1/\lambda$ is called the **canonical threshold** of X with respect to the linear system $\Phi_{\text{homaloidal}}^{-1}|H_{X'}|$.

(iii) e: **the number of crepant exceptional divisors**

When $\lambda > 0$, we define

$$e = \#\left\{E_k; a_k - \frac{1}{\lambda}b_k = 0 \text{ or equivalently } \frac{b_k}{a_k} = \lambda\right\}$$
$$= \#\left\{E; E \text{ is exceptional over } X \text{ and } a\left(E; X, \frac{1}{\lambda}\mathcal{H}_X\right) = 0\right\}.$$

(Exercise! Check the second equality, which implies that the number e is independent of the choice of a common resolution V.)

When $\lambda = 0$, i.e., when $\Phi_{\text{homaloidal}}^{-1}|H_{X'}|$ has no base points, all the exceptional divisors E_k have $b_k/a_k = 0 = \lambda$ for any common resolution V, and e is not well-defined. But when $\lambda = 0$, we are always in the case $\lambda \leq \mu$ in the algorithm of the Sarkisov program. The termination of untwistings in this case does not involve the invariant e, and we leave e undefined in this case.

This completes the definition of the triplet (μ, λ, e) associated to Φ with reference to $\phi' : X' \to Y'$ after fixing μ' and A'. We consider these triplets in lexicographical order. The next criterion tells us when Φ is an isomorphism of Mori fiber spaces, in terms of the Sarkisov degree (μ, λ, e) and the canonical divisor K_X of X.

In the following the notation $\mathcal{H}_{X'}$ always refers to a *general* member of $|H_{X'}|$, and $\mathcal{H}_X$ to its homaloidal transform.

Proposition 13-1-3 (Noether–Fano–Iskovskikh Criterion). *Suppose*

$$\lambda \leq \mu \quad and \quad K_X + \frac{1}{\mu}\mathcal{H} \text{ is nef on } X.$$

Then Φ is an isomorphism of Mori fiber spaces, i.e., there exists a commutative diagram

$$
\begin{array}{ccc}
X & \overset{\Phi}{\underset{\sim}{\longrightarrow}} & X' \\
\downarrow{\scriptstyle \phi} & & \downarrow{\scriptstyle \phi'} \\
Y & \overset{\sim}{\longrightarrow} & Y'
\end{array}
$$

PROOF.

Step 1. Show that $\mu = \mu'$.

First we claim that

$$\mu = \mu'.$$

We take a common resolution as before:

$$
\begin{array}{ccc}
 & V & \\
{\scriptstyle \sigma}\swarrow & & \searrow{\scriptstyle \sigma'} \\
X & \overset{\Phi}{\dashrightarrow} & X' \\
\downarrow{\scriptstyle \phi} & & \downarrow{\scriptstyle \phi'} \\
Y & & Y'.
\end{array}
$$

We write the log ramification formulas for the log pairs $(X, \frac{1}{\mu'}\mathcal{H}_X)$ and $(X', \frac{1}{\mu'}\mathcal{H}_{X'})$:

$$
\begin{aligned}
K_V + \frac{1}{\mu'}\mathcal{H}_V &= \sigma^*\left(K_X + \frac{1}{\mu'}\mathcal{H}_X\right) + \Sigma r_{k,\mu'} E_k \\
&= \sigma'^*\left(K_{X'} + \frac{1}{\mu'}\mathcal{H}_{X'}\right) + \Sigma r'_{j,\mu'} E'_j,
\end{aligned}
$$

where the E_k (respectively E'_j) are the exceptional divisors for σ (respectively σ'). Note that since $\mathcal{H}_V = \sigma'^* \mathcal{H}_{X'}$ and since X' has terminal singularities,

$$r'_{j,\mu'} = a'_j > 0 \quad \forall j,$$

where the usual ramification formula for X' reads

$$K_V = \sigma'^* K_{X'} + \Sigma a'_{j'} E'_{j'}.$$

Now if we take a general curve F in the general fiber of ϕ avoiding all $\sigma(E_k)$ and not contained in any of $\sigma(E'_j)$, then F can be considered to lie also on V, and we compute

$$\left(K_X + \frac{1}{\mu'} \mathcal{H}_X \right) \cdot F = \left\{ \sigma^* \left(K_X + \frac{1}{\mu'} \mathcal{H}_X \right) + \Sigma r_{k,\mu'} E_k \right\} \cdot F$$

$$= \left\{ \sigma'^* \left(K_{X'} + \frac{1}{\mu'} \mathcal{H}_{X'} \right) + \Sigma r'_{j,\mu'} E'_j \right\} \cdot F$$

$$= \left(\frac{1}{\mu'} \phi'^* A' + \Sigma r'_{j,\mu'} E'_j \right) \cdot F \geq 0.$$

(Note that we can take the general curve F to be the intersection of a general fiber $\phi^{-1}(p)$ and $\dim \phi^{-1}(p)$ general members of a very ample divisor on X.)
    This implies

$$\mu \geq \mu'.$$

Note that this inequality always holds without any extra assumption.
    In order to see the inequality in the opposite direction, we write down a similar ramification formula, replacing $1/\mu'$ by $1/\mu$:

$$K_V + \frac{1}{\mu} \mathcal{H}_V - \sigma^* \left(K_X + \frac{1}{\mu} \mathcal{H}_X \right) + \Sigma r_{k,\mu} E_k$$

$$= \sigma'^* \left(K_{X'} + \frac{1}{\mu} \mathcal{H}_{X'} \right) + \Sigma r'_{j,\mu} E'_j.$$

This time, since $\lambda \leq \mu$ by assumption, we have

$$r_{k,\mu} \geq 0 \quad \forall k.$$

Now if we take a general curve F' in the general fiber of ϕ' avoiding all $\sigma'(E'_j)$ and not contained in any of $\sigma'(E_k)$, then F' can be considered to lie also on V, and we compute

$$\left(K_{X'} + \frac{1}{\mu} \mathcal{H}_{X'} \right) \cdot F' = \left\{ \sigma'^* \left(K_{X'} + \frac{1}{\mu} \mathcal{H}_{X'} \right) + \Sigma r'_{j,\mu} E'_j \right\} \cdot F'$$

$$= \left\{ \sigma^* \left(K_X + \frac{1}{\mu} \mathcal{H}_X \right) + \Sigma r_{k,\mu} E_k \right\} \cdot F' \geq 0,$$

since $K_X + \frac{1}{\mu} \mathcal{H}_X$ is nef. This implies

$$\mu \leq \mu'.$$

Therefore, we finally conclude that

$$\mu = \mu'.$$

This completes Step 1.

We write down the ramification formulas

$$K_V = \sigma^* K_X + R,$$
$$K_V = \sigma'^* K_{X'} + R',$$
$$K_V + \frac{1}{\mu}\mathcal{H}_V = \sigma^*\left(K_X + \frac{1}{\mu}\mathcal{H}_X\right) + \Sigma r_k E_k,$$
$$K_V + \frac{1}{\mu}\mathcal{H}_V = \sigma'^*\left(K_{X'} + \frac{1}{\mu}\mathcal{H}_{X'}\right) + \Sigma r'_j E'_j.$$

Note that

$$R \geq \Sigma r_k E_k \geq 0, \text{ since } \lambda \leq \mu,$$

and that

$$R' = \Sigma r'_j E'_j, \text{ since } \mathcal{H}_V = \sigma'^* \mathcal{H}_{X'}.$$

Step 2. Show that in the obvious equality

$$\sigma^*\left(K_X + \frac{1}{\mu}\mathcal{H}_X\right) + \Sigma r_k E_k = \sigma'^*\left(K_{X'} + \frac{1}{\mu}\mathcal{H}_{X'}\right) + \Sigma r'_j E'_j$$

we have

$$\sigma^*\left(K_X + \frac{1}{\mu}\mathcal{H}_X\right) = \sigma'^*\left(K_{X'} + \frac{1}{\mu}\mathcal{H}_{X'}\right),$$

which is equivalent to

$$\Sigma r_k E_k = \Sigma r'_j E'_j.$$

We introduce the notation

$$(\cup E_k) \cup (\cup E'_j) = \cup F_l$$

and write

$$\Sigma r_k E_k = \Sigma_{l \in S_\sigma} r_l F_l + \Sigma_{l \in S_{\sigma,\sigma'}} r_l F_l,$$
$$\Sigma r'_j E'_j = \Sigma_{l \in S_{\sigma,\sigma'}} r'_l F_l + \Sigma_{l \in S_{\sigma'}} r'_l F_l,$$

where

the divisor F_l is $\begin{cases} \sigma\text{-exceptional but } not \ \sigma'\text{-exceptional,} \\ \sigma\text{-exceptional and } \sigma'\text{-exceptional,} \\ not \ \sigma\text{-exceptional but } \sigma'\text{-exceptional,} \end{cases}$ iff $\begin{cases} l \in S_\sigma, \\ l \in S_{\sigma,\sigma'}, \\ l \in S_{\sigma'}. \end{cases}$

Then by equalities

$$\Sigma_{l \in S_\sigma} r_l F_l + \Sigma_{l \in S_{\sigma,\sigma'}} (r_l - r'_l) F_l$$

$$\equiv -\sigma^*\left(K_X + \frac{1}{\mu}\mathcal{H}_X\right) + \sigma'^*\left(K_{X'} + \frac{1}{\mu}\mathcal{H}_{X'}\right) + \Sigma_{l\in S_{\sigma'}}r_l'F_l,$$

$$\Sigma_{l\in S_{\sigma,\sigma'}}(r_l' - r_l)F_l + \Sigma_{l\in S_{\sigma'}}r_l'F_l$$

$$\equiv -\sigma'^*\left(K_{X'} + \frac{1}{\mu}\mathcal{H}_{X'}\right) + \sigma^*\left(K_X + \frac{1}{\mu}\mathcal{H}_X\right) + \Sigma_{l\in S_\sigma}r_lF_l,$$

and by the following negativity lemma applied to σ over X and to σ' over X', we conclude that

$$r_l \le 0 \text{ if } l \in S_\sigma,$$
$$r_l - r_l' \le 0 \text{ if } l \in S_{\sigma,\sigma'},$$
$$r_l' - r_l \le 0 \text{ if } l \in S_{\sigma,\sigma'},$$
$$r_l' \le 0 \text{ if } l \in S_{\sigma'},$$

and hence

$$r_l = 0 \text{ if } l \in S_\sigma,$$
$$r_l = r_l' \text{ if } l \in S_{\sigma,\sigma'},$$
$$r_l' = 0 \text{ if } l \in S_{\sigma'}.$$

Therefore, we have

$$\Sigma r_k E_k = \Sigma_{l\in S_{\sigma,\sigma'}}r_lF_l = \Sigma_{l\in S_{\sigma,\sigma'}}r_l'F_l = \Sigma r_j'E_j'.$$

Lemma 13-1-4 (Negativity Lemma). *Let $f : V \to T$ be a birational morphism from a nonsingular projective variety V to another projective variety T. Suppose that a divisor $\Sigma\alpha_i E_i$ consisting of the divisors exceptional for f is numerically of the form*

$$\Sigma\alpha_i E_i \equiv N + G,$$

where N is a $\mathbb{Q}$-divisor that is f-nef (i.e., $N \cdot C \ge 0$ for any curve $C \subset V$ with $f(C) = \text{pt.}$) and where G is an effective $\mathbb{Q}$-divisor none of whose components are exceptional for f. Then

$$\alpha_i \le 0 \quad \forall i.$$

PROOF. Take very ample divisors A and B on T and V, respectively. We take any exceptional divisor E_i with $d_i = \dim f(E_i)$. Then by restricting the equality to the surface $S = \cap_{i=1}^{d_i} f^*A \cap \cap_{j=1}^{\dim V - d_i - 2} B_i$, the nonpositivity of α_i follows from the negativity lemma in dimension 2 (cf. Lemma 1-8-10). □

Step 3. Show that Φ induces an isomorphism of the Mori fiber spaces.

First we claim that Φ induces a morphism $\tau : Y \to Y'$ such that $\tau \circ \phi = \phi' \circ \Phi$. By Lemma 1-8-1 we have only to show that

$$(\phi' \circ \sigma')((\phi \circ \sigma)^{-1}(p)) \text{ is a point} \quad \forall p \in Y.$$

Since $(\phi \circ \sigma)_* \mathcal{O}_V = \mathcal{O}_Y$ (Exercise! Why?), $(\phi \circ \sigma)^{-1}(p)$ is connected by Zariski's Main Theorem (cf. Theorem 1-2-17) and hence so is $(\phi' \circ \sigma')((\phi \circ \sigma)^{-1}(p))$. Thus we have only to show that

$$\dim(\phi' \circ \sigma')((\phi \circ \sigma)^{-1}(p)) = 0.$$

Suppose $\dim(\phi' \circ \sigma')((\phi \circ \sigma)^{-1}(p)) > 0$ for some $p \in Y$. Then there exists a curve $C \subset (\phi \circ \sigma)^{-1}(p)$ such that $(\phi' \circ \sigma')(C)$ is not a point. (Exercise! Find such a curve C.) Then by the equality

$$\sigma^* \left(K_X + \frac{1}{\mu} \mathcal{H}_X \right) = \sigma'^* \left(K_{X'} + \frac{1}{\mu} \mathcal{H}_{X'} \right)$$

we compute

$$
\begin{aligned}
0 &= \sigma^* \left(K_X + \frac{1}{\mu} \mathcal{H}_X \right) \cdot C \\
&= \sigma'^* \left(K_{X'} + \frac{1}{\mu} \mathcal{H}_{X'} \right) \cdot C \\
&= \frac{1}{\mu} (\phi' \circ \sigma')^* A' \cdot C \\
&= \frac{1}{\mu} A' \cdot (\phi' \circ \sigma')_*(C) > 0,
\end{aligned}
$$

a contradiction!

Therefore, we have such a morphism $\tau : Y \to Y'$.

Observe that the inequality and equalities

$$R \geq \Sigma r_k E_k = \Sigma r'_j E'_j = R'$$

by Step 2 imply that all the exceptional divisors for σ' are also exceptional for σ. In dimension 2 we could immediately conclude from this that Φ^{-1} is a morphism. In dimension 3 or higher, though the final conclusion is the same, we have to struggle technically a little more.

Lemma 13-1-5 (Homaloidal Transform Preserves Numerical Equivalence).
Let

$$X \xleftarrow{\sigma} V \xrightarrow{\sigma'} X'$$

be as above. Let D'_1, D'_2 be two $\mathbb{Q}$-divisors on X' that are numerically equivalent, i.e.,

$$D'_1 \equiv D'_2 \text{ on } X'.$$

Then their homaloidal transforms are also numerically equivalent on X, i.e.,

$$D_1 = \sigma_* \sigma'^* D'_1 \equiv \sigma_* \sigma'^* D'_2 = D_2 \text{ on } X'.$$

PROOF.   We can write

$$\sigma'^* D'_1 = \sigma^* D_1 + \Sigma \alpha_k E_k,$$

$$\sigma'^* D_2' = \sigma^* D_2 + \Sigma \beta_k E_k,$$

where the E_k are the exceptional divisors for σ.

Since obviously

$$\sigma'^* D_1' \equiv \sigma'^* D_2',$$

we have

$$\Sigma(\alpha_k - \beta_k)E_k \equiv \sigma^*(D_2 - D_1) \equiv 0 \text{ over } X,$$
$$\Sigma(\beta_k - \alpha_k)E_k \equiv \sigma^*(D_1 - D_2) \equiv 0 \text{ over } X.$$

Thus by the negativity lemma applied to $\sigma : V \to X$ we conclude that

$$\Sigma(\alpha_k - \beta_k)E_k = 0$$

and hence

$$\sigma^* D_1 \equiv \sigma^* D_2.$$

Therefore, we conclude that

$$D_1 \equiv D_2. \qquad \square$$

Now we go back to the discussion of Step 3.

We claim that Φ^{-1} is an isomorphism in codimension one.

If not, then there would be a divisor $E_k \subset R$ such that E_k is exceptional for σ but not for σ'. We observe that $\sigma'(E_k)$ is a divisor that is ϕ'-ample, i.e., for any curve F' in a fiber of ϕ' we have

$$\sigma'(E_k) \cdot F > 0.$$

In fact, if not, then $\phi'(\sigma'(E_k))$ is a divisor on Y' such that $\sigma'(E_k) = \phi'^{-1}(\phi'(\sigma'(E_k)))$, since $\rho(X'/Y') = 1$. (Exercise! Why?) Since E_k is σ-exceptional, the set $(\tau \circ \phi)^{-1}(\phi'(\sigma'(E_k)))$ contains a divisor G whose strict transform on V is different from E_k. On the other hand, since $\sigma'(\sigma^{-1}(G)) \subset \phi'^{-1}(\phi'(\sigma'(E_k))) = \sigma'(E_k)$, the strict transform of G on V is σ'-exceptional but not σ-exceptional. By Step 2, there is no such divisor on V, a contradiction.

Therefore, $\sigma'(E_k)$ is ϕ'-ample, and hence there exist $e, a \in \mathbb{N}$ such that

$$D_1' = e\sigma'(E_k) + a\phi'^* A' \text{ is very ample on } X'.$$

Then for a curve C in the general fiber of ϕ we have

$$D_1 \cdot C = (\sigma_* \sigma'^* D_1') \cdot C = a\phi^* \tau^* A' \cdot C = 0.$$

On the other hand, since $e\sigma'(E_k) + a\phi'^* A'$ is very ample, for a general member $D_2' \in |e\sigma'(E_k) + a\phi'^* A'|$ we have

$$D_2 \cdot C = (\sigma_* \sigma'^* D_2') \cdot C > 0,$$

contradicting $D_1 \equiv D_2$, a consequence of Lemma 13-1-5.

Therefore, Φ^{-1} is an isomorphism in codimension one and hence so is Φ. Applying Lemma 13-1-5 to the homaloidal transforms $\sigma_* \sigma'$ and $\sigma'_* \sigma^*$, which are nothing but the strict transforms by Φ^{-1} and Φ, we conclude that they induce the isomorphism

$$N^1(X) \underset{\sigma_* \sigma'^*}{\overset{\sigma'_* \sigma^*}{\underset{\sim}{\longleftrightarrow}}} N^1(X').$$

Therefore, the inclusions

$$N^1(Y') \overset{\tau^*}{\hookrightarrow} N^1(Y) \hookrightarrow N^1(X) \cong N^1(X')$$

and

$$1 = \rho(X'/Y') = \dim N^1(X') - \dim N^1(Y')$$

imply

$$N^1(Y') = N^1(Y).$$

This implies that $\tau : Y \to Y'$ is a finite morphism and hence an isomorphism, since $\mathbb{C}(Y')$ is algebraically closed in $\mathbb{C}(X') = \mathbb{C}(X)$, i.e.

$$\tau : Y \overset{\to}{\sim} Y'.$$

Now if we take a (relatively) very ample divisor G' on X' over Y', then its strict transfrom G on X is also relatively ample over Y. Thus we finally conclude that

$$X = Proj_Y \oplus_{m \geq 0} \phi_* \mathcal{O}_X(mG) \overset{\sim}{\underset{\Phi}{\to}} Proj_{Y'} \oplus_{m \geq 0} \phi'_* \mathcal{O}_{X'}(mG') = X'.$$

This completes the proof of Proposition 13-1-3.   □

Flowchart for the Sarkisov Program

In the following we present a flowchart to untwist a birational map via the Sarkisov program,

$$
\begin{array}{ccc}
X & \overset{\Phi}{\underset{\text{birat}}{\dashrightarrow}} & X \\
\downarrow \phi & & \downarrow \phi' \\
Y & & Y'
\end{array}
$$

between two Mori fiber spaces. (See Flowchart 13-1-9 for an illustration of the entire algorithm.)
We

$$Start.$$

The first question to ask is:

$$\lambda \leq \mu \,?$$

According to whether the answer to this question is *yes* or *no*, we proceed separately into the case $\lambda \leq \mu$ or into the case $\lambda > \mu$.

Case: $\lambda \leq \mu$

If $\lambda \leq \mu$, then the next question to ask is:

$$K_X + \frac{1}{\mu}\mathcal{H}_X \text{ nef?}$$

If the answer to this question is *yes*, then $K_X + \frac{1}{\mu}\mathcal{H}_X$ is nef and $\lambda \leq \mu$ by the case assumption. Thus the Noether–Fano–Iskovskikh criterion applies to the situation to conclude that Φ is an isomorphism of Mori fiber spaces. This leads to an

End.

If $K_X + \frac{1}{\mu}\mathcal{H}_X$ is not nef, then we construct as follows a normal projective variety T dominated by a morphism with connected fibers $Y \to T$ s.t. $K_X + \frac{1}{\mu}\mathcal{H}_X$ is not relatively nef over T and $\rho(X/T) = 2$, so that we run $K + \frac{1}{\mu}\mathcal{H}$-MMP over T to have an untwisting link.

In fact, we construct such a morphism $Y \to T$ as follows. We pick a $K_X + \frac{1}{\mu}\mathcal{H}_X$-negative extremal ray P of $\overline{NE}(X/\operatorname{Spec} \mathbb{C})$ so that the span $F := P + R$ is a 2-dimensional extremal face, where R is the K_X-negative but $K_X + \frac{1}{\mu}\mathcal{H}_X$-trivial extremal ray giving the Mori fiber space $\phi : X \to Y$. (Exercise! Show why we can pick such an extremal ray P.)

Figure 13-1-6.

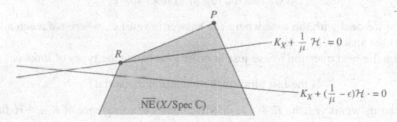

The face F is $K_X + (\frac{1}{\mu} - \epsilon)\mathcal{H}_X$-negative for $0 < \epsilon \ll 1$; thus we have the contraction morphism $\operatorname{cont}_F : X \to T$ to obtain T (cf. Theorem 8-1-3 and Exercise 8-1-4). Since $F \supset R$, the contraction cont_F factors through $\operatorname{cont}_R = \phi : X \to Y$, and by construction T satisfies all the required conditions.

2-Ray Game

Now we run $K + \frac{1}{\mu}\mathcal{H}$-MMP over T. It is a special kind of MMP, which we call the **2-ray game**. Since $\rho(X/T) = 2$, the cone $\overline{NE}(X/T)$ has two extremal rays (edges), one $R = R_\alpha^0$ for the original Mori fiber space $\phi = \operatorname{cont}_R : X \to Y (\to T)$ and the other $P = R_\beta^0$ on $X = X^0$. Note that we identify $\overline{NE}(X/T)$ with the face $P + R = F$ through the inclusion $N_1(X/T) \hookrightarrow N_1(X/\operatorname{Spec} \mathbb{C})$ (cf. Example-Exercise 3-5-1).

Figure 13-1-7.

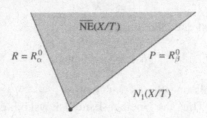

The ray R_α^0 is $K + \frac{1}{\mu}\mathcal{H}$-trivial, so in order to run $K + \frac{1}{\mu}\mathcal{H}$-MMP we just have to look at the other ray R_β^0.

If the contraction is either of divisorial type or fibering type, then the $K + \frac{1}{\mu}\mathcal{H}$-MMP necessarily comes to an end. If it is of (log) flipping type, then we continue the 2-ray game with the flip $X^1 = (X^0)^+$. The ray $R_\alpha^1 = (R_\alpha^0)^+$ is necessarily $K + \frac{1}{\mu}\mathcal{H}$-positive, so in order to run $K + \frac{1}{\mu}\mathcal{H}$-MMP we just have to look at the other ray R_β^1. This process must come to an end, since there is no infinite sequence of log flips, and the MMP terminates.

We reach at the end either a log minimal model or a log Mori fiber space (with respect to $K + \frac{1}{\mu}\mathcal{H}$ and over T).

So we ask the next question:

Do we reach a log Mori fiber space?

First we deal with the case where the answer is *yes*, i.e., where we reach a log Mori fiber space $X_1 \to Y_1$.

Then the next question to ask just in order to separate the types of links is:

Is the last birational contraction divisorial?

If the answer is *yes*, the $K + \frac{1}{\mu}\mathcal{H}$-MMP consists of a sequence of $K + \frac{1}{\mu}\mathcal{H}$-flips $X \dashrightarrow Z'$ followed by a $K + \frac{1}{\mu}\mathcal{H}$-negative divisorial contraction $Z' \to X_1$. Since $\rho(X_1/T) = 1$ and by the case assumption, $\phi_1 : X_1 \to Y_1 = T$ is a log Mori fiber space with respect to $K_{X_1} + \frac{1}{\mu}\mathcal{H}_{X_1}$, i.e., ϕ_1 is a $K_{X_1} + \frac{1}{\mu}\mathcal{H}_{X_1}$-negative and thus K_{X_1}-negative fiber space:

$$
\begin{array}{ccc}
X & \dashrightarrow & Z' \\
\downarrow & & \searrow \\
Y & & X_1 \\
& \searrow & \downarrow \\
& T & \xrightarrow{\sim} \; Y_1
\end{array}
$$

If the answer is *no*, then the $K + \frac{1}{\mu}\mathcal{H}$-MMP consists of a sequence of $K + \frac{1}{\mu}\mathcal{H}$-flips $X \dashrightarrow Z'$ followed by a $K_{X_1} + \frac{1}{\mu}\mathcal{H}_{X_1}$-negative and thus K_{X_1}-negative fibering contraction $\phi_1 : Z' = X_1 \to Y_1$. Since $\rho(X_1/T) = \rho(X/T) = 2$, we have

$\rho(Y_1/T) = 1$:

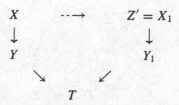

We claim in both cases that X_1 has only $\mathbb{Q}$-factorial and terminal singularities. First, $\mathbb{Q}$-factoriality of X_1 is automatic from the construction via the (log) MMP (cf. Section 8-2). In order to see that X_1 has only terminal singularities, let I be the locus of indeterminacy of the birational map $X_1 \dashrightarrow X$. If E is a discrete valuation whose center on X_1 is not contained in I (and has codimension ≥ 2), then

$$a(E; X_1, \emptyset) = a(E; X, \emptyset) > 0.$$

If the center of E on X_1 is contained in I, then

$$a(E; X_1, \emptyset) \geq a(E; X_1, \frac{1}{\mu}\mathcal{H}_{X_1}) > a(E; X, \frac{1}{\mu}\mathcal{H}_X) \geq 0.$$

We note that the second inequality follows from the logarithmic version of Proposition 8-2-1 (ii) and the observation of Shokurov in Chapter 9 that the discrepancy does not decrease under the process of the (log) MMP and it strictly increases if the center of the exceptional divisor is on the locus that is modified by the process of the (log) MMP. We also note that the third inequality follows from the case assumption $\lambda \leq \mu$.

Thus we have the claim.

Therefore, we have a link of Type (III) in the former case and a link of Type (IV) in the latter.

Moreover, since $K_{X_1} + \frac{1}{\mu}\mathcal{H}_{X_1}$ is negative over Y_1, we conclude in both cases that

$$\mu_1 < \mu.$$

Therefore, after untwisting Φ by a link of Type (III) or Type (IV), we go back to *Start* with a strictly decreased quasi-effective threshold.

Second, we deal with the case where the answer to the question; Do we reach a log Mori fiber space? is *no*, i.e., where we reach a log minimal model over T.

Then just in the previous case of reaching a log Mori fiber space the next question to ask is:

Is the last birational contraction divisorial?

If the answer is *yes*, the $K + \frac{1}{\mu}\mathcal{H}$-MMP consists of a sequence of $K + \frac{1}{\mu}\mathcal{H}$-flips $X \dashrightarrow Z'$ followed by a $K + \frac{1}{\mu}\mathcal{H}$-negative divisorial contraction $Z' \to X_1$. We will see below that $K_{X_1} + \frac{1}{\mu}\mathcal{H}_{X_1}$ is trivial over $T = Y_1$. Since $\rho(X_1/Y_1) = 1$,

$\phi_1 : X_1 \rightarrow Y_1$ is a K_{X_1}-negative fiber space:

$$
\begin{array}{ccc}
X & \dashrightarrow & Z' \\
\downarrow & & \searrow \\
Y & & X_1 \\
& \searrow & \downarrow \\
& T & \tilde{\rightarrow} \quad Y_1
\end{array}
$$

If the answer is *no*, then the $K + \frac{1}{\mu}\mathcal{H}$-MMP consists of a sequence of $K + \frac{1}{\mu}\mathcal{H}$-flips $X \dashrightarrow Z' = X_1$. In this case we will see that there is an extremal ray of $\overline{NE}(X_1/T)$ that is $K_{X_1} + \frac{1}{\mu}\mathcal{H}_{X_1}$-trivial and K_{X_1}-negative and that is of fibering type. Let $\phi_1 : X_1 \rightarrow Y_1$ be the contraction of the extremal ray and $Y_1 \rightarrow T$ the induced morphism with $\rho(Y_1/T) = 1$:

$$
\begin{array}{ccc}
X & \dashrightarrow & Z' = X_1 \\
\downarrow & & \downarrow \\
Y & & Y_1 \\
& \searrow \quad \swarrow & \\
& T &
\end{array}
$$

Just as before it follows that in both cases X_1 has only $\mathbb{Q}$-factorial and terminal singularities.

Now we verify the claim that in both cases there is a curve F_1 on X_1 mapping to a point on T such that

$$
\left(K_{X_1} + \frac{1}{\mu}\mathcal{H}\right) \cdot F_1 = 0 \quad \text{and} \quad K_{X_1} \cdot F_1 < 0.
$$

In fact, if we take a general curve F_1 in the general fiber of the morphism $X_1 \rightarrow T$ away from the locus of indeterminacy of the birational map $X_1 \dashrightarrow X$ (i.e., in the first case the union of the image of the exceptional divisor of the divisorial contraction and all the flipped curves and in the second case all the flipped curves) and not contained in the base locus of $\mathcal{H}_{X_1}$, then F_1 can be considered to lie also on X, and thus

$$
0 \geq \left(K_X + \frac{1}{\mu}\mathcal{H}\right) \cdot F_1 = \left(K_{X_1} + \frac{1}{\mu}\mathcal{H}_{X_1}\right) \cdot F_1 \geq 0,
$$

$$
0 > \left(K_X + \left(\frac{1}{\mu} - \epsilon\right)\mathcal{H}_X\right) \cdot F_1 = \left(K_{X_1} + \left(\frac{1}{\mu} - \epsilon\right)\mathcal{H}_{X_1}\right) \cdot F_1 \geq K_{X_1} \cdot F_1.
$$

When the last birational contraction is divisorial, since $\rho(X_1/Y_1) = 1$, the claim implies that $K_{X_1} + \frac{1}{\mu}\mathcal{H}_{X_1}$ is trivial over Y_1.

When the last birational contraction is flipping and we reach a log minimal model, $K_{X_1} + \frac{1}{\mu}\mathcal{H}_{X_1}$ is nef over T and cannot be trivial over T. (If it were trivial, then so would be $K_X + \frac{1}{\mu}\mathcal{H}_X$ over T, a contradiction!) Thus the claim implies the existence of an extremal ray containing $[F_1]$ of $\overline{NE}(X_1/T)$ which is $K_{X_1} + \frac{1}{\mu}\mathcal{H}_{X_1}$-

trivial and K_{X_1}-negative. The extremal ray is also of fibering type, since F_1 is a general curve in the general fiber of the morphism $X_1 \to T$.

Therefore, we have a link of Type (III) in the former case and a link of Type (IV) in the latter. We observe that in either case of untwisting by a link of Type (III) or (IV) reaching a log minimal model, the morphism $Y \to T$ is birational. In fact, suppose that the morphism $Y \to T$ is not birational, i.e., $\dim Y > \dim T$. Then F_1 being a general curve as before, $\phi(F_1)$ becomes a curve (not a point) under the assumption. But then

$$0 > \left(K_X + \frac{1}{\mu} \mathcal{H}_X \right) \cdot F_1 = \left(K_{X_1} + \frac{1}{\mu} \mathcal{H}_{X_1} \right) \cdot F_1 = 0,$$

a contradiction! (The first strict inequality follows from the fact that $K_X + \frac{1}{\mu} \mathcal{H}_X$ is nonpositive on $\overline{\mathrm{NE}}(X/T)$ and that it is zero only on the extremal ray $R \not\ni [F_1]$.) Note that this observation implies that the case $\lambda \leq \mu$ reaching a log minimal model after running $K + \frac{1}{\mu} \mathcal{H}$-MMP is impossible in dimension 2. (cf. Section 1-8.)

The morphism $Y \to T$ being birational implies that both links are "square." (See the end of this section for the definition of a "square" link.) In fact, say that the morphism $Y \to T$ is an isomorphism over a dense open subset $Y \supset U \subset T$. In the case of a link of Type (III), the $K + \frac{1}{\mu} \mathcal{H}$-MMP does not affect the locus over U (Exercise! Why?) Therefore, $\phi : X \to Y$ and $\phi_1 : X_1 \to Y_1 = T$ are identical over U. In the case of a link of Type (IV), the $K + \frac{1}{\mu} \mathcal{H}$-MMP does not affect the locus over U and the last contraction of fibering type contracts all the curves in fibers over U. Therefore, again $\phi : X \to Y$ and $\phi_1 : X_1 \to Y_1$ are identical over U.

Moreover, since $K_{X_1} + \frac{1}{\mu} \mathcal{H}_{X_1}$ is trivial over Y_1, we conclude in both cases that

$$\mu_1 = \mu.$$

Here we don't quite have control over the Sarkisov degree. (It is the subtle point we have to face when we try to verify "termination of Sarkisov program" in the next section.) But since we start from a canonical pair $K_X + \frac{1}{\mu} \mathcal{H}_X$ running $K + \frac{1}{\mu} \mathcal{H}$-MMP over T to reach X_1, we conclude that $K_{X_1} + \frac{1}{\mu} \mathcal{H}_{X_1}$ is also canonical. (Note that in general a canonical pair (X, B) may not stay canonical when we contract a component of the boundary B through $K + B$-MMP. But in our case, the boundary $\mathcal{H}$ is the strict transform of a general member of one unique base point free system, and thus we may assume that the contracted divisor, if any, is not contained in $\mathcal{H}$. Therefore, all the pairs in the process stay canonical.) Thus we conclude that

$$\lambda_1 \leq \mu = \mu_1.$$

Therefore, after untwisting Φ by a link of Type (III) or Type (IV), we stay in the case $\lambda \leq \mu$.

$$\boxed{\text{Case: } \lambda > \mu}$$

In this case we take what we call a **maximal divisorial blowup** $p : Z \to X$ with respect to $K_X + \frac{1}{\lambda}\mathcal{H}_X$, i.e., p is a projective birational morphism from Z with only $\mathbb{Q}$-factorial and terminal singularities such that

(i) $\rho(Z/X) = 1$,
(ii) the exceptional locus of p is a prime divisor E,
(iii) p is $K + \frac{1}{\lambda}\mathcal{H}$-crepant, i.e.,

$$K_Z + \frac{1}{\lambda}\mathcal{H}_Z = p^* \left(K_X + \frac{1}{\lambda}\mathcal{H}_X \right).$$

Proposition 13-1-8 (Existence of a Maximal Divisorial Blowup). *A maximal divisorial blowup $p : Z \to X$ with respect to $K_X + \frac{1}{\lambda}\mathcal{H}_X$ exists.*

Before giving a proof of the proposition, we make a remark. Note that the exceptional divisor E of p is necessarily one of the $K + \frac{1}{\lambda}\mathcal{H}$-crepant divisors $\{E_1, E_2, \ldots, E_e\}$ counted for the number e in the triplet (μ, λ, e) defining the Sarkisov degree. As long as we require Z to have only terminal singularities, we can't specify which E_i is to appear as the exceptional divisor for the maximal divisorial blowup p. However, if we allow Z to have canonical singularities, for each E_i we can construct a maximal divisorial blowup $p_i : Z_i \to X$ with the exceptional divisor being E_i. Moreover, once we fix E_i, such a maximal divisorial blowup $p_i : Z_i \to X$ is unique. (Exercise! Verify the assertions in this remark.)

PROOF.   Take a resolution $V \to X$ from a nonsingular projective variety V s.t.

(a) the exceptional locus is a divisor with only normal crossings,
(b) V dominates X', so that the strict transform $\mathcal{H}_V$ coincides with the total transform of $\mathcal{H}_{X'}$ and a general member $\mathcal{H}_V$ is smooth and crosses normally with the exceptional locus.

We run the $K + \frac{1}{\lambda}\mathcal{H}$-MMP over X to get a log minimal model $f : (Z', \frac{1}{\lambda}\mathcal{H}_{Z'}) \to (X, \frac{1}{\lambda}\mathcal{H}_X)$. As before, it is easy to see that Z' has only $\mathbb{Q}$-factorial and terminal singularities. Since both Z' and X are $\mathbb{Q}$-factorial, the exceptional locus of f is purely one-codimensional (cf. Shafarevich [1], Theorem 2 in Section 4 of Chapter II and the remark in the proof of Theorem 4-1-3). An easy application of the negativity lemma shows that the exceptional locus is actually $\cup_{i=1}^{e} E_i$ and that f is $K + \frac{1}{\lambda}\mathcal{H}$-crepant, i.e., $K_{Z'} + \frac{1}{\lambda}\mathcal{H}_{Z'} = f^*(K_X + \frac{1}{\lambda}\mathcal{H}_X)$.

Now we run the K-MMP starting from Z' over X ending necessarily with a divisorial contraction $p : Z \to X$. It is immediate that $p : Z \to X$ is a maximal divisorial blowup with respect to $K_X + \frac{1}{\lambda}\mathcal{H}_X$. (If we want to specify the exceptional divisor to be E_i allowing Z to have canonical singularities, then we run $K + \frac{1}{\lambda}\mathcal{H} + \epsilon \Sigma_{j \neq i} E_j$-MMP starting from Z' over X for sufficiently small $\epsilon > 0$ instead, ending necessarily with a minimal model $p_i : Z_i \to X$, which is the maximal divisorial blowup with the exceptional divisor E_i and Z_i having only canonical singularities.) ☐

We go back to the discussion of the flowchart in the case $\lambda > \mu$.

After taking a maximal divisorial blowup $p : Z \to X$, we run $K + \frac{1}{\lambda}\mathcal{H}$-MMP on Z over Y.

A priori we reach either a log minimal model or a log Mori fiber space (with respect to $K + \frac{1}{\lambda}\mathcal{H}$ over Y). So we ask the following question:

Do we reach a log Mori fiber space?

First we show that it is impossible to have the answer *no* to the above question, i.e., it is impossible to reach a log minimal model.

Suppose we did.

Then according to whether the last birational contraction is divisorial or not we should have two different diagrams reaching a log minimal model $X_1 \to Y_1 = Y$:

$$
\begin{array}{ccc}
Z & \cdots\!\longrightarrow & Z' \\
{\scriptstyle p}\swarrow & & \searrow{\scriptstyle q} \\
X & & X_1 \\
\downarrow & & \downarrow \\
Y & \xleftarrow{\;\sim\;} & Y_1
\end{array}
$$

$$
\begin{array}{ccc}
Z & \cdots\!\longrightarrow & X_1 \\
\swarrow & & \\
X & & \downarrow \\
\downarrow & & \\
Y & \xleftarrow{\;\sim\;} & Y_1
\end{array}
$$

We take a general curve F on X in the general fiber of ϕ, away from $p(E)$ (and thus can be considered to lie on Z) and away from all the flipping curves (and thus can be considered to lie on Z').

In the first case, we have

$$
\begin{aligned}
0 &\leq \left(K_{X_1} + \frac{1}{\lambda}\mathcal{H}_{X_1} \right) \cdot q_* F \\
&= \left\{ \left(K_{Z'} + \frac{1}{\lambda}\mathcal{H}_{Z'} \right) - a E_q \right\} \cdot F \; (a > 0) \\
&\leq \left(K_X + \frac{1}{\lambda}\mathcal{H}_X \right) \cdot F \\
&< \left(K_X + \frac{1}{\mu}\mathcal{H}_X \right) \cdot F = 0,
\end{aligned}
$$

a contradiction!

In the second case, we have

$$
0 \leq \left(K_{X_1} + \frac{1}{\lambda}\mathcal{H}_{X_1} \right) \cdot F
$$

$$= \left(K_Z + \frac{1}{\lambda} \mathcal{H}_Z \right) \cdot F$$

$$< \left(K_X + \frac{1}{\mu} \mathcal{H}_X \right) \cdot F = 0,$$

again a contradiction! (Note that in each of the above cases the last strict inequality follows from the case assumption $\lambda > \mu$.)

Second, we deal with the case where the answer to the question, Do we reach a log Mori fiber space? is *yes*, i.e., where we reach a log Mori fiber space $X_1 \to Y_1$.

Then the next question to ask just in order to separate the types of links is:

Is the last birational contraction divisorial?

If the answer is *yes*, then the $K + \frac{1}{\lambda} \mathcal{H}$-MMP consists of a sequence of $K + \frac{1}{\lambda} \mathcal{H}$-flips $Z \dashrightarrow Z'$ followed by a $K + \frac{1}{\lambda} \mathcal{H}$-negative contraction $Z' \to X_1$. Since $\rho(X_1/Y) = 1$, the morphism $\phi_1 : X_1 \to Y_1 = Y$ is a $K_{X_1} + \frac{1}{\lambda} \mathcal{H}_{X_1}$-negative and thus K_{X_1}-negative fiber space:

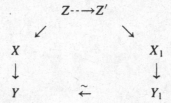

We note that the exceptional divisors E and E_q are distinct, since otherwise X and X_1 are isomorphic in codimension one, which would imply that X and X_1 are indeed isomorphic over $Y = Y_1$. (Exercise! Why?). But while the divisor E_q is *not* $K_{X_1} + \frac{1}{\lambda} \mathcal{H}_{X_1}$-crepant, the divisor E is $K_X + \frac{1}{\lambda} \mathcal{H}_X$-crepant; absurd!

If the answer is *no*, then the $K + \frac{1}{\lambda} \mathcal{H}$-MMP consists of a sequence of $K + \frac{1}{\lambda} \mathcal{H}$-flips $Z \dashrightarrow Z'$ followed by a $K_{X_1} + \frac{1}{\lambda} \mathcal{H}_{X_1}$-negative and thus K_{X_1}-negative fibering contraction $Z' = X_1 \to Y_1$. Since $\rho(X_1/Y) = \rho(Z/Y) = 2$, we have $\rho(Y_1/T) = 1$:

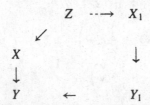

In both cases X_1 has only $\mathbb{Q}$-factorial and terminal singularities, and thus we have a link of Type (II) or a link of Type (I), respectively.

Now we study how the Sarkisov degree (μ, λ, e) changes after untwisting by a link of Type (II) or Type (I).

We claim that

$$\mu_1 \leq \mu$$

with equality holding only if either $\dim Y_1 > \dim Y$ or $\dim Y_1 = \dim Y$ and ψ_1 is square. Here ψ_1 is defined to be square if in the diagram

$$
\begin{array}{ccc}
X & \xrightarrow[\text{birat}]{\psi_1} & X' \\
\downarrow{\phi} & & \downarrow{\phi_1} \\
Y & \xleftarrow{\pi} & Y'
\end{array}
$$

there exists a birational map π that makes the diagram commute and $\psi_\eta : X_\eta \dashrightarrow (X_1)_\eta$ an isomorphism, where η is the generic point of Y.

First, by definition of λ and the assumption of this case $\lambda > \mu$, it follows that

$$
p^* \left(K_X + \frac{1}{\mu} \mathcal{H}_X \right) = K_Z + \frac{1}{\mu} \mathcal{H}_Z + bE
$$

for some $b > 0, b \in \mathbb{Q}$.

We take a general curve F_1 in the general fiber of $\phi_1 : X_1 \to Y_1$ away from the locus of indeterminacy of the birational map $X_1 \dashrightarrow Z$ (i.e., in the case of a link of Type (II) the union of $q(E_q)$ and all the flipped curves and in the case of a link of Type (I) the union of all the flipped curves) and not contained in the strict transform of E. Then F_1 can be considered to lie on Z and

$$
\begin{aligned}
0 &= \left(K_X + \frac{1}{\mu} \mathcal{H}_X \right) \cdot p_* F_1 \\
&= \left(K_Z + \frac{1}{\mu} \mathcal{H}_Z + bE \right) \cdot F_1 \\
&\geq \left(K_Z + \frac{1}{\mu} \mathcal{H}_Z \right) \cdot F_1 \\
&= \left(K_{X_1} + \frac{1}{\mu} \mathcal{H}_{X_1} \right) \cdot F_1,
\end{aligned}
$$

which implies

$$
\mu_1 \leq \mu.
$$

Moreover, if $\mu_1 = \mu$ and $\dim Y = \dim Y_1$ (which implies that $\pi : Y_1 \to Y$ is a birational morphism, since both field extensions $k(X)/k(Y)$ and $k(X_1) = k(X)/k(Y_1)$ are algebraically closed), then we necessarily have $E \cdot F_1 = 0$, which is equivalent to saying that ϕ_1 (the strict transform of E) is not equal to Y_1. Thus $\phi(p(E)) \neq Y$. It also follows easily that the process of $K + \frac{1}{\lambda} \mathcal{H}$-MMP only modifies the locus over $\phi(p(E))$. Therefore, ψ_1 is square.

We also claim that

$$
\lambda_1 \leq \lambda
$$

and

$$
\text{if } \lambda_1 = \lambda \text{ then } e_1 < e.
$$

First, $(X_1, \frac{1}{\lambda}\mathcal{H}_{X_1})$ is canonical, since it is obtained from a canonical pair $(Z, \frac{1}{\lambda}\mathcal{H}_Z)$ through $K + \frac{1}{\lambda}\mathcal{H}$-MMP. Thus $\lambda_1 \leq \lambda$. (See also the note at the end of the discussion of the case $\lambda \leq \mu$.)

Moreover, if $\lambda_1 = \lambda$, then in the case of untwisting by a link of Type (II) the ramification formula

$$K_{Z'} + \frac{1}{\lambda}\mathcal{H}_{Z'} = q^* \left(K_{X_1} + \frac{1}{\lambda}\mathcal{H}_{X_1} \right) + aE_q \qquad (a > 0)$$

implies that E_q is not a $K_{X_1} + \frac{1}{\lambda}\mathcal{H}_{X_1}$-crepant divisor (and E is a divisor on X_1 and thus not exceptional) and thus

$$e_1 \leq e - 1 < e.$$

In the case of untwisting by a link of Type (I) E is a divisor on X_1 (and thus not exceptional), and hence we have the same conclusion (cf. Lemma 9-1-3 and its logarithmic version).

Therefore, after untwisting by a link of Type (II) or Type (I), we go back to the *Start* with strictly decreased Sarkisov degree.

We summarize and visualize the algorithm in Flowchart 13-1-9.

Flowchart 13-1-9.

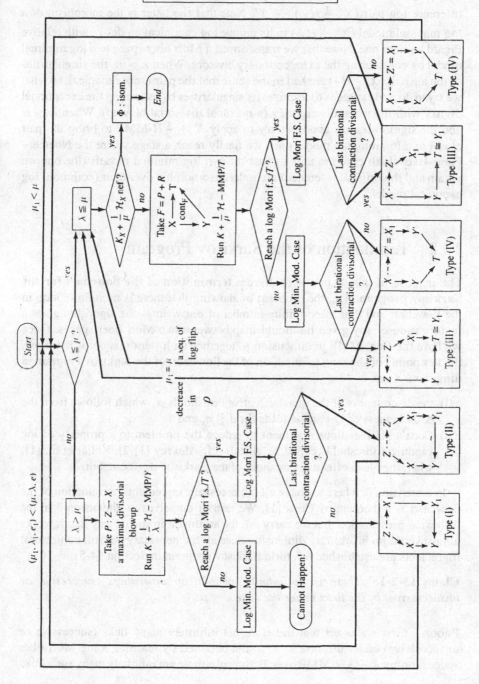

Sarkisov Program in Dimension 3

Remark 13-1-10 (A Logarithmic Viewpoint to Interpret What the Sarkisov Program Is Doing to Untwist a Birational Map). Roughly speaking, the Sarkisov program tries to bring the log pair $(X, \frac{1}{\mu}\mathcal{H}_X) \rightarrow Y$ closer and closer to the reference log pair $(X', \frac{1}{\mu'}\mathcal{H}_{X'}) \rightarrow Y'$. Note that the latter is the morphism of a log minimal model $(X', \frac{1}{\mu'}\mathcal{H}_{X'})$ to its unique log canonical model Y' with relative Picard number one. Note that we transformed a Mori fiber space to a log minimal model by considering the extra boundary divisors. When $\lambda > \mu$, the singularities of the log pair $(X, \frac{1}{\mu}\mathcal{H}_X)$ are bad in the sense that the pair is not canonical, so what we try to do in this case is to improve its singularities by extracting the exceptional divisor with the worst discrepancy (a maximal divisorial blowup). When $\lambda \leq \mu$ and the sigularities are good, we try to apply $K + \frac{1}{\mu}\mathcal{H}$-MMP to bring the pair closer to a log minimal model until we finally reach a stage where the Noether–Fano–Iskovskikh criterion tells us that the two log minimal models (the one we obtain and the original reference) are indeed isomorphic over their (common) log canonical model.

13.2   Termination of the Sarkisov Program

The purpose of this section is to discuss **termination of the flowchart for the Sarkisov program,** i.e., the problem of showing that there is no infinite loop in the flowchart and thus after a finite number of untwistings the algorithm gives a factorization of any given birational map between two Mori fiber spaces. Once we have the (log) MMP in dimension n, together with (termination of log flips), the key points of showing termination of the flowchart of the Sarkisov Program in dimension n are:

(i) the discreteness of the quasi-effective thresholds μ, which follows from the boundedness of $\mathbb{Q}$-Fano d-folds for $d \leq n$, and

(ii) Corti's [1] ingenious argument to reduce the problem to a property of log canonical thresholds called S_n (local) (cf. Alexeev [1][2], Kollár et al. [1]) when the quasi-effective threshold of the Sarkisov degree stabilizes.

In dimension 3, where we have all the necessary ingredients, termination of the flowchart is a theorem by Corti [1]. We restrict ourselves to dimension 3 in the following presentation, but we carry out the argument in such a way that it works almost verbatim in arbitrary dimension once all the necessary but still conjectural ingredients are established (cf. toric Sarkisov program in Section 14-5).

Claim 13-2-1. *There is no infinite number of untwistings (successive or unsuccessive) by the links under the case $\lambda \leq \mu$.*

PROOF.   First we assert that there are not infinitely many links (successive or unsuccessive) under the case $\lambda \leq \mu$ and obtained by reaching a log Mori fiber space running $K + \frac{1}{\mu}\mathcal{H}$-MMP over T. Suppose there are infinitely many such links (successive or unsuccessful):

$$
\begin{array}{ccc}
X_i & \overset{\psi_i}{\underset{\text{birat}}{\dashrightarrow}} & X_{i+1} \\
\downarrow \phi_i & & \downarrow \phi_{i+1} \\
Y_i & & Y_{i+1}.
\end{array}
$$

Note that in the case $\lambda \leq \mu$ we have dim $Y_i \geq 1$ (unless Φ_i becomes an isomorphism of Mori fiber spaces via the NFI criterion). When dim $Y_i = 2$, the general fiber l of ϕ_i is a rational curve, and hence we have

$$K_{X_i} \cdot l = -2,$$
$$(\mu K_{X_i} + \mathcal{H}_{X_i}) \cdot l = 0,$$

which implies

$$\mu \in \frac{1}{2} \mathbb{N}.$$

When dim $Y_i = 1$, we can take a rational curve l in the general fiber, which is a Del Pezzo surface s.t. (cf. Theorem 1-4-8 classification theorem of extremal rays in dimension 2)

$$K_{X_i} \cdot l = -1, -2, \text{ or } -3,$$
$$(\mu K_{X_i} + \mathcal{H}_{X_i}) \cdot l = 0,$$

which implies

$$\mu \in \frac{1}{3!} \mathbb{N}.$$

Since after any link under the case $\lambda \leq \mu$ and obtained by reaching a log Mori fiber space running $K + \frac{1}{\mu} \mathcal{H}$-MMP over T the quasi-effective threshold strictly decreases and since it does not increase after any link in any other case, an infinite sequence of links of the assumed type would lead to a strictly decreasing sequence in $\frac{1}{3!} \mathbb{N}$

$$\mu = \mu_0 > \mu_1 > \mu_2 > \cdots > 0,$$

a contradiction!

In general, we have only to use boundedness of $\mathbb{Q}$-Fano d-folds for $d \leq n - 1$ to derive the discreteness of μ and thus a contradiction to establish the first assertion. (Exercise! Show that the above-mentioned boundedness would imply boundedness of the indices and hence that via Theorem 10-3-1 it would imply the discreteness of the quasi-effective thresholds.)

Note that after untwisting by a link under the case $\lambda \leq \mu$ and obtained by reaching a log minimal model running $K + \frac{1}{\mu} \mathcal{H}$-MMP over T we will always go back to the case $\lambda \leq \mu$, though we lose track of the Sarkisov degree (μ, λ, e), not being able to control λ. (See the case analysis in the flowchart of Section 13-1.) We argue in the following way. The first assertion and the above note imply that in order for us to have an infinite number of untwisting by the links under the case $\lambda \leq \mu$ (successive or unsuccessive) we would have to have an infinite and successive sequence of untwisting by the links under the case $\lambda \leq \mu$ and obtained by reaching log minimal models running $K + \frac{1}{\mu}\mathcal{H}$-MMP over T:

$$
\begin{array}{ccccccc}
X = X_0 \dashrightarrow X_1 \dashrightarrow X_2 \dashrightarrow & \cdots & \dashrightarrow X_i \overset{\psi_i}{\dashrightarrow} X_{i+1} \cdots \\
\downarrow \phi \quad \downarrow \phi_1 \quad \downarrow \phi_2 & \cdots & \downarrow \phi_i \quad \downarrow \phi_{i+1} \cdots \\
Y = Y_0 \quad Y_1 \quad Y_2 & \cdots & Y_i \quad Y_{i+1} \cdots
\end{array}
$$

If the link ψ_i is of Type (III), then the Picard number drops by one:

$$\rho(X_{i+1}) = \rho(X_i) - 1.$$

If the link ψ_i is of Type (IV), then the Picard number stays the same:

$$\rho(X_{i+1}) = \rho(X_i).$$

Thus there are not infinitely many untwistings of Type (III). Thus we may assume that it is an infinite sequence consisting purely of links of Type (IV). But then the sequence

$$X_0 \dashrightarrow X_1 \dashrightarrow X_2 \dashrightarrow \cdots \dashrightarrow X_i \dashrightarrow X_{i+1} \dashrightarrow \cdots$$

would be an infinite sequence of $K + \frac{1}{\mu}\mathcal{H}$-flips, which contradicts termination of log flips (cf. Shokurov [2], Kawamata [15], Kollár et al. [1]). (Note that each transformation $X_i \dashrightarrow X_{i+1}$ is a sequence of log flips over T_i but also a sequence of log flips over Spec $\mathbb{C}$.)

Therefore, there is no infinite number of untwisting (successive or unsuccessive) by the links under the case $\lambda \leq \mu$.

This completes the proof of Claim 13-2-1. □

Claim 13-2-2. *There is no infinite (successive) sequence of untwisting by the links under the case $\lambda > \mu$ with stationary quasi-effective threshold.*

PROOF. This is the heart of the ingenious argument by Corti [1]. Suppose there is such an infinite sequence

$$
\begin{array}{ccccccc}
X = X_0 \dashrightarrow X_1 \dashrightarrow X_2 \dashrightarrow & \cdots & \dashrightarrow X_i \overset{\psi_i}{\dashrightarrow} X_{i+1} \cdots \\
\downarrow \phi \quad \downarrow \phi_1 \quad \downarrow \phi_2 & \cdots & \downarrow \phi_i \quad \downarrow \phi_{i+1} \cdots \\
Y = Y_0 \leftarrow Y_1 \leftarrow Y_2 \leftarrow & \cdots & \leftarrow Y_i \leftarrow Y_{i+1} \cdots
\end{array}
$$

Since $\mu_i = \mu_{i+1}$ for each i by the assumption, we have (see the case analysis in the flowchart of Section 13-1) either dim $Y_{i+1} >$ dim Y_i or dim $Y_{i+1} =$ dim Y_i and ψ_i is square. The first cannot happen infinitely many times, and thus we may assume that we are in the second case for all i. Note that dim $Y_i \geq 1$, since if dim $Y_i =$ dim $Y_{i+1} = 0$, then ψ_i being square would imply that ψ_i is an isomorphism of Mori fiber spaces, which is absurd!

We also know that since $\{\lambda_i\}$ is a nonincreasing sequence and since if $\lambda_i = \lambda_{i+1}$, then $e_{i+1} < e_i$, the value of λ_i cannot be stationary. Therefore, we have a sequence

$$\left\{\frac{1}{\lambda_i}\right\}, \quad \frac{1}{\lambda_i} < \frac{1}{\mu_i} = \frac{1}{\mu_0},$$

which accumulates from below to (but never equals) the value α with the property

$$\alpha \le \frac{1}{\mu_0}.$$

Step 1. We claim that the log pair $(X_i, \alpha \mathcal{H}_{X_i})$ and $(Z_i, \alpha \mathcal{H}_{Z_i})$ $(p_i : Z_i \to X_i$ is a maximal divisorial blowup with respect to $K + \frac{1}{\lambda_i}\mathcal{H}$ extracting the exceptional divisor E_i) have only log canonical singularities for i sufficiently large (and thus we may assume that this holds for all i).

Let α_i be the **log canonical threshold** of the pair X_i with respect to $\mathcal{H}_i$, i.e.,

$$\alpha_i = \max\{c \in \mathbb{Q}_{>0}; (X_i, c\mathcal{H}_{X_i}) \text{ is log canonical}\},$$

where the log pair $(X_i, c\mathcal{H}_{X_i})$ being **log canonical** means by definition (cf. Definition 4-4-2)

$$a(E; X_i, c\mathcal{H}_{X_i}) \ge -1 \quad \forall E \text{ exceptional over } X_i.$$

If $\alpha > \alpha_i$ $(> \frac{1}{\lambda_i})$ for infinitely many i's (i.e., if the above claim fails to hold), then there is a strictly increasing subsequence $\{\alpha_l\}$ of log canonical thresholds accumulating to α. This contradicts S_3(local) proved by Alexeev [1][2]. The same argument applies to $(Z_i, \alpha \mathcal{H}_{Z_i})$.

Theorem 13-2-3 (A Special Case of S_3 (local) by Alexeev [1][2]). *The log canonical thresholds of 3-folds with terminal singularities with respect to linear systems without base components satisfy the ascending chain condition.*

PROOF. We refer the reader to Kollár et al. [1] and the original papers of Alexeev [1][2]. □

Every link $X_i \dashrightarrow X_{i+1}$ is an outcome of $K + \frac{1}{\lambda_i}\mathcal{H}$-MMP over Y_i (after taking a maximal divisorial blowup $p_i = p_i^0 : Z_i \to X_i$) consisting of a finite number of $K + \frac{1}{\lambda_i}\mathcal{H}$-flips

$$Z_i = Z_i^0 \overset{t^0}{\dashrightarrow} Z_i^1 \overset{t^1}{\dashrightarrow} \cdots \overset{t^{m-2}}{\dashrightarrow} Z_i^{m-1} \overset{t^{m-1}}{\dashrightarrow} Z_i^m,$$

possibly followed by a divisorial contraction $q_i^m : Z_i^m \to X_i^{m+1} = X_{i+1}$ (otherwise, $Z_i^m = X_i^{m+1}$).

Step 2. We claim that every step

$$
\begin{array}{ccc}
Z_i^k & \dashrightarrow & Z_i^{k+1} \\
q_i^k \searrow & & \swarrow p_i^{k+1} \\
& X_i^{k+1} &
\end{array}
$$

is a step of $K + \alpha \mathcal{H}$-MMP.

We prove this by induction on k.

First note that since $\alpha > 1/\lambda_i$, we have

$$K_{Z_i} + \alpha \mathcal{H}_{Z_i} = p_i^{0*}(K_{X_i} + \alpha \mathcal{H}_{X_i}) - a E_i \ (a > 0).$$

Therefore, we have

$$(K_{Z_i^0}, +\alpha \mathcal{H}_{Z_i^0}) \cdot P_i^0 > 0,$$

where P_i^0 is the extremal ray corresponding to the morphism p_i^0.

Suppose we have

$$(K_{Z_i^k}, +\alpha \mathcal{H}_{Z_i^k}) \cdot P_i^k > 0,$$

where P_i^k is the extremal ray corresponding to the morphism p_i^k.

Note that $K_{Z_i^k} + \alpha \mathcal{H}_{Z_i^k}$ is never relatively nef over Y_i. We see this fact as follows: First, $\alpha \leq 1/\mu_i = 1/\mu_0$ by the assumption of Claim 13-2-2. Suppose $\alpha = 1/\mu_i$. Then

$$K_{Z_i^k} + \alpha \mathcal{H}_{Z_i^k} \equiv_{Y_i} -a \text{ (the strict transform of } E_i) \quad (a > 0)$$

is never relatively nef over Y_i. Suppose $\alpha < 1/\mu_i$. Then by taking a general curve F_i in the general fiber of $\phi_i : X_i \to Y_i$ away from the locus of indeterminacy of the birational map $X_i \dashrightarrow Z_i^k$ (thus the curve can be considered to lie on Z_i^k) we have

$$(K_{Z_i^k} + \alpha \mathcal{H}_{Z_i^k}) \cdot F_i = (K_{X_i} + \alpha \mathcal{H}_{X_i}) \cdot F_i$$

$$< (K_{X_i} + \frac{1}{\mu_i} \mathcal{H}_{X_i}) \cdot F_i = 0.$$

Therefore, $K_{Z_i^k} + \alpha \mathcal{H}_{Z_i^k}$ is never relatively nef over Y_i.

Since one of the two extremal rays, namely P_i^k, of $\overline{NE}(Z_i^k/Y_i)$ has positive intersection with $K_{Z_i^k} + \alpha \mathcal{H}_{Z_i^k}$, this implies that the other extremal ray, Q_i^k, has negative intersection with $K_{Z_i^k} + \alpha \mathcal{H}_{Z_i^k}$:

$$(K_{Z_i^k} + \alpha \mathcal{H}_{Z_i^k}) \cdot Q_i^k < 0.$$

This proves the claim.

A consequence of this claim is that (cf. Shokurov's observation, i.e., Lemma 9-1-3 and its logarithmic version in Chapter 9)

$$a(\nu; X_0, \alpha \mathcal{H}_{X_0}) \leq a(\nu; X_i, \alpha \mathcal{H}_{X_i})$$

for any discrete valuation ν of the function field $k(X)$, and strict inequality holds iff ψ_l is not an isomorphism at the center of ν on X_l for some $l < i$.

Step 3. We claim that the log pair $(X_i, \alpha \mathcal{H}_{X_i})$ has purely log terminal singularities for i sufficiently large (and thus we may assume that this holds for all i).

Assume to the contrary that there exist infinitely many indices i such that $(X_i, \alpha \mathcal{H}_{X_i})$ is not purely log terminal. Since $(X_i, \alpha \mathcal{H}_{X_i})$ is log canonical by Step 1, the assumption is equivalent to saying that for each of these i there exists a valuation ν_i of $k(X)$ with

$$a(\nu_i; X_i, \alpha\mathcal{H}_{X_i}) = -1,$$

which implies by the consequence that

$$a(\nu_i; X_0, \alpha\mathcal{H}_{X_0}) = -1$$

and that at the center $z(\nu_i, X_0)$ of ν_i on X_0, the birational map $\psi_{i-1} \circ \cdots \circ \psi_1 \circ \psi_0 : X_0 \dashrightarrow X_i$ is an isomorphism. Thus the local (w.r.t. the Zariski topology) canonical thresholds are the same:

$$c(z(\nu_i, X_i); X_i, \mathcal{H}_{X_i}) = c(z(\nu_i, X_0); X_0, \mathcal{H}_{X_0}).$$

On the other hand, by definition $1/\lambda_i$ is the global canonical threshold and hence is less than or equal to the local canonical threshold

$$\frac{1}{\lambda_i} \leq c(z(\nu_i, X_i), X_i, \mathcal{H}_{X_i}).$$

Since $K_{X_i} + \alpha\mathcal{H}_{X_i}$ is not canonical at the center $z(\nu_i, X_i)$, we have

$$c(z(\nu_i, X_i); X_i, \mathcal{H}_i) < \alpha.$$

Therefore,

$$\frac{1}{\lambda_i} \leq c(z(\nu_i, X_1); X_1, \mathcal{H}_{X_1}) < \alpha.$$

But $\{\frac{1}{\lambda_i}\}$ is a nondecreasing and nonstationary sequence converging to α, and it is easy to see (Exercise! Why?) that the set $\{c(x; X_0, \mathcal{H}_{X_0}); x \in X_0 = X\}$ is finite, a contradiction!

This proves the claim of Step 3.

Step 4. Conclusion of the argument.

We remark that the valuations of $k(X)$ corresponding to the divisors E_i extracted by the maximal divisorial blowups are all distinct. In fact, suppose that E_i and E_j coincide and hence that Z_i and Z_j are isomorphic in a neighborhood of the generic points of E_i and E_j. This would imply

$$a(E_i; X_i, \alpha\mathcal{H}_{X_i}) = a(E_j; X_j, \alpha\mathcal{H}_{X_j}).$$

On the other hand, from Step 2 we have

$$a(E_i; X_i, \alpha\mathcal{H}_{X_i}) < a(E_j; X_j, \alpha\mathcal{H}_{X_j}),$$

a contradiction!

Finally, we conclude the proof of Claim 13-2-2 as follows: From Step 3 we may assume that $(X_0, \alpha\mathcal{H}_{X_0})$ has only purely log terminal singularities. But on the other hand, for infinitely many E_i with distinct corresponding discrete valuations,

$$a(E_i; X_0, \alpha\mathcal{H}_{X_0}) \leq a(E_i; X_i, \alpha\mathcal{H}_{X_i}) < 0,$$

contradictiong the following lemma.   □

Lemma 13-2-4. *Let (X, D) be a log pair with only purely log terminal singularities. Then there are only finitely many discrete valuations v with negative log discrepancy $a(v, X, D) < 0$.*

PROOF.   The proof is easy and is left to the reader as an exercise. ·   □

In order to prove Claim 13-2-2 in dimension n, we need only S_n(local), which is still a missing ingredient when $n > 3$.

Remark 13-2-5. If we knew that the set of canonical thresholds satisfied the ascending chain condition, the proof of Claim 13-2-2 would be immediate by looking at the sequence $\{\frac{1}{\lambda_i}\}$. However, the behavior of the set of canonical thresholds seems to be much harder to grasp than that of the set of log canonical thresholds (cf. Remark 14-5-5).

Claim 13-2-6. *There is no infinite (successive) sequence of untwisting by the links under the case $\lambda > \mu$ with nonstationary quasi-effective thresholds.*

Suppose there is such an infinite sequence with nonstationary quasi-effective thresholds

$$X = X_0 \dashrightarrow X_1 \dashrightarrow X_2 \dashrightarrow \quad \cdots \quad \dashrightarrow X_i \overset{\psi_i}{\dashrightarrow} X_{i+1} \cdots$$
$$\downarrow\phi \qquad \downarrow\phi_1 \quad \downarrow\phi_2 \quad \cdots \qquad \downarrow\phi_i \qquad \downarrow\phi_{i+1} \cdots$$
$$Y = Y_0 \leftarrow Y_1 \leftarrow Y_2 \leftarrow \cdots \leftarrow Y_i \leftarrow Y_{i+1} \cdots$$

Case: For some i_0, we have dim $Y_{i_0} \geq 1$.

In this case, for all $i \geq i_0$ we have dim $Y_i \geq 1$ and hence

$$\mu_i \in \frac{1}{3!}\mathbb{N}$$

as before, and $\{\mu_i\}$ is a nonstationary and nonincreasing infinite sequence $\mu_0 \geq \mu_i > 0$, a contradiction!

In order to prove Claim 13-2-6 in dimension n and in the above case assumption, we need only the boundedness of $\mathbb{Q}$-Fano d-folds for $d \leq n - 1$.

Finally, we consider

Case: For all i, we have dim $Y_i = 0$.

In this case, the X_i's are all $\mathbb{Q}$-Fano varieties of dimension n with $\rho(X_i) = 1$. In dimension 3, we quote the following result of Kawamata [11].

Theorem 13-2-7 (Boundedness of $\mathbb{Q}$-Fano 3-Folds with Picard Number 1).
The family of $\mathbb{Q}$-Fano 3-folds (normal projective 3-folds with only $\mathbb{Q}$-factorial and terminal singularities having ample anticanonical divisors) with Picard number equal to one is bounded.

Therefore, there exists $r \in \mathbb{N}$ such that $r K_{X_i}$ is Cartier $\forall i$. (Exercise! Why?) Then Theorem 10-3-1 on the lengths of the extremal rational curves says that there exists a rational curve L_i on X_i s.t. $0 < -K_{X_i} \cdot L_i \le 2 \dim X_i$, which implies

$$\mu_i \in \frac{1}{(r \cdot 2 \dim X)! q} \mathbb{N}.$$

Again $\{\mu_i\}$ is a nonstationary and nonincreasing infinite sequence with $\mu_0 \ge \mu_i > 0$, a contradiction!

We remark that this last step is the only place where we use boundedness of $\mathbb{Q}$-Fano n-folds (with $\rho = 1$).   □

Claims 13-2-1, 13-2-2, and 13-2-6 show that there is no infinite loop in the flowchart of the Sarkisov program.

This completes the discussion of termination of the flowchart.

13.3 Applications

In this section we discuss some applications of the Sarkisov program. While it is expected that the Sarkisov program should shed new light on the birational structure of Mori fiber spaces in higher dimensions (e.g., in dimension 3), little is known or has been done at the moment (except for a marvelous recent work of Corti–Puklikov–Reid [1] on the structure of some special $\mathbb{Q}$-Fano 3-folds), and much is left for future research. Here we will present **Takahashi [3]'s work deriving the classical theorem of Jung** on the structure of the group of automorphisms $\mathrm{Aut}(\mathbb{A}^2)$ of the affine 2-space $\mathbb{A}^2$ **via the logarithmic Sarkisov program in dimension 2**.

First we will go over the logarithmic version of the Sarkisov program (called the log Sarkisov program for short) in dimension 2 very quickly, as the reader should have no difficulty modifying the argument in the previous sections into the language of the logarithmic category. Second, we will show how Jung's theorem can be derived by decomposing given birational maps into the links via the log Sarkisov program.

Theorem 13-3-1 (Log Sarkisov Program in Dimension 2 with Log Terminal Singularities) (cf. Bruno–Matsuki [1], Takahashi [1][2]). *Let*

$$(X, B_X) \cdots \overset{\Phi}{\longrightarrow} (X', B_{X'})$$
$$\downarrow \phi \qquad\qquad \downarrow \phi'$$
$$Y \qquad\qquad Y'$$

be a birational map between two log Mori fiber spaces in dimension 2 with only log terminal singularities, inducing a log MMP relation between them (cf. Definition 2-2-9 or Definition 11-3-1). Let (W, B_W) be a log pair consisting of a nonsingular projective surface W and a boundary divisor B_W with only normal crossings, which gives the log MMP relation. That is to say, the birational map Φ can be *decomposed as* $\Phi = \tau' \circ \tau^{-1}$ *where* τ *(respectively* τ'*) is a process of* $K + B$-*MMP starting from* (W, B_W) *finishing with the end result* (X, B_X)*: (respectively* $(X', B_{X'})$*):*

$$(W, B_W)$$

$$K + B\text{--MMP} \swarrow \tau \qquad\qquad \tau' \searrow K + B\text{--MMP}$$

$$(X, B_X) \qquad\qquad \overset{\Phi}{\dashrightarrow} \qquad\qquad (X', B_{X'})$$

$$\downarrow \phi \qquad\qquad\qquad\qquad\qquad\qquad\qquad\qquad \downarrow \phi'$$

$$Y \qquad\qquad\qquad\qquad\qquad\qquad\qquad\qquad Y'.$$

Take the log birational category $\mathcal{D}_{(W, B_W)}$ of our choice to be the one consisting of all the log pairs obtained from (W, B_W) via $K + B$-MMP.

The log birational category $\mathfrak{D}_{(W, B_W)}$ satisfies the following properties:

(i) $\mathfrak{D}_{(W, B_W)}$ contains the original two log Mori fiber spaces we consider.

(ii) For any finite family $\{(X_l, B_{X_l})\}_{l \in L} \subset \mathfrak{D}_{(W, B_W)}$ there exists an object $(V, B_V) \in \mathfrak{D}_{(W, B_W)}$ such that (V, B_V) dominates each (X_l, B_{X_l}) by a projective birational morphism $(V, B_V) \to (X_l, B_{X_l})$ that is a process of $K + B$-MMP over X_l (and hence over Spec $\mathbb{C}$).

(iii) Any $K + B$-MMP over Spec$\mathbb{C}$ starting from an object in $\mathfrak{D}_{(W, B_W)}$ stays inside of the category $\mathfrak{D}_{(W, B_W)}$, and so does any $K + B + c\mathcal{H}$-MMP over Spec$\mathbb{C}$ starting from an object in $\mathcal{D}_{(W, B_W)}$, where $\mathcal{H}$ is a linear system without base components (i.e., no divisor is contained in the base locus of the linear system) and c is a positive rational number.

As a consequence, there is an algorithm to decompose Φ into a composite of the four types of links (as described in the Sarkisov program) in the log birational category $\mathfrak{D}_{(W, B_W)}$, where the Noether-Fano-Iskovskikh criterion and termination also hold.

PROOF. It is straightforward to check properties (i) and (ii) for the log birational category $\mathfrak{D}_{(W, B_W)}$. (In checking property (ii) note that any process of $K + B + c\mathcal{H}$-MMP must also be a process of $K + B$-MMP, since a linear system without base components is nef, a feature unique to dimension 2.)

Thus the general mechanism of the log Sarkisov program works in this log birational category. (Exercise! Check that in order for the general mechanism of the (log) Sarkisov program in Sections 13-1 and 13-2 to work we need only these 3 properties (i) (ii) (iii).)

The definition of the log Sarkisov degree is more subtle (than the general mechanism) and is explained below.

The quasi-effective threshold is defined just as before.

The maximal multiplicity λ of the linear system of homaloidal transforms is defined as

$$\frac{1}{\lambda} = \max\{c \in \mathbb{Q}_{>0}; (K_X + B_X) + c\mathcal{H}_X \text{ is } \mathfrak{D}_{(W, B_W)}\text{-canonical}\},$$

where $(K_X + B_X) + c\mathcal{H}_X$ being $\mathfrak{D}_{(W,B_W)}$-canonical means by definition

$$a(E; X, B_X + c\mathcal{H}_X) \geq a(E; W, B_W) \quad \forall E \text{ exceptional over } X.$$

Now, in order to define the number e of the $\mathfrak{D}_{(W,B_W)}$-crepant exceptional divisors we consider only those exceptional divisors that appear as divisors on W, i.e., only those exceptional divisors that have codimension-one centers on W. (If we consider all the exceptional divisors, then the number could be infinite and not well-defined.)

The Noether–Fano–Iskovskikh criterion holds without any change.

Finally, we check termination of the program. Claim 13-2-1 follows easily, since the general fibers of the fibering morphisms ϕ_i with $\dim Y_i = 1$ are all $\mathbb{P}^1$ and we have the discreteness of the quasi-effective thresholds.

The proof for Claim 13-2-2 goes without change for Steps 1 and 2 replacing K with $K + B$. We disregard Step 3. (Note that the assertion of Step 3 was used to guarantee the finiteness of the divisors extracted by the maximal divisorial blowups, which follow easily in our case.) In Step 4 we conclude the argument as follows. First note that the discrete valuations of the function field corresponding to the exceptional divisors E_i of the maximal divisorial blowups are all distinct as before. Moreover, all the E_i are divisors on W. On the other hand,

$$a(E_i; X_0, B_{X_0} + \alpha\mathcal{H}_{X_0}) \leq a(E_i; X_i, B_{X_i} + \alpha\mathcal{H}_{X_i}) < 0.$$

But there are only finitely many divisors on W with negative discrepancies with respect to the boundary divisor $B_{X_0} + \alpha\mathcal{H}_{X_0}$, a contradiction!

In order to show Claim 13-2-5 for the case with $\dim Y_{i_o} = 1$ for some i_o (and thus for all $i \geq i_o$), we show the discreteness of the quasi-effective thresholds, again noting that the general fibers of ϕ_i are all $\mathbb{P}^1$. For the case $\dim Y_i = 0 \quad \forall i$, we note that the X_i are all log Del Pezzo surfaces (normal projective surfaces with quotient singularities having ample anticanonical divisors) that are dominated by one fixed nonsingular projective surface W. Therefore, it is easy to see that the X_k belong to a bounded family, from which fact the discreteness of the quasi-effective thresholds follows just as before.

This completes the discussion for termination of the log Sarkisov program in the log birational category $\mathfrak{D}_{(W,B_W)}$.

This completes the proof of Theorem 13-3-1. $\qquad\qquad\qquad\qquad\qquad$ □

Now we discuss the application by Takahashi [3] of the log Sarkisov program in dimension 2 to the following classical theorem of Jung:

Theorem 13-3-2 (Structure of Aut($\mathbb{A}^2$)). *Any automorphism of $\mathbb{A}^2 = \text{Spec}\,\mathbb{C}[x, y]$ is a composition of linear (affine) transformations and de Jonquiéres transformations, where the latter consist of the transformations of type*

$$X = x,$$

$$Y = y + f(x), \quad where \ f(x) \ is \ a \ polynomial \ in \ x.$$

The rest of the section will be spent on the proof of this theorem in the framework of our log Sarkisov program in dimension 2 with log terminal singularities, following Takahashi.

Take an automorphism

$$\sigma \in \text{Aut}(\mathbb{A}^2).$$

Then taking a compactification $\mathbb{P}^2$ of $\mathbb{A}^2$ by adding the hyperplane H at infinity we realize that σ induces a self-birational map Φ of the log Mori fiber space

$$(\mathbb{P}^2, H) \to \text{Spec } \mathbb{C}.$$

Therefore, by setting

$$(X, B_X) = (\mathbb{P}^2, H), \qquad (X', B_{X'}) = (\mathbb{P}^2, H),$$
$$Y = \text{Spec } \mathbb{C}, \qquad \text{and} \qquad Y' = \text{Spec } \mathbb{C},$$

we obtain the diagram

$$
\begin{array}{ccc}
(X, B_X) & \overset{\Phi}{\dashrightarrow} & (X', B_{X'}) \\
\downarrow \phi & & \downarrow \phi' \\
Y & & Y'.
\end{array}
$$

We take a common resolution

$$(X, B_X) \overset{\pi}{\leftarrow} (W, B_W) \overset{\pi'}{\to} (X', B_{X'})$$

so that it induces the isomorphisms

$$\mathbb{A}^2 \cong X - B_X \cong W - B_W \cong X' - B_{X'} \cong \mathbb{A}^2$$

and so that

$$\pi^{-1}(B_X)_{\text{red}} = B_W = \pi'^{-1}(B_{X'})_{\text{red}}$$

is a divisor with only normal crossings. It is straightforward to check that both π and π' are processes of $K + B$-MMP starting from (W, B_W). Thus Φ induces a log MMP relation, and we take the log birational category $\mathcal{D}_{(W, B_W)}$ as in Theorem 13-3-1.

Our strategy now is to use the untwisting of the birational map Φ to deduce the decomposition of the automorphism σ.

> **(Takahashi's Main Idea)**
> We describe each link in terms of the toric geometry by choosing some appropriate coordinate system.
> The coordinate system chosen depends on each link.
> The difference between the chosen coordinate systems for the adjacent links is described by a linear (affine) transformation or by a de Jonquiéres transformation.

We use the notation $T(k, l)$ for the toric projective surface defined by the complete fan (in $N_{\mathbb{R}} = N \otimes \mathbb{R} = \mathbb{Z}^2 \otimes \mathbb{R}$) whose 1-dimensional cones are defined by (k, l), $(0, 1)$, and $(-1, -1)$ (where $k > l \geq 0$ and $\gcd(k, l) = 1$ unless

$(k, l) = (1, 0))$, and the notation $B(k, l)$ for the (closed) divisor corresponding to the 1-dimensional cone spanned by (k, l).

Figure 13-3-3.

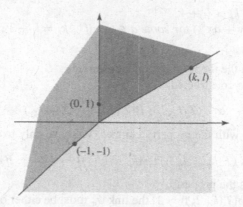

Note that

$$T(k, l) - B(k, l) = \text{the toric variety associated to the fan}$$

spanned by the vectors $(0, 1)$ and $(-1, -1)$

$$\cong \mathbb{A}^2.$$

We fix the last isomorphism and choose the coordinates x and y for $\mathbb{A}^2$ such that $x = (-1, 1)$ and $y = (-1, 0)$ in the dual space $M = \text{Hom}_{\mathbb{Z}}(N, \mathbb{Z})$. According to this coordinate system we have $B(0, 1) = \{x = 0\}$ and $B(-1, -1) = \{y = 0\}$.

Theorem 13-3-4 (after Takahashi [3]). *Let*

$$\Phi = \psi_{N-1} \circ \cdots \circ \psi_1 \circ \psi_0$$

be the untwisting of the birational map

$$\Phi : (X, B_X) = (X_0, B_{X_0}) \dashrightarrow (X_N, B_{X_N}) = (X', B_{X'})$$

by the links

$$\psi_i : (X_i, B_{X_i}) \dashrightarrow (X_{i+1}, B_{X_{i+1}}) \text{ for } i = 0, 1, \ldots, N - 1$$

in the log birational category $\mathfrak{D}_{(W, B_W)}$ via the process of the log Sarkisov program. Then there exist isomorphisms of log pairs

$$\alpha_i : (T(k_i, l_i), B(k_i, l_i)) \xrightarrow{\sim} (X_i, B_{X_i}) \text{ for } i = 0, 1, \ldots, N - 1$$

such that

$$\alpha_{i+1}^{-1} \circ \psi_i \circ \alpha_i : (T(k_i, l_i), B(k_i, l_i)) \dashrightarrow (T(k_{i+1}, l_{i+1}), B(k_{i+1}, l_{i+1}))$$

$$\text{for } i = 0, 1, \ldots, N - 1$$

induces an automorphism

$$\alpha_{i+1}^{-1} \circ \psi_i \circ \alpha_i : \mathbb{A}^2 \cong Y(k_i, l_i) - B(k_i, l_i) \to Y(k_{i+1}, l_{i+1}) - B(k_{i+1}, l_{i+1}) \cong \mathbb{A}^2,$$

which is described in terms of the precribed coordinate system as

(a) $(x, y) \mapsto (x, y + ax)$ *for some* $a \in \mathbb{C}$ *or* $(x, y) \mapsto (y, x)$ *if* $(k_i, l_i) = (1, 0)$,

(b) *identity   if* $k_i > l_i + 1$,

(c) $(x, y) \mapsto (x, y + ax^{k_i})$ *for some* $a \in \mathbb{C}$ *if* $k_i = l_i + 1 > 1$.

PROOF.   We prove the theorem by induction on i.

Suppose we have constructed

$$\alpha_i : (T(k_i, l_i), B(k_i, l_i)) \overset{\sim}{\to} (X_i, B_{X_i})$$

for $i = 0, 1, \ldots, n$ with the property claimed. We have only to construct

$$\alpha_{n+1} : (T(k_{n+1}, l_{n+1}), B(k_{n+1}, l_{n+1})) \overset{\sim}{\to} (X_{n+1}, B_{X_{n+1}}),$$

which also satisfies the property.

Since $\rho(X_n) = \rho(T(k_n, l_n)) = 1$, the link ψ_n must be either of Type (I) or Type (II) via some maximal divisorial blowup

$$p_n : (Z_n, B_{Z_n}) \to (X_n, B_n)$$

with the exceptional divisor E_n extracted by p_n. Furthermore, since by induction

$$X_0 - B_{X_0} \overset{\psi_0}{\cong} X_1 - B_{X_1} \overset{\psi_1}{\cong} \ldots \overset{\psi_{n-1}}{\cong} X_{n-1} - B_{X_{n-1}} \overset{\psi_{n-1}}{\cong} X_n - B_{X_n} \cong X_N - B_{X_N},$$

we conclude that the image of E_n, which is in the base locus of the homaloidal transforms of a very ample divisor on X_N, must be on the boundary, i.e.,

$$p_n(E_n) \in B_{X_n}.$$

Let

$$V_{0,n} \to X_n$$

be the minimal resolution and

$$V_{m,n} \to V_{m-1,n} \to \cdots \to V_{0,n}$$

the succession of blowups at the centers of E_n until E_n shows up as a divisor on $V_{m,n}$.

Observe that the log pair (Z_n, B_{Z_n}), where $B_{Z_n} = p_{n*}^{-1}(B_{X_n}) + E_n$, has only log terminal singularities. We recall one fact, which follows from the classification of log terminal singularities in dimension 2 (cf. Exercise-Theorem 4-6-30).

Fact 13-3-5.   *Let* $p \in (S, D)$ *be a germ of a log terminal singularity in dimension 2, where the boundary divisor* D *consists of two irreducible components* D_1 *and* D_2 *with coefficient 1 intersecting at* p. *Then* p *is necessarily a smooth point of the surface* S, *while* D_1 *and* D_2 *intersect transversally at* p.

Since $p_{n*}^{-1}(B_{X_n}) \cong B_{X_n} \cong \mathbb{P}^1$, we conclude by the fact above that $p_{n*}^{-1}(B_n)$ intersects $E_n = p_n^{-1}(p_n(E_n))$ at one point only and the intersection point is a smooth point of Z_n. Therefore, noting that the minimal resolution of Z_n would factor through $V_{m,n}$, we conclude that the center of each blowup $V_{j+1,n} \rightarrow V_{j,n}$ must be the point of intersection of the strict transform of B_{X_n} and the exceptional locus of $V_{j,n} \rightarrow X_n$.

Actually, this completely determines the way we obtain (Z_n, B_{Z_n}) from (X_n, B_{X_n}) as given by the following recipe:

(i) Get the minimal resolution

$$V_{0,n} \rightarrow X_n.$$

(ii) Keep blowing up the intersection point of the strict transform of B_{X_n} and the exceptional locus of $V_{j,n} \rightarrow X_n$ until E_n appears as a divisor.

(iii) Contract all the exceptional divisors other than E_n to get Z_n, with B_{Z_n} the union of the strict transform of B_{X_n} and E_n.

Now we analyze and construct α_{n+1} case by case looking at the toric description of

$$(X_n, B_n) \overset{\alpha_n^{-1}}{\tilde{\rightarrow}} (T(k_n, l_n), B(k_n, l_n))$$

according to the recipe.

Case (a) $(k_n, l_n) = (1, 0)$.

In this case, $X_n \cong T(k_n, l_n)$ is nonsingular and $V_{0,n} = X_n$ is the minimal resolution.

Subcase (a.1) $p_n(E_n) = B(1, 0) \cap B(0, 1)$.

In this subcase, according to the recipe (ii) we successively blowup the points

$$P_0 = B(0, 1) \cap B(1, 0),$$
$$P_1 = B(1, 1) \cap B(1, 0),$$
$$\cdots$$
$$P_{k-1} = B(k-1, 1) \cap B(1, 0),$$

until E_n shows up as the divisor $B(k, 1)$ and then blow down all the exceptional divisors other than $B(k, 1)$ to obtain Z_n.

In terms of the cones in $N_{\mathbb{R}}$, the procedure can be illustrated as in Figure 13-3-6.

Figure 13-3-6.

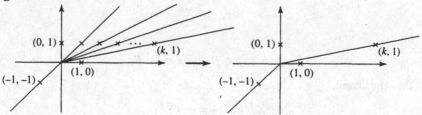

Finally, we blow down $B(1, 0)$ to obtain X_{n+1} by the contraction morphism $q_n : Z_n \to X_{n+1}$. (We show in Claim 13-3-7 that it is impossible to have $(k, 1) = (1, 1)$, in which case there would be no birational contraction blowing down all the exceptional divisors other than $B(k, 1)$.)

This way we obtain the link

$$\psi_n : (X_n, B_{X_n}) \overset{\alpha_n^{-1}}{\underset{\to}{\sim}} (T(1, 0), B(1, 0))$$

$$\overset{p_n}{\leftarrow} (Z_n, B_{Z_n})$$

$$\overset{q_n}{\to} (T(k_{n+1} = k, l_{n+1} = 1), B(k_{n+1}, l_{n+1})) \overset{\alpha_{n+1}}{\underset{\to}{\sim}} (X_{n+1}, B_{X_{n+1}})$$

and

$$\alpha_{n+1}^{-1} \circ \psi_n \circ \alpha_n = \text{identity},$$

mapping $(x, y) \mapsto (x, y + 0 \cdot x)$.

Claim 13-3-7. *Suppose $(k_n, l_n) = (1, 0)$ and $p_n(E_n) = B(0, 1) \cap B(1, 0)$ as above. Then the last ray $(k, 1)$ to add to obtain E_n as a divisor cannot be $(1, 1)$. That is to say,*

$$E_n \neq B(1, 1).$$

PROOF.   If $(k, 1) = (1, 1)$, then the log Mori fiber space $(X_{n+1}, B_{X_{n+1}}) \to Y_{n+1}$ is nothing but $(\mathbb{F}_1, \sigma_1 + F_1) \to \mathbb{P}^1$ from the Hirzebruch surface $\mathbb{F}_1$, where $\sigma_1 = B(1, 1)$ is the unique (-1)-curve and $F_1 = B(1, 0)$ is a fiber of the ruling.

Since the log Mori fiber space $\phi_n : (X_{n+1}, B_{X_{n+1}}) \to Y_{n+1}$ does not coincide with the reference log Mori fiber space $\phi_N : (X_N, B_{X_N}) \to Y_N = \phi' : (X', B_{X'}) \to Y'$, we have to have another link ψ_{n+1}.

Now, the link ψ_{n+1} cannot be of Type (III). In fact, a link of Type (III) would contract σ_1, which brings us back to (X_n, B_{X_n}), a contradiction.

Therefore, it must be of Type (II) via the maximal divisorial blowup

$$p_{n+1} : (Z_{n+1}, B_{Z_{n+1}}) \to (X_{n+1}, B_{X_{n+1}})$$

with the exceptional divisor E_{n+1}.

(Note also that the link ψ_{n+1} must be in the case where we reach a log Mori fiber space running $K + \frac{1}{\mu}\mathcal{H}$-MMP when $\lambda \leq \mu$ and running $K + \frac{1}{\lambda}\mathcal{H}$-MMP when $\lambda > \mu$. See Flowchart 13-1-9.)

Since

$$X_{n+1} - B_{X_{n+1}} \subset X' - B_{X'},$$

we have

$$p_{n+1}(E_{n+1}) \in B_{X_{n+1}} = \sigma_1 + F_1.$$

On the other hand,

$$p_{n+1}(E_{n+1}) \neq \sigma_1 \cap F_1,$$

since $\sigma_1 \cap F_1$ is a point that is log terminal but not purely log terminal of the logarithmic pair $(X_{n+1}, B_{X_{n+1}})$. (Recall that (W, B_W) dominates both $(Z_{n+1}, B_{Z_{n+1}})$ and $(X_{n+1}, B_{X_{n+1}})$ by a process of the log MMP.) We can also see this fact by observing via direct computation that if $p_{n+1}(E_{n+1}) = \sigma_1 \cap F_1$, then $K_{Z_{n+1}} + B_{Z_{n+1}}$ would not be p_{n+1}-negative, contradicting the property of the maximal divisorial blowup. We also see that

$$p_{n+1}(E_{n+1}) \notin F_1 - \sigma_1,$$

since otherwise $K_{Z_{n+1}} + B_{Z_{n+1}}$ would not be q_{n+1}-negative via direct computation, where

$$q_{n+1} : (Z_{n+1}, B_{Z_{n+1}}) \to (X_{n+2}, B_{n+2})$$

is the contraction of the strict transform of F_1.

Thus

$$p_{n+1}(E_{n+1}) \in \sigma_1 - F_1,$$

in which case

$$q_{n+1} : (Z_{n+1}, B_{Z_{n+1}}) \to (X_{n+2}, B_{n+2})$$

is the contraction of the strict transform of the ruling passing through $p_{n+1}(E_{n+1})$. Then $X_{n+2} \cong \mathbb{F}_2$, the Hirzebruch surface with the unique negative section σ_2, and

$$B_{X_{n+2}} = \sigma_2 + F_1 + F_2,$$

where F_1 and F_2 are rulings.

Arguing in the same way, we see that the log Mori fiber space $(X_{n+i}, B_{X_{n+i}}) \to Y_i$ for $i \geq 1$, which is nothing but $(\mathbb{F}_i, \sigma_i + F_1 + F_2 + \cdots + F_i) \to \mathbb{P}^1$ satisfying

$$X_i - B_{X_i} = \mathbb{F}_i - (\sigma_r + F_1 + F_2 + \cdots + F_i) \subset X' - B_{X'},$$

where $\mathbb{F}_i$ is the Hirzebruch surface having the unique section with the minimum self-intersection $\sigma_i^2 = -i$ and where $F_1 + F_2 + \cdots + F_i$ are rulings, would be followed only by a link of Type (II) ending with a log Mori fiber space of the same type. (It cannot be followed by a link of Type (III), since it would contract σ_r, which is not $K_{X_i} + B_{X_i}$-negative. The rest of the reasoning is exactly the same as above.)

This would produce an infinite sequence of links, contradicting termination of the log Sarkisov program.

Therefore, the ray $(k, 1)$ cannot be equal to $(1, 1)$.

This completes the proof of Claim 13-3-7.   □

Subcase (a.2) $p_n(E_n) = B(1, 0) \cap B(-1, -1)$.

In this case, we take an automorphism

$$\tau_n^{-1} : (T(k_n, l_n), B(k_n, l_n)) \to (T(k_n, l_n), B(k_n, l_n)),$$

which is induced by the automorphism of the lattice N switching the points $(0, 1)$ and $(-1, -1)$. Then $\alpha_n \circ \tau_n^{-1}$ is the one in Subcase (a.1), and hence we can find

$$\alpha_{n+1} : (T(k_{n+1}, l_{n+1}), B(k_{n+1}, l_{n+1})) \xrightarrow{\sim} (X_{n+1}, B_{X_{n+1}}) \text{ such that}$$

$$\alpha_{n+1}^{-1} \circ \psi_n \circ (\alpha_n \circ \tau_n^{-1}) = \text{identity}.$$

Therefore,

$$\alpha_{n+1}^{-1} \circ \psi_n \circ \alpha_n = \tau_n,$$

mapping $(x, y) \mapsto (y, x)$.

Subcase (a.3) $p_n(E_n) \in B(1, 0) - \{B(0, 1) \cup B(-1, -1)\}$.

Observe that in this subcase y/x gives the affine coordinate system for $B(1, 0) - \{B(0, 1) \cup B(-1, -1)\} \cong \mathbb{P}^1 - \{0, \infty\} \cong \mathbb{C}^*$. Say $p_n(E_n) = \{\frac{y}{x} = a \in \mathbb{C}^*\}$. We take an automorphism

$$\tau_n^{-1} : (T(k_n, l_n), B(k_n, l_n)) \rightarrow (T(k_n, l_n), B(k_n, l_n))$$

that sends $(x, y) \in \mathbb{A}^2 = T(k_n, l_n) - B(k_n, l_n)$ to $(x, y - ax) \in \mathbb{A}^2 = T(k_n, l_n) - B(k_n, l_n)$. Then $\alpha_n \circ \tau_n^{-1}$ is the one in Subcase (a.1), and hence we can find $\alpha_{n+1} :$ $(T(k_{n+1}, l_{n+1}), B(k_{n+1}, l_{n+1})) \xrightarrow{\sim} (X_{n+1}, B_{X_{n+1}})$ such that

$$\alpha_{n+1}^{-1} \circ \psi_n \circ (\alpha_n \circ \tau_n^{-1}) = \text{identity}.$$

Therefore,

$$\alpha_{n+1}^{-1} \circ \psi_n \circ \alpha_n = \tau_n$$

mapping $(x, y) \mapsto (x, y + ax)$.

> Case (b) (k_n, l_n) with $k_n > l_n + 1$.

In this case, $T(k_n, l_n)$ has two singular points that are the closed orbits corresponding to the 2-dimensional cones generated by (k_n, l_n) and $(0, 1)$, and by (k_n, l_n) and $(-1, -1)$. They are on the divisor $B(k_n, l_n)$.

Claim 13-3-8. *Suppose* $(X_n, B_n) \overset{\alpha_n^{-1}}{\sim}{\rightarrow} (T(k_n, l_n), B(k_n, l_n))$ *with* $k_n > l_n + 1$. *Then* $p_n(E_n)$ *must be one of the two singular points.*

PROOF.   Suppose not. Then after taking the maximal divisorial blowup

$$p_n : (Z_n, B_{Z_n} = p_{n\,*}^{-1}(B_{X_n}) + E_n) \rightarrow (X_n, B_{X_n}),$$

we have two extremal rays on Z_n with $\rho(Z_n) = 2$, one, which corresponds to the contraction of E_n, and the other, which corresponds to the contraction of $p_{n*}^{-1}(B_{X_n})$ as

$$\{p_{n*}^{-1}(B_{X_n})\}^2 \leq B(k_n, l_n)^2 - 1 = \frac{1}{k_n(k_n - l_n)} - 1 < 0$$

to obtain

$$q_n : (Z_n, B_{Z_n}) \rightarrow (X_{n+1}, B_{X_{n+1}}).$$

But then $K_{Z_n} + B_{Z_n}$ is not q_n-negative, since

$$(K_{Z_n}+B_{Z_n})\cdot p_{n*}^{-1}(B_{X_n}) = (K_{T(k_n,l_n)}+B(k_n, l_n))\cdot B(k_n, l_n)+1 = \frac{-2k_n + l_n}{k_n(k_n - l_n)}+1 > 0,$$

contradicting the construction of links of Type (I) or Type (II) (since we are always in the case where we reach a log Mori fiber space no matter whether $\lambda \leq \mu$ or $\lambda > \mu$). (Exercise! Check the computation of the intersection numbers on the toric surface above.) □

Subcase (b.1) $p_n(E_n)$ is the singularity corresponding to the 2-dimensional cone generated by (k_n, l_n) and $(0, 1)$.

We follow the recipe to construct (Z_n, B_{Z_n}) from (X_n, B_{X_n}) explained right before the case-by-case study.

(i) We take the Newton polygon of the cone generated by (k_n, l_n) and $(0, 1)$ to obtain the minimal resolution.

(ii) Keep taking the subdivision between (k_n, l_n) and the adjacent lattice point until E_n shows up as a divisor corresponding to (k_{n+1}, l_{n+1}). By construction, we have $k_{n+1} > l_{n+1} > 0$ with $l_{n+1}/k_{n+1} > l_n/k_n$.

(iii) Contract all the exceptional divisors other than E_n. This corresponds to obtaining the toric variety associated to the complete fan whose 1-dimensional cones are generated by the vectors $(0, 1)$, $(-1, -1)$, (k_n, l_n), and (k_{n+1}, l_{n+1}), respectively.

Figure 13-3-9.

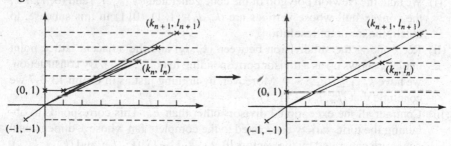

Finally, by eliminating the 1-dimensional cone generated by (k_n, l_n) we obtain

$$\alpha_{n+1} : (T(k_{n+1}, l_{n+1}), B(k_{n+1}, l_{n+1})) \xrightarrow{\sim} (X_{n+1}, B_{X_{n+1}})$$

such that

$$\alpha_{n+1}^{-1} \circ \psi_n \circ \alpha_n = \text{identity}.$$

Subcase (b.2) $p_n(E_n)$ is the singularity corresponding to the 2-dimensional cone generated by (k_n, l_n) and $(-1, -1)$.

We follow the recipe to construct (Z_n, B_{Z_n}) from (X_n, B_{X_n}) explained right before the case-by-case study.

(i) We take the Newton polygon of the cone generated by (k_n, l_n) and $(-1, -1)$ to obtain the minimal resolution.

(ii) Keep taking the subdivision between (k_n, l_n) and the adjacent lattice point until E_n shows up as a divisor corresponding to (k_{n+1}, l_{n+1}). By construction, we have $k_{n+1} > l_{n+1} > 0$ with $l_{n+1}/k_{n+1} < l_n/k_n$.

(iii) Contract all the exceptional divisors other than E_n. This corresponds to obtaining the toric variety associated to the complete fan whose 1-dimensional cones are generated by the vectors $(0, 1), (-1, -1), (k_n, l_n)$, and (k_{n+1}, l_{n+1}), respectively.

Finally, by eliminating the 1-dimensional cone generated by (k_n, l_n) we obtain

$$\alpha_{n+1} : (T(k_{n+1}, l_{n+1}), B(k_{n+1}, l_{n+1})) \xrightarrow{\sim} (X_{n+1}, B_{X_{n+1}})$$

such that

$$\alpha_{n+1}^{-1} \circ \psi_n \circ \alpha_n = \text{identity}.$$

Case (c) (k_n, l_n) with $k_n = l_n + 1 > 1$.

Subcase (c.1) $p_n(E_n) = B(k_n, l_n) \cap B(0, 1)$.

In this subcase, we proceed as in the subcases of the previous case.

We follow the recipe to construct (Z_n, B_{Z_n}) from (X_n, B_{X_n}) explained right before the case-by-case study.

(i) We take the Newton polygon of the cone generated by (k_n, l_n) and $(0, 1)$, i.e., the convex hull whose vertices are $(k_n, l_n), (1, 1), (0, 1)$ in this subcase, to obtain the minimal resolution.

(ii) Keep taking the subdivision between (k_n, l_n) and the adjacent lattice point until E_n shows up as a divisor corresponding to (k_{n+1}, l_{n+1}). By construction, we have $k_{n+1} \geq l_{n+1} > 0$. Moreover, by arguing just as in Claim 13-3-7 we conclude that $(k_n, l_n) \neq (1, 1)$.

(iii) Contract all the exceptional divisors other than E_n. This corresponds to obtaining the toric variety associated to the complete fan whose 1-dimensional cones are generated by the vectors $(0, 1), (-1, -1), (k_n, l_n)$, and (k_{n+1}, l_{n+1}), respectively.

Finally, by eliminating the 1-dimensional cone generated by (k_n, l_n) we obtain

$$\alpha_{n+1} : (T(k_{n+1}, l_{n+1}), B(k_{n+1}, l_{n+1})) \xrightarrow{\sim} (X_{n+1}, B_{n+1})$$

such that

$$\alpha_{n+1}^{-1} \circ \psi_n \circ \alpha_n = \text{identity}.$$

Subcase (c.2) $p_n(E_n) \in B(k_n, l_n) \cap B(-1, -1)$.

We follow the recipe to construct (Z_n, B_{Z_n}) from (X_n, B_{X_n}) explained right before the case-by-case study.

(i) In this subcase the cone generated by (k_n, l_n) and $(-1, -1)$ already corresponds to a nonsingular affine surface.

(ii) Keep taking the subdivision between (k_n, l_n) and the adjacent lattice point until E_n shows up as a divisor corresponding to (k_{n+1}, l_{n+1}). By construction, we have $k_{n+1} > l_{n+1} \geq 0$.

(iii) Contract all the exceptional divisors other than E_n. This corresponds to obtaining the toric variety associated to the complete fan whose 1-dimensional cones are generated by the vectors $(0, 1)$, $(-1, -1)$, (k_n, l_n), and (k_{n+1}, l_{n+1}), respectively.

Finally, by eliminating the 1-dimensional cone generated by (k_n, l_n) we obtain

$$\alpha_{n+1} : (T(k_{n+1}, l_{n+1}), B(k_{n+1}, l_{n+1})) \xrightarrow{\sim} (X_{n+1}, B_{n+1})$$

such that

$$\alpha_{n+1}^{-1} \circ \psi_n \circ \alpha_n = \text{identity}.$$

Subcase (c.3) $p_n(E_n) \in B(k_n, l_n) - \{B(0, 1) \cup B(-1, -1)\}$.

Observe that in this subcase y/x^{k_n} gives the affine coordinates for $B(k_n, l_n) - \{B(0, 1) \cup B(-1, -1)\} \cong \mathbb{P}^1 - \{0, \infty\} \cong \mathbb{C}^*$. Say $p_n(E_n) = \{y/x^{k_n} = a \in \mathbb{C}^*\}$. We take an automorphism

$$\tau_n^{-1} : (T(k_n, l_n), B(k_n, l_n)) \rightarrow (T(k_n, l_n), B(k_n, l_n))$$

that sends $(x, y) \in \mathbb{A}^2 = T(k_n, l_n) - B(k_n, l_n)$ to $(x, y - ax^{k_n}) \in \mathbb{A}^2 = T(k_n, l_n) - B(k_n, l_n)$. Then $\alpha_n \circ \tau_n^{-1}$ is the one in Subcase (c.1), and hence we can find $\alpha_{n+1} : (T(k_{n+1}, l_{n+1}), B(k_{n+1}, l_{n+1})) \xrightarrow{\sim} (X_{n+1}, B_{X_{n+1}})$ such that

$$\alpha_{n+1}^{-1} \circ \psi_n \circ (\alpha_n \circ \tau_n^{-1}) = \text{identity}.$$

Therefore,

$$\alpha_{n+1}^{-1} \circ \psi_n \circ \alpha_n = \tau_n,$$

mapping $(x, y) \mapsto (x, y + ax^{k_n})$.

This completes the proof of Theorem 13-3-4.   □

In order to see Theorem 13-3-2 from Theorem 13-3-4,

$$\beta_0 : (T(1, 0), B(1, 0)) \xrightarrow{\sim} (X_0, B_{X_0}) = (X, B_X),$$

$$\beta_N : (T(1, 0), B(1, 0)) \xrightarrow{\sim} (X_N, B_{X_N}) = (X', B_{X'}),$$

being the isomorphisms such that

$$\sigma = \beta_{N+1}^{-1} \circ \Phi \circ \beta_0,$$

we observe that

$$\sigma = (\beta_N^{-1} \circ \alpha_N) \circ (\alpha_N^{-1} \circ \psi_{N-1} \circ \alpha_{N-1}) \circ \cdots \circ (\alpha_{i+1}^{-1} \circ \psi_i \circ \alpha_i) \circ \cdots \circ (\alpha_1^{-1} \circ \psi_0 \circ \alpha_0) \circ (\alpha_0^{-1} \circ \beta_0),$$

where

$$\alpha_{i+1}^{-1} \circ \psi_i \circ \alpha_i \text{ for } i = 0, 1, \ldots, N - 1$$

are de Jonquiéres transformations or linear (affine) transformations by Theorem 13-3-4 and where

$$\beta_N^{-1} \circ \alpha_N \text{ and } \alpha_0^{-1} \circ \beta_0$$

are linear (affine) transformations of $\mathbb{A}^2$, since they are automorphisms of $T(1, 0) \cong \mathbb{P}^2$ fixing the plane at infinity $B(1, 0)$.

This completes the presentation of Takahashi's work showing Jung's theorem on the structure of $\text{Aut}(\mathbb{A}^2)$ as an application of the log Sarkisov program in dimension 2.   □

Remark 13-3-10. For an application of the Sarkisov program in dimension 3 or higher to study the birational properties of the $\mathbb{Q}$-Fano varieties via the structure of their self-birational maps, we refer the reader, e.g., to the recent paper of Corti–Pukhlikov–Reid [1]. See also Example 3-2-13 to taste the flavor of their methods.

CHAPTER 14

Birational Geometry of Toric Varieties

This chapter is intended as a coffee break after the previous thirteen chapters of hard work. We will just play around with the **toric varieties** and see all the ingredients of the Mori program at work in terms of the concrete geometry of convex cones, following the paper Reid [5]. It is more of my personal note to his beautiful paper, only to "draw legs on the picture of a snake."

We refer the reader to Danilov [1], Fulton [2], Kempf–Knudsen–Mumford–Saint–Donat [1], or Oda [2] for the basics of toric geometry. We also follow their notation for our presentation.

14.1 Cone Theorem and Contraction Theorem for Toric Varieties

Let $N \cong \mathbb{Z}^n$ be a lattice of rank n and $M = \mathrm{Hom}_{\mathbb{Z}}(N, \mathbb{Z})$ the dual lattice. A toric variety $X(\Delta)$ is associated to a fan $\Delta = \{\sigma\}$, a collection of convex cones $\sigma \subset N_{\mathbb{R}} = N \otimes_{\mathbb{Z}} \mathbb{R}$ satisfying:

(i) Each convex cone σ is rational polyhedral in the sense there are finitely many $v_1, \ldots, v_s \in N \subset N_{\mathbb{R}}$ such that

$$\sigma = \{r_1 v_1 + \cdots + r_s v_s; r_i \geq 0\} = \langle v_1, \ldots, v_s \rangle,$$

and it is strongly convex in the sense

$$\sigma \cap -\sigma = \{0\}.$$

(ii) Each face τ of a convex cone $\sigma \in \Delta$ is again an element in Δ.
(iii) The intersection of two cones in Δ is a face of each.

Accordingly we have a collection of affine schemes

$$\{U_\sigma = \operatorname{Spec} \mathbb{C}[\sigma^\vee \cap M]\}, \sigma^\vee = \{u \in M_\mathbb{R} = M \otimes_\mathbb{Z} \mathbb{R}; u(v) \geq 0 \quad \forall v \in \sigma\},$$

where any two affine schemes U_σ and U_τ are to be patched together along their common principal open subset $U_{\sigma \cap \tau}$ to obtain the toric variety $X(\Delta)$.

A toric variety $X(\Delta)$ has an intrinsic characterization as an equivariant embedding of a torus $T_N \cong (G_m)^n$: A toric variety $X(\Delta)$ contains a torus T_N as an open subset such that the translation of T_N extends to an action of T_N on $X(\Delta)$. We will, however, stick to the extrinsic description of the toric varieties in terms of the fans of convex cones in $N_\mathbb{R}$ and its dual $M_\mathbb{R}$.

In our presentation all the toric varieities are assumed to be **complete** and $\mathbb{Q}$-**factorial** unless specified otherwise. We sometimes further assume that they are projective, where the projectivity condition is given in terms of the fan, e.g., as in Fulton [2], Section 3.4. The condition for $\mathbb{Q}$-factoriality is given by the following lemma.

Lemma 14-1-1. *A toric variety $X(\Delta)$ is $\mathbb{Q}$-factorial if and only if each $\sigma \in \Delta$ is simplicial, i.e., there exist exactly $s = \dim \sigma$ lattice points $v_1, \ldots, v_s$ such that $\sigma = \{r_1 v_1 + \cdots + r_s v_s; r_i \geq 0\}$. It is factorial if and only if it is nonsingular, i.e., $v_1, \ldots, v_s$ form a (part of a) $\mathbb{Z}$-basis of N.*

PROOF.   Since the assertion is local, we have to prove the criteria only on each cone $\sigma \in \Delta$ and the corresponding affine toric variety U_σ. Without loss of generality, we may assume $\dim \sigma = n$. (Exercise! Why?)

First we check the criterion for $\mathbb{Q}$-factoriality.

Suppose σ is not simplicial. Then there exist two 1-dimensional faces τ, τ' of σ such that *no* 2-dimensional face of σ contains both of them. The two irreducible divisors $V(\tau)$ and $V(\tau')$ corresponding to τ and τ', respectively, have the orbit decompositions

$$V(\tau) = \cup_{\tau \subset \gamma, \gamma \in \Delta} O(\gamma) \text{ and } V(\tau') = \cup_{\tau' \subset \gamma, \gamma \in \Delta} O(\gamma).$$

Therefore,

$$V(\tau) \cap V(\tau') = \cup_{\tau + \tau' \subset \gamma, \gamma \in \Delta} O(\gamma),$$

which is not empty, since $\tau + \tau' \subset \sigma$, and which has codimension at least 3, since $O(\gamma)$ has at least codimension 3, because $\dim \gamma$ is at least 3 if γ contains both τ and τ' by the assumption. On the other hand, if U_σ is $\mathbb{Q}$-factorial, then (some multiples of) $V(\tau)$ and $V(\tau')$ are both locally factorial, implying that the codimension of $V(\tau) \cap V(\tau')$ is exactly 2 by Krull's principal ideal theorem, a contradiction. Thus U_σ is not $\mathbb{Q}$-factorial.

Suppose σ is simplicial, generated by $v_1, \ldots, v_n \in N$. Then we set N' to be the sublattice generated by $v_1, \ldots, v_n$, and σ' to be the same convex cone as σ but considered sitting over the sublattice N' instead of N. Then we have a finite morphism $\phi : U_{\sigma'} \to U_\sigma$ associated to the inclusion $\varphi : N' \to N$ of the lattices of finite index. Then for a divisor $D \subset U_\sigma$, locally

$$\deg \phi \cdot D = \phi_*(\phi^* D) = \mathrm{div}(\mathrm{Norm}(f)),$$

where f is the defining equation for $\phi^* D = \mathrm{div}(f)$ on a nonsingular $U_{\sigma'}$. (See below for the reason why $U_{\sigma'}$ is nonsingular. Note also that the pullback $\phi^* D$ is well-defined for any Weil divisor D on U_σ, since ϕ is a finite morphism between normal varieties.) Thus U_σ is $\mathbb{Q}$-factorial.

Second, we check the criterion for factoriality, which turns out to be equivalent to nonsingularity.

Observe that in order for U_σ to be factorial it has to be simplicial from the first part of the criterion for $\mathbb{Q}$-factoriality. If σ is simplicial, then its dual $\sigma^\vee$ is also simplicial. The primitive generators $v_1, \ldots, v_n$ of σ form a $\mathbb{Z}$-basis of N if and only if the primitive generators $w_1, \ldots, w_n$ of the dual $\sigma^\vee$ form a $\mathbb{Z}$-basis of M. It is easy to see that this condition is equivalent to nonsingularity of U_σ. (See, e.g., Fulton [2], Section 2.1.) Note that $w_1, \ldots, w_n$ taken to be primitive as above represent irreducible elements in the local ring $\mathbb{C}[\sigma^\vee \cap M]_{m_0}$, where m_0 is the maximal ideal generated by $\sigma^\vee \cap M$. If U_σ is not nonsingular, then it is not nonsingular at m_0 (Exercise! Why?), and there exists $w \in \sigma^\vee \cap M$ other than $w_1, \ldots, w_n$ such that w is not a summation of any other two elements in $\sigma^\vee \cap M$, since $\dim m_0/m_0^2 > n$. Then w represents an irreducible element in the same local ring. On the other hand, there would exist $l \in \mathbb{Z}_{>0}$ and $r_1, \ldots, r_n \in \mathbb{Z}_{\geq 0}$ such that

$$lw = r_1 w_1 + \cdots + r_n w_n,$$

i.e.,

$$w^l = w_1{}^{r_1} \cdots w_n{}^{r_n} \text{ in } \mathbb{C}[\sigma^\vee \cap M]_{m_0},$$

giving two different ways of factoring one element into a product of irreducible elements. Thus the local ring is not UFD and hence U_σ is not factorial. The converse, i.e., U_σ nonsingular implies factoriality, is obvious.

Now we start discussing the

Minimal Model Program for Toric Varieties

If we stick to the category over the base $\mathrm{Spec}\,\mathbb{C}$ and run MMP over $\mathrm{Spec}\,\mathbb{C}$, then starting from a toric variety we would never see a minimal model and always end up with a Mori fiber space, since a toric variety contains a torus T_N and is thus rational (cf. Theorem 3-2-4 (easy dichotomy theorem) and Exercise 3-2-7). In order to avoid this prejudiced polarization toward Mori fiber spaces, we will also consider the category of $\mathbb{Q}$-factorial toric varieties $X(\Delta)$ projective (or only proper) over another fixed toric variety $S(\Delta_S)$ (which we don't assume to be $\mathbb{Q}$-factorial), where we will have relative minimal models over $S(\Delta_S)$ as well as relative Mori fiber spaces over $S(\Delta_S)$ as the end results of MMP over $S(\Delta_S)$. We assume that the morphisms $X(\Delta) \to S(\Delta_S)$ are equivariant, i.e., induced from the $\mathbb{Z}$-linear maps of lattices.

The first ingredient of MMP for toric varieties is the cone theorem, just as MMP for general varieties. The next proposition is a baby version of Mori's bend and break technique realized not by the hard modulo p reduction method but by the easy action of 1-parameter subgroups unique to toric varieties.

Proposition 14-1-2 (Baby Version of Mori's Bend and Break for Toric Varieties). *Let $X(\Delta)$ be a complete toric variety. Then every irreducible curve $C \subset X(\Delta)$ is rationally equivalent to a sum of rational curves that are just closed orbits corresponding to the $(n-1)$-dimensional cones in Δ, i.e.,*

$$C \sim \Sigma_{w \in \Delta, \dim w = n-1} a_w V(w), \quad a_w \in \mathbb{Z}_{\geq 0}.$$

In the relative setting of $\phi : X(\Delta) \to S(\Delta_S)$ a toric variety proper over another, the w in the linear relation above run over those $(n-1)$-dimensional cones with $\phi(V(w)) = a$ point.

PROOF.   For an irreducible curve C, take the smallest closed orbit $V(\sigma)$ that contains C. Let w be an $(n-1)$-dimensional cone that has σ as a face. If $v \in N$ is contained in the relative interior of w, then

$$\lim_{z \to 0} \lambda_v(z) = x_w \subset O(w),$$

where x_w is the distinguished point in the orbit $O(w)$ (cf. Section 2.3 in Fulton [2]). Regard $V(\sigma)$ as a toric variety containing a torus

$$T_{N(\sigma)} = O(\sigma) = \text{Hom}(M(\sigma), \mathbb{C}^*),$$

where $M(\sigma) = \sigma^{\perp} \cap M$ is dual to the quotient lattice $N(\sigma) = N/N_\sigma$ (cf. Fulton [2], Section 3.1). Then for a point $c \in C \cap O(\sigma)$ we have

$$\lim_{z \to 0}(\lambda_v(z) \cdot c) = \lim_{z \to 0}(\lambda_{\overline{v}}(z) \cdot c) = c \cdot \lim_{z \to 0} \lambda_{\overline{v}}(z) = c \cdot x_w,$$

where $\overline{v}$ is the image of v in $N(\sigma)$ and where in the third and fourth terms in the above equation c is considered as an element of the torus $T_{N(\sigma)}$.

Observe (Exercise! Why?) that

$$\cap_{w \in \Delta, \dim w = n-1, w \supset \sigma} \text{Stab}(x_w) = \text{a finite multiplicative group} \subset T_{N(\sigma)}$$

where "Stab" denotes the stabilizer

$$\text{Stab}(x_w) = \{t' \in T_{N(\sigma)}; t' \cdot x_w = x_w\}.$$

This implies that there exists an $(n-1)$-dimensional cone w_o that has σ as a face such that for any $v \in N$ contained in the relative interior of w_o we have

$$\left\{ \lim_{z \to 0}(\lambda_v(z) \cdot c); c \in C \cap O(\sigma) \right\} = O(w_o).$$

Choose $v \in N$ from the relative interior of w_o and let $\lambda_v : G_m \to T_N$ be the corresponding 1-parameter subgroup. Consider the action of G_m on C via this 1-parameter subgroup,

$$G_m \times C \to V(\sigma)$$

$$z \times c \mapsto \lambda_v(z) \cdot c,$$

and regard it as a rational map (via the inclusion $G_m \times C \subset \mathbb{A}^1 \times C$)

$$\mathbb{A}_1 \times C \dashrightarrow V(\sigma).$$

We extend this rational map to a morphism from a nonsingular surface W (which maps onto $\mathbb{A}^1 \times C$ by a proper birational morphism) to $V(\sigma)$ and obtain the commutative diagram

$$
\begin{array}{ccc}
W & \xrightarrow{f} & V(\sigma) \\
\downarrow & & \| \\
\mathbb{A}^1 \times C & \dashrightarrow & V(\sigma) \\
\downarrow{p_1} & & \\
\mathbb{A}^1 & &
\end{array}
$$

Now observe that

$$C = f_*(p_1^{-1}(1)) \sim f_*(p_1^{-1}(0)) = a_{w_o} V(w_o) + C'$$

by the choice of w_o, where C' is, a sum of irreducible curves. If we assume that $X(\Delta)$ is projective, then $\deg C' < \deg C$, where "deg" is the degree taken with respect to some fixed ample line bundle on $X(\Delta)$. Now the induction on deg gives the assertion.

For a general complete toric variety $X(\Delta)$ we take an equivariant birational morphism $X(\Delta') \to X(\Delta)$ from a projective toric variety $X(\Delta')$ using a toric version of Chow's lemma (cf. Oda [2], Proposition 2.17) and apply the above argument to $X(\Delta')$. The assertion for $X(\Delta)$ immediately follows from the one for $X(\Delta')$.

We leave the verification of the relative case to the reader. $\qquad\square$

Remark 14-1-3. (i) In the above proof the original curve C is deformed along a 1-parameter subgroup parametrized by $G_m \subset \mathbb{A}^1$, whereas in Mori's argument in Theorem 10-1-1 the original curve C (having negative intersection with K_X) is deformed along a subspace of the deformation space parametrized by $D \subset \Delta$ (bend). As a result, in the above proof the curve $V(w_o)$ arises as a component in the degeneration, whereas in Mori's argument the exceptional curve for the elimination of indeterminacy arises as a component in the degeneration (break).

(ii) The above proof is taken from (cf. Reid [5], Proposition (1.6)), who concludes that

$$f(p_1^{-1}(0)) \subset V(\sigma) - O(\sigma)$$

(taking $v \in N$ to be contained in the relative interior of some cone $\sigma' \overset{\neq}{\supset} \sigma$) and uses the induction on the dimension of $V(\sigma)$ without referring to the projectivity of $X(\Delta)$. Unfortunately, his last argument leading to this conclusion does not make sense to the author. (The set $g(T_0)$, which is claimed to be a union of substrata of Y on page 399 of Reid [5], is actually all of Y.) This is why we provide a slight modification of the argument with the projectivity condition.

(iii) We could give a more direct proof if we used the exact sequence (cf. Fulton [1], Proposition 1.8)

$$A_1(X(\Delta) - T_N) \to A_1(X(\Delta)) \to A_1(T_N) \to 0$$

and the facts

$$A_1(T_N) = 0,$$

$$X(\Delta) - T_N = \cup_{\sigma \in \Delta, \dim \sigma = 1} V(\sigma),$$

together with induction on the dimension of the complete toric varieties applied to the $V(\sigma)$. Note, however, that in the proof of $A_1(T_N) = 0$ we can also observe a form of the bend and break method (cf. Fulton [1], Proposition 1.9).

(iv) The same argument shows that any divisor $D \subset X(\Delta)$ is linearly equivalent to a sum of the invariant divisors, which are just the closed orbits corresponding to the 1-dimensional cones $\langle v_i \rangle$ in Δ,

$$D \sim \Sigma_{\langle v_i \rangle \in \Delta, \dim \langle v_i \rangle = 1} a_{\langle v_i \rangle} V(\langle v_i \rangle),$$

a fact that follows immediately from the exact sequence

$$A_{n-1}(X(\Delta) - T_N) = \oplus \mathbb{Z} \cdot V(\langle v_i \rangle) \to A_{n-1}(X(\Delta)) \to A_{n-1}(T_N)(= 0) \to 0.$$

As an immediate consequence of the above proposition, we obtain the cone theorem for toric varieties.

Theorem 14-1-4 (Cone Theorem for Toric Varieties). *Let $X(\Delta)$ be a complete toric variety. Then*

$$\overline{NE}(X(\Delta)) = NE(X(\Delta)) = \Sigma_{w \in \Delta, \dim w = n-1} \mathbb{R}_+[V(w)],$$

shaping the polyhedral cone.

In the relative setting of dealing with the category of toric varieties $\phi : X(\Delta) \to S(\Delta_S)$ proper over another fixed toric variety, we consider the intersection pairing between $Z_1(X(\Delta)/S(\Delta_S))$, the space of 1-cycles on $X(\Delta)$ contracted to points on $S(\Delta_S)$, and $\text{Pic}(X(\Delta))$, the group of line bundles on $X(\Delta)$:

$$\text{Pic}(X(\Delta)) \times Z_1(X(\Delta)/S(\Delta_S)) \to \mathbb{Z},$$

which induces the nondegenerate pairing

$$N^1(X(\Delta)/S(\Delta_S)) \times N_1(X(\Delta)/S(\Delta_S)) \to \mathbb{R}.$$

Then we have

$$\overline{NE}(X(\Delta)/S(\Delta_S)) = NE(X(\Delta)/S(\Delta_S))$$

$$= \Sigma_{w \in \Delta, \dim w = n-1, \dim \phi(V(w)) = 0} \mathbb{R}_+[V(w)].$$

Our next task is to establish the contraction theorem. In toric geometry, contracting a 1-dimensional closed orbit $V(w)$ corresponding to an $(n-1)$-dimensional wall w should be described as "removing" this wall from the fan Δ. Then the issue is this:

After removing all the walls giving 1-dimensional closed orbits in the extremal ray do we still get a fan consisting of convex cones and thus forming a toric variety?

In order to settle this issue, we study the geometry of those $(n-1)$-dimensional cones w with $\mathbb{R}_+[V(w)]$ on the edge of $\overline{NE}(X(\Delta))$, i.e., extremal in the sense that the following claim holds:

If $z_1, +z_2 \in \mathbb{R}_+[V(w)]$ with $z_1, z_2 \in \overline{NE}(X(\Delta))$ then $z_1, z_2 \in \mathbb{R}_+[V(w)]$.

The study will yield the contraction theorem for toric varieties almost immediately.

Proposition 14-1-5 (Study of the Extremal 1-Dimensional Closed Orbits $V(w)$).
Let $X(\Delta)$ be a complete and $\mathbb{Q}$-factorial toric variety, $V(w)$ the 1-dimensional closed orbit of $X(\Delta)$ corresponding to an $(n-1)$-dimensional cone

$$w = \langle v_1, \ldots, v_{n-1} \rangle$$

generated by primitive vectors $v_1, \ldots, v_{n-1} \in N$. Let $v_n, v_{n+1} \in N$ be the two primitive vectors such that they together with w generate the two n-dimensional cones that belong to Δ:

$$\tau_{n+1} = \langle v_1, \ldots, v_{n-1}, v_n \rangle \in \Delta,$$
$$\tau_n = \langle v_1, \ldots, v_{n-1}, v_{n+1} \rangle \in \Delta.$$

Suppose $\mathbb{R}_+[V(w)]$ is extremal on $\overline{NE}(X(\Delta))$. (In the relative setting of ϕ : $X(\Delta) \to S(\Delta_S)$ a $\mathbb{Q}$-factorial toric variety proper over another, we assume that $\phi(V(w))$ is a point on $S(\Delta_S)$ and that $\mathbb{R}_+[V(w)]$ is extremal on $\overline{NE}(X(\Delta)/S(\Delta_S))$.)
Then the following two statements hold:

(i) *For every divisor $V(\langle v_i \rangle)$ having positive intersection with the extremal ray*

$$V(\langle v_i \rangle) \cdot V(w) > 0,$$

we have

$$\tau_i = \langle v_1, \ldots, \overset{\vee}{v_i}, \ldots, v_{n-1}, v_n, v_{n+1} \rangle \in \Delta,$$

and the two $(n-1)$-dimensional cones

$$w_{i,n+1} = \langle v_1, \ldots, \overset{\vee}{v_i}, \ldots, v_{n-1}, v_n, \overset{\vee}{v_{n+1}} \rangle,$$
$$w_{i,n} = \langle v_1, \ldots, \overset{\vee}{v_i}, \ldots, v_{n-1}, \overset{\vee}{v_n}, v_{n+1} \rangle$$

give rise to two 1-dimensional closed orbits both of which belong to the extremal ray

$$V(w_{i,n+1}), V(w_{i,n}) \in \mathbb{R}_+[V(w)].$$

(ii) *For every divisor $V(\langle v_i \rangle)$ having zero intersection with the extremal ray*

$$V(\langle v_i \rangle) \cdot V(w) = 0,$$

there exists a unique primitive vector v_i' such that

(a) $\langle v_i' \rangle \in \Delta$;

(b) *the following four n-dimensional cones all belong to Δ:*

$$\tau_{n+1} \in \Delta,$$

$$\tau_n \in \Delta,$$

$$\tau_{n+1}' := \langle v_1, \ldots, v_i', \ldots, v_{n-1}, v_n, \overset{\vee}{v}_{n+1} \rangle \in \Delta,$$

$$\tau_n' := \langle v_1, \ldots, v_i', \ldots, v_{n-1}, \overset{\vee}{v}_n, v_{n+1} \rangle \in \Delta.$$

(c) *The $(n-1)$-dimensional cone*

$$w' = \langle v_1, \ldots, v_i', \ldots, v_{n-1} \rangle$$

gives rise to the 1-dimensional closed orbit that belongs to the extremal ray

$$V(w') \in \mathbb{R}_+[V(w)].$$

(In the relative setting for statement (ii), such a vector v_i' may not exist at all. If that is the case, then the two $(n-1)$-dimensional cones $w_{i,n}$ and $w_{i,n+1}$ are on the boundary of Δ.)

The above two statements are illustrated in Figure 14-1-6.

Figure 14-1-6.

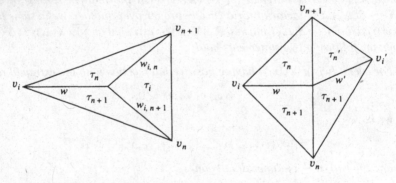

In the figure, v_i represents one of the primitive vectors for $i = 1, \ldots, n-1$ with the specified features as in (i) or (ii).

PROOF.   First we make an important remark about the relation between the intersection number

$$V(\langle v_i \rangle) \cdot V(w)$$

and the coefficient of v_i in the nontrivial linear equation

$$a_1 v_1 + \cdots + a_i v_i + \cdots + a_{n-1} v_{n-1} + a_n v_n + a_{n+1} v_{n+1} = 0 \text{ with } a_{n+1} = 1.$$

We make the following observation in the form of a lemma.

Lemma 14-1-7.

$$V(\langle v_i \rangle) \cdot V(w) \underset{<}{\overset{\geq}{=}} 0 \iff a_i \underset{<}{\overset{\geq}{=}} 0$$

$$\iff \tau_n \cup \tau_{n+1} \text{ is } \begin{matrix} \text{strictly concave} \\ \text{flat} \\ \text{strictly convex} \end{matrix} \text{ along } \sigma_i = \langle v_1, \ldots, \overset{\vee}{v_i}, \ldots, v_{n-1} \rangle.$$

Figure 14-1-8.

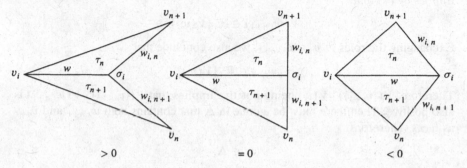

PROOF.   Proof of Lemma 14-1-7 In fact, if we take the dual basis $(v_1^*, \ldots, v_i^*, \ldots, v_{n-1}^*, v_n^*)$ to $(v_1, \ldots, v_i, \ldots, v_{n-1}, v_n)$ and $c \in \mathbb{N}$ such that $m = cv_i^* \in M$, then

$$0 \sim \operatorname{div}(x^m) = \Sigma_{v \in N, v \text{ primitive}, \langle v \rangle \in \Delta} m(v) V(\langle v \rangle)$$
$$= c\{V(\langle v_i \rangle) + v_i^*(v_{n+1}) V(\langle v_{n+1} \rangle) + \cdots\},$$

where $\cdots$ indicates the summation of divisors disjoint from $V(w)$. Therefore,

$$V(\langle v_i \rangle) \cdot V(w) = -v_i^*(v_{n+1}) V(\langle v_{n+1} \rangle) \cdot V(w)$$
$$= a_i V(\langle v_{n+1} \rangle) \cdot V(w) \underset{<}{\overset{\geq}{=}} 0$$
$$\iff a_i \underset{<}{\overset{\geq}{=}} 0.$$

The assertion about the concavity (flatness or convexity) along the $(n-2)$-dimensional cone σ_i follows easily.   □

Now we go back to the proof of Proposition 14-1-5.

Take $m = cv_i^* \in M$ as above. Note that since $V(\langle v_i \rangle) \cdot V(w) \geq 0$ in (i) or (ii), which implies that $\tau_n \cup \tau_{n+1}$ is concave or flat along σ_i, we conclude that w is the only $(n-1)$-dimensional cone in Δ that contains σ_i and whose relative interior is contained in the half-space $\{z \in N_\mathbb{R}; m(z) > 0\}$.

Consider the 2-dimensional closed orbit $V(\sigma_i)$. Since $m \in M \cap \sigma_i^\perp$, meaning $V(\sigma_i)$ is not contained in either poles or zeros of the rational function x^m, the restriction $x^m|_{V(\sigma_i)}$ is a well-defined rational function on $V(\sigma_i)$. Now

$$0 \sim \operatorname{div}(x^m|_{V(\sigma_i)}) = c_w V(w) + \Sigma_{w_j \in \Delta, \dim w_j = n-1, \sigma_i \subset w_j, w_j \neq w} c_{w_j} V(w_j)$$

implies

$$(*) \qquad V(w) \equiv \Sigma_{w_j \in \Delta, \dim w_j = n-1, \sigma_i \subset w_j, w_j \neq w} - \frac{c_{w_j}}{c_w} V(w_j),$$

where $-c_{w_j}/c_w \geq 0$ for all w_j appearing in the summation by the conclusion above and $-c_{w_j}/c_w > 0$ if the relative interior of w_j is in the half-space $\{z \in N_\mathbb{R}; m(z) < 0\}$.

In the situation of (i), the relative interior of $w_{i,n}$ is in the half-space $\{z \in N_\mathbb{R}; m(z) < 0\}$ and so are the relative interiors of the other $(n-1)$-dimensional faces containing σ_i, except for w and $w_{i,n+1}$. Thus $\mathbb{R}_+[V(w)]$ being extremal implies by $(*)$ that

$$V(w_{i,n+1}) \in \mathbb{R}_+[V(w)].$$

Exchanging the roles of v_n and v_{n+1}, we also conclude that

$$V(w_{i,n}) \in \mathbb{R}_+[V(w)].$$

Therefore, $V(\langle v_{n+1}\rangle) \cdot V(w)$ being positive implies that $V(\langle v_{n+1}\rangle) \cdot V(w_{i,n+1})$ is also positive. Then there must be a cone in Δ that contains both $w_{i,n+1}$ and v_{n+1} as faces. Therefore,

$$\tau_i \in \Delta.$$

In the situation of (ii), by the condition for Δ to be a complete fan, there exists an $(n-1)$-dimensional convex cone

$$w' = \langle \sigma_i, v_i'\rangle \in \Delta$$

whose relative interior is in the half-space $\{z \in N_\mathbb{R}; m(z) < 0\}$. Again by equation $(*)$, $\mathbb{R}_+[V(w)]$ being extremal implies

$$V(w') \in \mathbb{R}_+[V(w)].$$

Therefore, the intersection numbers $V(\langle v_n\rangle) \cdot V(w)$ and $V(\langle v_{n+1}\rangle) \cdot V(w)$ being positive implies that $V(\langle v_n\rangle) \cdot V(w')$ and $V(\langle v_{n+1}\rangle) \cdot V(w')$ are also positive. As before, we conclude that

$$\tau'_{n+1} = \langle \sigma_i, v_i', v_n\rangle \in \Delta,$$
$$\tau'_n = \langle \sigma_i, v_i', v_{n+1}\rangle \in \Delta.$$

The uniqueness of such w' and hence that of v_i' follows immediately from the above.

We leave the verification of the relative case to the reader.

This completes the proof of Proposition 14-1-5.   □

Theorem 14-1-9 (Contraction Theorem for Toric Varieties). *Let $X(\Delta)$ be a complete and $\mathbb{Q}$-factorial toric variety, $R \in \overline{NE}(X(\Delta))$ an extremal ray. Then there exists an equivariant surjective morphism with connected fibers*

$$\varphi_R : X(\Delta) \to Y(\Delta_Y)$$

such that for an $(n-1)$-dimensional cone $w \in \Delta$ whether the 1-dimensional orbit $V(w)$ is contracted by φ_R or not is determined by whether $[V(w)]$ is in the extremal ray R or not, i.e.,

$$\varphi_R(V(w)) = \text{a point on } Y(\Delta_Y) \iff [V(w)] \in R.$$

Accordingly, Δ_Y is obtained from Δ by "removing" all the $(n-1)$-dimensional cones $w \in \Delta$ with $[V(w)] \in R$.

Moreover, if $X(\Delta)$ is projective, then $Y(\Delta_Y)$ is also projective, and for a curve $C \subset X(\Delta)$,

$$\varphi_R(C) = \text{ a point on } Y(\Delta_Y) \Longleftrightarrow [C] \in R.$$

(In the relative setting of $\phi : X(\Delta) \to S(\Delta_S)$ being a $\mathbb{Q}$-factorial toric variety proper over $S(\Delta_S)$ and $R \in \overline{NE}(X(\Delta)/S(\Delta_S))$, the morphism φ_R is over $S(\Delta_S)$.)

PROOF.   Let $\{\tau^n\}$ be the collection of n-dimensional convex cones in Δ. We construct a collection of n-dimensional convex bodies $\{\tau_Y^n\}$ by pasting together those τ^n's that share $(n-1)$-dimensional faces w with $[V(w)] \in R$. In the nondegenerate case where each τ_Y^n is strictly convex, then we have only to set Δ_Y to be the collection of all τ_Y^n's and their faces to obtain the desired equivariant morphism

$$\varphi_R : X(\Delta) \to Y(\Delta_Y).$$

In the degenerate case where there exists a k-dimensional subspace $U \subset N_{\mathbb{R}}$ such that each τ_Y^n is of the form $\tau_Y^n = \tau_Y^{n-k} \times U$ for a strictly convex $(n-k)$-dimensional cone τ_Y^{n-k} and for a fixed linear subspace $U \subset N_1$, they are considered to lie in the space $N_{\mathbb{R}}/U$ with the lattice $N/U \cap N$. We have only to set Δ_Y to be the collection of all τ^{n-k}'s and their faces to obtain the desired equivariant morphism

$$\varphi_R : X(\Delta) \to Y(\Delta_Y).$$

First we use condition (i) of Proposition 14-1-5 to observe that each τ_Y^n is convex. Suppose that one n-dimensional body τ_Y^n is not convex. Then there exist two $(n-1)$-dimensional faces $\omega, \omega' \in \Delta$ of τ_Y^n such that τ_Y^n is concave along the $(n-2)$-dimensional face $\sigma = \omega \cap \omega'$. By construction the 1-dimensional orbits corresponding to ω and ω' do not belong to the extremal ray, i.e., $[V(\omega)], [V(\omega')] \notin R$. Let $w_1, \ldots, w_k \in \Delta$ be the $(n-1)$-dimensional cones that lie inside of τ_Y^n such that ω and w_1 are faces of the n-dimensional cone $\tau = \omega + \langle v_\tau \rangle$ in Δ and that so are ω' and w_k of the n-dimensional cone $\tau' = \omega' + \langle v_{\tau'} \rangle$. By construction, the 1-dimensional orbits corresponding to $\omega_1, \ldots, \omega_k$ belong to the extremal ray, i.e., $V(w_1), \ldots, V(w_k) \in R$.

Figure 14-1-10.

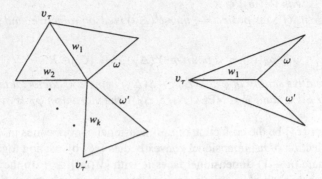

On the other hand, if $k > 1$, then

$$V(\langle v_\tau \rangle) \cdot V(\omega) > 0 \quad \text{and} \quad V(\langle v_\tau \rangle) \cdot V(\omega') = 0,$$

a contradiction. If $k = 1$, then we are in the situation described by (i) of Proposition 14-1-5, which implies $V(\omega)$, $V(\omega') \in R$, a contradiction.

Thus each τ_Y^n is convex.

Suppose some τ_Y^n is convex but not strictly convex. Then it is easy to see from condition (ii) of Proposition 14-1-5 that there exists a k-dimensional linear subspace $U \subset N_{\mathbb{R}}$ such that each τ_Y^n is of the form $\tau_Y^n = \tau_Y^{n-k} \times U$ for a strictly convex $(n - k)$-dimensional cone τ_Y^{n-l}, leading us to the degenerate case.

By construction $X(\Delta) \to Y(\Delta_Y)$ corresponds to the homomorphism of the lattices

$$(0 \to U \cap N \to)N \to \text{Coker} \to 0,$$

where Coker does not have any torsion. Thus the field extension is algebraically closed, and Zariski's Main Theorem implies that the fibers are connected. (In Section 14-2 we will see more explicitly what the fibers are.)

This completes the proof of the first part of the contraction theorem.

For the second part of the contraction theorem with the projectivity condition, we quote the following base point freeness theorem for toric varieties (cf. Theorem 8-1-3 and its proof).

Lemma 14-1-11 (Base Point Freeness Theorem for Toric Varieties). *Let $X(\Delta)$ be a complete toric variety. Then a line bundle $\mathcal{L} \in \text{Pic}(X(\Delta))$ is base point free (i.e., generated by global sections) if and only if it is nef. (In the relative setting of having an equivariant morphism $f : X(\Delta) \to S(\Delta_S)$, a line bundle $\mathcal{L}$ is f-generated, i.e., $f^* f_* \mathcal{L} \to \mathcal{L}$ is surjective if and only if $\mathcal{L}$ is f-nef.)*

PROOF.   Note that a line bundle $\mathcal{L}$ on $X(\Delta)$ is of the form $\mathcal{O}_{X(\Delta)}(D)$ for some T_N-invariant Cartier divisor $D = \Sigma_{\langle v_i \rangle \in \Delta, \dim\langle v_i \rangle = 1} a_i V(\langle v_i \rangle)$. Let ψ_D be the piecewise linear function on Δ associated with D such that

$$\psi_D(v_i) = -a_i \text{ for the primitive vector } v_i \in \langle v_i \rangle \in \Delta.$$

Then it is easy to see that $\mathcal{L} = \mathcal{O}_{X(\Delta)}(D)$ is nef if and only if ψ_D is a convex function, which implies that $|D|$ is base point free by Fulton [2], Section 3.4. If $|D|$ is base point free, $\mathcal{L} = \mathcal{O}_{X(\Delta)}(D)$ is obviously nef. This proves the lemma. (We leave the verification of the relative case to the reader.) □

We go back to the discussion of the contraction theorem.

If $X(\Delta)$ is projective, then $\overline{NE}(X(\Delta))$ must be polyhedral and strictly convex, i.e., no line is contained (either by Kleiman's criterion for ampleness or by Fulton [2], Section 3.4). Therefore, there exists an (equivariant) line bundle $\mathcal{L} \in \text{Pic}(X(\Delta))$ such that $\mathcal{L}$ is nef and

$$R = \overline{NE}(X(\Delta)) \cap \mathcal{L}^\perp.$$

Then $\mathcal{L}$ is base point free by the lemma above and the equivariant morphism

$$\Phi_{|\mathcal{L}|} : X(\Delta) \to Y$$

is onto a normal (toric) projective variety Y with connected fibers (by replacing $\mathcal{L}$ with its high multiple if necessary, cf. Proposition 1-2-16) such that

$$\Phi_{|\mathcal{L}|}(C) = \text{ a point on } Y \text{ iff } [C] \in R,$$

which implies

$$\Phi_{|\mathcal{L}|}(V(w)) = \text{ a point on } Y \text{ iff } [V(w)] \in R.$$

Therefore, we conclude $\Phi_{|\mathcal{L}|} = \varphi_R$ and hence $Y = Y(\Delta_Y)$ is projective.

We remark that actually the condition that for a curve $C \subset X(\Delta)$

$$\varphi_R(C) = \text{ a point on } Y(\Delta_Y) \Longleftrightarrow [C] \in R$$

follows without the projectivity assumption. (Exercise! Check this remark.)

Again we leave the verification of the relative case to the reader as an exercise. This completes the proof of Theorem 14-1-9. □

14.2   Toric Extremal Contractions and Flips

The purpose of this section is to give a detailed study of **contractions of extremal rays for toric varieties**, whose existence and general features were described in Section 14-1. The study will yield the existence of flips for toric varieties in arbitrary dimension, which leads to the minimal model program for toric varieties together with easy termination arguments.

We start with a closer look at the operation of "removing" all the walls w with $[V(w)] \in R$ for an extremal ray R described in Theorem 14-1-9.

We recall the situation considered in Proposition 14-1-5. Let $X(\Delta)$ be a complete and $\mathbb{Q}$-factorial toric variety, $V(w)$ the 1-dimensional closed orbit of $X(\Delta)$

corresponding to an $(n-1)$-dimensional cone

$$w = \langle v_1, \ldots, v_{n-1} \rangle$$

generated by primitive vectors $v_1, \ldots, v_{n-1} \in N$. Let $v_n, v_{n+1} \in N$ be the two primitive vectors such that they together with w generate the two n-dimensional cones that belong to Δ:

$$\tau_{n+1} = \langle v_1, \ldots, v_{n-1}, v_n \rangle \in \Delta,$$
$$\tau_n = \langle v_1, \ldots, v_{n-1}, v_{n+1} \rangle \in \Delta.$$

Suppose the half-line $\mathbb{R}_+[V(w)]$ is extremal on $\overline{NE}(X(\Delta))$. (In the relative setting, we assume that $V(w)$ contracts to a point on $S(\Delta_S)$ and that $\mathbb{R}_+[V(w)]$ is extremal on $\overline{NE}(X(\Delta)/S(\Delta_S))$.) Take the nontrivial linear relation

$$a_1 v_1 + \cdots + a_i v_i + \cdots + a_{n-1} v_{n-1} + a_n v_n + a_{n+1} v_{n+1} = 0 \text{ with } a_{n+1} = 1.$$

By reordering the v_i's we may assume that

$$a_i \begin{cases} < 0 \text{ for } 1 \le i \le \alpha, \\ = 0 \text{ for } \alpha + 1 \le i \le \beta, \\ > 0 \text{ for } \beta + 1 \le i \le n+1, \end{cases}$$

where $0 \le \alpha \le \beta \le n - 1$, since $a_n > 0$.

Proposition 14-2-1. *Let the situation be as above. Then*

$$\tau(w) := \tau_n + \tau_{n+1} = \langle v_1, \ldots, v_{n-1}, v_n, v_{n+1} \rangle$$

is an n-dimensional cone in Δ_Y, which is obtained from Δ by removing all the walls corresponding to the extremal ray R. The cone $\tau(w)$ has a decomposition into the union of the n-dimensional convex cones

$$\tau(w) = \cup_{j=\beta+1}^{n+1} \tau_j,$$

where

$$\tau_j = \langle v_1, \ldots, \overset{\vee}{v_j}, \ldots, v_{n-1}, v_n, v_{n+1} \rangle$$

with

$$\tau_j \in \Delta \text{ for } j = \beta + 1, \ldots, n+1.$$

Set

$$U(w) = \langle v_1, \ldots, v_\alpha, v_{\beta+1}, \ldots, v_{n+1} \rangle.$$

Then $\tau(w)$ is degenerate, i.e., it contains some nonzero linear subspace, if and only if $\alpha = 0$, i.e., there is no negative coefficient $a_i < 0$ in the nontrivial linear relations among the v_i (with $a_{n+1} = 1$).

(i) *When $\alpha = 0$, we have*

$$\tau(w) = \tau(w)_Y^\beta \times U(w),$$

where

$$\tau(w)_Y^\beta = \langle v_1, \ldots, v_\beta \rangle$$

is a strictly convex β-dimensional cone and

$$U(w) = \langle \pm v_{\beta+1}, \ldots, \pm v_n \rangle$$

is an $n - \beta$-dimensional vector space.

(ii) *When $\alpha > 0$, the cone $\tau(w)$ has another decomposition*

$$\tau(w) = \cup_{j=1}^\alpha \tau_j$$

with

$$\tau_j \notin \Delta \text{ for } 1 \le j \le \alpha.$$

Moreover, $U(w)$ is the smallest face of $\tau(w)$ that contains $\langle v_1, \ldots, v_\alpha \rangle$.

PROOF.   If $\alpha = 0$, then

$$v_{n+1} = -(a_{\beta+1} v_{\beta+1} + \cdots + a_n v_n),$$

where

$$a_{\beta+1}, \ldots, a_n > 0.$$

Therefore,

$$\tau(w) \supset U(w) = \langle \pm v_{\beta+1}, \ldots, \pm v_n \rangle$$

is degenerate.

If $\alpha \neq 0$, then any point $x \in \tau(w)$ is of the form

$$x = x_1 v_1 + \cdots + x_n v_n + x_{n+1} v_{n+1}$$
$$\text{where } x_i \ge 0 \quad \forall i$$
$$= x_1 v_1 + \cdots + x_n v_n + x_{n+1}(-a_1 v_1 - \cdots - a_n v_n)$$
$$= \{x_1 + x_{n+1}(-a_1)\} v_1 + \cdots$$
$$\text{where } x_1 + x_{n+1}(-a_1) \ge 0 \text{ and } > 0 \text{ if } x_{n+1} \neq 0.$$

Therefore, if $\tau(w)$ contains a line L, then for any point $x = \Sigma x_i v_i \in L$ the coefficient $x_{n+1} = 0$, since $-x \in L$. But then

$$L \subset \tau(w) \cap \{x_{n+1} = 0\} = \langle v_1, \ldots, v_n \rangle,$$

a contradiction!

Thus $\tau(w)$ is degenerate if and only if $\alpha = 0$.

Let

$$x = x_1 v_1 + \cdots + x_n v_n + x_{n+1} v_{n+1} \in \tau(w)$$

be a point in $\tau(w)$, where

$$x_i \ge 0 \quad \forall i.$$

Take $\beta + 1 \leq j \leq n + 1$ such that

$$\frac{x_j}{a_j} = \min\left\{\frac{x_i}{a_i}; i = \beta + 1, \dots, n + 1\right\}.$$

Then

$$x = (x_1 v_1 + \cdots + x_n v_n + x_{n+1} v_{n+1}) - \frac{x_j}{a_j}(a_1 v_1 + \cdots + a_\alpha v_\alpha + a_{\beta+1} v_{\beta+1})$$

$$= \left(x_1 - \frac{x_j}{a_j}a_1\right)v_1 + \cdots + \left(x_\alpha - \frac{x_j}{a_j}a_\alpha\right)v_\alpha$$

$$+ x_{\alpha+1}v_{\alpha+1} + \cdots + x_\beta v_\beta$$

$$+ \left(x_{\beta+1} - \frac{x_j}{a_j}a_{\beta+1}\right)v_{\beta+1} + \cdots + \left(x_{n+1} - \frac{x_j}{a_j}a_{n+1}\right)v_{n+1}$$

with
$$\begin{cases} x_i - \dfrac{x_j}{a_j}a_i \geq 0 \text{ since } a_i < 0 & \text{for } 1 \leq i \leq \alpha, \\[2mm] x_i \geq 0 & \text{for } \alpha + 1 \leq i \leq \beta, \\[2mm] x_i - \dfrac{x_j}{a_j}a_i \geq 0 \text{ by the choice of } \dfrac{x_j}{a_j} & \text{for } \beta + 1 \leq i \leq n + 1, \end{cases}$$

$$\in \langle v_1, \dots, v_\alpha, v_{\alpha+1}, \dots, v_\beta, v_{\beta+1}, \dots, \overset{\vee}{v_j}, \dots, v_{n+1}\rangle = \tau_j.$$

This gives the decomposition

$$\tau(w) = \cup_{j=\beta+1}^{n+1}\tau_j.$$

Now by Proposition 14-1-5 (i) the cones τ_j belong to Δ, i.e.,

$$\tau_j \in \Delta \text{ for } j = \beta + 1, \dots, n + 1,$$

since

$$V(\langle v_j\rangle) \cdot V(w) > 0, \text{ because } a_j > 0 \text{ for } j = \beta + 1, \dots, n + 1.$$

Also by (the repeated use of) the second part of Proposition 14-1-5 (i), we see that for all the $(n - 1)$-dimensional walls in the interior of $\tau(w)$,

$$w_{i,j} = \langle v_1, \dots, v_\beta, v_{\beta+1}, \dots, \overset{\vee}{v_i}, \dots, \overset{\vee}{v_j}, \dots, v_{n+1}\rangle \text{ for } \beta + 1 \leq i < j \leq n + 1$$

the corresponding 1-dimensional closed orbits belong to the extremal ray

$$[V(w_{ij})] \in R.$$

Moreover, for the $(n - 1)$-dimensional walls on the boundary of $\tau(w)$,

$$w_{ij} = \langle v_1, \dots, \overset{\vee}{v_i}, \dots, \overset{\vee}{v_j}, \dots, v_{n+1}\rangle \text{ for } 1 \leq i \leq \beta, i < j \leq n + 1,$$

the corresponding 1-dimensional closed orbits do not belong to the extremal ray

$$[V(w_{ij})] \notin R,$$

since

$$V(\langle v_i\rangle) \cdot V(w_{ij}) > 0, \text{ while } V(\langle v_i\rangle) \cdot V(w) \leq 0.$$

This proves that $\tau(w)$ is an n-dimensional cone in Δ_Y obtained by removing all the walls corresponding to the extremal ray.

(i) When $\alpha = 0$, we have the description

$$\tau(w) = \{x = x_1 v_1 + \cdots + x_n v_n + x_{n+1} v_{n+1}; x_i \geq 0\}$$
$$= \{x = x_1 v_1 + \cdots + x_\beta v_\beta + x_{\beta+1} v_{\beta+1} + \cdots + x_n v_n;$$
$$x_1, \ldots, x_\beta \geq 0, x_{\beta+1}, \ldots, x_n \in \mathbb{R}\}$$
$$= \tau(w)_Y^\beta \times U(w).$$

(ii) When $\alpha > 0$, for a point

$$x = x_1 v_1 + \cdots + x_n v_n + x_{n+1} v_{n+1} \in \tau(w)$$

take $1 \leq j \leq \alpha$ such that

$$-\frac{x_j}{a_j} = \min\left\{-\frac{x_i}{a_i}; i = 1, \ldots, \alpha\right\}.$$

Then

$$x = (x_1 v_1 + \cdots + x_n v_n + x_{n+1} v_{n+1}) - \frac{x_j}{a_j}(a_1 v_1 + \cdots + a_n v_n + a_{n+1} v_{n+1})$$

$$= \left(x_1 - \frac{x_j}{a_j} a_1\right) v_1 + \cdots + \left(x_\alpha - \frac{x_j}{a_j} a_\alpha\right) v_\alpha$$
$$+ x_{\alpha+1} v_{\alpha+1} + \cdots + x_\beta v_\beta$$
$$+ \left(x_{\beta+1} - \frac{x_j}{a_j} a_{\beta+1}\right) v_{\beta+1} + \cdots + \left(x_{n+1} - \frac{x_j}{a_j} a_{n+1}\right) v_{n+1}$$

with $\begin{cases} x_i - \dfrac{x_j}{a_j} a_i \geq 0 \text{ by the choice of } \dfrac{x_j}{a_j} & \text{for } 1 \leq i \leq \alpha, \\[2mm] x_i \geq 0 & \text{for } \alpha + 1 \leq i \leq \beta, \\[2mm] x_i - \dfrac{x_j}{a_j} a_i \geq 0 \text{ since } a_i > 0 & \text{for } \beta + 1 \leq i \leq n+1, \end{cases}$

$$\in \langle v_1, \ldots, \overset{\vee}{v_j}, \ldots, v_\alpha, v_{\alpha+1}, \ldots, v_\beta, v_{\beta+1}, \ldots, v_{n+1}\rangle = \tau_j.$$

This gives the decomposition

$$\tau(w) = \cup_{j=1}^\alpha \tau_j.$$

The cones in the above decomposition do not belong to Δ, i.e.,

$$\tau_j \notin \Delta \quad j = 1, \ldots, \alpha,$$

since they are distinct from any of the cones τ_j, $\beta + 1 \leq j \leq n + 1$, that belong to Δ.

Let

$$v_1^*, \ldots, v_\alpha^*, v_{\alpha+1}^*, \ldots, v_\beta^*, v_{\beta+1}^*, \ldots, v_n^*$$

be the dual basis of

$$v_1, \ldots, v_\alpha, v_{\alpha+1}, \ldots, v_\beta, v_{\beta+1}, \ldots, v_n.$$

Let

$$y = y_1 v_1^* + \cdots + y_n v_n^* \in M_{\mathbb{R}}.$$

Suppose that

$$y = 0 \text{ on } \langle v_1, \ldots, v_\alpha \rangle$$

and that

$$y \geq 0 \text{ on } \tau(w) = \langle v_1, \ldots, v_n, v_{n+1} \rangle.$$

The first condition implies

$$y_1 = \cdots = y_\alpha = 0,$$

while the second together with these equalities implies

$$y_{\beta+1} = \cdots = y_n = 0.$$

Therefore, $y = 0$ on $\langle v_{n+1} = -\Sigma a_i v_i \rangle$ and hence

$$y = 0 \text{ on } \langle v_1, \ldots, v_\alpha, v_{\beta+1}, \ldots, v_n, v_{n+1} \rangle.$$

It is also easy to see that all such linear functions $y \in M_{\mathbb{R}}$ will cut out as the zero locus the face

$$\langle v_1, \ldots, v_\alpha, v_{\beta+1}, \ldots, v_n, v_{n+1} \rangle \subset \tau(w).$$

This proves the "moreover" part of (ii) and completes the proof of Proposition 14-2-1. □

The analysis of Proposition 14-2-1 leads immediately to the following detailed description of the extremal contractions.

Corollary 14-2-2 (Description of Contractions of Extremal Rays of Fibering, Divisorial and Flipping Type for Toric Varieties). *Let $X(\Delta)$ be a complete and $\mathbb{Q}$-factorial toric variety (respectively over $S(\Delta_S)$), $R \in \overline{NE}(X(\Delta))$ (respectively $R \in \overline{NE}(X(\Delta)/S(\Delta_S)))$ an extremal ray. Let*

$$\varphi_R : X(\Delta) \to Y(\Delta_Y)$$

be the contraction of R (cf. Theorem 14-1-9). Then the contraction morphism φ_R is classified into one of the following three types:

(i) *Contraction of Fibering Type:* $\dim Y(\Delta_Y) < \dim X(\Delta)$.
   In this case, for every $(n-1)$-dimensional face $w \in \Delta$ with $[V(w)] \in R$ we have

$$\alpha = \text{codim}(Exc(\varphi_R)) = 0,$$
$$\beta = \dim Y(\Delta_Y).$$

   (See the description right before Proposition 14-2-1 for the definitions of the numbers α and β.)
   The fan Δ_Y (after taking the quotient by the vector space $U(w)$) is simplicial, implying that $Y(\Delta_Y)$ is $\mathbb{Q}$-factorial (cf. Lemma 14-1-1).

(ii) *Contraction of Divisorial Type:* $\dim Y(\Delta_Y) = \dim X(\Delta)$ *and* $\mathrm{codim Exc}(\varphi_R) = 1$. *In this case, for every* $(n-1)$*-dimensional face* $w \in \Delta$ *with* $[V(w)] \in R$ *we have*

$$\alpha = \mathrm{codim}(Exc(\varphi_R)) = 1,$$

$$\beta \qquad \textit{independent of } w.$$

The fan Δ_Y *is simplicial, since so is*

$$\tau(w) = \tau_1 = \langle v_2, \ldots, v_{n-1}, v_n, v_{n+1} \rangle \textit{ for every } w \textit{ with } [V(w)] \in R,$$

implying that $Y(\Delta_Y)$ *is* $\mathbb{Q}$*-factorial. The exceptional locus consists of a unique divisor* $\mathrm{Exc}(\varphi_R) = V(\langle v_1 \rangle)$.

(iii) *Contraction of Flipping Type:* $\dim Y(\Delta_Y) = \dim X(\Delta)$ *and* $\mathrm{codim Exc}(\varphi_R) \geq 2$. *In this case, for every* $(n-1)$*-dimensional face* $w \in \Delta$ *with* $[V(w)] \in R$,

$$\alpha = \mathrm{codim Exc}(\varphi_R) \geq 2,$$

$$\beta \qquad \textit{independent of } w.$$

The fan Δ_Y *is not simplicial, implying that* $Y(\Delta_Y)$ *is not* $\mathbb{Q}$*-factorial.*

In all of the cases above, the exceptional locus $E = \mathrm{Exc}(\varphi_R)$ is irreducible and coincides with the closed orbit $V(\langle v_1, \ldots, v_\alpha \rangle)$ of dimension $n - \alpha$ in $X(\Delta)$ and maps onto the closed orbit $V(U(w))$ of dimension $\beta - \alpha$ in $Y(\Delta_Y)$,

$$\varphi_R : E = V(\langle v_1, \ldots, v_\alpha \rangle) \to F = V(\langle v_1, \ldots, v_\alpha, v_{\beta+1}, \ldots, v_{n+1} \rangle).$$

There exists a toric $\mathbb{Q}$-Fano variety G of dimension $n - \beta$ with Picard number 1 such that

$$\varphi_R^{-1}(P)_{\mathrm{red}} \cong G \quad \forall P \in V(U(w)).$$

More precisely, $\tau(w)_Y$ being the image of $\tau(w)$ in $N_\mathbb{R}/\langle \pm v_1, \ldots, \pm v_\alpha, \pm v_{\beta+1}, \ldots, \pm v_{n+1} \rangle$, there exists a finite cover

$$f_w : U'_{\tau(w)_Y} \to U_{\tau(w)_Y} \subset F$$

such that the pullback by f_w becomes (locally) a trivial G-bundle

$$E \times_F U'_{\tau(w)_Y} \cong G \times U'_{\tau(w)_Y},$$

where G has the description of a toric variety

$$G \cong Z(\Delta_Z),$$

Δ_Z being a fan in

$$\langle \pm v_1, \ldots, \pm v_\alpha, \pm v_{\beta+1}, \ldots, \pm v_{n+1} \rangle / \langle \pm v_1, \ldots, \pm v_\alpha \rangle \subset N_\mathbb{R}/\langle \pm v_1, \ldots, \pm v_\alpha \rangle$$

with the 1-dimensional edges $\langle v_{\beta+1} \rangle, \ldots, \langle v_{n+1} \rangle$.

PROOF. In all of the cases above, α is the number of the T-invariant divisors $V(\langle v_i \rangle)$ that have negative intersection with R (cf. the proof of Proposition 14-1-5).

Thus it is independent of w with $[V(w)] \in R$. Now,

$$\text{Exc}(\varphi_R) \subset \cap_{i=1}^{\alpha} V(\langle v_i \rangle) = V(\langle v_1, \ldots, v_\alpha \rangle),$$

since each curve C with $[C] \in R$ has negative intersection with $V(\langle v_i \rangle)$ for $1 \le i \le \alpha$. On the other hand,

$$\varphi_R(V(\langle v_1, \ldots, v_\alpha \rangle)) = V(\langle v_1, \ldots, v_\alpha, v_{\beta+1}, \ldots, v_{n+1} \rangle) = V(U(w))$$

by (ii) in Lemma 14-2-1. Since $\dim V(\langle v_1, \ldots, v_\alpha \rangle) > \dim V(U(w))$, we conclude that

$$\text{Exc}(\varphi_R) = V(\langle v_1, \ldots, v_\alpha \rangle).$$

Thus

$$\dim \varphi_R(\text{Exc}(\varphi_R)) = \dim V(U(w)) = \beta - \alpha$$

and β is independent of w.

The Δ_Y being simplicial in the case of the contraction of fibering type or divisorial type is an immediate consequence of (i) and (ii) of Proposition 14-2-1. In the case of the contraction of flipping type, if Δ_Y were simplicial, i.e., $\tau(w) = \langle v_1, \ldots, v_n, v_{n+1} \rangle$ were simplicial, then one of the primitive vectors $v_1, \ldots, v_n, v_{n+1}$ must be a nonnegative linear combination of the others. But this would contradict the fact that in the unique nontrivial linear relation

$$a_1 v_1 + \cdots + a_i v_i + \cdots + a_n v_n + a_{n+1} v_{n+1}$$

we have more than 1 strictly negative coefficients $a_1, \ldots, a_\alpha$ and more than 1 strictly positive coefficients a_n, a_{n+1}. Thus the fan Δ_Y is not simplicial.

For the last assertion, note that the equivariant morphism $\varphi_R : E \to F$ corresponds to the exact sequence of vector spaces

$$0 \to \langle \pm v_1, \ldots, \pm v_\alpha, \pm v_{\beta+1}, \ldots, \pm v_{n+1} \rangle / \langle \pm v_1, \ldots, \pm v_\alpha \rangle$$
$$\to N_\mathbb{R} / \langle \pm v_1, \ldots, \pm v_\alpha \rangle$$
$$\to N_\mathbb{R} / \langle \pm v_1, \ldots, \pm v_\alpha, \pm v_{\beta+1}, \ldots, \pm v_{n+1} \rangle \to 0,$$

where the convex cone

$$\tau(w)_Y \subset N_\mathbb{R} / \langle \pm v_1, \ldots, \pm v_\alpha, \pm v_{\beta+1}, \ldots, \pm v_{n+1} \rangle$$

has a "lifting"

$$\langle v_1, \ldots, v_\beta \rangle \subset N_\mathbb{R} / \langle \pm v_1, \ldots, \pm v_\alpha \rangle.$$

Moreover, the "composite" of $\langle v_1, \ldots, v_\beta \rangle$ with the fan in $\langle \pm v_1, \ldots, \pm v_\alpha, \pm v_{\beta+1}, \ldots, \pm v_{n+1} \rangle / \langle \pm v_1, \ldots, \pm v_\alpha \rangle$ generated by the 1-dimensional edges $v_{\beta+1}, \ldots, v_{n+1}$ gives the decomosition of $\tau(w)$ in $N_\mathbb{R} / \langle \pm v_1, \ldots, \pm v_\alpha \rangle$ into the images of cones in Δ. Then the assertion is an immediate consequence of, e.g., Fulton [2], exercise on page 41. It is not usually a locally trivial fiber bundle, and we have to take a finite morphism, since the lattice in the lifting could map only onto the sublattice of finite index and not onto the full lattice of $\tau(w)_Y$.

This completes the proof.   □

Remark 14-2-3. The general fiber G of the exceptional locus $\varphi_R : E \to F$ is not a weighted projective space in general, as opposed to what is claimed in Reid [5], (2.6). Let $\hat{N}$ be the lattice for $\langle \pm v_1, \ldots, \pm v_\alpha, \pm v_{\beta+1}, \ldots, \pm v_{n+1} \rangle / \langle \pm v_1, \ldots, \pm v_\alpha \rangle$ induced from the quotient lattice for $N_{\mathbb{R}} / \langle \pm v_1, \ldots, \pm v_\alpha \rangle$ of $N \subset N_{\mathbb{R}}$. Let $N' \subset \hat{N}$ be the sublattice generated by the primitive vectors $v_{\beta+1}, \ldots, v_{n+1}$ (considered as elements in $\hat{N}$). There exist $d_0, \ldots, d_{n-\beta} \in \mathbb{N}$ such that

$$d_0 v_{\beta+1} + d_1 v_{\beta+2} + \cdots + d_{n-\beta} v_{n+1} = 0.$$

Let $N'' \subset N'$ be the sublattice generated by the vectors $d_0 v_{\beta+1}, \ldots, d_{n-\beta} v_{n+1}$. We take $\Delta_{Z''} = \Delta_{Z'} = \Delta_Z$ to be same fan generated by the 1-dimensional edges $\langle v_{\beta+1} \rangle, \ldots, \langle v_{n+1} \rangle$ identifying $N''_{\mathbb{R}} = N'_{\mathbb{R}} = \hat{N}_{\mathbb{R}}$. Then we get the corresponding finite morphisms

$$Z''(\Delta_{Z''}) \longrightarrow Z'(\Delta_{Z'}) \longrightarrow Z(\Delta_Z)$$
$$\| \qquad\qquad \| \qquad\qquad \|$$
$$\mathbb{P}^{n-\beta} \longrightarrow \mathbb{P}(d_0 : \cdots : d_{n-\beta}) \longrightarrow G$$

where the last finite morphism from the weighted projective space $\mathbb{P}(d_0 : \cdots : d_{n-\beta})$ to G is in general not an isomorphism.

In fact, take the complete fan Δ_Z to be the one having four 1-dimensional edges generated by the primitive vectors below in $N_{\mathbb{R}} \cong \mathbb{R}^3$ (cf. Borisov-Borisov [1]):

$$(1, 0, 1), (-2, 1, 1), (1, -2, 0), (0, 1, -2).$$

Then $G = Z(\Delta_Z)$ is a $\mathbb{Q}$-Fano 3-fold with only $\mathbb{Q}$-factorial and terminal singularities of Picard number 1. But it is not isomorphic to any weighted projective space of dimension 3. This can be checked easily, as Professor Włodarczyk communicated to us, if we use the above observation and a theorem of Danilov–Demushkin (cf. Danilov [1] Demushkin [1]), which claims; If two complete toric varieties $W = X(\Delta_W)$ and $V = X(\Delta_V)$ are isomorphic as abstract varieties (i.e., an isomorphim between them may not be equivariant), then there exists an isomorphism of lattices between N_W and N_V, which induces an isomorphism of fans between Δ_W and Δ_V.

The existence of (log) flip also follows immediately from the description of the change of decompositions of the cone $\tau(w)$ under the extremal contraction of the extremal ray $R = \mathbb{R}_+[V(w)]$ in Proposition 14-2-1.

Proposition 14-2-4 (Existence of (Log) Flip for Toric Varieties). *Let $X(\Delta)$ be a complete and $\mathbb{Q}$-factorial toric variety, $R \in \overline{\text{NE}}(X(\Delta))$ an extremal ray. Suppose that the contraction of the extremal ray R*

$$\varphi_R : X(\Delta) \to Y(\Delta_Y)$$

is of flipping type with $\text{codimExc}(\varphi_R) \geq 2$. *Then there exists an equivariant birational morphism* $\varphi_R^+ : X^+(\Delta^+) \to Y(\Delta_Y)$ *called the flip of* φ_R *such that*

(i) $X^+(\Delta^+)$ *is a complete and* $\mathbb{Q}$*-factorial toric variety,*

(ii) $\mathrm{codim}\,\mathrm{Exc}(\varphi_R^+) \geq 2$, *and*

(iii) *for a* $(n-1)$*-dimensional cone* $w^+ \in \Delta^+$,

$$\varphi_R^+(V(w^+)) = a\ point\ on\ Y(\Delta_Y) \Longleftrightarrow V(w^+) \in -R,$$

where we identify $N^1(X(\Delta))$ *with* $N^1(X^+(\Delta^+))$ *by taking the strict transfroms of divisors and hence identify* $N_1(X(\Delta))$ *with* $N_1(X^+(\Delta^+))$ *as their duals.*
Properties (i) (ii) *and* (iii) *characterize the flip* $\varphi_R^+ : X^+(\Delta^+) \to Y(\Delta_Y)$, *and thus it is unique.*
Moreover, when $X(\Delta)$ *is projective,* $X^+(\Delta^+)$ *is projective and condition* (iii) *can be replaced by*

(iii) *for* $C \subset X^+(\Delta^+)$

$$\varphi_R^+(C) = a\ point\ on\ Y(\Delta_Y) \Longleftrightarrow C \in -R.$$

(In the relative setting of $\phi : X(\Delta) \to S(\Delta_S)$ *a* $\mathbb{Q}$*-factorial toric variety proper over another and* $R \in \overline{\mathrm{NE}}(X(\Delta)/S(\Delta_S))$), *the flip* $\varphi_R^+ : X^+(\Delta^+) \to Y(\Delta_Y)$ *is over* $S(\Delta_S)$.)

PROOF.   We take the simplicial subdivision Δ^+ of Δ_Y given by the decomposition

$$\tau(w) = \cup_{j=1}^\alpha \tau_j$$

for each $\tau(w)$ as in (ii) of Proposition 14-2-1. Then conditions (i) and (ii) are clearly satisfied. For condition (iii) observe that any $(n-1)$-dimensional face $w^+ \in \Delta^+$ with $\varphi^+(V(w^+)) = $ a point on $Y(\Delta_Y)$ is of the form

$$w^+ = w_{i,j} = \langle v_1, \ldots, \overset{\vee}{v_i}, \ldots, \overset{\vee}{v_j}, \ldots, v_n, v_{n+1} \rangle,$$

where $1 \leq i < j \leq \alpha$. Recalling that the nontrivial linear relation for w is written

$$a_1 v_1 + \cdots + a_i v_i + \cdots + a_j v_j + \cdots + a_n v_n + a_{n+1} v_{n+1} = 0 \text{ with } a_{n+1} = 1$$

and that the coefficients for v_i and v_j are both negative $a_i, a_j < 0$, we conclude that reversing the sign (and then divide by $-a_i$ or $-a_j$) will give the nontrivial linear relation for w^+. This implies (cf. Lemma 14-1-7 and its proof) that there exists a positive constant $c > 0$ such that

$$V(\langle v_i \rangle) \cdot V(w) = -c V(\langle v_i \rangle) \cdot V(w^+) \text{ for } i = 1, \ldots, n, n+1.$$

Therefore, we conclude (Exercise! Why?) that for any divisor D on $X(\Delta)$,

$$D \cdot V(w) \gtreqless 0 \Longleftrightarrow D^+ \cdot V(w^+) \lesseqgtr 0,$$

where D^+ is the strict transfrom of D. Thus we have

$$[V(w^+)] \in -R.$$

We leave it as an exercise for the reader to verify whether for an $(n-1)$-dimensional cone $w^+ \in \Delta^+$ we have $[V(w^+)] \in -R$, then $\varphi_R^+(V(w^+)) = $ a point.

Conversely, suppose $X^+(\Delta^+)$ satisfies conditions (i), (ii), and (iii). Conditions (i) and (ii) imply that Δ^+ must be a simplicial subdivisoin of Δ_Y without adding any extra 1-dimensional edge. By condition (iii) we conclude that any new $(n-1)$-dimensional face we are adding is of the form $w_{i,j}$ as above. We have to add all the $w_{i,j}$ as above, since if we omit some, then Δ^+ would not be simplicial. Therefore, Δ^+ must be given by the subdivision

$$\tau(w) = \cup_{j=1}^{\alpha} \tau_j,$$

proving the uniqueness.

When $X(\Delta)$ is projective, so is $Y(\Delta_Y)$ by Theorem 14-1-9. If A is a relatively ample divisor over $Y(\Delta_Y)$ such that

$$A \cdot V(w) > 0 \text{ for every } w \in \Delta \text{ with } [V(w)] \in R,$$

then

$$(-A^+) \cdot V(w^+) > 0 \text{ for every } w^+ \in \Delta^+ \text{ with } [V(w^+)] \in -R.$$

Thus $-A^+$ is relatively ample over $Y(\Delta_Y)$, and hence $X^+(\Delta^+)$ is projective. The last assertion follows easily from Proposition 14-1-2.

We remark that condition (iii) actually holds without the projectivity assumption. (Exercise! Check this remark.)

The verification of the relative case is left to the reader.   □

We give some examples of toric flipping contractions. We refer the reader to the next section for the details of the criterion for a toric variety to have only terminal singularities.

Example-Claim 14-2-5 (Classification of Some Special Cases of Contractions of Flipping Type for Toric 3-Folds with Only $\mathbb{Q}$-factorial and Terminal Singularities). *Let*

$$\varphi_R : X(\Delta) \to Y(\Delta_Y)$$

be the contraction morphism of an extremal ray R with $K_{X(\Delta)} \cdot R < 0$ of flipping type from a complete toric 3-fold with only $\mathbb{Q}$-factorial and terminal singularities. We further assume that the (unique) rational curve that is contracted passes through only one singular point of $X(\Delta)$.

Then we have the following description of the flipping contraction:

There exist two 3-dimensional cones

$$\tau_4 = \langle v_1, v_2, v_3 \rangle \in \Delta,$$
$$\tau_3 = \langle v_1, v_2, v_4 \rangle \in \Delta,$$

sharing the 2-dimensional wall

$$w = \langle v_1, v_2 \rangle$$

such that

$$[V(w)] \in R$$

and such that for some $\mathbb{Z}$-coordinate of N,

$$
\begin{array}{llll}
v_1 = (1, 0, 0), & & v_1 = (1, 0, 0), \\
v_2 = (0, 1, 0), & & v_2 = (0, 1, 0), \\
v_3 = (0, 0, 1), & or & v_3 = (0, 0, 1), \\
v_4 = (a, r - a, -r), & & v_4 = (a, 1, -r),
\end{array}
$$

where

$$
0 < a < r \text{ and } \gcd(r, a) = 1
$$

to satisfy the nontrivial relation

$$
(-a) \cdot v_1 + (-(r - a)) \cdot v_2 + r \cdot v_3 + 1 \cdot v_4 = 0
$$

or

$$
(-a) \cdot v_1 + 1 \cdot v_2 + r \cdot v_3 + 1 \cdot v_4 = 0,
$$

respectively.

PROOF. `Because of the condition that the unique rational curve $V(w)$ that is contracted passes through only one singular point of $X(\Delta)$, we may assume that one of the 3-dimensional cones (sharing the 2-dimensional wall w), say $\tau_4 = \langle v_1, v_2, v_3 \rangle$, is nonsingular. Choose the $\mathbb{Z}$-coordinate such that

$$
\begin{aligned}
v_1 &= (1, 0, 0), \\
v_2 &= (0, 1, 0), \\
v_3 &= (0, 0, 1),
\end{aligned}
$$

and

$$
v_4 = \langle -a_1, -a_2, -a_3 \rangle
$$

to satisfy the nontrivial linear relation

$$
a_1 v_1 + a_2 v_2 + a_3 v_3 + v_4 = 0 \text{ with } a_1, a_2 < 0 \text{ and } a_3 > 0.
$$

Note that since $K_{X(\Delta)} \cdot V(w) < 0$, we have (cf. Section 14-3)

$$
(-a_1) + (-a_2) + (-a_3) < 1.
$$

Case: $(-a_1) + (-a_2) + (-a_3) = 0$

In this case, set

$$
r = a_3, \quad a = -a_1, \quad r - a = -a_2.
$$

Then $0 < a < r$ and $\gcd(r, a) = 1$, since the vector v_4 is primitive. We have

$$
v_4 = (a, r - a, -r)
$$

as desired.

Case: $(-a_1) + (-a_2) + (-a_3) < 0$

In this case, set

$$r = a_3, \quad a = -a_1, \quad b = -a_2.$$

Then we have

$$0 < a, \quad b < a + b < r,$$

and

$$\gcd(r, a) = \gcd(r, b) = 1.$$

In fact, $\gcd(r, a) = 1$ (respectively $\gcd(r, b) = 1$), since the triangle with vertices 0, v_2, v_4 (respectively 0, v_1, v_4) contains no integral lattice points other than the vertices themselves.

Since the 3-simplex with vertices 0, v_1, v_2, v_4 contains no integral lattice points other than the vertices themselves (see Section 14-3 for the condition on the 3-simplex so that the toric variety has only terminal singularities), there is *no* triple λ_1, λ_2, λ_3 satisfying

$$0 < \lambda_1 < 1,$$
$$0 < \lambda_2 < 1,$$
$$0 < \lambda_3 < 1,$$
$$0 < \lambda_1 + \lambda_2 + \lambda_3 \leq 1,$$
$$\lambda_1 v_1 + \lambda_2 v_2 + \lambda_3 v_4 = (\lambda_1 + a\lambda_3, \lambda_2 + b\lambda_3, -\lambda_3 r) \in N.$$

Therefore, if we set

$$\lambda_3 = \frac{k}{r}, \quad k = 1, \ldots, r - 1,$$
$$\lambda_1 = \{-a\lambda_3\},$$
$$\lambda_2 = \{-b\lambda_3\}$$

(where $\{x\}$ denotes the fractional part $x - [x]$ of a rational number x),

then

$$0 < \lambda_1 < 1,$$
$$0 < \lambda_2 < 1,$$
$$0 < \lambda_3 < 1,$$
$$\lambda_1 v_1 + \lambda_2 v_2 + \lambda_3 v_4 = (\lambda_1 + a\lambda_3, \lambda_2 + b\lambda_3, -\lambda_3 r) \in N,$$

and hence

$$\lambda_1 + \lambda_2 + \lambda_3 > 1.$$

That is to say, we have

$$\left\{-a\frac{k}{r}\right\} + \left\{-b\frac{k}{r}\right\} + \left\{1\frac{k}{r}\right\} > 1 \text{ for } k = 1, \ldots, r - 1.$$

Now we use the following terminal lemma of White–Frumkin (cf. White [1] and the terminal lemma in Oda [2], Section 1.6).

Proposition 14-2-6 (Terminal Lemma). *Suppose*

$$\left\{\alpha\frac{k}{q}\right\} + \left\{\beta\frac{k}{q}\right\} + \left\{\gamma\frac{k}{q}\right\} > 1 \; for \; k = 1, \ldots, q - 1,$$

where α, β, γ, and $q > 0$ are integers with

$$\gcd(\alpha, q) = \gcd(\beta, q) = \gcd(\gamma, q) = 1.$$

Then at least one of $\alpha + \beta$, $\beta + \gamma$, and $\gamma + \alpha$ is divisible by q.

Applying the terminal lemma to

$$\alpha = -a, \quad \beta = -b, \quad \gamma = 1, \quad q = r,$$

we conclude (via the inequalities at the beginning) that

$$a = 1 \quad (\text{or } b = 1).$$

Thus

$$v_4 = (a, 1, -r)$$

as desired.   □

Remark 14-2-7. (i) The essence of the analysis above is the classification of the 3-simplices that give rise to terminal singularities in dimension 3 via the terminal lemma. We refer the reader to the terminal lemma in Oda [2], Section 1.6, for more discussion of this subject.

(ii) There are examples of a flipping contraction from a complete toric 3-fold with only $\mathbb{Q}$-factorial and terminal singularities where the extremal rational curve passes through 2 singular points, e.g., take the cones τ_4, τ_3, w as in the example but with the primitive vectors replaced by

$$v_1 = (1, 0, 0),$$
$$v_2 = (0, 1, 0),$$
$$v_3 = (0, 1, 5),$$
$$v_4 = (1, -1, -17).$$

We leave the analysis of the examples where the extremal rational curve passes through 2 singular points to the reader.

In Kawamata–Matsuda–Matsuki [1] at the end of Example 5-2-5 there is a slightly misleading statement: "The morphisms given in Examples 5-2-4 and 5-2-5 (which are the examples given in our Example-Claim 14-2-5) are the only contractions of flipping type from $\mathbb{Q}$-factorial terminal toric varieties of dimension 3 by the theorem of White–Frumkin." This is, however, true only under the assumption that the extremal rational curve passes only one singular point.

Example-Claim 14-2-8 (Classification of Contractions of Flipping or Flopping Type for Toric Smooth 4-Folds). *Let*

$$\varphi_R : X(\Delta) \to Y(\Delta_Y)$$

be the contraction morphism of an extremal ray R with $K_{X(\Delta)} \cdot R < 0$ of flipping type from a complete nonsingular toric 4-fold. Then there exist two 4-dimensional cones

$$\tau_5 = \langle v_1, v_2, v_3, v_4 \rangle \in \Delta,$$
$$\tau_4 = \langle v_1, v_2, v_3, v_5 \rangle \in \Delta,$$

sharing the 3-dimensional wall

$$w = \langle v_1, v_2, v_3 \rangle$$

such that

$$[V(w)] \in R$$

and such that for some $\mathbb{Z}$-coordinate of N:

$$v_1 = (1, 0, 0, 0),$$
$$v_2 = (0, 1, 0, 0),$$
$$v_3 = (0, 0, 1, 0),$$
$$v_4 = (0, 0, 0, 1),$$
$$v_5 = (1, 1, -1, -1),$$

to satisfy the nontrivial relation

$$(-1) \cdot v_1 + (-1) \cdot v_2 + 1 \cdot v_3 + 1 \cdot v_4 + 1 \cdot v_5 = 0.$$

If φ_R is of flopping type with $K_{X(\Delta)} \cdot R = 0$, then

$$
\begin{array}{llll}
v_1 = (1, 0, 0, 0), & & v_1 = (1, 0, 0, 0), \\
v_2 = (0, 1, 0, 0), & & v_2 = (0, 1, 0, 0), \\
v_3 = (0, 0, 1, 0), & \text{or} & v_3 = (0, 0, 1, 0), \\
v_4 = (0, 0, 0, 1), & & v_4 = (0, 0, 0, 1), \\
v_5 = (1, 1, 0, -1), & & v_5 = (1, 2, -1, -1),
\end{array}
$$

respectively, to satisfy the nontrivial relation

$$(-1) \cdot v_1 + (-1) \cdot v_2 + 0 \cdot v_3 + 1 \cdot v_4 + 1 \cdot v_5 = 0$$

or

$$(-1) \cdot v_1 + (-2) \cdot v_2 + 1 \cdot v_3 + 1 \cdot v_4 + 1 \cdot v_5 = 0.$$

PROOF. Let

$$\tau_5 = \langle v_1, v_2, v_3, v_4 \rangle \in \Delta,$$
$$\tau_4 = \langle v_1, v_2, v_3, v_5 \rangle \in \Delta,$$

be the two 4-dimensional cones in Δ sharing the 3-dimensional wall

$$w = \langle v_1, v_2, v_3 \rangle \text{ with } [V(w)] \in R.$$

Since Δ is nonsingular, we may take a $\mathbb{Z}$-coordinate of N such that

$$v_1 = (1, 0, 0, 0),$$
$$v_2 = (0, 1, 0, 0),$$
$$v_3 = (0, 0, 1, 0),$$
$$v_4 = (0, 0, 0, 1),$$

and

$$v_5 = (-a_1, -a_2, -a_3, -1),$$

where $a_1, a_2 < 0$, since $\alpha \geq 2$.

Now, since $K_{X(\Delta)} \cdot V(w) < 0$, we have the condition

$$(-a_1) + (-a_2) + (-a_3) + (-1) < 1.$$

This implies $a_3 > 0$. Thus by Proposition 14-1-5 (i) we conclude that

$$\tau_3 = \langle v_1, v_2, v_4, v_5 \rangle \in \Delta.$$

Again, the nonsingularity of Δ implies $a_3 = 1$ and hence $a_1 = a_2 = -1$. The case for φ_R being of flopping type is similar and left to the reader as an exercise.

This completes the proof.   □

Remark 14-2-9. We remark that the exceptional locus $E = \mathrm{Exc}(\varphi_R)$ for the flipping contraction as above is the closed orbit $V(\langle v_1, v_2 \rangle)$, which is the toric variety $\mathbb{P}^2$ corresponding to the fan generated by the 1-dimensional edges $\langle v_3 \rangle, \langle v_4 \rangle, \langle v_5 \rangle$ in $N_\mathbb{R}/\langle \pm v_1, \pm v_2 \rangle$. The normal bundle is given by

$$\mathcal{N}_{E/X(\Delta)} \cong \mathcal{O}_{\mathbb{P}^2}(-1) \oplus \mathcal{O}_{\mathbb{P}^2}(-1).$$

The flip $\varphi_R^+ : X^+(\Delta^+) \to Y(\Delta_Y)$ is a morphism from a smooth 4-fold with the exceptional locus $\cong \mathbb{P}^1$.

Actually, the above is just a toric description of the example given in Example 8-3-8 (ii) in dimension 4.

The exceptional locus for the first flopping contraction is a $\mathbb{P}^1$-bundle over a smooth curve. It is a 1-dimensional family of Atiyah's flop in Example 3-3-3. The flopped variety $X^+(\Delta^+)$ is again smooth.

The exceptional locus for the second flopping contraction is isomorphic to $\mathbb{P}^2$ with normal bundle $\mathcal{O}_{\mathbb{P}^2}(-1) \oplus \mathcal{O}_{\mathbb{P}^2}(-2)$. The flop $\varphi_R^+ : X^+(\Delta^+) \to Y(\Delta_Y)$ is a morphism from a *singular* 4-fold with the exceptional locus $\cong \mathbb{P}^1$.

We leave the verification of the assertions in this remark to the reader as an exercise.

Exercise 14-2-10. (i) Show that there is no contraction morphism of an extremal ray R with $K_{X(\Delta)} \cdot R < 0$ of flipping type from a complete *smooth* toric 3-fold $X(\Delta)$.

(ii) Classify all the contraction morphisms of extremal rays of flopping type (i.e., $K_{X(\Delta)} \cdot R = 0$ and codimExc(φ_R) ≥ 2) from complete *smooth* toric 3-folds $X(\Delta)$.

Proposition 14-2-11 (Termination of (Log) Flips for Toric Varieties). *There is no infinite sequence of (log) flips with respect to some fixed (log) canonical divisor for complete toric varieties with $\mathbb{Q}$-factorial singularities.*

PROOF. Our first argument for termination is cheap. Observe that a (log) flip does not add or eliminate any 1-dimensional edges from the fan Δ (cf. Proposition 14-2-4). There are only finitely many fans with the same 1-dimensional edges. If there exists an infinite sequence of (log) flips, we have to have the same fan repeated at least twice in the sequence, which is impossible by the discrepancy consideration of Lemma 9-1-3.

Our second argument, following Reid [5], is more sophisticated and suggests a connection with the invariant "difficulty" and/or minimal discrepancy argument of Shokurov [1] [2]. We consider the case of a sequence of flips with respect to the canonical divisor. Let

$$w = \langle v_1, \ldots, v_{n-1} \rangle$$

be an $(n-1)$-dimensional cone with $[V(w)] \in R$, where R is the extremal ray of flipping type. Let

$$\tau_{n+1} = \langle v_1, \ldots, v_{n-1}, v_n \rangle \in \Delta,$$
$$\tau_n = \langle v_1, \ldots, v_{n-1}, v_{n+1} \rangle \in \Delta,$$

be the two n-dimensional cones in Δ containing w, as discussed at the beginning of this section. Observe that $K_{X(\Delta)} \cdot V(w) < 0$, and hence $\Theta_{\tau_{n+1}} \cup \Theta_{\tau_n}$ is strictly convex along σ_w (see lemma 14-3-4), where Θ_τ for $\tau \in \Delta$ is the simplex that is the convex hull of the origin and the primitive vectors generating τ and where σ_w is the convex hull of the primitive vectors generating w. If $X^+(\Delta^+)$ is the flipped variety with the new $(n-1)$-dimensional cone $w^+ \in \Delta^+$, then $K_{X^+(\Delta^+)} \cdot V(w^+) > 0$, and hence the two n-simplices sandwiching w^+ are strictly concave along σ_{w^+}. This implies the strict inequality for the volumes (cf. Figure 14-2-12)

$$\text{Vol}(\cup_{\tau \in \Delta, \dim \tau = n} \Theta_\tau) > \text{Vol}(\cup_{\tau^+ \in \Delta^+, \dim \tau^+ = n} \Theta_{\tau^+}).$$

On the other hand, the values of both volumes are in $\frac{1}{n!}\mathbb{N}$. Therefore, this decrease of volumes cannot happen infinitely many times, excluding the possibility of an infinite sequence of flips. The argument for the general case of a sequence of log flips with respect to some fixed log canonical divisor is similar and left to the reader as an exercise.   □

Figure 14-2-12.

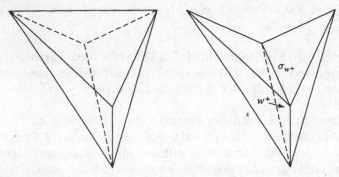

14.3   Toric Canonical and Log Canonical Divisors

So far, the analysis of toric varieties in Sections 14-1 and 14-2 did not refer to their canonical or log canonical divisors (except for the distinction between flip and flop in terms of the intersection number with the canonical divisor in Example-Claim 14-2-8 and except for the argument for termination of (log)flips with respect to some fixed (log) canonical divisor in Proposition 14-2-11). In this section we describe the **toric canonical and log canonical divisors** and give some **criteria for toric singularities to be terminal, canonical, log terminal, or log canonical** in terms of the corresponding geometry of convex cones.

Proposition 14-3-1 (Toric Terminal and Canonical Singularities).   *Let $X(\Delta)$ be a toric variety. Then the canonical divisor $K_{X(\Delta)}$ has the description*

$$K_{X(\Delta)} \sim -\Sigma_{\langle v_i \rangle \in \Delta, \dim\langle v_i \rangle = 1} V(\langle v_i \rangle).$$

Moreover,

(i) *the canonical divisor $K_{X(\Delta)}$ is $\mathbb{Q}$-Cartier if and only if for every $\tau \in \Delta$ there exists $m_\tau \in M \otimes \mathbb{Q}$ such that*

$$m_\tau(v_i) = 1 \quad \forall v_i \in N \text{ primitive with } \langle v_i \rangle \in \Delta, \langle v_i \rangle \subset \tau, \text{ and}$$

(ii) *when $K_{X(\Delta)}$ is $\mathbb{Q}$-Cartier, the criteria for $X(\Delta)$ to have terminal or canonical singularities are given as follows:*

$$terminal \Longleftrightarrow N \cap \tau \cap \{m_\tau \leq 1\}$$
$$= \{0\} \cup \{v_i \in N; \text{ primitive with } \langle v_i \rangle \in \Delta \text{ and } \langle v_i \rangle \subset \tau\} \forall \tau \in \Delta,$$
$$canonical \Longleftrightarrow N \cap \tau \cap \{m_\tau < 1\} = \{0\} \quad \forall \tau \in \Delta.$$

PROOF.   Take a subdivision Δ_Z of Δ such that

$$f : Z(\Delta_Z) \to X(\Delta)$$

provides a resolution of singularities. If the first assertion holds for $Z(\Delta_Z)$, then it holds for $X(\Delta)$:

$$K_{X(\Delta)} = f_* K_{Z(\Delta_Z)} = f_*(-\Sigma_{\langle v_j \rangle \in \Delta_Z, \dim \langle v_j \rangle = 1} V(\langle v_j \rangle))$$
$$= -\Sigma_{\langle v_i \rangle \in \Delta, \dim \langle v_i \rangle = 1} V(\langle v_i \rangle).$$

Therefore, we may assume that $X(\Delta)$ is nonsingular.
Set

$$B = \Sigma_{\langle v_i \rangle \in \Delta, \dim \langle v_i \rangle = 1} V(\langle v_i \rangle).$$

Note that for any (n-dimensional) $\tau \in \Delta$ with its dual (to the generators of τ) $\mathbb{Z}$-basis $y_1, \ldots, y_n \in M$ the affine space U_τ is isomorphic to $\mathbb{A}^n$ with its coordinates $y_1, \ldots, y_n$ and the boundary divisor $B|_{U_\tau} = \cup_{i=1}^n \{y_i = 0\}$.

Therefore, fixing linearly independent $x_1, \ldots, x_n \in M$ we have

$$\mathcal{O}_{X(\Delta)}(K_{X(\Delta)} + B)|_{U_\tau} = \mathcal{O}_{U_\tau} \cdot \frac{dy_1}{y_1} \wedge \cdots \wedge \frac{dy_n}{y_n}$$
$$= \mathcal{O}_{U_\tau} \cdot d\log y_1 \wedge \cdots \wedge d\log y_n$$
$$= \mathcal{O}_{U_\tau} \cdot d\log x_1 \wedge \cdots \wedge d\log x_n$$
$$= \mathcal{O}_{U_\tau} \cdot \frac{dx_1}{x_1} \wedge \cdots \wedge \frac{dx_n}{x_n};$$

hence globally,

$$\mathcal{O}_{X(\Delta)}(K_{X(\Delta)} + B) = \mathcal{O}_{X(\Delta)} \cdot \frac{dx_1}{x_1} \wedge \cdots \wedge \frac{dx_n}{x_n} \cong \mathcal{O}_{X(\Delta)}.$$

Therefore,

$$K_{X(\Delta)} + B \sim 0, \text{ i.e., } K_{X(\Delta)} \sim -B.$$

If you prefer a more down-to-earth but less illuminating explanation, then take two cones sharing a face (the cones are assumed to be n-dimensional and their common face $(n-1)$-dimensional for the sake of simplicity of presentation):

$$\tau = \langle v_1, \ldots, v_{n-1}, v_n \rangle,$$
$$\tau' = \langle v_1, \ldots, v_{n-1}, v_n' = -v_n + \Sigma_{i \neq n} a_i v_i \rangle.$$

$\mathcal{O}_{X(\Delta)}(-B)$ has the generators

$$\mathcal{O}_{X(\Delta)}(-B)|_{U_\tau} = \mathcal{O}_{U_\tau} \cdot x_1 \cdots x_{n-1} x_n,$$
$$\mathcal{O}_{X(\Delta)}(-B)|_{U_{\tau'}} = \mathcal{O}_{U_{\tau'}} \cdot x_1' \cdots x_{n-1}' x_n',$$
$$= \mathcal{O}_{U_\tau} \cdot (x_1 x_n^{a_1}) \cdots (x_{n-1} x_n^{a_{n-1}})(x_n^{-1}).$$

On the other hand,

$$\mathcal{O}_{X(\Delta)}(K_{X(\Delta)})|_{U_\tau} = \mathcal{O}_{U_\tau} \cdot dx_1 \wedge \cdots \wedge dx_{n-1} \wedge dx_n,$$
$$\mathcal{O}_{X(\Delta)}(K_{X(\Delta)})|_{U_{\tau'}} = \mathcal{O}_{U_{\tau'}} \cdot dx_1' \wedge \cdots dx_{n-1}' \wedge dx_n'$$
$$= \mathcal{O}_\tau \cdot x_n^{a_1} dx_1 \wedge \cdots \wedge x_{n-1}^{a_{n-1}} dx_{n-1} \wedge d(x_n^{-1}).$$

Thus $\mathcal{O}_{X(\Delta)}(-B)$ and $\mathcal{O}_{X(\Delta)}(K_{X(\Delta)})$ have the same transition function (up to a constant)

$$\frac{x_1' \cdots x_{n-1}' x_n'}{x_1 \cdots x_n} = x_n^{a_1} \cdots x_n^{a_{n-1}} x_n^{-2} = -\frac{dx_1' \wedge \cdots \wedge dx_{n-1}' \wedge dx_n'}{dx_1 \wedge \cdots \wedge dx_{n-1} \wedge dx_n},$$

and hence they are isomorphic.

Criterion (i) follows easily from, e.g., Fulton [2], Section 3.3.

In order to see criterion (ii), take a resolution of singularities $f : Z(\Delta_Z) \to X(\Delta)$ by choosing a subdivision Δ_Z.

Take a primitive $v_j \in N$ with $\langle v_j \rangle \in \Delta_Z$, $\dim \langle v_j \rangle = 1$.

Suppose $\langle v_j \rangle \subset \tau \in \Delta$. Then the discrepancy at $V(\langle v_j \rangle)$ is

$$\mathrm{ord}_{V(\langle v_j \rangle)}\{K_{Z(\Delta_Z)} - f^* K_{X(\Delta)}\} = \mathrm{ord}_{V(\langle v_j \rangle)}\{(-B_Z) - f^*(-B)\}$$
$$= -1 + m_\tau(v_j).$$

Moreover, the divisor $V(\langle v_j \rangle)$ is exceptional with respect to f if and only if $\langle v_j \rangle \notin \Delta$. Therefore,

$$\text{terminal} \iff -1 + m_\tau(v_j) > 0 \quad \forall V(\langle v_j \rangle) \text{ exceptional for } f \quad \forall \tau \in \Delta$$
$$\iff N \cap \tau \cap \{m_\tau \leq 1\}$$
$$= \{0\} \cup \{v_i \in N; \text{ primitive with } \langle v_i \rangle \in \Delta \text{ and } \langle v_i \rangle \subset \tau\} \quad \forall \tau \in \Delta.$$

The criterion for toric canonical singularities is similar and immediate. $\qquad\square$

We leave the verification of the following criterion for toric log terminal and log canonical singularities and its corollary as an exercise for the reader.

Proposition 14-3-2 (Toric Log Terminal and Log Canonical Singularities).
Let $X(\Delta)$ be a toric variety and $D = \Sigma_{\langle v_i \rangle \in \Delta, \dim \langle v_i \rangle = 1} d_i V(\langle v_i \rangle)$, $0 \leq a_i \leq 1$, a boundary $\mathbb{Q}$-divisor consisting of invariant divisors.

(i) *$K_{X(\Delta)} + D$ is $\mathbb{Q}$-Cartier if and only if for every $\tau \in \Delta$ there exists $m_\tau \in M \otimes \mathbb{Q}$ such that*

$$m_\tau(v_i) = 1 - d_i \quad \forall \langle v_i \rangle \subset \tau,$$

(ii) *when $K_{X(\Delta)} + D$ is $\mathbb{Q}$-Cartier the criteria for $X(\Delta)$ to have log terminal or log canonical singularities are given as follows:*

log terminal and $\mathbb{Q}$-factorial $\iff$ for each $\tau \in \Delta$ τ is simplicial and
$$\{v_i; \langle v_i \rangle \in \Delta, \qquad \langle v_i \rangle \subset \tau, d_i = 1\} \text{ form a (part of a) } \mathbb{Z}\text{-basis}$$
log canonical $\iff$ no condition!

That is to say, every log pair $(X(\Delta), D)$ consisting of a toric variety and an invariant boundary $\mathbb{Q}$-divisor, with $K_{X(\Delta)} + D$ being $\mathbb{Q}$-Cartier, has at most log canonical singularities.

We also remark that without $\mathbb{Q}$-factoriality the criterion for $X(\Delta)$ to have only log terminal singularities becomes more subtle.

Our final task of this section is to describe the contractions of extremal rays in reference to the toric (log) canonical divisors.

Theorem 14-3-3 (Contractions of Extremal Rays for Toric Varieties in Reference to the (Log) Canonical Divisor). *Let $X(\Delta)$ be a complete and $\mathbb{Q}$-factorial toric variety, $R \in \overline{NE}(X(\Delta))$ an extremal ray. Let*

$$\varphi_R : X(\Delta) \to Y(\Delta_Y)$$

be the contraction of R.

(i) *Suppose that $X(\Delta)$ has only terminal (respectively canonical) singularities and that $K_{X(\Delta)} \cdot R < 0$. Then*

   (a) *Contraction of Fibering Type*
      $Y(\Delta_Y)$ has only log terminal singularities.
   (b) *Contraction of Divisorial Type*
      $Y(\Delta_Y)$ has terminal (respectively canonical) singularities.
   (c) *Contraction of Flipping Type*
      $X^+(\Delta^+)$ has terminal (respectively canonical) singularities.

(ii) *Suppose $(X(\Delta), D)$ has only log terminal (respectively log canonical) singularities for a boundary $\mathbb{Q}$-divisor $D = \Sigma_{\langle v_i \rangle \in \Delta, \dim \langle v_i \rangle = 1} d_i V(\langle v_i \rangle)$, $0 \leq d_i, < 1$ and that $(K_{X(\Delta)} + D) \cdot R < 0$. Then*

   (a) *Contraction of Fibering Type*
      $Y(\Delta_Y)$ has only log terminal singularities.
   (b) *Contraction of Divisorial Type*
      $(Y(\Delta_Y), D_Y)$ has log terminal (respectively log canonical) singularities.
   (c) *Contraction of Flipping Type*
      $(X(\Delta^+), D^+)$ has log terminal (respectively log canonical) singularities.

Moreover, for any extremal ray $R \in \overline{NE}(X(\Delta))$, there exists some invariant boundary $\mathbb{Q}$-divisor

$$D = \Sigma_{\langle v_i \rangle \in \Delta, \dim \langle v_i \rangle = 1} d_i V(\langle v_i \rangle) \text{ with } 0 \leq d_i < 1$$

such that $(X(\Delta), D)$ has only log terminal singularities and

$$(K_{X(\Delta)} + D) \cdot R < 0.$$

In the relative setting of $\phi : X(\Delta) \to S(\Delta_S)$ a $\mathbb{Q}$-factorial toric variety proper over another and $R \in \overline{NE}(X(\Delta)/S(\Delta_S))$, the contraction morphism $\varphi_R : X(\Delta) \to Y(\Delta_Y)$ is over $S(\Delta_S)$, and the identical statements hold about their singularities.

PROOF.   Let

$$w = \langle v_1, \ldots, v_{n-1} \rangle \in \Delta$$

be an $(n-1)$-dimensional cone generated by the primitive vectors $v_1, \ldots, v_{n-1} \in N$ such that $[V(w)] \in R$. Let $v_n, v_{n+1} \in N$ be the two primitive vectors such that they generate the two n-dimensional cones that belong to Δ:

$$\tau_{n+1} = \langle v_1, \ldots, v_{n-1}, v_n \rangle \in \Delta,$$
$$\tau_n = \langle v_1, \ldots, v_{n-1}, v_{n+1} \rangle \in \Delta.$$

Then we observe the following relation between the intersection number $K_{X(\Delta)} \cdot V(w)$ and how the two hyperplanes $\{m_{\tau_n} = 1\}$ and $\{m_{\tau_{n+1}} = 1\}$ meet along w.

Lemma 14-3-4.

$$K_{X(\Delta)} \cdot V(w) \overset{\geq}{\underset{<}{=}} 0 \iff \{m_{\tau_n} = 1\} \text{ and } \{m_{\tau_{n+1}} = 1\} \text{ meet } \begin{array}{c} \text{strictly concave} \\ \text{flat} \\ \text{strictly convex} \end{array} \text{ along } w.$$

PROOF.   Since

$$0 \sim_{\mathbb{Q}} \mathrm{div}(x^{m_{\tau_{n+1}}}) = \Sigma_{i=1}^n V(\langle v_i \rangle) + m_{\tau_{n+1}}(v_{n+1})V(\langle v_{n+1} \rangle) + G,$$

where $V(w) \cap G = \emptyset$, we have

$$
\begin{aligned}
K_{X(\Delta)} \cdot V(w) &= (-B) \cdot V(w) \\
&= (-B + \mathrm{div}(x^{m_{\tau_{n+1}}})) \cdot V(w) \\
&= (m_{\tau_{n+1}}(v_{n+1}) - 1)V(\langle v_{n+1} \rangle) \cdot V(w) \overset{\geq}{\underset{<}{=}} 0
\end{aligned}
$$

$$\iff \{m_{\tau_n} = 1\} \text{ and } \{m_{\tau_{n+1}} = 1\} \text{ meet } \begin{array}{c} \text{strictly concave} \\ \text{flat} \\ \text{strictly convex} \end{array} \text{ along } w. \quad \square$$

Now we go back to the proof of Theorem 14-3-3.

(i) Suppose $X(\Delta)$ has terminal singularities.

(a) Contraction of Fibering Type

By Corollary 14-2-2 Δ_Y is simplicial, since so is Δ. Thus by Proposition 14-3-2 $Y(\Delta_Y)$ has only log terminal singularities. (Note that we do not use the assumption of $X(\Delta)$ having only terminal singularities here.)

(b) Contraction of Divisorial Type

Let $w \in \Delta$ be an $(n-1)$-dimensional face with $[V(w)] \in R$, and let τ_n, τ_{n+1} be as above (cf. Proposition 14-2-1). Since $K_{X(\Delta)} \cdot V(w) < 0$, the hyperplanes $\{m_{\tau_n} = 1\}$ and $\{m_{\tau_{n+1}} = 1\}$ meet strictly convex along w by Lemma 14-3-4. This holds for all the other $(n-1)$-dimensional faces $w' \in \Delta$ in the interior of $\tau(w)$ (which necessarily implies $[V(w')] \in R$). Since $X(\Delta)$ has only terminal singularities, for all n-dimensional cones $\tau \subset \tau(w)$ we have

$$N \cap \tau \cap \{m_\tau \leq 1\} = \{0\} \cup \{v_i; \langle v_i \rangle \subset \tau\}.$$

Together with convexity along all the $(n-1)$-dimensional faces, we conclude (say, the divisor $V(\langle v_1 \rangle)$ is contracted by the contraction morphism) that

$$N \cap \tau(w) \cap \{m_{\tau(w)} \leq 1\} = \{0\} \cup \{v_2, \ldots, v_n, v_{n+1}\}.$$

Thus $Y(\Delta_Y)$ has only terminal singularities.

(c) Contraction of Flipping Type

The same argument as in the case of contraction of divisorial type via Lemma 14-3-4 yields that $X^+(\Delta^+)$ has terminal singularities. The case for canonical singularities is identical.

We leave the verification of the logarithmic case (ii) to the reader.

In order to see the last assertion, note that there always exists an invariant divisor $V(\langle v_o \rangle)$ such that $V(\langle v_o \rangle) \cdot R > 0$ in each case of the contraction of an extremal ray R, since $R = \mathbb{R}_+[V(w)]$ for an $(n-1)$-dimensional cone w. Then set

$$D = \Sigma_{\langle v_i \rangle \in \Delta, \dim\langle v_i \rangle = 1, \langle v_i \rangle \neq \langle v_o \rangle} (1 - \epsilon) V(\langle v_i \rangle) + (1 - \eta) V(\langle v_o \rangle)$$

for some $\epsilon, \eta \in \mathbb{Q}$ with $0 < \epsilon \ll \eta \ll 1$. Then

$$(K_{X(\Delta)} + D) \cdot R < 0,$$

and the log pair $(X(\Delta), D)$ has only log terminal singularities by Proposition 14-3-2 (ii). □

14.4 Toric Minimal Model Program

As a summary of the results in the previous sections, we have established the **toric (log) minimal model program**.

Theorem 14-4-1 (Toric (Log) Minimal Model Program). *The minimal model program works in the category of complete toric varieties $X(\Delta)$ with only $\mathbb{Q}$-factorial and terminal singularities. The log minimal model program works in the category of logarithmic pairs $(X(\Delta), D)$ consisting of complete toric varieties $X(\Delta)$ and invariant boundary $\mathbb{Q}$-divisors D with only $\mathbb{Q}$-factorial and log terminal (or log canonical) singularities.*

They also work in the relative setting of the category of $\mathbb{Q}$-factorial toric varieties proper over another fixed toric variety.

We may also add the projectivity condition on the category.

PROOF. As in the proof of Theorem 3-1-16 in order to establish the minimal model program, we have only to check the following requirements on the category:

(a) Each object in the category has a well-defined $\mathbb{Q}$-Cartier canonical divisor. This requirement clearly checks.
(b) The cone theorem holds for the objects in the category. We checked this in Theorem 14-1-4.
(c) The contraction theorem holds for the objects in the category. We checked this in Theorem 14-1-9.
(d) We require that when the contraction $\phi = \text{cont}_R : X(\Delta) \to Y(\Delta_Y)$ is divisorial, the resulting variety $Y(\Delta_Y)$ be again in the category and that the

canonical divisor of $Y(\Delta_Y)$ be the pushforward of the canonical divisor of $X(\Delta)$. This was checked in Theorem 14-3-3. We also require that when $\phi = cont_R : X(\Delta) \to Y(\Delta_Y)$ is small,

$$\oplus_{m \geq 0} \phi_*(mK_{X(\Delta)})$$

be a finitely generated $\mathcal{O}_{Y(\Delta_Y)}$-algebra and that

$$\phi^+ : X^+(\Delta^+) = Proj \oplus_{m \geq 0} \phi_*(mK_{X(\Delta)}) \to Y(\Delta)$$

be again small with $X^+(\Delta^+)$ in the category and that the canonical divisor of $X^+(\Delta^+)$ be the strict transform of the canonical divisor of $X(\Delta)$. This was checked in Proposition 14-2-4 and Theorem 14-3-3.

(e) Finally, we require that the program terminate after finitely many steps. Every time we have a divisorial contraction, we lose one 1-dimensional edge, while a flip does not add or eliminate 1-dimensional edges from the fan Δ. Thus we cannot have infinitely many divisorial contractions. There is no infinite sequence of flips by Proposition 14-2-11. Therefore, there is no infinite loop in the program, and it terminates after finitely many steps.

The verification of the logarithmic case is almost identical and is left to the reader as an exercise..   □

Remark 14-4-2. The log minimal model program works without assuming that the boundary $\mathbb{Q}$-divisors are invariant.

Proposition 14-4-3 ((Log) Abundance Theorem for Toric Varieties). *Let $X(\Delta)$ (respectively $(X(\Delta), D)$) be a complete and $\mathbb{Q}$-factorial toric variety (respectively a log pair consisting of a complete and $\mathbb{Q}$-factorial toric variety $X(\Delta)$ and a boundary divisor D). Assume that $K_{X(\Delta)}$ (respectively $K_{X(\Delta)} + D$) is nef. Then $|mK_{X(\Delta)}|$ (respectively $|m(K_{X(\Delta)} + D)|$) is base point free for sufficiently divisible $m \in \mathbb{N}$, and we have the canonical model (respectively log canonical model)*

$$Proj \oplus_{m \geq 0} H^0(X(\Delta), \mathcal{O}_{X(\Delta)}(mK_{X(\Delta)})),$$
$$(respectively \ Proj \oplus_{m \geq 0} H^0(X(\Delta), \mathcal{O}_{X(\Delta)}(m(K_{X(\Delta)} + D)))).$$

In the relative setting of $\phi : X(\Delta) \to S(\Delta_S)$ a toric variety proper over another, the canonical divisor (respectively log canonical divisor) being ϕ-nef implies that the sheaf $\mathcal{O}_{X(\Delta)}(mK_{X(\Delta)})$ (respectively $\mathcal{O}_{X(\Delta)}(m(K_{X(\Delta)} + D))$) is ϕ-generated, and we have the relative canonical model (respectively log canonical model) over $S(\Delta_S)$.

PROOF.   This is a direct consequence of Lemma 14-1-11.   □

Remark 14-4-4. The toric minimal model program can be considered an algorithm to provide factorization of an equivariant birational map between toric varieties into divisorial contractions and flips. The varieties in the category we work with are not necessarily nonsingular, or rather, we inevitably introduce singularities in the process of the minimal model program (cf. Solution 3-1-4 to Obstacle 3-1-2).

The classical factorization problem asks whether a given birational map between two nonsingular varieties can be factored into smooth blowups and smooth blowdowns. In the toric case, after the work of Danilov [2] (cf. Ewald [1]) in dimension 3, this problem has been recently settled affirmatively in a series of papers Włodarczyk [1], Morelli [1][2], and Abramovich–Matsuki–Rashid [1][2].

14.5 Toric Sarkisov Program

The purpose of this section is to discuss the **toric Sarkisov program**, providing an algorithm to untwist equivariant birational maps among toric Mori fiber spaces into links. The Sarkisov program in general is established only in dimension 3, by Corti [1], leaving the higher-dimensional case conjectural. The toric case, however, has all the necessary ingredients established, and we can state the program in arbitrary dimensions as a theorem.

Theorem 14-5-1 (Toric Sarkisov Program). *The Sarkisov program works in the category of projective toric varieties with $\mathbb{Q}$-factorial and terminal singularities, i.e., there exists an algorithm to decompose an equivariant birational map*

$$
\begin{array}{ccc}
X & \overset{\Phi}{\dashrightarrow} & X' \\
\downarrow{\scriptstyle\phi} & & \downarrow{\scriptstyle\phi'} \\
Y & & Y'
\end{array}
$$

between two toric Mori fiber spaces in dimension n into a composite of the 4 types of links of Type (I), (II), (III), or (IV) as in Theorem 13-1-1 in the category of projective toric n-folds with only $\mathbb{Q}$-factorial and terminal singularities, where all the birational maps are required to be equivariant.

PROOF.

General Mechanism

Recall that the general mechanism of the Sarkisov program as described in Section 13-1 requires only the (log) MMP. Since the (log) MMP for projective toric varieties with $\mathbb{Q}$-factorial and terminal, canonical, log terminal, or log canonical singularities works, we have all the ingredients for the general mechanism in order for the toric Sarkisov program to work (cf. Chapter 13).

Termination

We have to check whether if the program terminates, in order to guarantee that the program untwists the given birational maps after finitely many steps.

Recall that according to the arguments in Section 13-2 we have to establish the following three claims:

Claim 13-2-1: There is no infinite number of untwisting (successive or unsuccessive) by the links under the case $\lambda \leq \mu$.

Claim 13-2-2: There is no infinite (successive) sequence of untwisting by the links under the case $\lambda > \mu$ with stationary quasi-effective threshold.

Claim 13-2-6: There is no infinite (successive) sequence of untwisting by the links under the case $\lambda > \mu$ with nonstationary quasi-effective threshold.

In order to check Claim 13-2-1 and Claim 13-2-6 we have only to show that the set of quasi-effective thresholds is discrete. For this purpose it is enough to show that the general fibers G of toric Mori fiber spaces, which are toric $\mathbb{Q}$-Fano varieties of dimension less than or equal to n with Picard number 1 having only $\mathbb{Q}$-factorial and terminal singularities (cf. Corollary 14-2-2), form a bounded family. (See the remarks in Section 13-2.)

Proposition 14-5-2 (Boundedness of Toric $\mathbb{Q}$-Fano Varieties with Picard Number 1 (cf. Borisov–Borisov [1]). *There are only finitely many toric $\mathbb{Q}$-Fano varieties $G = Z(\Delta_Z)$ of fixed dimension d with Picard number 1 having only $\mathbb{Q}$-factorial and terminal singularities.*

PROOF.   The condition for a toric variety $G = Z(\Delta_Z)$ to be a $\mathbb{Q}$-Fano variety of dimension d with Picard number 1 is nothing but that the fan Δ_Z be generated by $d + 1$ of 1-dimensional edges $\langle v_0 \rangle, \langle v_1 \rangle, \ldots, \langle v_d \rangle$, where $v_0, v_1, \ldots, v_d$ are primitive vectors in the lattice $N \cong \mathbb{Z}^d$. The condition for G to have only terminal singularities (cf. Proposition 14-3-1) is that the d-simplex $\diamond$ whose vertices are $v_0, v_1, \ldots, v_d$ contain no lattice points other than the vertices themselves and the origin, i.e.,

$$\diamond \cap N = \{0, v_0, v_1, \ldots, v_d\}.$$

We give a proof of the proposition above, extracting the simple and elegant ideas from Borisov–Borisov [1].

Let $N' \subset N$ be the sublattice generated by the primitive vectors $v_0, v_1, \ldots, v_d$.

Case: $N' = N$. (This is the case where $G = Z(\Delta_Z)$ is a weighted projective space. cf. Remark 14-2-3.)

First we change the coordinate system by an affine transformation: Pick one of the primitive vectors, say v_0, and bring v_0 to the origin of the new coordinate system

$$0 = v_0 - v_0$$

and set

$$V_1 = v_1 - v_0 = (1, 0, \ldots, 0),$$
$$V_2 = v_2 - v_0 = (0, 1, \ldots, 0),$$
$$\cdots$$
$$V_d = v_d - v_0 = (0, 0, \ldots, 1),$$

and the previous origin

$$W = 0 - v_0.$$

Note that according to this new coordinate system, the lattice N is generated by $V_1, \ldots, V_d$ and W.

Figure 14-5-3.

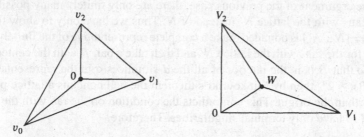

We have only to prove there are only finitely many choices for the position of the transformed origin W. Suppose not. Then there would be an infinite sequence $\{W_l\}_{l \in \mathbb{N}}$ (of the transformed origins of the fans representing $\mathbb{Q}$-Fano varieties G_l), which we may assume to be all distinct, accumulating to a point $A \in \diamond$.

If A coincides with one of the vertices of the simplex $\diamond$, by replacing v_0 by that vertex vector, we may assume $A = 0$, i.e., $\{W_l\}_{l \in \mathbb{N}}$ accumulates to the origin. But then for $l \gg 0$, the point

$$W_l' = 2W_l$$

is a lattice point in N and W_l' is different from any of $0, V_1, \ldots, V_d$ or W, contradicting the condition for G_l to have only terminal singularities.

If A does not coincide with any of the vertices of the simplex $\diamond$, then take the minimal face of the simplex $\diamond$ that contains A in its relative interior. By choosing v_0 again and renumbering the vectors if necessary, we may assume that the face has the vertices $0, V_1, \ldots, V_t$ $(t \le d)$. Let ϵ be the distance from A to the boundary of the face. Choose a sufficiently divisible integer $k \in \mathbb{N}$ and a point L in the sublattice generated by $V_1, \ldots, V_t$ such that

$$|kA - L| < \frac{\epsilon}{3}.$$

Take $l \in \mathbb{N}$ so large that

$$|W_l - A| < \frac{\epsilon}{3} \quad \text{and} \quad |kA - kW_l| < \frac{\epsilon}{3}.$$

Since all the points in the sequence $\{W_l\}_{l \in \mathbb{N}}$ are distinct, we may also assume

$$L - kW_l \ne 0.$$

But then the point

$$W_l' = W_l + (L - kW_l)$$

is a lattice point in N, which is in the relative interior of the face, since

$$|W_l' - A| \le |W_l - A| + |L - kA| + |kA - kW_l| < \epsilon.$$

Thus W'_l is a point in the simplex $\diamond$ and is different from any of 0, $V_1, \ldots, V_d$ or W_l, contradicting the condition for G_l to have only terminal singularities.

Case: $N' \not\subset N$.

By the argument of the previous case, there are only finitely many possibilities for the fans with the lattice N' instead of N. Thus we have only to show that the index $i = [N : N']$ is bounded. Take a complete representative of the finitely many choices for the fans with the lattice N' and then take a ball B with the center at the origin so that B is in the interior of all the d-simplices $\diamond$ in the representative. If $i \cdot \text{vol}(B) \geq 2^d$, then by Minkowski's theorem the ball contains a lattice point in N other than the origin. This contradicts the condition on the fan with the lattice N for G to have only terminal singularities. Therefore,

$$i < \frac{2^d}{\text{Vol}(B)},$$

and we complete the proof of Proposition 14-5-2.   □

Proposition 14-5-2 implies the discreteness of the set of quasi-effective thresholds and hence Claim 13-2-1 and Claim 13-2-6.

In order to check Claim 13-2-2 we have only to show a special case of S_n(local) for toric varieties, applied to the linear system $\Phi_{\text{homaloidal}}^{-1}|H_{X'}|$ on X, consisting of the homaloidal transforms of the members of the linear system $|H_{X'}|$ for an invariant very ample divisor $H_{X'}$ on X'.

Proposition 14-5-4 (A Special Case of S_n (local) for Toric Varieties). *The set of log canonical thresholds for toric n-folds with $\mathbb{Q}$-factorial singularities with respect to invariant linear systems (not necessarily complete) satisfies the ascending chain condition.*

PROOF.   Let $V \subset |D|$ be an invariant linear system on a toric variety with only $\mathbb{Q}$-factorial singularities. Since V is invariant and finite-dimensional, it is generated by a finite number of invariant divisors associated to semi-invariant sections for the action of the torus. The question is also local. Therefore, we have only to consider the following situation:

Let $\{D_k; D_k = \Sigma a_{ki} V(\langle v_i \rangle)\}$ be a finite set of invariant effective divisors (corresponding to the semi-invariant sections generating V) on an affine toric variety $X(\Delta)$, where Δ is the fan corresponding to an n-dimensional cone τ spanned by primitive vectors $v_1, \ldots, v_n$. The log canonical threshold is defined to be

$$\text{lc}(X(\Delta), D) := \max\{c \in \mathbb{Q}_{>0}; a(E; X(\Delta), cD) \geq -1 \quad \forall E \text{ exceptional divisors}\},$$

where D is a general member of V.

Recall that if we take a resolution of singularities $f : V(\Delta_V) \to X(\Delta)$ given by the subdivision Δ_V of Δ, then

$$K_{V(\Delta_V)} + f_*^{-1}(cD) = \Sigma(ca_i - 1)V(\langle v_i \rangle) + \Sigma(-1) \cdot V(\langle v_j \rangle)$$
$$= f^*(K_{X(\Delta)} + cD) + \Sigma a(E_j, X(\Delta), cD)E_j,$$

where the E_j are the exceptional divisors corresponding to the new 1-dimensional edges $\langle v_j \rangle$ (v_j being the primitive vectors in N) that we add to obtain Δ_V from Δ. Take $m_\tau, m_{D_k} \in M \otimes \mathbb{Q}$ such that

$$m_\tau(v_i) = 1 \quad \forall i,$$
$$m_{D_k}(v_i) = a_{ki} \quad \forall i.$$

Note that for a general member D we have

$$\mathrm{ord}_{E_j}(f^*D) = \min_k \{m_{D_k}(v_j)\}.$$

Therefore, if we set the function g on the cone τ to be the minimum among the m_{D_k}'s,

$$g(v) = \min_k \{m_{D_k}(v)\} \text{ for } v \in \tau,$$

then we interpret the conditions

$$a(E_j; X(\Delta), cD) \geq -1 \quad \forall v_j$$
$$\Longleftrightarrow$$
$$(m_\tau - cg)(v_j) \geq 0 \quad \forall v_j$$
$$\Longleftrightarrow$$
$$1 - cg(v) \geq 0 \quad \forall v \in \tau_1,$$

where τ_1 is the face of the simplex generated by $v_1, \ldots, v_n$,

$$\tau_1 := \{x_1 v_1 + \cdots + x_n v_n ; x_1 + \cdots + x_n = 1, x_1 \geq 0, \ldots, x_n \geq 0\}.$$

Therefore, we obtain the formula

$$\mathrm{lc}(X(\Delta), D) = \frac{1}{\max_{v \in \tau_1}\{g(v)\}}.$$

Now, in dimension $n = 2$ the ascending chain condition can be checked in the following manner: Note first that the maximum value of g in τ_1 is attained at one of the two vertices or at a point where two of the functions m_{D_k} coincide, i.e.,

$$\max_{v \in \tau_1}\{g(v)\} \in \cup_k \{m_{D_k}(v_i); i = 1, 2\} \cup_k \{m_{D_k}(v); v \in \tau_1,$$

$$m_{D_k}(v) = m_{D_{k'}}(v) \text{ for some } k' \neq k\}.$$

The values on the right-hand side of the above inclusion can be computed explicitly in terms of the integers a_{ik}. Therefore, we see that the set of log canonical thresholds is contained in the following set:

$$\left\{ \frac{\alpha + \beta}{\alpha\beta + \gamma\beta + \delta\alpha}; \alpha, \beta, \gamma, \delta \in \mathbb{N} \cup \{0\} \right\}.$$

It is easy to see that the only accumulation points of the set above are

$$\left\{ \frac{1}{n}; n \in \mathbb{N} \right\} \cup \{0\},$$

where points in the set accumulate only from the above and hence it satisfies the ascending chain condition. (Exercise! Check these assertions.)

The verification of the ascending chain condition in dimension $n > 2$ is similar and is left to the reader as an exercise.   □

Remark 14-5-5. (i) If we consider the set of the log canonical thresholds $\{lc(X(\Delta), D)\}$ where the D represent not general members of invariant linear systems but fixed effective invariant divisors, it has a more direct description

$$\left\{\frac{1}{n}; n \in \mathbb{N}\right\}.$$

(ii) The behavior of the set of canonical thresholds

$$c(X(\Delta), D) := \max\{c \in \mathbb{Q}_{>0}; a(E; X(\Delta), cD) \geq 0 \quad \forall E \text{ exceptional divisors}\}$$

seems much more subtle than that of the set of log canonical thresholds even for toric varieties. The reason for this difference, at least in the toric case, may be interpreted that while the equation for the log canonical threshold is given by

$$1 - cg(v) \geq 0 \quad \forall v \in \tau_1,$$

which is a problem in "linear algebra," the equation for the canonical threshold is given by

$$(m_\tau - cg)(v_j) \geq 1 \quad \forall v_j \in \tau \cap N \text{ and primitive,}$$

which is a problem in "number theory" involving the behavior of primitive points in the lattice.

The author does not know the explicit description of this set or whether the set satisfies the ascending chain condition.

This completes the argument for termination and hence establishes the toric Sarkisov program.   □

References

Abhyankar, S.
[1] *On the valuations centered in a local domain*, Amer. J. Math. **78** (1956), 321–348.

Abramovich, D. and De Jong, A.J.
[1] *Smoothness, semistability and toroidal geometry*, alg-geom/9604024.

Abramovich, D. and Karu, K.
[1] *Weak semistable reduction in characteristic zero*, alg-geom/9707012 (1997).

Abramovich, D., Matsuki, K. and Rashid, S.
[1] *A note on the factorization theorem of torpic birational maps after Morelli and its toroidal extension*, Tohoku Math. J. **51** (1999), 489–537.
[2] *Erratum to the above paper*, preprint (2000).

Abramovich, D., Karu, K., Matsuki, K. and Wlodarczyk, J.
[1] *Torification and factorization of birational maps*, preprint (1999).

Abramovich, D. and Wang, J.
[1] *Equivariant resolution of singularities in characteristic zero*, Math. Res. Letters (to appear) (1997).

Alexeev, V.
[1] *Two two-dimensional terminations*, Duke. Math. Journal **69** (1993).
[2] *Boundedness of K^2 for long surfaces*, alg-geom 9402007 (1994).
[3] *Log canonical singularities and complete moduli of stable pairs*, alg-geom 9608013 (1996).

Altman, A. and Kleiman, S.
[1] *Introduction to Grothendieck Duality Theory*, Lecture Notes in Math. **146**, Springer-Verlag (1970).

456   References

Ando, T.

[1] *On extremal rays of the higher dimensional varieties*, Invent. Math. **81** (1985), 347–357.

Angehrn, U. and Siu, Y.-T.

[1] *Effective freeness and point separation for adjoint bundles*, Invent. Math. **122** (1995), 291–308.

Arbarello, E., Cornalba, M., Griffiths, P.A., and Harris, J.

[1] *Geometry of Algebraic Curves Volume I*, Grundlehren der mathematischen Wissenschaften **267**, Springer-Verlag (1984).

Artin, M.

[1] *Some numerical criteria for contractibility of curves on algebraic surfaces*, Amer. J. Math. **84** (1962), 485–496.

[2] *On isolated rational singularities of surfaces*. Amer. J. Math. **88** (1966), 129–136.

[3] *On the solution of analytic equations*, Invent. Math. **5** (1968), 277–291.

[4] *Algebraic approximation of structures over complete local rings*, Publ. Math. IHES **36** (1969), 23–58.

[5] *Coverings of the Rational Double Points in Characteristic p*, Complex Analysis and Algebraic Geometry, a collection of papers dedicated to K. Kodaira, Iwanami Shoten, Publishers Cambridge University Press (1977).

Artin, M. and Mumford, D.

[1] *Some elementary example of uniruled varieties which are not rational*, Proc. London Math. Soc. **25** (1972), 75–95.

Atiyah, M.

[1] *On analytic surfaces with double points*, Proc. Royal Soc. London **A-247** (1958), 237–244.

Barth, W., Peters, C. and Van de Ven, A.

[1] *Compact Complex Surfaces*, Ergebnisse der Mathematik und ihrer Grenzgebiete **4**, Springer-Verlag (1984).

Beauville, A.

[1] *Comples Algebraic Surfaces*, London Mathematical Society Lecture Note Series **68** (1983).

Benveniste, X.

[1] *Sur l'anneau canonique de certaines variétés de dimension 3*. Invent. Math. **73** (1983), 157–164.

[2] *Sur les applications pluricanoniques des variété de type très général en dimension 3*, preprint (1984).

Bierstone, E. and Milman, P.D.

[1] *Canonical desingularization in characteristic zero by blowing up the maximum strata of a local invariant*, Invent. Math. **128** (1997), 207–302.

Bombieri, E.

[1] *Canonical models of surfaces of general type*, Publ. Math. IHES **42** (1973), 171–219.

Bogomolov, F.A. and Pantev, T.G.

[1] *Weak Hironaka Theorem*, Math. Research Letters **3** (1996), 299–308.

Bombieri, E. and Mumford, D.

[1] *Enriques's classification of surfaces in characteristic p, II*, Complex Analysis and Algebraic Geometry, a collection of papers dedicated to K. Kodaira, W. Bailey and T. Shioda edtd., Iwanami Shoten (1977), 23–42.

Borisov, A.

[1] *Boundedness theorem for Fano log-threefold*, J. Alg. Geom. **5** (1996), 119–133.

[2] *Minimal discrepancies of toric singularities*, To appear in Manuscripta Math. (1994).

Borisov, A., and Borisov, L.

[1] *Singular toric Fano varieties*, Russian Acad. Sbornik Math. No 1 **75** (1993), 277–283.

Brieskorn, E.

[1] *Rationale Singularitäten komplexer Flächen*, Invent. Math. **14** (1968), 336 358.

Bruno, A. and Matsuki, K.

[1] *Log Sarkisov Program*, Internat. J. Math. **8** (1997), 451–494.

Christensen, C.

[1] *Strong domination/weak factorization of three dimensional regular local rings*, Journal of the Indian Math. Soc. **45** (1981), 21–47.

Clemens, H. and Griffiths, G.

[1] *The intermediate Jacobian of a cubic threefold*, Annals of Math, **95** (1972), 281–356.

Clemens, H., Kollár, J. and Mori, S.

[1] *Higher Dimensional Complex Geometry*, Astérisque **166** (1988).

Corti, A.

[1] *Factorizing birational maps of threefolds after Sarkisov*, J. Alg. Geom. **4** (1995), 23–254.

[2] *Del Pezzo surfaces over Dedekind schemes*, Ann. Math. **144** (1996), 641–683.

Corti, A., Pukhikov, A.V. and Reid, M.

[1] *Fano 3-fold hypersurfaces*, Explicit Birational Geometry of 3-Folds, Cambridge Univ. Press (2000).

Cutkosky, S.D.

[1] *Elementary contractions of Gorenstein threefolds*, Math. Ann. **280** (1988), 521–525.

[2] *Local Factorization of Birational Maps*, Advances in Math. **132** (1997), 167–315.

[3] *Local factorization and monomialization of morphisms*, math.AG/9803078 (1998).

Danilov, V.

[1] *Birational geometry of toric varieties*, Math. USSR-Izv. **21** (1983), 269–280.

[2] *The birational geometry of toric 3-folds*, Math. USSR-Izv. **21** (1983), 269–280.

De Jong, A.J.

[1] *Smoothness, semistability, and alterations*, Publ. math. I.H.E.S. **83** (1996), 51–93.

Deligne, P.

[1] *Théorie de Hodge I*, Actes ICM, Niece, Gauthier-Villas, t. I (1970), 425–430.

[2] *Théorie de Hodge II*, Publ. Math. Inst. Hautes Etud. Sci. **40** (1971), 5–58.

[3] *Théorie de Hodge III*. Pbul. Math. Inst. Hautes Etud. Sci. **44** (1975), 5–77.

Demushkin, A.S.

[1] *Combinatorial invariance of toric singularities (Russian)*, Vestnik Moskov. Univ. Ser. I No. **2** Mat. Mekh. (1982), 80–87.

Dolgachev, I. and Gross, M.

[1] *Elliptic threefolds I: Ogg-Shafarevich theory*, J. Alg. Geom. **3** (1994), 39–80.

Du Val, P.

[1] *On isolated singularities of surfaces which do not affect the conditions of adjunction I-II*, Proc. Camb. Soc. **30** (1934), 453–465, 483–491.

Elin, L. and Lazarsfeld, R.

[1] *Global generation of pluricanonical and adjoint linear series on smooth projective threefolds*, Journal AMS **6** (1993), 875–903.

[2] *Singularities of theta divisors and the birational geometry of irregular varieties*, J. of Amer. Math. Soc. **10** (1997), 243–258.

Eisenbud, D.

[1] *Commutative Algebra with a view toward Algebraic Geometry*, Graduate Texts in Mathematics **150** (1995).

Elkik, R.

[1] *Rationalité des singularités canoniques*, Invent. Math **64** (1981), 1–6.

Encinas, S. and Villamayor, O.

[1] *A course on constructive desingularization and equivariance*, preprint (1999).

Esnault, H. and Viehweg, E.

[1] *Lectures on Vanishing Theorems*, DMV Seminar Band **20** (1992).

Ewald, G.

[1] *Blow-ups of smooth toric 3-varieties*, Abh. Math. Sem. Univ. Hamburg **57** (1987), 193–201.

Faltings, G.

[1] *Entlichkeitssätze für abelsche Varietäten über Zahlkörpern*, Invent. Math. **73** (1983), 349–366.

Francia, P.

[1] *Some remarks on minimal models. I*, Compositio Math. **40** (1980), 301–313.

Fujita, T.

[1] *Semipositive line bundles*, J. Fac. Sci. Tokyo Univ. **30** (1983), 353–378.

[2] *Fractionally logarithmic canonical rings of algebraic surfaces*, J. Fac. Sci. Univ. Tokyo Sect. IA **30** (1984), 685–696.

[3] *A relative version of Kawamata-Viehweg's vanishing theorem*, preprint, Tokyo Univ. (1985).

[4] *Zariski-decomposition and canonical rings of elliptic threefolds*, J. Math. Soc. Japan **38** (1986), 19–37.

Fulton, W.

[1] *Intersection Theory*, Ergebnisse der Mathematik und ihrer Grenzgebiete **2** (1984), Springer-Velag.

[2] *Introduction to Toric Varieties*, Annals of Mathematics Studies **131** (1993), Princeton University Press.

Grassi, A.

[1] *On minimal models of elliptic threefolds*, Math. Ann. **290** (1991), 287–301.

Grauert, H.

[1] *Über Modifikationen und exzeptionelle analytische Mengen*, Math. Ann. **146** (1962), 331–368.

Grauert, H. and Remmert, R.

[1] *Coherent analytic sheaves*, Springer-Verlag (1984).

Grauert, H. and Riemenschneider, O.

[1] *Verschwindungssätze für analytische Kohomologiegruppen auf komplexen Räumen*, Invent. Math. **11** (1970), 263–292.

Griffiths, P. and Harris, J.

[1] *Principles of Algebraic Geometry*, Pure & Applied Mathematics, Wiley Interscience (1978).

Grothendieck, A.

[1] *Fondements de Géométrie Algébrique*, Séminaire Bourbaki 1957–62, Secrétariat Math. Paris (1965).

[2] *On the de Rham cohomology of algebraic varieties*, Publ. Math. Inst. Hautes Etud. Sci. **29** (1966), 95–103.

Grothendieck, A. and Dieudonné, J.

[1] *Eléments de Géométrie Algébrique, Le langage des schémas*, Publ. Math. Inst. Hautes Etud. Sci. **4** (1960).

[2] *Eléments de Géométrie Algébrique II, Étude globale élémentaire de quelques classes de morphismes*, Publ. Math. Inst. Hautes Etud. Sci. **8** (1961).

[3] *Eléments de Géométrie Algébrique III, Étude cohomologique des faisceaux cohérents*, Publ. Math. Inst. Hautes Etud. Sci. **11** (1960), **17** (1963).

[4] *Eléments de Géométrie Algébrique IV, Étude locale des schémas et des morphismes de schémas*, Publ. Math. Inst. Hautes Etud. Sci. **20** (1964), **24** (1965), **28** (1966), **32** (1967).

[5] *Eléments de Géométrie Algébrique I*, Grundlehren der mathematischen Wissenschaften **166**, Springer-Verlag (1971).

Harris, J.

[1] *Algebraic Geometry*, Graduate Texts in Mathematics **133**, Springer-Verlag (1992).

Harris, J. and Morrison, I.

[1] *Moduli of Curves*, Graduate Texts in Math, **187**, Springer-Verlag (1998).

Hartshorne, R.

[1] *Residues and Duality*, Lecture notes in Math. **20**, Springer-Verlag (1966).

[2] *Ample Subvarieties of Algebraic Varieties*, Lecture Notes in Math. **156**, Springer-Verlag (1970).

[3] *Algebraic Geometry*, Graduate Texts in Math. **52**, Springer-Verlag (1977).

Hassett, B.

[1] *Special cubic hypersurfaces of dimension 4*, thesis, Harvard University (1996).

[2] *Special cubic fourfolds*, preprint (1997).

Hironaka, H.

[1] *On the theory of birational blowing-up*, thesis, Harvard University (1960).

[2] *Resolution of singularities of an algebraic variety over a field of characteristic zero*, Annals of Math. **79** (1964), 109–326.

[3] *Flattening theorem in complex analytic geometry*, Amer. J. of Math. **97** (1975), 503–547.

Hironaka, H. and Rossi, H.

[1] *On the equivalence of embeddings of exceptional complex spaces*, Math. Ann. **156** (1964), 313–368.

Hirzebruch, F.

[1] *Topological Methods in Algebraic Geometry*, Grundlehren **131**, Springer-Verlag (1966).

[2] *The Hilbert modular groups, resolution of the singularities at the cusps and related problems*, Lecture Notes in Math. **244**, Springer-Verlag (1971), 275–288.

Humphreys, J.

[1] *Reflection Groups and Coxeter Groups*, Cambridge University Press (1990).

Iitaka, S.

[1] *On D-dimensions of algebraic varieties*, J. Math. Soc. Japan **23** (1971), 356–373.

[2] *Logarithmic forms of algebraic varieties*, J. Fac. Sci. Univ. Tokyo **23** (1976), 525–544.

[3] *On logarithmic Kodaira dimension of algebraic varieties*, Complex Analysis and Algebraic Geometry, a collection of papers dedicated to K. Kodaira, W. Bailey and T. Shioda ed., Iwanami Shoten, Tokyo (1977), 123–127.

[4] *Birational geometry of algebraic varieties*, Proc. Internat. Congr. Math. (Warsaw, 1883), Polish Scientific Publishers, Warsaw (1984), 727–732.

[5] *Algebraic Geometry (An Introduction to Birational Geometry of Algebraic Varieties)*, Graduate Texts in Math. **76**, Springer-Verlag (1981).

Iskovskikh, V.A.

[1] *Fano 3-folds I*, Math. USSR-Izv. **11** (1977), 485–527.

[2] *Fano 3-folds II*, Math. USSR-Izv. **12** (1978), 496–506.

Iskovskikh, V.A., and Manin, Yu.I.

[1] *Three-dimensional quartics and counterexamples to the Lüroth problem*, Math. USSR Sbornik **15** (1971), 141–166.

Karu, K.

[1] *Minimal models and boundedness of stable varieties*, preprint (1998).

Kawamata, Y.

[1] *On the classification of open algebraic surfaces*, Springer Lecture Notes **732** (1979), 51–55.

[2] *On Singularities in the Classification Theory of Algebraic Varieties*, Math. Ann. **251** (1980), 51–55.

[3] *A generalization of Kodaira-Ramanujam's vanishing theorem*, Math Ann. **261** (1982), 43–46.

[4] *On the finiteness of generators of the pluri-canonical ring for a threefold of general type*, Amer. J. Math. **106** (1984), 1503–1512.

[5] *The cone of curves of algebraic varieties*, Annals of Math. **119** (1984), 603–633.

[6] *Pluricanonical systems on minimal algebraic varieties*, Invent. Math. **79** (1985), 567–588.

[7] *On the pluri-genera of minimal algebraic threefolds with $K \equiv 0$*, Math. Ann. **275** (1986), 539–546.

[8] *Minimal models and the Kodaira dimension of algebraic fiber spaces*, Journal für die reine und angewandte Mathematik (1985), 1–46.

[9] *Crepant blowing-ups of three dimensional canonical singularities and its application to degeneration of surfaces*, Ann. of Math. **127** (1988), 93–163.

[10] *Small Contractions of Four Dimensional Algebraic manifolds*, UTYO-MATH 88-23, 1–13.

[11] *Boundedness of $\mathbb{Q}$-Fano threefolds*, Proc. Int. Conf. Algebra. Contemp. Math. **131** (1989), 439–445.

[12] *Abundance theorem for minimal threefolds*, Invent. Math. **108** (1991), 229–246.

[13] *On the length of an extremal rational curve*, Invent. Math. **105** (1991), 609–611.

[14] *Log Canonical Models of Algebraic 3-Folds*, Internat. J. Math. **3** (1992), 351–357.

[15] *Termination of log flips for algebraic 3-Folds*, Internat. J. Math. **3** (1992), 653–660.

[16] *Semistable minimal models of threefolds in positive or mixed characteristic*, J. Alg. Geom. **3** (1994), 463–491.

[17] *On the cone of divisors of Calabi-Yau fiber spaces*, Internat. J. Math. **8** (1997), 665–687.

[18] *On Fujita's freeness conjecture for 3-folds and 4-folds*, Math. Ann. **308** (1997), 491–505.

[19] *Index-1 cover of log terminal surface singularities*, math.AG/9802044 (1998).

Kawamata, Y. and Matsuki, K.

[1] *The number of minimal models for 3-folds of general type is finite.*, Math, Ann. **276** (1987), 595–598.

Kawamata, Y., Matsuda, K. and Matsuki, K.

[1] *Introduction to minimal model problem*, Adv. Stud. Pure Math. **10**, Alg. Geom., Sendai, T. Oda ed. (1985), 283–360.

Keel, S., Matsuki, K. and McKernan, J.

[1] *Log abundance theorem for threefolds*, Duke Math. Journal **75** (1994), 99–119.

Keel, S., and McKernan, J.

[1] *Rational curves on log Del Pezzo surfaces*, preprint (1994).

Kempf, G., Knudsen, F., Mumford, D. and Saint-Donat, B.

[1] *Toroidal Embeddings I*, Lecture Notes in Mathematics **339**, Springer-Verlag (1973).

Kleiman, S.

[1] *Toward a numerical theory of ampleness*, Annals of Math. **84** (1996), 293–344.

Kodaira, K.

[1] *On a differential-geometric method in the theory of analytic stacks*, Proc. Natl. Acad. Sci. USA **39** (1953), 1268–1273.

[2] *On compact complex analytic surfaces I*, Annals of Mathematics **71** (1960), 111–152.

[3] *On compact complex analytic surfaces II*, Annals of Mathematics **77** (1963), 563–626.

[4] *On compact complex analytic surfaces III*, Annals of Mathematics **78** (1963), 1–40.

[5] *Collected Works Vol. I, II, III*, Iwanami Shoten and Princeton University Press (1975).

[6] *Complex Manifolds and Deformations of Complex Structures*, Grundlehren der mathematishen Wissenschaften **283**, translated by K. Akao, Springer-Verlag (1986).

Kollár, J.

[1] *The cone theorem: Note to a paper of Y. Kawamata*, Annals of Math. **120** (1984), 1–5.

[2] *Toward moduli of singular varieties*, Comp. Math. **56** (1985), 369–398.

[3] *Higher direct images of dualizing sheaves I*, Annals of Mathematics **123** (1986), 11–42.

[4] *Higher direct images of dualizing sheaves II*, Annals of Mathematics **124** (1986), 171–202.

[5] *The structure of algebraic threefolds—an introduction to Mori's program*, Bull. AMS **17** (1987), 211–273.

[6] *Subadditivity of the Kodaira dimension: Fibers of general type*, Adv. Stud. Pure Math. **10**, Proc. Sympos. Algebraic Geom., Sendai (1985), 361–398.

[7] *Vanishing Theorems for Cohomology Groups*, Proc. Symp. Pure math. **46** Algebraic Geometry Bowdoin 1985 (1987), 233–243.

[8] *Flops*, Nagoya Math. J. **113** (1989), 15–36.

[9] *Projectivity of Complete Moduli*, J. Differential Geometry **32** (1990), 235–268.

[10] *Extremal rays on smooth threefolds*, Ann. Sci. ENS **24** (1991), 339–361.

[11] *Flips, Flops, Minimal Models, etc.*, Surv. in Diff. Geom. **1** (1991), 113–199.

[12] *Cone theorems and bug-eyed covers*, J. Alg. geom. **1** (1992), 293–323.

[13] *Log surfaces of general type; some conjectures*, Contemp. Math. **162** (1994), 261–275.

[14] *Nonrational hypersurfaces*, Journal AMS **8** (1995), 241–249.

[15] *Rational Curves on Algebraic Varieties*, Ergebnisse der Mathematik und ihrer Grenzgebiete **32**, Springer-Verlag (1996).

[16] *Singularities of pairs*, Algebraic Geometry, Santa-Cruz 1995, Proc. Symp. Pure Math. **62** (1997), 221–287.

Kollár, J. et al.

[1] *Flips and abundance for algebraic threefolds, Summer Seminar Note at the University of Utah*, Astéridque **211**, J. Kollár ed. (1992)

Kollár, J. and Mori S.

[1] *Classification of Three-Dimensional Flips*, Journal of AMS **5** (1992), 533–703.

[2] *Birational geometry of algebraic varieties (with the collaboration of H. Clemens and A. Corti)*, Cambridge University Press (1998).

[3] *Soyurikikagaku (Kyoryoku H. Clemens·A. Corti)*, Gendaisugaku no Tenkai **16**, Iwanami-Shoten (1998).

Kollár, J. and Shepherd-Barron, N.

[1] *Threefolds and deformations of surface singularities*, Invent. Math. **91** (1988), 299–338.

Kovács, S.

[1] *The cone of curves of a K3 surface*, Math Ann. **300** (1994), 681–691.

[2] *Smooth families over rational and elliptic curves*, J. Alg. Geometry **5** (1996), 369–385.

Lamothke, K.

[1] *Regular Solids and Isolated Singularities*, Advanced Lectures in Mathematics (1986).

Laufer, H.

[1] *Normal two-dimensional singularities*, Ann. Math. Studies **71**, Princeton University Press (1971).

[2] *Taut two-dimensional singularities*, Math. Ann. **205** (1973), 131–164.

[3] *One minimally elliptic singularities*, American J. of Math. **99** (1977), 1257–1295.

Lazarsfeld, R.

[1] *Lectures on linear series*, Complex Algebraic Geometry-Park City **3**, Park City/IAS Math. Ser., AMS (1996), 161–220.

Lipman, J.

[1] *Rational singularities with applications to algebraic surfaces and unique factorization*, Publ. Math. IHES **36** (1969), 195–279.

464   References

Manin, Y.I.
[1] *CUBIC FORMS, Algebra, Geometry, Arithmetic*, North-Holland Mathe-matical Library **4** (1974).

Masek, V.
[1] *Minimal discrepancies of hypersurface singularities*, preprint (1998).

Matsuki, K.
[1] *On pluricanonical maps for 3-folds of general type*, J. Math. Soc. Japan **38** (1986), 339–359.
[2] *A criterion for the canonical bundle of a threefold to be ample*, Math. Ann. **276** (1987), 557–564.
[3] *An approach to the abundance conjecture for 3-folds*, Duke Math. J. **61** (1990), 207–220.
[4] *Termination of flops for 4-folds*, Amer. J. Math. **113** (1991), 835–859.
[5] *A note on the Sarkisov's program*, preprint (unpublished) (1992).
[6] *Weyl groups and birational transformations among minimal models*, Memoirs of AMS **557** (1995).
[7] *Surface log terminal singularities in positive characteristic*, in preparation (2001).
[8] *Lectures on Factorization of Birational Maps*, RIMS–1281 (2000).

Matsuki, K. and Wentworth, R.
[1] *Mumford-Thaddeus principle on the moduli space of vector bundles on an algebraic surface*, Internat. J. Math. **8** (1997), 97–148.

Matsumura, H.
[1] *Commutative Ring Theory*, Cambridge studies in advanced mathematics **8** (1989).

Migliorini, L.
[1] *A smooth family of minimal surfaces of general type over a curve of genus at most one is trivial*, J. Algebraic Geometry **4** (1995), 353–361.

Milnor, J.
[1] *Singular Points of Complex Hypersurfaces*, Annals of Mathematics Studies **61**, Princeton University Press (1968).
[2] *Morse Theory*, Annals of mathematics Studies **51**, Princeton University Press (1969).

Miyanishi, M. and Tsunoda, S.
[1] *The structure of open algebraic surfaces II*, in Classification of Algebraic and Analytic Manifolds (K. Ueno ed.), Progress in Math. **39**, Birkhaüser (1975), 499–544.

Miyaoka, Y.
[1] *The Chern classes and Kodaira dimension of a minimal variety*, Adv. Stud. Pure Math. **10**, Alg. Geom., Sendai, T. Oda ed. (1985), 449–476.
[2] *On the Kodaira dimension of minimal threefolds*, Math. Ann. **281** (1988), 325–332.
[3] *Abundance Conjecture for 3-folds: case $v = 1$*, Compositio Math. **68** (1988), 203–220.

Miyaoka, Y. and Mori, S.

[1] *A numerical criterion for uniruledness* Ann. of Math. **124** (1986), 65–69.

Morelli, R.

[1] *The birational geometry of toric varieties*, J. Alg. Geom. **5** (1996), 751–782.

[2] *Correction to "The birational geometry of toric varieties,"* home page at the Univ. of Utah (1997), 767–770.

Mori, S.

[1] *Projective manifolds with ample tangent bundles*, Annals of Mathematics **110** (1986), 593–606.

[2] *Threefolds whose canonical bundles are not numberically effective*, Ann. of Math. **116** (1986), 133–176.

[3] *Classification of Higher-Dimensional Varieties*, Proc. Sympos. Pure Math. **46** Part 1 (Summer Research Inst., Bowdoin 1985), 269–331.

[4] *On 3-dimensional terminal singularities*, Nagoya Math. J. **98** (1985), 43–66.

[5] *Flip theorem and the existence of minimal models for 3-folds*, J. Amer. Math. Soc. **1** (1988), 117–253.

Mori, S. and Mukai, S.

[1] *On Fano 3-Folds with $B_2 \geq 2$. Advanced Studies in Pure Mathematics* **1** (1983), 101–129.

Mukai, S.

[1] *Moduli Theory I*, Gendaisugaku no Tenkai **13**, Iwanami-Shoten (1998).

Mumford, D.

[1] *The canonical ring of an algebraic surface, Appendix to Zariski's paper "The theorem of Riemann–Roch for high multiples of a divisor."* Ann. of Math. **76** (2) (1962), 612–615.

[2] *Lectures on curves on an algebraic surface*, Princeton University Press (1966).

[3] *The topology of normal singularities of an algebraic surface and a criterion for simplicity*, Publ. Math. IHES **9** (1969), 5–22.

[4] *Enriques' classification of surfaces in characteristic p, I*, Global Analysis, papers in honour of K. Kodaira, Univ. of Tokyo Press, Tokyo and Princeton Univ. Press, Princeton N.J. (1969), 325–339.

[5] *Abelian Varieties*, Oxford University Press (1970).

[6] *Algebraic Geometry I Complex Projective Varieties*, Grundlehren der mathematischen Wissenschaften **221**, Springer-Verlag (1976).

Mumford, D., Fogarty, J. and Kirwan, F.

[1] *Geometric Invariant Theory*, Ergebnisse der Mathematik und ihrer Grenzgebiete **34** Third Enlarged Edition (1994).

Nagata, M.

[1] *On rational surfaces I-II*, Mem. Coll. Sci. Univ. Kyoto, Ser. A **32, 33** (1960), 351–370, 271–293.

466   References

Nakayama, N.

[1] *Invariance of the plurigenera of algebraic varieties under minimal model conjectures*, Topology **25** (1986), 237–251.

[2] *The Lower Semi-Continuity of the Plurigenera of Complex Varieties*, Advanced Studies in Pure Mathematics **10**, Algebraic geometry, Sendai (1985), 551-590.

[3] *The singularity of the canonical model of compact Kähler manifolds*, Math. Ann. **280** (1988), 509–512.

Nikulin, V.V.

[1] *Del pezzo surfaces with log terminal singularies III*, Math. USSR Izv. **35** (1990).

Oda, T.

[1] *Lectures on Torus Embeddings and Applications*, Based on joint work with Katsuya Miyake, Tata Inst. of Fund. Research **58** (1966), Springer-Verlag.

[2] *Convex Bodies and Algebraic Geometry*, Ergebnisse der Mathematik und ihrer Grengezbiete **15** (1987).

Paranjape, K.H.

[1] *Bogomolov-Pantev resolution—an expository account*, New trends in Algebraic Geometry, Warwick July 1996, Cambridge University Press (1998).

Prill, D.

[1] *Local classification of quotients of complex manifolds by discontinuous groups*, Duke Math. J. **34** (1967), 375–386.

Ramanujam, C.P.

[1] *Remarks on the Kodaira vanishing theorem*, J. Indian Math. Soc. **36** (1972), 41–51.

Reid, M.

[1] *Elliptic Gorenstein singularities of surfaces*, preprint (1976).

[2] *Canonical threefolds*, in Gemétrie Algébrique Angers 1979, A. Beauville ed. Sijthoff and Noordhoff (1980), 273–310.

[3] *Minimal models of canonical 3-folds*, Adv. Stud. in Pure math. **1** (1983), 131–180.

[4] *Projective morphisms according to Kawamata*, preprint, Univ. of Warwick (1983).

[5] *Decomposition of toric morphisms*, Arithmetic and Geometry II (M. Artin and J. Tate, eds.), Progress in Math. **36** Birkhäuser (1983), 395–418.

[6] *Birational geometry of 3-folds according to Sarkisov*, preprint (1991).

[7] *Young person's guide to canonical singularities* Proc. Pure Math. **46** Algebraic Geometry Bowdoin 1985 (1987), 345–416.

[8] *Tendencious survey of 3-folds*, Proc. Pure Math. **46** Algebraic Geometry Bowdoin 1985 (1987), 333–344.

[9] *Chapters on Algebraic Surfaces*, IAS/Park City Mathematics Series **3** (1997).

Sally, J.

[1] *Regular overrings of regular local rings*, Trans. Amer. Math. Soc. **171** (1972), 291–300.

Sarkisov, V.G.

[1] *Birational automorphisms of conic bundles*, Math. USSR Izv. **17** (1981), 177–202.

[2] *On the structures of conic bundles*, Math. USSR Izv. **120** (1982) 355–390.

[3] *Birational maps of standard ℚ-Fano fiberings*, I.V. Kurchatov Institute Atomic Energy preprint (1989).

Serre, J.P.

[1] *Géométrie algébrique et géomírie analytique*, Ann. Inst. Fourier **6** (1956), 1–42.

Shafarivich, I.R.

[1] *Basic Algebraic Geometry*, Springer-Verlag (1972).

Shannon, D.L.

[1] *Monoidal transforms*, Amer. J. Math. **45** (1973), 284–320.

Shokurov, V.V.

[1] *The Non-Vanishing Theorem*, Math. USSR Izv. **19** (1985), 591–604.

[2] *3-fold log flips*, Math. USSR Izv. **56** (1992), 105–203.

Silverman, J.

[1] *The arithmetic of elliptic curves*, Graduate Texts in Math. **106** (1986).

Smith, K.

[1] *Fujita's Freeness Conjecture in Terms of Local Cohomology*, To appear in Journal of Algebraic Geometry (1995).

Szabó, E.

[1] *Divisorial log terminal singularities*, J. Math Sci. Univ. Tokyo **1** (1995), 631–639.

Takagi, H.

[1] *3-flop log flips according to V.V. Shokurov*, math.AG/9803145 (1998).

Takahashi, N.

[1] *Sarkisov Program for Log Surfaces*, Tokyo University Master's Thesis (1995), 395–418.

[2] *An Application of Nöther-Fano Inequalities*, preprint (1995).

[3] *Decomposition of Automorphisms of the Affine Plane*, preprint (1995).

Tsunoda, S.

[1] *Structure of open algebraic surfaces I*, J. Math. Kyoto Univ. **32** (1983), 95–125.

Ueno, K.

[1] *Classification theory of algebraic varieties and compact complex spaces*, Lecture Notes in Math. **439**, Springer-Verlag (1975).

Viehweg, E.

[1] *Klassifikationstheorie algebrasischer Varietäten der dimension drei*, Compositio Math. **41** (1980), 361–400.

[2] *Vanishing theorems*, Jour. Reine Angew. Math. **335** (1982), 1–8.

[3] *Weak positivity and the additivity of the Kodaira dimension for certain fiber spaces*, Algebraic Varieties and Analytic Varieties (S. Iitaka, ed.), Adv Stud. in Pure Math. **1** (1983), 329–353.

[4] *Weak positivity and the additivity of the Kodaira dimension II—the local Torelli map*, Classification of Algebraic and Analytic Manifolds (K. Ueno ed.), Progress in Math. **39** (1983), 567–589.

[5] *Quasi-projective Moduli for Polarized Manifolds* Ergebnisse der mathematik und ihrer Grenzgebiete **30** (1995).

Villamayor, O.

[1] *Constructiveness of Hironaka's resolution*, Ann. Sci. École Norm. Sup. (4) **22** no. 1 (1989), 1–32.

Vojta, P.

[1] *A Higher Dimensional Mordell Conjecture*, Arithmetic Geometry, ed. by G. Cornell and J. Silverman, Springer-Verlag (1986), 341–353.

Wahl, J.

[1] Equations defining rational singularities, Ann. Sci. École Norm. Sup. (4) **10** (1977), 231–264.

Wells, R.O.

[1] *Differential Analysis on Complex Manifolds*, Graduate Texts in Math. **65**, Springer-Verlag (1980).

White, G.K.

[1] *Lattice tetrahedra*, Canad. J. Math. **16** (1964), 389–396.

Wilson, P.M.H.

[1] *Towards birational classification of algebraic varieties*, Bull. London Math. Soc. **19** (1987), 1–48.

Wlodarczyk, J.

[1] *Decomposition of Birational Toric Maps in Blow-Ups and Blow-Downs. A Proof of the Weak Oda Conjecture*, Transaction of the AMS **349** (1997), 373–411.

[2] *Combinatorial structures on toroidal varieties. A proof of the weak Factorization Theorem*, preprint (1999).

Zariski, O.

[1] *Algebraic surfaces. 2nd edition*, Ergebnisse der Math. **61**, Springer-Verleg (1970).

[2] *The theorem of Riemann–Roch for high multiples of an effective divisor on an algebraic surface*, Annals of Math. **76** (2) (1962), 560–615.

Index

Universitext   *(continued)*